AF213511

Springer Monographs in Mathematics

Editors-in-Chief

Minhyong Kim, School of Mathematics, Korea Institute for Advanced Study, Seoul, South Korea, International Centre for Mathematical Sciences, Edinburgh, UK

Katrin Wendland, School of Mathematics, Trinity College Dublin, Dublin, Ireland

Series Editors

Sheldon Axler, Department of Mathematics, San Francisco State University, San Francisco, USA

Maria Chudnovsky, Department of Mathematics, Princeton University, Princeton, USA

Tadahisa Funaki, Department of Mathematics, University of Tokyo, Tokyo, Japan

Isabelle Gallagher, Département de Mathématiques et Applications, Ecole Normale Supérieure, Paris, France

Sinan Güntürk, Courant Institute of Mathematical Sciences, New York University, New York, USA

Claude Le Bris, CERMICS, Ecole des Ponts ParisTech, Marne la Vallée, France

Pascal Massart, Département de Mathématiques, Université de Paris-Sud, Orsay, France

Alberto A. Pinto, Department of Mathematics, University of Porto, Porto, Portugal

Gabriella Pinzari, Department of Mathematics, University of Padova, Padova, Italy

Ken Ribet, Department of Mathematics, University of California, Berkeley, USA

René Schilling, Institute for Mathematical Stochastics, Technical University Dresden, Dresden, Germany

Panagiotis Souganidis, Department of Mathematics, University of Chicago, Chicago, USA

Endre Süli, Mathematical Institute, University of Oxford, Oxford, UK

Shmuel Weinberger, Department of Mathematics, University of Chicago, Chicago, USA

Boris Zilber, Mathematical Institute, University of Oxford, Oxford, UK

This series publishes advanced monographs giving well-written presentations of the "state-of-the-art" in fields of mathematical research that have acquired the maturity needed for such a treatment. They are sufficiently self-contained to be accessible to more than just the intimate specialists of the subject, and sufficiently comprehensive to remain valuable references for many years. Besides the current state of knowledge in its field, an SMM volume should ideally describe its relevance to and interaction with neighbouring fields of mathematics, and give pointers to future directions of research.

Zdeněk Dvořák

Graph Minors

Theory and Applications

Springer

Zdeněk Dvořák
Computer Science Institute
Charles University
Prague, Czech Republic

ISSN 1439-7382 ISSN 2196-9922 (electronic)
Springer Monographs in Mathematics
ISBN 978-3-031-87468-0 ISBN 978-3-031-87469-7 (eBook)
https://doi.org/10.1007/978-3-031-87469-7

© The Editor(s) (if applicable) and The Author(s), under exclusive license to Springer Nature Switzerland AG 2025

This work is subject to copyright. All rights are solely and exclusively licensed by the Publisher, whether the whole or part of the material is concerned, specifically the rights of translation, reprinting, reuse of illustrations, recitation, broadcasting, reproduction on microfilms or in any other physical way, and transmission or information storage and retrieval, electronic adaptation, computer software, or by similar or dissimilar methodology now known or hereafter developed.
The use of general descriptive names, registered names, trademarks, service marks, etc. in this publication does not imply, even in the absence of a specific statement, that such names are exempt from the relevant protective laws and regulations and therefore free for general use.
The publisher, the authors and the editors are safe to assume that the advice and information in this book are believed to be true and accurate at the date of publication. Neither the publisher nor the authors or the editors give a warranty, expressed or implied, with respect to the material contained herein or for any errors or omissions that may have been made. The publisher remains neutral with regard to jurisdictional claims in published maps and institutional affiliations.

This Springer imprint is published by the registered company Springer Nature Switzerland AG
The registered company address is: Gewerbestrasse 11, 6330 Cham, Switzerland

If disposing of this product, please recycle the paper.

Preface

The theory of graph minors has a rich history starting with Kuratowski's characterization of planar graphs, intertwines with the graph coloring through Hadwiger's conjecture, and plays an important role in algorithmic graph theory. The development of the structural characterization of graphs avoiding a fixed minor by Robertson and Seymour is arguably the most important foundational result in structural graph theory. And, if any of this is news to you, this book is probably not for you.

There are quite a few books that present the basic concepts of the theory. They will explain to you the importance of forbidden minors in relation to graphs on surfaces and other proper minor-closed classes, the notion of treewidth and its applications in design of efficient algorithms, and Hadwiger's conjecture and its relationship to the Four Color Theorem. They will speak about well-quasi-ordering, Wagner's conjecture, and the amazing fact that every minor-monotone property is characterized by only finitely many forbidden minors. They will maybe state the Minor Structure Theorem with all its technical and slightly mysterious ingredients, and if you are lucky, actually explain to you why the ingredients are needed and how the theorem can be used. And they will let you hanging there, aware that there is a lot to the theory that you do not see.

Admittedly, this is already enough for you to appreciate many of the applications of the theory, or even to use it in your research. But if you want to learn more, there are not that many options available. You can go hardcore and actually read the series of papers by Robertson and Seymour, or some of the newer developments of the theory. This is indeed possible, but of course journal papers need to show all the details of every argument, no matter how distracting from the important ideas. Adding to the difficulty, it needs to be admitted that many of the papers on graph minors theory are not written in an especially reader-friendly way. And, in the end, you may end up knowing a lot of theory without actually being able to use it.

So perhaps you could go the other way, start by studying various applications of the theory, gain an understanding of what is important in this way, and gradually learn the details as needed. This seems to be the most popular avenue, usually with the help of a mentor who can point you to the right materials on each particular aspect. Unsurprisingly, this top-down approach leaves you with a great overall understanding, but somewhat shaky foundations.

Or, you could give this book a try. As far as I am aware, this is the first book to offer a somewhat comprehensive treatment of the intermediate and advanced aspects of the theory of graph minors. The book is split into three parts:

- Part I discusses in detail the ideas and notions that are used in the proof of the Minor Structure Theorem and its applications. It starts with the development of the theory of tree decompositions and dual objects such as tangles and brambles. A significant attention is given to graphs drawn on surfaces, especially in connection the metric arising from tangles that respect the drawings and the corresponding sufficient conditions for the presence of minors. Building on these results, we prove the Flat Wall Theorem and explain how it is used in the algorithm to test whether the input graph contains a fixed minor. Finally, we give a detailed outline of the proof of the Minor Structure Theorem.
- Part II showcases applications of the Minor Structure Theorem of various complexity. We take this as an opportunity to introduce several important properties of graphs from proper minor-closed classes (existence of low-treewidth colorings and large grid minors), and to talk about recent progress on the structural theory of topological minors. This part also introduces and motivates several strengthenings of the Theorem that simplify handling of apices and vortices, the main technical ingredients.
- Part III is devoted to results on graph minors that avoid the Structure Theorem, on topics such as sublinear separators, density, isomorphism testing, ...

My hope is that this book can help you to learn the more advanced aspects of theory of graph minors. Still, it is not strictly speaking a textbook. I make an effort to explain overall ideas, but the theory is somewhat infamous for its technicality and I do not think I can provide a detailed guidance for every single argument. Moreover, given the intended audience, the text assumes a lot of maturity of the reader.

Let me acknowledge that there are important topics that are not covered in the book. Given the richness of the theory, it is of course not possible to talk about everything, and while the book can be useful as a reference, it is not its intended purpose. A significant omission I would like to mention is the theory of well-quasi-ordering. This may seem rather strange; after all, the proof of Wagner's conjecture on minor well-quasi-ordering was the guiding force behind the ultimate goal of the development of the Minor Structure Theorem. However, I feel the subject is covered elsewhere (e.g., in [1]) to a satisfactory depth, and that it is difficult to go beyond the basic ideas showcased on trees and graphs on surfaces without getting bogged down in technical details. And admittedly, this is one of the parts of the theory which my own research did not touch and consequently I lack the depth of understanding needed to give it justice.

Finally, let me note that although I try to mention many of the more recent developments in the structural theory of graph minors, including the new improved proof of the Minor Structure Theorem by Kawarabayashi, Thomas, and Wollan [2], the presentation mostly follows the "classical" development of the theory in the series of papers by Robertson and Seymour, which I find more instructive.

Kladno, Czech Republic Zdeněk Dvořák
January 2025

References

1. R. Diestel, *Graph Theory*. Graduate Texts in Mathematics, vol. 173, 5th edn. (Springer, Heidelberg, 2016)
2. K. Kawarabayashi, R. Thomas, P. Wollan, Quickly excluding a non-planar graph (2020). ArXiv, 2010.12397

Declarations

Competing Interests The author has no competing interests to declare that are relevant to the content of this manuscript.

Contents

1 Introduction ... 1
1.1 Notation ... 2
1.2 Minors: Definitions and Basic Results ... 5
References ... 7

Part I Understanding the Structure Theorem

2 Tree Decompositions and Treewidth ... 11
2.1 The Basic Concepts ... 12
2.2 Statement of the Minor Structure Theorem ... 16
2.3 Treewidth ... 18
2.4 Balanced Separators and Unbreakable Sets ... 19
2.5 Tangles ... 22
2.6 Brambles ... 26
2.7 Grid Minors and Walls ... 28
2.8 Which Obstruction Is the Best One? ... 32
2.9 Operations on Tangles, Rank, and Freedom ... 34
2.10 Proof of the Grid Theorem (⤳) ... 39
2.10.1 Pregrids ... 39
2.10.2 Cleaning Up Path Systems ... 41
2.10.3 Building the Base ... 43
2.10.4 Constructing the Grid ... 45
2.11 Erdős–Pósa for Planar Minors ... 46
2.12 Treewidth in Graphs on Surfaces ... 51
2.13 Respectful Tangles ... 60
2.13.1 Slopes ... 63
2.13.2 Pretangles and Slopes ... 66
2.13.3 Tangles and Slopes ... 68
2.14 Related Notions (⤳) ... 70
2.14.1 Branch Decompositions and Related Parameters ... 70
2.14.2 Pathwidth ... 75

2.14.3 More Restrictive Parameters 77
References 80

3 **Linkedness** 83
3.1 Exploiting a Clique Minor 84
3.2 Linkedness in Graphs of High Connectivity 87
3.3 Unique Linkage Theorem 90
3.4 Linked Tree Decompositions 93
References 100

4 **Graphs on Surfaces** 101
4.1 Linkage Across a Disk or Cylinder 102
4.2 Metric from Respectful Tangles 106
4.2.1 The Definition of the Metric 107
4.2.2 The Structure of Balls 109
4.2.3 Clearing a Zone 115
4.3 Linking Through a Cylindrical Grid 122
4.4 Linking Across a Surface 129
4.5 Constructions of Minors 134
References 151

5 **Towards the Structure Theorem** 153
5.1 Disjoint Crossed Paths 153
5.2 Flat Wall Theorem 163
5.3 Testing the Presence of a Minor 169
5.4 Structure Around a Crossing 176
5.5 Local Form of the Minor Structure Theorem 182
5.6 Arranging a Graph on a Surface 188
5.7 Finishing the Proof 195
References 202

6 **Pointers and Sources** 203
6.1 Sources 204
References 205

Part II Using the Structure Theorem

7 **Low-Treewidth Colorings** 209
7.1 Treewidth-Fragility of Minor-Closed Classes 210
7.2 Applications of Treewidth-Fragility 215
7.3 Approximation of the Chromatic Number 219
7.4 Degeneracy-Treewidth Partitioning 222
References 224

8 **Tighter Grid Theorem** 225
8.1 Linear-Sized Grid Minors 225
8.2 Linear-Sized Grid Contractions 228
References 232

9 Topological Minors 233
9.1 Untangling the Cuts 236
9.2 Structure in Minor-Free Graphs 239
9.3 Dealing with a Clique Minor 246
9.4 Local to Global 248
References 253

10 Minors in Large Connected Graphs 255
10.1 Localizing Apices 256
10.2 Taming Vortices 260
10.2.1 From Vortical Decompositions to Strong Vortical Decompositions 261
10.2.2 Guardrails Around a Vortex 263
10.2.3 Enlarging a Vortex 264
10.2.4 Establishing a Linkage 269
10.3 Improved Structure Theorem 271
10.4 Finding Large Bipartite Minors 273
10.5 Coloring K_t-Minor-Free Graphs 276
References 282

11 Sources 283
References 284

Part III Avoiding the Structure Theorem

12 Sublinear Separators 287
12.1 Existence of Small Separators 288
12.2 Applications 290
References 294

13 Chordal Partitions 295
13.1 Fractional Chromatic Number 296
13.2 Degeneracy and Its Generalizations 301
References 305

14 Chromatic Number 307
14.1 Density 308
14.2 Beyond the Density 312
14.3 Relaxed Colorings 314
References 318

15 Product Structure 321
15.1 Product Structure in Graphs on Surfaces 322
15.2 Beyond Embedded Graphs 326
15.3 Applications 330
References 337

16 Iterated Layerings ... 339
16.1 Iterated BFS Layerings ... 339
16.2 Baker-amenable Classes ... 347
References ... 357

17 Isomorphism Testing ... 359
17.1 Weisfeiler-Lehman Algorithm ... 360
17.2 Small Cuts from Stable Colorings ... 362
17.3 Handling the Small Cuts ... 366
17.4 Group Theoretic Ingredients ... 368
17.5 Isomorphism Testing ... 373
References ... 375

18 Sources ... 377
References ... 377

Name Index ... 379

Index ... 381

Chapter 1
Introduction

The goal of this book is to give the reader a solid foundation in advanced aspects of graph minor theory, and studying it "cover to cover" should achieve that. However, not everyone has the time to study the whole book in detail. Since the book focuses on the development of a theory, the later chapters in general rely on the material presented so far. Still, it is sometimes possible to skip some parts or even to jump directly to a section one is interested in.

For a reader interested in the proof of the Minor Structure Theorem, I would recommend to simply read Part I in full; it gradually builds the tools and intuitions necessary for this task. On the other hand, if one is interested only in the applications of the Theorem that we give in Part II, much of this first part can be skipped. My recommendation would be to:

- Skim over Chap. 2. This chapter introduces important notions concering tree decompositions and the dual concepts (in particular, tangles) which are then used throughout the book. The reader should already be familiar with many of them, but I would recommend to pay attention to Sect. 2.13, which studies tangles in embedded graphs.
- Read Chap. 3; the connectivity notions and the methods introduced in this chapter are often useful in the applications.
- Chapter 4 is only needed for the applications in Chaps. 9 and 10.
- From Chap. 5, only the Sect. 5.5 introducing the local forms of the Minor Structure Theorem should be needed.

The applications showcased in Part II are independent and each chapter can be read separately. Finally, Part III is almost entirely independent on the first two parts, with the exception of the basic concepts introduced in Chap. 2 and a few motivational references.

The structural theory of graph minors is somewhat infamous for being quite technical. I generally try to avoid technical arguments that in my opinion do not contribute much in terms of exposition or interesting proof ideas, and refer the

© The Author(s), under exclusive license to Springer Nature Switzerland AG 2025
Z. Dvořák, *Graph Minors*, Springer Monographs in Mathematics,
https://doi.org/10.1007/978-3-031-87469-7_1

reader to the relevant papers if needed. However, occasionally I felt it suitable to include them, usually because of the importance of the overall result or discussed concept. As an example, I give a proof of the Grid Theorem in Sect. 2.10, even though its ideas are not needed anywhere in the book and to my knowledge do not have many applications elsewhere, either. To make it clear that a certain section or proof is optional and safe to skip, I mark them by the (⤼) symbol.

The sources on which the presentation is based are listed at the end of each part. The notation used throughout the book is fairly standard, familiar to anyone with interest in modern graph theory. Nevertheless, there are many slight variations used by different authors, and thus we prefer to go over the fundamentals anyway.

1.1 Notation

Graphs are generally allowed to have parallel edges and loops, unless they are explicitly specified to be *simple*. When we contract an edge, the edge that we contracted is eliminated, but we only suppress parallel edges or loops when we work with simple graphs. In particular, if there are $k \geq 2$ edges between two vertices and we contract one of them, this results in $k - 1$ new loops. The graph obtained from a graph G by contracting an edge e is denoted by G/e. We say that a graph H is a ***minor*** of G if a graph isomorphic to H can be obtained from a subgraph of G by contracting edges. A graph is *H-minor-free* if G does not contain H as a minor. If $\mathcal{P}$ is a partition of $V(G)$, then $G/\mathcal{P}$ denotes the graph obtained by contracting each of the parts to a single vertex; in this case, the loops are always eliminated and parallel edges suppressed.

For a graph G and a vertex $v \in V(G)$, $N_G(v)$ is the set of vertices of G adjacent to v (excluding v when there is a loop on v); the vertices in $N_G(v)$ are the *neighbors* of v. We let $N_G[v] = N_G(v) \cup \{v\}$. For $X \subseteq V(G)$, we let $N_G(X)$ be the set of the vertices in $V(G) \setminus X$ with a neighbor in X. When the graph G is clear from the context, we drop the subscript and use $N(v)$, $N[v]$, and $N(X)$ for brevity. We use $|G|$ and $\|G\|$ to refer to the number of vertices and edges of G, respectively. For graphs G_1 and G_2, by $G_1 \cap G_2$ we mean the graph with vertex set $V(G_1) \cap V(G_2)$ and the edge set $E(G_1) \cap E(G_2)$, and similarly, the graph $G_1 \cup G_2$ has vertex set $V(G_1) \cup V(G_2)$ and edge set $E(G_1) \cup E(G_2)$.

A ***vertex separation*** of a graph G is a pair (A, B) of subsets of $V(G)$ such that $V(G) = A \cup B$ and G has no edge with one end in $A \setminus B$ and the other end in $B \setminus A$; i.e., both $A \setminus B$ and $B \setminus A$ induce unions of connected components of $G - (A \cap B)$. The size of the set $A \cap B$ is the *order* of the vertex separation (A, B). The vertex separation is *non-trivial* if $A \not\subseteq B$ and $B \not\subseteq A$. Note that a graph is k-connected if and only if it has no non-trivial vertex separation of order less than k.

A ***separation*** of a graph G is a pair (A, B) of edge-disjoint subgraphs of G such that $G = A \cup B$, and the *order* of the separation is $|A \cap B|$. Note that $(V(A), V(B))$ is a vertex separation of G of the same order. Separations provide a finer decomposition of G into two parts on a vertex cut than vertex separations, since they explicitly specify how the edges with both ends in the cut are partitioned. This is occasionally useful in applications, though also somewhat more cumbersome.

For any (vertex) separations (A, B) and (C, D) of the same graph, $(A \cap C, B \cup D)$ and $(A \cup C, B \cap D)$ are also (vertex) separations of the same graph, and their orders satisfy the following equality:

$$|A \cap B| + |C \cap D| = |A \cap C \cap (B \cup D)| + |(A \cup C) \cap B \cap D| \tag{1.1}$$

A *tree* is a connected graph without cycles. A *rooted tree* is a tree T with a specified vertex $r \in V(T)$ called the *root* of T. Note that for each vertex $v \in V(T)$, there exists a unique path P_v in T from v to r. If $u \in V(P_v) \setminus \{v\}$, we say that u is an *ancestor* of v, and v is a *descendant* of u. The descendants of v that are adjacent to it are the *children* of v, and for $v \neq r$, the unique ancestor of v that is adjacent to it is the *parent* of v. The *subtree of T rooted in v* is the subtree T' of T induced by v and its descendants, with v being the root of T'. The *depth* of a rooted tree T is the maximum number of vertices of a path in T starting in the root.

We often work with graphs drawn on surfaces. A ***surface*** is a compact 2-dimensional manifold, without boundary unless explicitly specified otherwise. It is well known that every surface is homeomorphic to one obtained from the sphere by adding finitely many handles and crosscaps. Sometimes, it is also convenient to allow the usage of crosshandles, though these are equivalent to adding two crosscaps. Surfaces with boundary additionally contain finitely many holes. The ***Euler genus*** g of the surface is twice the number of handles plus the number of crosscaps used to obtain it. For a surface Σ with boundary, we use $\mathrm{bd}(\Sigma)$ to denote the boundary. The basic surfaces are the *sphere*, the *projective plane* (the sphere with one crosscap), the *torus* (the sphere with a handle), and the *Klein bottle* (the sphere with a crosshandle or two crosscaps). The basic surfaces with boundary are the *disk* (sphere with one hole) and the *cylinder* (sphere with two holes). The *open disk* is the interior of the disk; in contrast, we sometimes use the term *closed disk* to refer to the disk.

A ***drawing*** of a graph G on a surface Σ assigns to each vertex $v \in V(G)$ a distinct point $p(v) \in \Sigma$ and to each edge $e = uv \in E(G)$ a simple curve $p(e)$ in Σ with ends $p(u)$ and $p(v)$ such that

(i) $p(e)$ intersects $p(V(G))$ only in its endpoints, and
(ii) for any other edge $e' \in E(G)$, $p(e)$ intersects $p(e')$ only in their shared endpoints (if any).

In case that e is a loop, $p(e)$ is a simple closed curve, instead. Sometimes, we are going do need to consider drawings with crossings; in that case, (ii) is replaced by the condition that $p(e) \cap p(e')$ is finite and no point in $\Sigma \setminus p(V(G))$ is contained in more than two curves representing edges. The *crossings* are the points in $\Sigma \setminus p(V(G))$ contained in two curves representing edges. Throughout the book, drawings are only allowed to contains crossings if this is explicitly specified.

We generally only consider one fixed drawing of the graph on a surface; thus, to simplify the notation, we are not going to distinguish between the graph G or its parts (vertices, edges, ...) and the subsets of Σ representing them. Thus, for instance, if Δ is a subset of Σ, we are going to write $v \in \Delta$ instead of $p(v) \in \Delta$, $e \subseteq \Delta$ instead of saying that all the points of the curve $p(e)$ are contained in Δ, etc. Conversely, in case that every edge $e \in E(G)$ satisfies $e \subseteq \Delta$ or is disjoint from Δ except possibly for its endpoints, by $G \cap \Delta$ we mean the subgraph of G consisting of vertices and edges whose drawings are contained in Δ, as well as the drawing of this subgraph in Δ.

The ***faces*** of the drawing of G on a surface Σ are the maximal arcwise-connected subsets of $\Sigma \setminus G$. A face of a graph drawn on a surface is *cellular* if it is homeomorphic to an open disk (and in particular, its boundary is connected). A drawing of a graph is ***cellular*** if all its faces are cellular. Note that a plane drawing of a graph G is cellular if and only if G is connected. If g is the Euler genus of Σ and f is the number of faces of G, then the ***generalized Euler's formula*** states that

$$\|G\| \leq |G| + f + g - 2,$$

with equality exactly if the drawing of G is cellular. A well-known consequence that we often use is that

$$\|G\| \leq 3|G| + 3(g - 2)$$

if G is simple and $|G| \geq 3$.

A cycle C in G is *contractible* if the closed curve created by composing the curves representing the edges of C is contractible, i.e., can be continuously deformed to a single point. A curve in a surface is ***G-normal*** if it only intersects the drawing of G in the points representing vertices. The ***representativity*** of a drawing of G on a surface other than the sphere is the minimum number of intersections of G with a non-contractible closed curve; clearly, this minimum number of intersections is achieved by a G-normal closed curve. The ***edgewidth*** of the drawing of G is the length of the shortest non-contractible cycle in G. As our focus is not on the topological aspects of graphs on surfaces, we are going to treat topology on a somewhat "intuitive" level; e.g., we are going to use claims such as the following ones without worrying about their proofs.

- If G is a graph drawn on a surface Σ and every cycle in G is contractible, then there exists an open disk $\Lambda \subseteq \Sigma$ with $G \subseteq \Lambda$.
- If C is a contractible cycle drawn on Σ, then C has exactly two faces, at least one of which is an open disk.

We let $\mathbb{N}$ denote the set of all non-negative integers, i.e., $0 \in \mathbb{N}$. For $n \in \mathbb{N}$, we let $[n]$ denote the set $\{1, \ldots, n\}$; i.e., $[0] = \emptyset$. For a function $f : A \to B$, we let $\mathrm{dom}(f) = A$ and $\mathrm{img}(f) = \{f(a) : a \in A\}$; the latter is taken as a set, not as a multiset.

In some of the proofs, we adopt the following convention to ensure that the ideas are not obscured by technical computations: By $a \ll b \ll c$, we mean that b should be chosen large enough compared to a so that any of the inequalities arising in the proof involving only a and b are satisfied; and then c should be chosen large enough so that any of the inequalities involving a, b, and c are satisfied.

1.2 Minors: Definitions and Basic Results

There may be several non-equivalent ways how to obtain a graph H as a minor of another graph G, and it is often needed to specify one. A ***model*** μ of a graph H in a graph G is a function that

(i) maps vertices of H to pairwise vertex-disjoint connected subgraphs of G,
(ii) maps edges of H to pairwise different edges of G not contained in the subgraphs assigned to vertices of H, and such that
(iii) for each edge $e = uv \in E(H)$, the edge $\mu(e)$ has one end in $\mu(u)$ and the other end in $\mu(v)$.

Clearly H is a minor of G if and only if there exists a model of H in G. Note that the definition becomes slightly simpler when H is a simple graph, as then there is no need to say in (ii) that μ cannot assign the same edge of G to distinct edges of H, and that edges belonging to subgraphs assigned to vertices cannot be assigned to edges of H, as (iii) implies that the former is only a concern for parallel edges and the latter only for loops.

The ***Hadwiger number*** $\mathrm{Had}(G)$ of a graph G is the maximum integer k such that the clique K_k is a minor of G. Famous conjecture of Hadwiger [3] states that the chromatic number $\chi(G)$ of a graph G is always at most its Hadwiger number.

Conjecture 1.1 For every positive integer t, every K_{t+1}-minor-free graph is t-colorable.

It is known that K_k-minor-free graphs have bounded average degree. Indeed, we have the following asymptotically tight result.

Theorem 1.2 (Thomason [5]) *For every positive integer k, let $f(k)$ be the minimum integer such that every simple graph of average degree at least $f(k)$ contains K_k as a minor. Then*

$$f(k) = (0.638\ldots + o(1))k\sqrt{\log k}.$$

Here, the constant $0.638\ldots$ is the solution to a specific equation. Consequently, every K_k-minor-free graph has a vertex with most $O(k\sqrt{\log k})$ neighbors, and thus the greedy algorithm shows that the chromatic number of a K_k-minor-free graph is $O(k\sqrt{\log k})$. For more than 30 years, this was the best partial result towards Hadwiger's conjecture; only recently, Delcourt and Postle [2] improved the bound to $O(k \log\log k)$. Although many people (including myself) doubt the validity of Hadwiger's conjecture, it undeniably is one of the most influential questions in both structural and chromatic graph theory, and we discuss it in more detail in Chap. 14.

We say that a graph H is a ***topological minor*** of G if a subdivision of H is a subgraph of G. More precisely, there exists a function π mapping vertices of H to pairwise different vertices of G (the ***branch vertices*** of the topological minor) and each edge $e = uv$ of H to a path $\pi(e)$ of G with ends $\pi(u)$ and $\pi(v)$ and with at least one edge such that for distinct edges e and e', the paths $\pi(e)$ and $\pi(e')$ intersect exactly in their common endpoints (if any). Note that the condition that $E(\pi(e)) \neq \emptyset$ is only relevant when e is a loop. We say that π is a ***topological model*** of H in G, and we let $\pi(H) = \pi(V(H)) \cup \bigcup \pi(E(H))$ be the corresponding subgraph of G forming a subdivision of H. Note that if H is a topological minor of G, then it is also a minor of G (obtained by contracting all but one edge of each of the paths forming the topological minor), but the converse is false; for example, note that a graph of maximum degree Δ can only contain graphs of maximum degree at most Δ as topological minors, but no such restriction holds for ordinary minors when $\Delta \geq 3$. However, let us note the following important special case.

Observation 1.3 *If H is a graph of maximum degree at most three and G contains H as a minor, then it contains H as a topological minor as well.*

For an injective function ρ from a set $\mathrm{dom}(\rho) \subseteq V(H)$ to $V(G)$, we say that a model μ of H in G is *ρ-rooted* if $\rho(v) \in V(\mu(v))$ for each $v \in \mathrm{dom}(\rho)$. When H has a ρ-rooted model in G, we say that H is a ***ρ-rooted minor*** of G. Informally, we can obtain H from a subgraph of G by contracting edges so that the vertex $\rho(v)$ is contracted into v for every root $v \in \mathrm{dom}(\rho)$. To simplify the notation, in case that the vertices of $M = \mathrm{dom}(\rho)$ are shared by H and G and ρ is the identity function, we say that H is an *M-rooted minor* of G. As an important example, if H is a matching and $\mathrm{dom}(\rho) = V(H)$, then H is a ρ-rooted minor of a graph G if and only if G contains $\|H\|$ pairwise vertex-disjoint paths $\{P_e : e \in E(H)\}$, where for $e = uv \in V(H)$, the path P_e connects $\rho(u)$ to $\rho(v)$. Let us remark that *ρ-rooted topological models* (and ρ-rooted topological minors) are defined analogously, as topological models π such that π restricted to $\mathrm{dom}(\rho)$ is equal to ρ.

For a set $S \subseteq V(G)$, we say that a graph G is *(S, k)-connected* if G has no non-trivial vertex separation (A, B) of order less than k with $S \subseteq A$. As another basic result, let us note the following sufficient condition for the presence of a rooted triangle minor.

Lemma 1.4 *Suppose ρ is a function mapping vertices of a triangle K_3 to distinct vertices of a simple graph G. If G is $(\mathrm{img}(\rho), 3)$-connected, then*

- *$V(G) = \mathrm{img}(\rho)$ and $\|G\| \leq 2$, or*
- *$V(G) \setminus \mathrm{img}(\rho)$ consists of exactly one vertex v and $E(G)$ consists exactly of the three edges between v and $\mathrm{img}(\rho)$, or*
- *K_3 is a ρ-rooted minor of G.*

Proof The claim is clear if $V(G) = \mathrm{img}(\rho)$, and thus suppose that there exists a vertex $v \in V(G) \setminus \mathrm{img}(\rho)$. Since G is $(\mathrm{img}(\rho), 3)$-connected, Menger's theorem implies that there exist three paths P_1, P_2, and P_3 from v to $\mathrm{img}(\rho)$ pairwise intersecting exactly in v. If each vertex of $\mathrm{img}(\rho)$ is in a different component of $G - v$, then since G is $(\mathrm{img}(\rho), 3)$-connected, we conclude that each of the components consists of a single vertex and that the second outcome of the lemma holds. Hence, suppose that $G - v$ contains a path R between $P_1 - v$ and $P_2 - v$. Contracting all edges of P_3, all edges of P_1 and P_2 except for those incident with v, and all but one edge of R gives K_3 as a ρ-rooted minor of G.

A class $\mathcal{G}$ of graphs is ***minor-closed*** if for every $G \in \mathcal{G}$, all minors of G also belong to $\mathcal{G}$. A ***forbidden minor*** for $\mathcal{G}$ is a graph $F \notin \mathcal{G}$ such that all proper minors of F (i.e., all minors of F except for F itself) belong to $\mathcal{G}$. For a set $\mathcal{F}$ of graphs, we say that a graph G is *$\mathcal{F}$-minor-free* if G does not contain any graph in $\mathcal{F}$ as a minor.

Observation 1.5 *Let $\mathcal{G}$ be a minor-closed class of graphs and let $\mathcal{F}$ be the set of forbidden minors for $\mathcal{G}$. Then a graph belongs to $\mathcal{G}$ if and only if it is $\mathcal{F}$-minor-free.*

For example, as shown by Wagner [6], the forbidden minors for planar graphs are K_5 and $K_{3,3}$. Perhaps the most important application of the graph minors theory of Robertson and Seymour (as well as the motivation for its development) is the proof of Wagner's conjecture on well-quasi-ordering property of the minor relation, stated in an equivalent form in the following theorem.

Theorem 1.6 (Robertson and Seymour [4]) *The set of forbidden minors for every minor-closed class of graphs is finite.*

As an example, consider any fixed surface Σ. Since the class of graphs drawn on Σ is minor-closed, there exists a finite set $\mathcal{F}$ such that a graph can be drawn on Σ if and only if it is $\mathcal{F}$-minor-free. Remarkably, there is only one non-trivial surface (projective plane) for which we know exactly what the forbidden minors are [1].

References

1. D. Archdeacon, A Kuratowski theorem for the projective plane. J. Graph Theory **5**, 243–246 (1981)
2. M. Delcourt, L. Postle, Reducing linear Hadwiger's conjecture to coloring small graphs (2021). ArXiv 2108.01633
3. H. Hadwiger, Über eine Klassifikation der Streckenkomplexe. Vierteljschr. Naturforsch. Ges. Zürich **88**, 133–143 (1943)
4. N. Robertson, P.D. Seymour, Graph Minors. XX. Wagner's conjecture. J. Combin. Theory Ser. B **92**, 325–357 (2004)
5. A. Thomason, The extremal function for complete minors. J. Combin. Theory Ser. B **81**, 318–338 (2001)
6. K. Wagner, Über eine Eigenschaft der ebenen Komplexe. Math. Ann. **114**, 570–590 (1937)

Part I
Understanding the Structure Theorem

In order to prove Theorem 1.6, Robertson and Seymour developed a rich theory of graph minors. The most prominent and best known part of this theory is the Minor Structure Theorem giving an approximate characterization of graphs avoiding a fixed minor. However, it also includes important arguments on topics such as graph decompositions, the existence of minors in graphs on surfaces, finding vertex-disjoint paths with prescribed ends, and many others. The goal of this part is to give you an understanding of all these ingredients and to show how they combine in the proof of the Minor Structure Theorem.

We start by developing the theory of tree decompositions, showing how they naturally arise (twice) in the context of the Minor Structure Theorem. We study treewidth and prove the famous Grid Theorem. We then give tools for showing the existence of minors in well-connected graphs and in graphs on surfaces, and use them to prove a weak version of the structure theorem, the Flat Wall Theorem. We study the algorithmic implications of this result, most importantly the polynomial-time algorithm to test the presence of a fixed minor. We describe the structural results about "highly non-planar" parts of a graph, which are needed to turn the proof of the Flat Wall Theorem into the proof of the Minor Structure Theorem. Finally, we describe the stronger "local" form of the Minor Structure Theorem and show how the results obtained so far combine in its proof.

Chapter 2
Tree Decompositions and Treewidth

Tree decompositions and related notions (brambles, tangles, grid minors, ...) play a key role in the formulation of the Minor Structure Theorem, its proof and applications, and in structural graph theory in general. In this chapter, we develop the theory of these concepts.

We expect the reader to be already somewhat familiar with graphs of bounded treewidth, especially from the algorithmic perspective; and in particular, to know the celebrated theorem of Courcelle showing that all problems expressible in the Monadic Second Order Logic can be solved efficiently on graphs of bounded treewidth. However, our focus will be on the structural properties of treewidth.

The Minor Structure Theorem takes the form of stating that every graph avoiding a fixed minor has a tree decomposition into "nice" pieces, which are either small or nearly drawn on a surface. It is important to understand what forces the presence of the large pieces, and how to determine which parts of the graph should belong to a particular piece. Thus, we are going to study various obstructions to small treewidth; the most important for us will be the notion of tangles—informally, a tangle is an orientation of small cuts towards one of such large, well connected pieces.

Next, we need to "nearly draw" these large pieces on a surface. Chapters 4 and 5 are largely devoted to this goal, but we are going to take first steps in this direction. We prove the well-known Grid Theorem, showing that a tangle of large order (equivalent to high treewidth) gives rise to a minor of a large grid. Such a grid minor can be seen as a drawing of a substantial part of the graph on the sphere. We also study the notion of *respectful tangles* arising from drawings of graphs on surfaces, which is very important when we consider which minors appear in these graphs.

Finally, we are going to give a brief overview of other relevant width parameters and their applications in the theory of graph minors.

© The Author(s), under exclusive license to Springer Nature Switzerland AG 2025

Z. Dvořák, *Graph Minors*, Springer Monographs in Mathematics,
https://doi.org/10.1007/978-3-031-87469-7_2

2.1 The Basic Concepts

Let us start by recalling basic concepts and results about graph decompositions which the reader may be already familiar with, in part to fix notation. Although we are mostly going to use tree decompositions, it is occasionally useful to use graphs other than trees to describe the global structure. Hence, for a graph G, we are going to define a more general notion of a decomposition consisting of another graph B and a function β from the vertices of B to subsets of $V(G)$. To decrease the chance of confusion, we call the elements of $V(B)$ *nodes*, while we refer to the elements of $V(G)$ as vertices.

A ***general decomposition*** of a graph G is a pair (B, β), where B is a graph and β assigns a subset of $V(G)$ (called a *bag*) to each node of B so that

(D1) for every edge $uv \in E(G)$, there exists a node $x \in V(B)$ such that $\{u, v\} \subseteq \beta(x)$, and
(D2) for every vertex $v \in V(G)$, the set $\{x \in V(B) : v \in \beta(x)\}$ of nodes whose bags contain v is non-empty and induces a connected subgraph of B.

For a subgraph F of B, let us define $\beta(F) = \bigcup_{v \in V(F)} \beta(v)$. Slightly abusing the notation, for a set $Z \subseteq V(G)$ we let $\beta^{-1}(Z) = \{x \in V(B) : Z \cap \beta(x) \neq \emptyset\}$; hence, (D2) says that for each $v \in V(G)$, $\beta^{-1}(\{v\})$ induces a non-empty connected subgraph of B. The most important special cases of this general concept are

- ***tree decompositions***, where B is a tree,
- ***path decompositions***, where B is a path,
- ***star decompositions***, where B is a star, and
- ***cycle decompositions***, where B is a cycle.

We define the *width* $\text{width}(B, \beta)$ of a general decomposition (B, β) as the maximum of $|\beta(x)| - 1$ over the nodes $x \in V(B)$. Let us remark that the somewhat unfortunate minus one in this definition is forced on us by historical circumstances: It was considered to be desirable for trees to have a tree decomposition of width one rather than two. The ***treewidth*** $\text{tw}(G)$ of a graph G is the minimum possible width of a tree decomposition of G, and the ***pathwidth*** $\text{pw}(G)$ is the minimum possible width of a path decomposition of G. Clearly, $\text{tw}(G) \leq \text{pw}(G)$. For an edge $e = xy$ of B, let $\beta(e) = \beta(x) \cap \beta(y)$. The ***adhesion*** of a general decomposition (B, β) is the maximum of $|\beta(e)|$ over all edges $e \in E(B)$. Importantly, edges of tree decompositions correspond to vertex cuts in the original graph.

Observation 2.1 *If (T, β) is a tree decomposition of a graph G, e is an edge of T, and T_1 and T_2 are the components of $T - e$, then $(\beta(T_1), \beta(T_2))$ is a vertex separation of G of order $|\beta(e)|$.*

In some instances it may be useful to define bags of a general decomposition (B, β) of a graph G not as vertex sets, but as pairwise edge-disjoint subgraphs of G, with property (D1) from the definition replaced by $\beta(B) = G$. In that case, Observation 2.1 gives a separation of G rather than a vertex separation.

General decompositions behave well with respect to minors, because of the following observation generalizing the condition (D2).

Lemma 2.2 *If* (B, β) *is a general decomposition of* G *and* F *is a connected subgraph of* G*, then* $\beta^{-1}(V(F))$ *induces a connected subgraph of* B.

Proof By induction on the number of vertices of F. The basic case $|F| = 1$ is the condition (D2) from the definition of a general decomposition. If $|F| > 1$, let v be a vertex of F such that $F - v$ is connected, i.e., a leaf of a spanning tree of F. By the induction hypothesis both $\beta^{-1}(V(F - v))$ and $\beta^{-1}(\{v\})$ induce connected subgraphs of B. Moreover, by (D1) applied to an edge incident with v, the subgraphs intersect. Therefore, $\beta^{-1}(V(F)) = \beta^{-1}(V(F-v)) \cup \beta^{-1}(\{v\})$ induces a connected subgraph of B. □

Let (B, β) be a general decomposition of a graph G and let μ be a model of H in G. Based on Lemma 2.2, we can turn (B, β) into a general decomposition (B, β_μ) of the minor H of G: For each node $x \in V(B)$, let $\beta_\mu(x) = \{v \in V(H) : V(\mu(v)) \cap \beta(x) \neq \emptyset\}$. Observe that for each $v \in V(H)$, we have $\beta_\mu^{-1}(\{v\}) = \beta^{-1}(V(\mu(v)))$, and thus $\beta_\mu^{-1}(\{v\})$ induces a connected subgraph of B by Lemma 2.2. Hence, (B, β_μ) satisfies the condition (D2), and it is also straightforward to check that it satisfies (D1). We say (B, β_μ) is the *decomposition of* H *induced by* (B, β) *and* μ. Observe that the width of (B, β_μ) is at most as large as the width of (B, β).

Let us also note the following important property of tree decompositions.

Lemma 2.3 *If* $K \subseteq V(G)$ *forms a clique in a graph* G *and* (T, β) *is a tree decomposition of* G*, then there exists a node* $x \in V(T)$ *such that* $K \subseteq \beta(x)$.

We postpone the proof till Sect. 2.6, where we show a more general statement of this form (see Lemma 2.26 and the discussion following it).

A tree decomposition of a graph can be viewed as a description of its construction from smaller graphs through clique-sums. A graph G is a ***clique-sum*** of graphs G_1 and G_2 if G has a vertex separation (A_1, A_2) such that for $i \in [2]$, G_i is isomorphic to the graph obtained from $G[A_i]$ by adding all edges with both ends in $A_1 \cap A_2$ that are not already present in $G[A_i]$. Informally, G is obtained by selecting cliques of the same size in G_1 and G_2, gluing them on these cliques, and possibly deleting some edges of the resulting clique. The ***torso*** of a node $x \in V(T)$ in a tree decomposition (T, β) of a graph G is the graph obtained from $G[\beta(x)]$ by, for each edge $xy \in E(T)$ incident with x, adding all edges with both ends in $\beta(x) \cap \beta(y)$ that are not already present in G. Lemma 2.3 has the following key consequence.

Lemma 2.4 *A graph* G *is obtained by clique-sums (in any order) from graphs* G_1, …, G_n *if and only if* G *has a tree decomposition whose torsos are* $G_1, \ldots, G_n$.

Proof It is easy to see that the backward implication holds. We prove the forward one by induction on n, with the basic case $n = 1$ being trivial. Suppose that G is obtained by clique-sums from graphs $G_1, \ldots, G_n$. As the last step of the process, G is a clique-sum of graphs H_1 and H_2 on cliques Z_1 and Z_2, where H_1 is obtained by clique-sums say from graphs $G_1, \ldots, G_k$ and H_2 from graphs $G_{k+1}, \ldots, G_n$. By the induction hypothesis, H_1 has a tree decomposition (T_1, β_1) with torsos $G_1, \ldots, G_k$ and H_2 has a tree decomposition (T_2, β_2) with torsos $G_{k+1}, \ldots, G_n$. By Lemma 2.3, there exist nodes $x_1 \in V(T_1)$ and $x_2 \in V(T_2)$ such that $Z_1 \subseteq \beta(x_1)$ and $Z_2 \subseteq \beta(x_2)$. Let T be the tree obtained from $T_1 \cup T_2$ by adding the edge x_1x_2 and let β be obtained from $\beta_1 \cup \beta_2$ by replacing the vertices of Z_1 and Z_2 by the corresponding vertices of G. Then (T, β) is a tree decomposition of G with torsos $G_1, \ldots, G_n$. □

By Lemma 2.4, a graph has treewidth at most k if and only if it can be obtained by clique-sums from graphs with at most $k+1$ vertices. Using this characterization, it is easy to see that a clique-sum of graphs G_1 and G_2 has treewidth at most $\max(\mathrm{tw}(G_1), \mathrm{tw}(G_2))$. Similarly, it is easy to see that clique-sums do not increase the chromatic number and the Hadwiger number.

Observation 2.5 *If G is a clique-sum of graphs G_1 and G_2, then*

$$\chi(G) \leq \max(\chi(G_1), \chi(G_2)),$$
$$\mathrm{tw}(G) \leq \max(\mathrm{tw}(G_1), \mathrm{tw}(G_2)), \textit{ and}$$
$$\mathrm{Had}(G) \leq \max(\mathrm{Had}(G_1), \mathrm{Had}(G_2)).$$

By Lemma 2.4, this has the following consequence.

Observation 2.6 *If G has a tree decomposition in which each torso has Hadwiger number at most k, then* $\mathrm{Had}(G) \leq k$.

However, let us remark that because of the edges added when forming torsos, the clique-sum perspective loses some of the information contained in G. In particular, is sometimes problematic that the torsos are not necessarily minors of G. In other words, if G is an H-minor-free graph, its tree decomposition does not necessarily give a way to express it as a clique-sum of smaller H-minor-free graphs. Even in the special case that H is a clique, the converse to Observation 2.6 does not hold. However, let us note the following useful partial converse. A graph G is ***essentially 4-connected*** if it is 3-connected and for every 3-cut S in G, S is an independent set and $G - S$ has exactly two components, one of which consists of a single vertex. Note that we define all cliques to be essentially 4-connected.

Lemma 2.7 *Every simple graph G has a tree decomposition whose torsos are essentially* 4*-connected minors of G.*

Proof It suffices to prove that if G is not essentially 4-connected, then it is a clique-sum of minors G_1 and G_2 of G with fewer vertices. The claim then follows by induction and Lemma 2.4.

Let (A, B) be a non-trivial vertex separation of G of smallest order and let $A \cap B = \{v_1, \ldots, v_k\}$. If (A, B) has order at most one, then G is a clique-sum of $G[A]$ and $G[B]$, and thus the claim holds. Hence, suppose that G is 2-connected. If (A, B) has order two, then since G is 2-connected, the graphs $G[A]$ and $G[B]$ are connected. By contracting $G[B]$ to a single edge, we conclude that $G[A]+v_1v_2$ is a minor of G, and similarly $G[B]+v_1v_2$ is a minor of G. Note that G is a clique-sum of $G[A]+v_1v_2$ and $G[B]+v_1v_2$. Hence, we can assume that G is 3-connected.

Since G is not essentially 4-connected, it follows that (A, B) has order three. Let ρ be a bijection between the vertices of a triangle K_3 and $\{v_1, v_2, v_3\}$, and let $K = \{v_1v_2, v_2v_3, v_1v_3\}$. If K_3 is a ρ-rooted minor of $G[B]$, then $G[A]+K$ is a minor of G. If additionally K_3 is a ρ-rooted minor of $G[A]$, then $G[B]+K$ is a minor of G and G is a clique-sum of $G[A]+K$ and $G[B]+K$. Suppose that K_3 is not a ρ-rooted minor of $G[A]$. Since G is 3-connected, $G[A]$ is $(A \cap B, 3)$-connected, and by Lemma 1.4, we conclude that $V(A \cap B)$ is an independent set and $A \setminus B$ consists of a single vertex adjacent to all three vertices of $V(A \cap B)$.

If G is not a clique-sum of minors with fewer vertices, then this conclusion holds for any non-trivial vertex separation of G of order three. Note that this implies that for every 3-cut S in G,

- the set S is independent in G,
- the graph $G - S$ has at most three components,
- if $G - S$ has two components then one of them is a single vertex, and
- if $G - S$ has three components then each of them is a single vertex.

Since we assumed that G is not essentially 4-connected, the last possibility holds, and thus $G = K_{3,3}$. In this case we observe that K_4 is a minor of G and that G is a clique-sum of three copies of K_4. □

Tree decompositions are also important from the algorithmic perspective; assuming that we can solve a problem efficiently for the torsos of the decomposition, it is often possible to extend the solution to the whole graph by dynamic programming. Consequently, many algorithmic problems which are hard in general graphs admit efficient algorithms when restricted to graphs of bounded treewidth. In particular, the famous meta-algorithmic result of Courcelle [8] shows this is the case for all problems expressible in the Monadic Second Order Logic.

Theorem 2.8 *For every graph property $\mathcal{P}$ expressible in the Monadic Second Order Logic and every integer t, there exists an algorithm that, given a graph G of treewidth at most t, decides in time $O(|G|)$ whether G satisfies the property $\mathcal{P}$.*

Recall that a property is expressible in the Monadic Second Order Logic if it is described by a logic formula with variables for vertices, edges, sets of vertices, and sets of edges, quantification over such variables, and predicates for adjacency, incidence of vertices and edges, and set membership. For example, the property that a graph G is 3-colorable can be expressed as

$$\begin{aligned}(\exists A_1, A_2, A_2 \subseteq V(G))\ ((\forall v)\ v \in A_1 \vee v \in A_2 \vee v \in A_3)\\ \wedge (\forall u, v \in A_1)\ uv \notin E(G)\\ \wedge (\forall u, v \in A_2)\ uv \notin E(G)\\ \wedge (\forall u, v \in A_3)\ uv \notin E(G).\end{aligned}$$

Thus, we can decide whether a graph of bounded treewidth is 3-colorable in linear time. In contrast, even determining if a planar graph is 3-colorable is NP-complete [15]. Note that Theorem 2.8 has many variations; e.g., if $\mathcal{P}$ is a property of subgraphs of G, the algorithm can return a largest subgraph satisfying the property or count the number of such subgraphs; we will somewhat loosely refer to Theorem 2.8 even for such generalizations.

2.2 Statement of the Minor Structure Theorem

Tree decompositions feature heavily in the Minor Structure Theorem. They appear on the global level, dividing the graph on small cuts to more complex, well connected pieces; this is justified by Observation 2.6. They also appear once more in the description of the pieces of decomposition, in the definition of *vortices*.

A ***bordered graph*** is a graph F with a subset ∂F of its vertices specified as the *boundary* of F. A ***society*** is a bordered graph where a cyclic ordering is specified for ∂F. A ***vortical decomposition*** of F is a cycle decomposition $(\partial F, \beta)$ of F, where we slightly abuse the notation and view ∂F as a cycle rather than as a cyclic sequence of vertices, such that

(V) $v \in \beta(v)$ for every $v \in \partial F$.

For example, consider the following societies:

- H_1 is the society such that $\partial H_1 = v_1 \dots v_{4k}$, $V(H_1) = \partial H_1$, and H_1 has edges $v_{4i-3}v_{4i-1}$ and $v_{4i-2}v_{4i}$ for $i \in [k]$; that is, H_1 consists of k consecutive crosses.
- H_2 is the society such that $\partial H_2 = v_1 \dots v_{4k}$, $V(H_2) = \partial H_2$, and H_2 has edges $v_i v_{i+2k}$ for $i \in [2k]$; that is, H_2 is a "crosscap".

Then H_1 has a vortical decomposition of width three, obtained by repeating the bag $\{v_{4i-3}, \dots, v_{4i}\}$ four times for each $i \in [k]$. On the other hand, although H_2 has a cycle (or even path) decomposition of width one, it is easy to see that the condition (V) prevents us from obtaining a vortical decomposition of H_2 of bounded width.

In the Minor Structure Theorem, vortices (societies with a vortical decomposition of bounded width) attach to a graph drawn on a surface through their boundaries. More precisely, let G_0 be a graph, and for $i \in [m]$, let F_i be a society obtained from a subgraph of G_0 by choosing a cyclically ordered boundary ∂F_i such that

- all neighbors (in G_0) of every vertex of $V(F_i) \setminus \partial F_i$ belong to F_i, and
- for distinct $i, j \in [m]$, the societies F_i and F_j are vertex-disjoint.

Let G be the subgraph of G_0 with the vertex set

$$V(G) = V(G_0) \setminus \bigcup_{i=1}^{m} (V(F_i) \setminus \partial F_i)$$

and the edge set $E(G) = E(G_0) \setminus \bigcup_{i=1}^{m} E(F_i)$. We say that G_0 is ***drawn on* Σ *up to outgrowths*** $F_1, \ldots, F_m$ if there exists a drawing of G on Σ and pairwise disjoint disks $\Delta_1, \ldots, \Delta_m \subset \Sigma$ such that for $i \in [m]$,

- the interior of Δ_i is disjoint from G, and
- the boundary $\mathrm{bd}(\Delta_i)$ of Δ_i intersects G exactly in the vertices of ∂F_i, and the cyclic ordering in which these vertices appear on $\mathrm{bd}(\Delta_i)$ matches the fixed cyclic ordering of ∂F_i (up to reversal).

We say that G is the ***surface part*** of G_0. We say that G_0 is ***drawn on* Σ *up to* m *vortices of width at most*** d if each of $F_1, \ldots, F_m$ has a vortical decomposition of width at most d.

For every surface Σ and integers m and d, there exists k such that K_k is not a minor of any graph drawn on Σ up to m vortices of width at most d.[1] Moreover, consider the society H_1 discussed above, consisting of many crossings, and suppose that it appears as one of the vortices. It should be clear that the crossings cannot in general be eliminated by drawing the graph on a surface of slightly larger genus. It also may not be possible to cut H_1 off from the rest of G_0 by a small separation, and thus we cannot get rid of H_1 using the global tree decomposition. This indicates that any structure theorem for K_k-minor-free graphs must necessarily deal with vortices of bounded width.

The last ingredient of the Minor Structure Theorem arises from the following observation: If a graph G if K_k-minor-free and a graph G' is obtained from G by adding a single vertex v adjacent to an arbitrary subset of $V(G)$, then G' is K_{k+1}-minor-free. As we have no control over the neighborhood of v, the structure theorem must necessarily allow for a bounded number of exceptional vertices, called ***apex vertices*** or *apices*.

With these observations in mind, we say that a graph G_1 can be ***(a, m, d)-nearly drawn on*** Σ if there exists a set $A \subseteq V(G_1)$ of size at most a such that $G_1 - A$ can be drawn on Σ up to at most m vortices of width at most d. Also, let us say a surface is ***H-avoiding*** if the graph H cannot be drawn on it. We now have all the ingredients necessary to state the best-known version of the Minor Structure Theorem.

Theorem 2.9 (Robertson and Seymour [24]) *For every graph H, there exist constants a, m, and d such that every H-minor-free graph G has a tree decomposition in which each torso with more than a vertices can be (a, m, d)-nearly drawn on an H-avoiding surface.*

By Lemma 2.4, this is equivalent to G being obtained by clique-sums from graphs with at most a vertices and (a, m, d)-nearly drawn graphs. Let us remark that we chose to specifically emphasize the case of the pieces of the decomposition having size at most a. This does not matter when H is non-planar, as in that case any

[1] Intuitively, a vortex of bounded width only enables us to incorporate crossings in a drawing of K_k on Σ that appear close to a face of the drawing, and it is not possible to draw K_k with all crossings close to just a few of its faces; for a less handwavy argument, see Corollary 7.6.

such piece is $(a, 0, 0)$-nearly drawn on the sphere, and the sphere is H-avoiding. However, if H is planar, then no surface is H-avoiding, and thus Theorem 2.9 in this case just states that $\mathrm{tw}(G) < a$. This is a result of separate interest and with substantially simpler (though still very much non-trivial) proof, see Sects. 2.7 and 2.10 for more details. Many other sources choose to make the statement of Theorem 2.9 more elegant by dropping the part concerning the torsos of size at most a; this is possible by either restricting the statement to non-planar graphs H, or by allowing for a null surface in which only the graph with no vertices can be drawn.

2.3 Treewidth

Let us now explore some basic properties of treewidth. Lemma 2.3 implies that $\mathrm{tw}(G) \geq \omega(G) - 1$. Another way to show this is by observing that the treewidth is an upper bound on the degeneracy of the graph. For a non-negative integer d, we say that a graph G is ***d-degenerate*** if every subgraph of G has a vertex of degree at most d. Equivalently, the vertices of G can be ordered in such a way that each vertex has at most d neighbors that appear after it in the ordering.

Lemma 2.10 *Every simple graph G is* $\mathrm{tw}(G)$*-degenerate.*

Proof If H is a subgraph of G, then $\mathrm{tw}(H) \leq \mathrm{tw}(G)$. Hence, it suffices to show that every graph G contains a vertex of degree at most $\mathrm{tw}(G)$. Let (T, β) be a tree decomposition of G of width $\mathrm{tw}(G)$ with the smallest number of nodes, and let x be a leaf of T. If $|T| > 1$ and y is the neighbor of x in T, then $\beta(x) \not\subseteq \beta(y)$, as otherwise $(T - x, \beta \restriction V(T - x))$ is a tree decomposition of G of width $\mathrm{tw}(G)$ with fewer nodes. Hence, by (D2) there exists a vertex $v \in \beta(x)$ that does not appear in any other bag of the decomposition. By (D1), all edges incident with v have both ends in $\beta(x)$, and thus $\deg v \leq |\beta(x)| - 1 \leq \mathrm{tw}(G)$. □

Recall that if μ is a model of H in G, then the induced tree decomposition (T, β_μ) of H has at most as large width as (T, β), giving the following key observation.

Observation 2.11 *If H is a minor of G, then* $\mathrm{tw}(H) \leq \mathrm{tw}(G)$. *In particular,* $\mathrm{tw}(G) \geq \mathrm{Had}(G) - 1$.

That is, having a large clique minor forces the treewidth to be large. Interestingly, clique minors are the only obstructions to treewidth at most three.

Lemma 2.12 *For any $k \leq 2$, a graph has treewidth at most k if and only if it does not contain K_{k+2} as a minor.*

Proof Without loss of generality, we can assume that G is a simple graph. If $\mathrm{tw}(G) \leq k$, then K_{k+2} is not a minor of G by Observation 2.11. Conversely, suppose that K_{k+2} is not a minor of G. By Lemma 2.7, there exists a tree decomposition (T, β) such that each torso H is 3-connected and a minor of G, and in particular K_{k+2} is not a minor of H. We claim that $|H| \leq k + 1$ for every torso H, and thus the width of (T, β) is at most k.

If $k = 0$, then since H is connected and K_2-minor-free (i.e., edgeless), we have $|H| = 1$. If $k = 1$, then since H is K_3-minor-free, H does not contain a cycle, and since H is 2-connected, we have $|H| \leq 2$. Finally, consider the case $k = 2$ and suppose for a contradiction that $|H| \geq 4$. Let C be a shortest (and thus necessarily induced) cycle in H. Since H is 3-connected, we have $H \neq C$, and thus there exists a vertex $v \in V(H) \setminus V(C)$. However, then Menger's theorem implies that there are three paths from v to C intersecting only in C, and these paths together with C give K_4 as a minor in H, which is a contradiction. □

Hence, for $k \leq 2$, K_{k+2} is the only forbidden minor for having treewidth at most k. However, this stops being true for $k = 3$. Indeed, there are planar graphs of arbitrarily large treewidth, but planar graphs are K_5-minor-free. The exact set of forbidden minors for treewidth at most three is known [1, 26], but for larger treewidth the number of forbidden minors grows fast. Moreover, it is unclear whether their knowledge would give any significant insights. Less precise but better structured obstructions to small treewidth turn out to be much more useful. We devote the following sections to describing the most important ones and the relationships between them.

2.4 Balanced Separators and Unbreakable Sets

Let w be an assignment of non-negative weights to vertices of a graph G. For a subgraph $D \subseteq G$, we let $w(D) = \sum_{v \in V(D)} w(v)$. Similarly, the weight $w(A)$ of a set $A \subseteq V(G)$ is $\sum_{v \in A} w(v)$. A set $S \subseteq V(G)$ is a ***w*-balanced cut** in a graph G if every component C of $G - S$ satisfies $w(C) \leq \frac{2}{3} w(G)$. A vertex separation (A, B) of a graph G is *w-balanced* if $w(A \setminus B) \leq \frac{2}{3} w(G)$ and $w(B \setminus A) \leq \frac{2}{3} w(G)$.

The constant $\frac{2}{3}$ in these definitions is a bit arbitrary; any positive number smaller than 1 would work, and often $\frac{1}{2}$ is a more natural choice. A motivation for the value $\frac{2}{3}$ comes from the following observation, which does not hold as an equivalence with other values.

Observation 2.13 *Let w be an assignment of non-negative weights to vertices of a graph G. A set $S \subseteq V(G)$ is a w-balanced cut if and only if G has a w-balanced vertex separation (A, B) such that $S = A \cap B$.*

Proof If (A, B) is a vertex separation, then every component of $G - (A \cap B)$ is contained in $A \setminus B$ or $B \setminus A$; hence, if (A, B) is a w-balanced vertex separation, then $A \cap B$ is a w-balanced cut.

Conversely, suppose S is a w-balanced cut and let C be the vertex set of a component of $G - S$ of maximum weight. If $w(C) \geq \frac{1}{3} w(G)$, then $(S \cup C, V(G) \setminus C)$

is the desired w-balanced vertex separation. If $w(C) < \frac{1}{3}w(G)$, then let A_0 be an inclusionwise-minimal union of components of $G - S$ such that $w(A_0) \geq \frac{1}{3}w(G)$. Since each component has weight less than $\frac{1}{3}w(G)$ and deleting any component from A_0 would decrease the weight of A_0 to less than $\frac{1}{3}w(G)$, we have $w(A_0) < \frac{2}{3}w(G)$. Hence, $(V(A_0) \cup S, V(G) \setminus V(A_0))$ is the desired w-balanced vertex separation. □

A key property of trees is that they have balanced cuts of size one.

Observation 2.14 *A tree T has a w-balanced cut of size one for every assignment w of non-negative weights.*

Proof Note that each edge e splits T to two subtrees T_1 and T_2; we direct e towards T_1 if $w(T_1) \geq w(T_2)$ and towards T_2 otherwise. Since $\|T\| < |T|$, there exists a vertex v of outdegree 0 in this orientation. The choice of the orientation implies that each component C of $T - v$ has weight at most $\frac{1}{2}w(T)$. □

And importantly, graphs of bounded treewidth have a similar property.

Lemma 2.15 *Let G be a graph of treewidth at most k. Then G has a w-balanced cut of size at most $k + 1$ for every assignment w of non-negative weights.*

Proof Let (T, β) be a tree decomposition of G of width at most k. For each vertex $v \in V(G)$, choose a node $\sigma(v) \in V(T)$ such that $v \in \beta(\sigma(v))$ arbitrarily. Let us assign weight $w'(x) = w(\sigma^{-1}(x))$ to each node $x \in V(T)$, and observe that $w'(T) = w(G)$. By Observation 2.14, there exists a node $x \in V(T)$ such that $\{x\}$ is a w'-balanced cut in T. By Lemma 2.2, for any component C of $G - \beta(x)$, the set $\beta^{-1}(V(C))$ induces a connected subtree T_C of T, and clearly $x \notin V(T_C)$. Hence, T_C is contained in a single component of $T - x$, and thus $w'(T_C) \leq \frac{2}{3}w'(T)$. Moreover, $\sigma(V(C)) \subseteq V(T_C)$, and thus

$$w(C) \leq w'(T_C) \leq \tfrac{2}{3}w'(T) = \tfrac{2}{3}w(G).$$

It follows that $\beta(x)$ is a w-balanced cut in G. □

Hence, if there exists an assignment w of non-negative weights such that G has no small w-balanced cut, then G has large treewidth. And indeed, large treewidth can be always certified in this way, even using a more explicit version of the notion. For a set $X \subseteq V(G)$, we say that a cut or a vertex separation is *X-balanced* if it is w-balanced for the characteristic function w of X, i.e., the function such that $w(v) = 1$ for $v \in X$ and $w(v) = 0$ for $v \in V(G) \setminus X$. We say X is ***k-unbreakable*** if G has no X-balanced cut of size less than k; that is, for every set $S \subseteq V(G)$ of size less than k, there exists a component of $G - S$ containing more than $\frac{2}{3}|X|$ vertices of X.

Lemma 2.16 *Let k be a positive integer. If G is a graph of treewidth at least $4k - 3$, then there exists a k-unbreakable set of size at most $3k - 2$ in G.*

Proof Suppose for a contradiction that G has an X-balanced cut of size less than k for every set $X \subseteq V(G)$ of size at most $3k-2$. Let us remark that this property is hereditary, i.e., it holds also for every induced subgraph G' of G, since for every set $X \subseteq V(G')$, if a set S is an X-balanced cut in G, then the set $S \cap V(G')$ is an X-balanced cut in G'.

We are going to prove the following claim:

($\star$) For every induced subgraph G' of G and for every set $R \subseteq V(G')$ of size at most $3k-2$, the graph G' has a tree decomposition (T, β) of width less than $4k-3$ such that $R \subseteq \beta(x)$ for a node $x \in V(T)$.

For $G' = G$, this contradicts the assumption that the treewidth of G is at least $4k-3$. We prove ($\star$) by induction on the number of vertices of G.

If $|G'| \leq 4k-3$, then we can simply let T be the single-node tree with bag $V(G')$. Hence, suppose that $|G'| > 4k-3$. We can without loss of generality assume that $|R| = 3k-2$, as otherwise we can add vertices to R, only making the problem harder. By the assumptions and Observation 2.13, there exists an R-balanced vertex separation (A, B) of G' of order less than k. Let $R_A = A \cap (R \cup B)$ and $R_B = B \cap (R \cup A)$. Note that

$$R_A \leq \tfrac{2}{3}|R| + k - 1 = \tfrac{2}{3}(3k-2) + k - 1 < 3k - 2,$$

and similarly $|R_B| < 3k-2$. In particular, $A \neq V(G') \neq B$, since neither A nor B contains all vertices of R.

For $Y \in \{A, B\}$, we apply the induction hypothesis to $G'[Y]$ with R_Y playing the role of R; this gives us a tree decomposition (T_Y, β_Y) of $G'[Y]$ of width less than $4k-3$ and a node $x_Y \in V(T_Y)$ such that $R_Y \subseteq \beta_Y(x_Y)$. Let T be the tree obtained from $T_A \cup T_B$ by adding a new node x adjacent to x_A and x_B. Let β match β_A on T_A and β_B on T_B, and let $\beta(x) = R \cup (A \cap B)$. Note that $|\beta(x)| \leq 3k-2+k-1 = 4k-3$, and thus the resulting tree decomposition (T, β) of G' has width less than $4k-3$. □

Let us remark that the proof method of Lemma 2.16 is quite important and appears in many arguments where we want to obtain a tree decomposition into (not necessarily small) parts with some property $\mathcal{P}$; an example is the Minor Structure Theorem, where $\mathcal{P}$ = nearly drawn on a surface. In this setting, the argument can get "stuck" if at an inductive step it reaches a k-unbreakable set R of size $3k-2$. To deal with this case, we need a separate result that gives a decomposition relative to this unbreakable set, of the following form: There exists a star decomposition (T_0, β_0) of G such that

- the torso of the center c of the star T_0 has the property $\mathcal{P}$, and
- for each ray r of T_0, $|\beta(r) \cap (R \cup \beta(c))| \leq 3k-3$.

We then use $\beta_0(c) \cup R$ as a bag of the resulting decomposition and process the subgraphs $G[\beta(r)]$ inductively, with $\beta(r) \cap (R \cup \beta(c))$ playing the role of R. This results in a tree decomposition (T, β) where for each node $x \in V(T)$, either

- $|\beta(x)| \leq 4k - 3$, or
- we can delete at most $3k - 2$ vertices from the torso of x to obtain a graph with property $\mathcal{P}$.

Lemma 2.16 shows that the property of having large treewidth can be certified by providing an unbreakable set, or equivalently, giving an assignment of weights with no small balanced cut. It is natural to ask whether it is necessary to consider weighted balance in order to capture the notion of treewidth: Let us say that a cut or a vertex separation in a graph G is *balanced* if it is $V(G)$-balanced, i.e., it splits the graph into parts with at most $\frac{2}{3}|G|$ vertices. Lemma 2.15 implies that a graph of treewidth at most k has a balanced cut of size at most $k + 1$. The converse is clearly false (consider e.g. a union of two large cliques of the same size, a graph of large treewidth but with a balanced cut of size zero). However, would it be enough to assume the existence of small balanced cuts for all induced subgraphs of G? Note that this is different from assuming that there is a small X-balanced cut for every $X \subseteq V(G)$, since $G[X]$ may have a balanced cut of size at most k even if X is k-unbreakable in G. However, Dvořák and Norin [11] proved that even this weaker assumption suffices.

Theorem 2.17 *If every induced subgraph of G has a balanced cut of size at most k, then* $\mathrm{tw}(G) \leq 15k$.

2.5 Tangles

Another way to phrase the definition of k-unbreakability of a set X is as follows: For every vertex separation (A, B) of G of order less than k, either $|X \setminus A| > \frac{2}{3}|X|$, or $|X \setminus B| > \frac{2}{3}|X|$. Clearly exactly one of these two options happens, and thus this uniquely defines an orientation of small vertex separations "towards" the set X.

More precisely, for a graph G, we say that a system $\mathcal{T}$ of (vertex) separations of G of order less than k is an ***orientation of (vertex) (< k)-separations*** if for each (vertex) separation (A, B) of G of order less than k, exactly one of $(A, B) \in \mathcal{T}$ or $(B, A) \in \mathcal{T}$ holds. For a k-unbreakable set X, the system

$$\mathcal{T}_X = \{(A, B) : (A, B) \text{ is a vertex separation, } |A \cap B| < k,\ |X \setminus A| > \tfrac{2}{3}|X|\}$$

is clearly an orientation of vertex $(<k)$-separations.

Note that it is possible to have distinct k-unbreakable sets X_1 and X_2 such that the orientations $\mathcal{T}_{X_1}$ and $\mathcal{T}_{X_2}$ are the same. Consider e.g. the case that $|X_1| = |X_2| \geq 3k$ and there are $|X_1|$ pairwise vertex-disjoint paths from X_1 to X_2. If $(A, B) \in \mathcal{T}_{X_1}$, then $|X_1 \setminus A| \geq \frac{2}{3}|X_1|$, and since at most $|A \cap B| < k$ of the paths from $X_1 \setminus A$ to X_2 may intersect the cut $A \cap B$, it follows that $|X_2 \setminus A| > \frac{2}{3}|X_1| - k \geq \frac{1}{3}|X_2|$. Hence, $|X_2 \setminus B| \leq |X_2 \cap A| < \frac{2}{3}|X_2|$, and since X_2 is k-unbreakable, we must have $|X_2 \setminus A| > \frac{2}{3}|X_2|$ and $(A, B) \in \mathcal{T}_{X_2}$ as well.

This suggests that "consistent enough" orientations of vertex separations might be a more fundamental object than k-unbreakable sets, and motivates the important notion of *tangles*. A ***(vertex) tangle*** of order θ in a graph G is an orientation of (vertex) $(<\theta)$-separations satisfying the following properties:

(T1) If $(A_1, B_1), (A_2, B_2), (A_3, B_3) \in \mathcal{T}$, then $G \neq A_1 \cup A_2 \cup A_3$ for tangles and $G \neq G[A_1] \cup G[A_2] \cup G[A_3]$ for vertex tangles.
(T2) For tangles, if $(A, B) \in \mathcal{T}$, then $V(A) \neq V(G)$.

Note that the analogue of (T2) for vertex tangles follows from (T1) applied to (A, B), (A, B), (A, B), and thus we can refer to this property even for vertex tangles. On the other hand, for tangles the condition (T2) is important to forbid the following degenerate case: Fix any edge e_0 of G, and let $\mathcal{T}_{e_0}$ consist of all separations (A, B) of order less than θ such that $e_0 \in E(B)$. The orientation $\mathcal{T}_{e_0}$ of $(<\theta)$-separations of G satisfies (T1) and exists regardless of the treewidth of G.

Occasionally, it is useful to work with orientations of $(<\theta)$-separations satisfying only the property (T1), but not necessarily (T2); we call such orientations ***pretangles***. Since the condition (T2) does not play any role for vertex tangles, vertex tangles and pretangles coincide. As an example, only (T1) is needed to prove the monotonicity of the orientation of $(<\theta)$-separations.

Observation 2.18 *Let $\mathcal{T}$ be a (vertex) pretangle of order θ in a graph G. If $(A, B) \in \mathcal{T}$ and (A', B') is a (vertex) separation of G of order less than θ with $A' \subseteq A$, then $(A', B') \in \mathcal{T}$.*

Proof Since $A' \subseteq A$ and $A' \cup B' = G$ (or $G[A'] \cup G[B'] = G$), we have $A \cup B' = G$ (or $G[A] \cup G[B'] = G$). Hence, (T1) applied to (A, B), (A, B), and (B', A') implies $(B', A') \notin \mathcal{T}$. □

It may seem that the notion of tangles as opposed to vertex tangles offers a finer gradation, allowing us to take the edges with both ends in $V(A \cap B)$ into account when deciding the orientation of the separation (A, B). However, this is not actually the case.

Lemma 2.19 *Let $\mathcal{T}$ be a tangle of order θ in a graph G. If $(A_1, B_1) \in \mathcal{T}$ and (A_2, B_2) is a separation of G such that $V(A_2) = V(A_1)$ and $V(B_2) = V(B_1)$, then $(A_2, B_2) \in \mathcal{T}$.*

Proof It suffices to consider the case that $E(A_2) = E(A_1) \cup \{e\}$ and $E(B_2) = E(B_1) \setminus \{e\}$ for a single edge e with both ends in $V(A_1 \cap B_1) = V(A_2 \cap B_2)$: Repeatedly applying this special case, we can show that $(A', B') \in \mathcal{T}$ for the separation (A', B') such that $V(A') = V(A_1)$, $V(B') = V(B_1)$ and all the edges with both ends in $V(A_1 \cap B_1)$ belong to A', and the lemma then follows from Observation 2.18.

Note that θ is necessarily greater than the number of ends of e (1 if e is a loop and 2 otherwise), since (A_1, B_1) and (A_2, B_2) are separations of order less than θ. Let S be the subgraph of G consisting of e and the vertices incident with e. By (T2), we have $(S, G - e) \in \mathcal{T}$. Note that $A_1 \cup S = A_2$, and thus $A_1 \cup S \cup B_2 = G$. Then

(T1) applied to (A_1, B_1), $(S, G - e)$, and (B_2, A_2) implies that $(B_2, A_2) \notin \mathcal{T}$, and thus $(A_2, B_2) \in \mathcal{T}$. □

Hence, the notions of tangles and vertex tangles are equivalent.

Corollary 2.20 *Let G be a graph and θ a positive integer, let $\mathcal{T}$ be an orientation of $(<\theta)$-separations and $\mathcal{T}'$ an orientation of vertex $(<\theta)$-separations of G.*

- *$\mathcal{T}$ is a tangle if and only if $\{(V(A), V(B)) : (A, B) \in \mathcal{T}\}$ is a vertex tangle.*
- *$\mathcal{T}'$ is a vertex tangle if and only if*

$$\{(A, B) : (A, B) \text{ is a separation of } G,\ (V(A), V(B)) \in \mathcal{T}'\}$$

is a tangle.

Therefore, we can use tangles and vertex tangles interchangeably, depending on what is notationally more convenient. Let us remark that the graph minors series of Robertson and Seymour only uses tangles, not vertex tangles.

Going back to the beginning of the section, it is easy to see that if X is a k-unbreakable set, then $\mathcal{T}_X$ is a vertex tangle of order k; indeed, if $(A_i, B_i) \in \mathcal{T}_X$ for $i \in [3]$, then $|X \cap A_i| < \frac{1}{3}|X|$, and thus $X \not\subseteq A_1 \cup A_2 \cup A_3$, implying (T1). Hence, Lemma 2.16 gives the following claim.

Corollary 2.21 *For any positive integer θ, if $\mathrm{tw}(G) \geq 4\theta - 3$, then there exists a tangle of order θ in G.*

On the other hand, the presence of a tangle of large order implies large treewidth, showing that treewidth can be approximately characterized in terms of tangles. To see that, we need the following consequence of the property (T1).

Lemma 2.22 *Suppose $\mathcal{T}$ is a (vertex) pretangle of order θ, n is a positive integer, and $(A_i, B_i) \in \mathcal{T}$ for $i \in [n]$. Let $A = \bigcup_{i=1}^n A_i$, $B = \bigcap_{i=1}^n B_i$, and $C = \bigcup_{i=1}^n (A_i \cap B_i)$. If $|C| < \theta$, then $(A, B) \in \mathcal{T}$.*

Proof By induction on n. The case $n = 1$ is trivial, and thus suppose that $n \geq 2$. Letting $A' = \bigcup_{i=1}^{n-1} A_i$ and $B' = \bigcap_{i=1}^{n-1} B_i$, we have $(A', B') \in \mathcal{T}$ by the induction hypothesis. Since $|C| < \theta$, (A, B) is a (vertex) separation of order less than θ, and thus either $(A, B) \in \mathcal{T}$ or $(B, A) \in \mathcal{T}$. However, we cannot have $(B, A) \in \mathcal{T}$ by (T1), since $G[A'] \cup G[A_n] \cup G[B] = G$ if $\mathcal{T}$ is a vertex pretangle and $A' \cup A_n \cup B = G$ if $\mathcal{T}$ is a pretangle. □

Let us now give a lower bound on treewidth in terms of the order of a tangle.

Lemma 2.23 *For a positive integer θ, if G contains a vertex tangle of order θ, then* $\mathrm{tw}(G) \geq \theta - 1$.

Proof Let $\mathcal{T}$ be a vertex tangle of order θ in G. Suppose for a contradiction that G has a tree decomposition (T, β) of width less than $\theta - 1$. Each edge e splits T to two subtrees T_1 and T_2, and by Observation 2.1, $(\beta(T_1), \beta(T_2))$ is a vertex separation of G of order less than θ. We orient e towards T_2 if $(\beta(T_1), \beta(T_2)) \in \mathcal{T}$

and towards T_1 otherwise. Since $\|T\| < |T|$, there exists a node x of outdegree 0 in this orientation. The choice of the orientation implies that for each edge $e = xy$, letting T_e and T'_e be the components of $T - e$ labeled so that $y \in V(T_e)$, we have $(\beta(T_e), \beta(T'_e)) \in \mathcal{T}$. Note that for each of these vertex separations, we have $\beta(T_e) \cap \beta(T'_e) = \beta(x) \cap \beta(y) \subseteq \beta(x)$. Moreover, we have $(\beta(x), V(G)) \in \mathcal{T}$ by (T2). Lemma 2.22 (where $C = \beta(x)$ has size less than θ) then implies that

$$\left(\beta(x) \cup \bigcup_{e=xy\in E(T)} \beta(T_e), \bigcap_{e=xy\in E(T)} \beta(T'_e)\right) \in \mathcal{T}.$$

However, $\beta(x) \cup \bigcup_{e=xy\in E(T)} \beta(T_e) = V(G)$, contradicting (T2). □

Let us remark that Lemma 2.23 is almost tight, as can be seen by considering grids. The $n \times m$ ***grid*** is the Cartesian product of an n-vertex and an m-vertex path, i.e., the graph with vertex set $\{(i, j) : i \in [n], j \in [m]\}$, with two vertices (i_1, j_1) and (i_2, j_2) adjacent if and only if $|i_1 - i_2| + |j_1 - j_2| = 1$. For any $i \in [n]$, the set $R_i = \{(i, j) : j \in [m]\}$ is the i-th *row* of the grid. For any $j \in [m]$, the set $C_j = \{(i, j) : i \in [n]\}$ is the j-th *column* of the grid. Let $\mathcal{T}_n$ consist of vertex separations (A, B) of the $n \times n$ grid of order less than n such that A does not contain any row of the grid.

Observation 2.24 *For any positive integer n, $\mathcal{T}_n$ is an orientation of vertex ($<n$)-separations of the $n \times n$ grid.*

Proof Let (A, B) be a vertex separation of the $n \times n$ grid of order less than n. Since the grid has n pairwise vertex-disjoint rows, there exists a row R_i disjoint from $A \cap B$, and since R_i induces a connected subgraph, it follows that $R_i \subseteq A \setminus B$ or $R_i \subseteq B \setminus A$. By symmetry, suppose that $R_i \subseteq B \setminus A$; it follows that $(B, A) \notin \mathcal{T}_n$. Similarly, there exists a column C_j of the grid disjoint from $A \cap B$. Since C_j induces a connected subgraph and intersects R_i, we conclude that $C_j \subseteq B \setminus A$. Since every row of the grid intersects C_j, no row is contained in A, and thus $(A, B) \in \mathcal{T}_n$. Therefore, $\mathcal{T}_n$ is indeed an orientation of vertex ($<n$)-separations of the $n \times n$ grid. □

A technical argument given in [22] shows that $\mathcal{T}_n$ satisfies (T1) and thus it is actually a vertex tangle of order n; we say that $\mathcal{T}_n$ is the *canonical vertex tangle* of the $n \times n$ grid. Moreover, it is easy to show (see the next section) that the $n \times n$ grid has treewidth exactly n.

On the other hand, Corollary 2.21 is far from best possible. A tight bound was given in [22].

Theorem 2.25 (Robertson and Seymour [22]) *Every graph G has a tangle of order $\lceil \frac{2}{3}(\mathrm{tw}(G) + 1) \rceil$.*

The tightness follows by considering complete graphs. The complete graph K_n has treewidth $n - 1$ by Lemma 2.3. The only vertex separations in K_n are of form $(A, V(K_n))$ and $(V(K_n), A)$ for $A \subseteq V(K_n)$. Hence, for a positive integer θ, the only orientation of vertex ($< \theta$)-separations of K_n satisfying (T2) is $\mathcal{T}_\theta = \{(A, V(K_n)) : A \subseteq V(K_n), |A| < \theta\}$. If $\theta > \lceil \frac{2}{3}n \rceil$, then $\mathcal{T}_\theta$ is not a vertex tangle, since it does not satisfy (T1); to see that, observe that there exist sets $A_1, A_2, A_3 \subseteq V(K_n)$ of size $\lceil \frac{2}{3}n \rceil$ such that every pair of vertices belongs to one of them.

Let us remark that $\mathcal{T}_\theta$ for $\theta \leq \lceil \frac{2}{3}n \rceil$ is a vertex tangle in K_n: Indeed, if $|A_1|, |A_2|, |A_3| < \frac{2}{3}n$, then there exists a vertex v which belongs to at most one of A_1, A_2, and A_3, say to A_1, and thus the edges from v to $V(K_n) \setminus A_1$ are not contained in $K_n[A_1] \cup K_n[A_2] \cup K_n[A_3]$. This implies that $\mathcal{T}_\theta$ satisfies the property (T1).

2.6 Brambles

Tangles have the nice interpretation of "pointing towards a part of the graph that forces high treewidth", but as we have seen in the previous section, they fail to characterize treewidth exactly. Seymour and Thomas [27] developed the theory of exact obstructions to treewidth, which they referred to as *screens* but nowadays are called *brambles*. A system $\mathcal{B}$ of subsets of vertices of a graph G is a ***bramble*** if

(B1) each set $B \in \mathcal{B}$ is non-empty and induces a connected subgraph of G, and
(B2) for distinct $B_1, B_2 \in \mathcal{B}$, either $B_1 \cap B_2 \neq \emptyset$ or G contains an edge with one end in B_1 and the other end in B_2.

Equivalently, $\mathcal{B}$ is a bramble if all its elements are non-empty and for any (not necessarily distinct) $B_1, B_2 \in \mathcal{B}$, the graph $G[B_1 \cup B_2]$ is connected. A *hitting set* for $\mathcal{B}$ is a subset of vertices of G intersecting all elements of $\mathcal{B}$, and the *order* of $\mathcal{B}$ is the size of the smallest hitting set for $\mathcal{B}$.

Let us give three important examples of brambles.

- If K is a clique in a graph G, then $\mathcal{B}_K = \{\{v\} : v \in K\}$ is a bramble of order $|K|$.
- For $n \geq 2$, let G_n be the $n \times n$ grid with rows R_i and columns C_i for $i \in [n]$. Let $R'_i = R_i \setminus C_n$ and $C'_i = C_i \setminus R_n$ be the rows and columns truncated by removing their last vertex, and let $\mathcal{B}_n = \{R'_i \cup C'_j : i, j \in [n-1]\} \cup \{R_n, C'_n\}$. That is, $\mathcal{B}_n$ consists of $(n-1)^2$ "truncated crosses", the full last row, and the truncated last column. Note that $\mathcal{B}_n$ is a bramble: The truncated cross $R'_i \cup C'_j$ with $i, j \in [n-1]$ intersects all other truncated crosses and there is an edge from it to R_n and an edge from it to C'_n, and there is an edge between R_n and C'_n.

  We claim that the order of this bramble is $n+1$. Indeed, any hitting set X must intersect either every row or every column of the $(n-1) \times (n-1)$ subgrid $G_n - (R_n \cup C_n)$, as otherwise there would be a truncated cross avoiding it. Moreover, X needs to have a vertex in R_n and a vertex in C'_n. We conclude that $|X| \geq (n-1) + 2 = n + 1$.
- If Z is a k-unbreakable set of vertices of G, then let $\mathcal{B}_Z$ consist of non-empty sets $B \subseteq V(G)$ such that $G[B]$ is connected and $|B \cap Z| > \frac{1}{2}|Z|$. Clearly any

two sets in $\mathcal{B}_Z$ intersect in a vertex belonging to Z, and thus $\mathcal{B}_Z$ is a tangle. If X is a hitting set for $\mathcal{B}_Z$, then $G - X$ does not have any component containing more than half of the vertices of Z, and thus X is a Z-balanced cut. Since Z is k-unbreakable, it follows that $|X| \geq k$, and thus $\mathcal{B}_Z$ has order at least k.

Tangles have the following key property of duality to tree decompositions.

Lemma 2.26 *If $\mathcal{B}$ is a bramble and (T, β) is a tree decomposition of a graph G, then there exists a node $x \in V(T)$ such that $\beta(x)$ is a hitting set for $\mathcal{B}$.*

Proof For contradiction, suppose that this is not the case. Hence, for every node $x \in V(T)$ there exists a set $B_x \in \mathcal{B}$ disjoint from $\beta(x)$. Since the graph $G[B_x]$ is connected, Lemma 2.2 implies that $\beta^{-1}(B_x)$ induces a connected subtree of T. Since $x \notin \beta^{-1}(B_x)$, there exists a unique neighbor y of x such that $\beta^{-1}(B_x)$ is a subset of vertices of the connected component T_{xy} of $T - xy$ that does not contain x. Let us define $f(x)$ as the edge xy.

Since $\|T\| < |T|$, there exists an edge xy of T such that $f(x) = xy = f(y)$. That is, $\beta^{-1}(B_x) \subseteq V(T_{xy})$ and $\beta^{-1}(B_y) \subseteq V(T_{yx})$, where T_{xy} and T_{yx} are the two components of $T - xy$. But then there is no node $z \in V(T)$ containing a vertex of B_x as well as a vertex of B_y. It follows that $B_x \cap B_y = \emptyset$, and moreover, the property (D1) from the definition of tree decomposition implies that there is no edge with one end in B_x and the other end in B_y. This is a contradiction, since $\mathcal{B}$ is a bramble. □

Let us remark that if K is a clique in a graph G, Lemma 2.26 applied to the bramble $\mathcal{B}_K$ implies that every tree decomposition has a bag containing K, giving us the proof of Lemma 2.3 which we promised earlier. More importantly, Lemma 2.26 implies that every tree decomposition has a bag whose size is at least the order of $\mathcal{B}$.

Corollary 2.27 *If a graph G contains a bramble of order at least k, then* $\mathrm{tw}(G) \geq k - 1$.

In particular, the tangle $\mathcal{T}_n$ constructed above shows that the $n \times n$ grid G_n has treewidth at least n. Let us remark that this bound is tight; in fact, $\mathrm{tw}(G_n) = \mathrm{pw}(G_n) = n$ since the $n \times n$ grid has the following natural "row by row and cell by cell" path decomposition: Let $v_1, \ldots, v_{n^2}$ be the vertices of G_n labeled row by row with the indices increasing from left to right in each row. Let P be the path with vertices $x_1, \ldots, x_{n^2-n}$, and for $i \in [n^2 - n]$, let $\beta(x_i) = \{i, i+1, \ldots, i+n\}$. Then (P, β) is a path decomposition of G_n of width n.

Conversely, since a k-unbreakable set Z corresponds to a bramble $\mathcal{B}_Z$ of order at least k, Lemma 2.16 shows that a graph of treewidth at least $4k - 3$ contains a bramble of order at least k, showing that brambles give another approximate characterization of treewidth. However, the connection between treewidth and brambles is actually precise!

Theorem 2.28 (Seymour and Thomas [27]) *A graph has treewidth at least $k - 1$ if and only if it contains a bramble of order at least k.*

2.7 Grid Minors and Walls

The obstructions that we considered so far are somewhat nebulous. A tangle points towards a reason for large treewidth, but it is not clear what this reason actually is. An unbreakable set is an explicit set of vertices, but the property of unbreakability relies on connectivity of this set ensured by parts of the graph external to the set. Brambles remedy this by fully capturing all the connections, but their elements are only loosely constrained, allowing them to have quite complicated structure.

A much more explicit obstruction to small treewidth is provided by grid minors. Combining the observation on the treewidth of the $n \times n$ grid from the previous section with Observation 2.11, we obtain the following lower bound.

Observation 2.29 *If a graph G contains the $n \times n$ grid as a minor, then* $\mathrm{tw}(G) \geq n$.

One of the cornerstones of the graph minors theory is that a weak converse to this statement holds.

Theorem 2.30 (Grid Theorem, Robertson and Seymour [21]) *There exists a function $f_{2.30} : \mathbb{N} \to \mathbb{N}$ such that for every positive integer n, every graph of treewidth at least $f_{2.30}(n)$ contains the $n \times n$ grid as a minor.*

Note that the Grid Theorem cannot be precise, or even give a linear approximation the way unbreakable sets or tangles do. Indeed, K_{n^2} has treewidth $n^2 - 1$, but clearly cannot contain a grid larger than the $n \times n$ one as a minor. Hence, $f_{2.30}(n) = \Omega(n^2)$. By considering random graphs, this lower bound can be improved to $f_{2.30}(n) = \Omega(n^2 \log n)$. The original proof of Robertson and Seymour [21] gives a very loose upper bound on $f_{2.30}$, so bad that they do not even bother to compute it. This was substantially improved by Robertson, Seymour, and Thomas [25], who proved $f_{2.30}(n) \leq 20^{2n^5}$. Despite several subsequent improvements, the question of whether the Grid Theorem holds for a polynomial function $f_{2.30}$ was open for more than 20 years, until it was eventually answered in positive by Chekuri and Chuzhoy [6]. The best bound at the time of the writing is $f_{2.30}(n) = O(n^9 \operatorname{polylog} n)$ given by Chuzhoy and Tan [7].

If we do not care about the magnitude of the dependency, the proof of Theorem 2.30 is not particularly difficult and it is somewhat instructive. However, it is not very short, and as there is more to say on the topic of obstructions to bounded treewidth, we prefer to postpone it till Sect. 2.10.

One important observation is that grids contain all planar graphs as minors. This should be intuitively obvious (imagine drawing the graph on a very fine grid paper, then slightly altering the edges to follow the lines of the grid). Making this argument rigorous is a simple though tedious exercise. We prefer to give the elegant argument from [25], which starts with the following special case.

Lemma 2.31 *If G is a simple planar graph with n vertices, G has a Hamiltonian cycle C, and e_0 is an edge of C, then $G - e_0$ is a minor of the $n \times n$ grid.*

Proof Without loss of generality, we can assume that G is a triangulation of the plane, as otherwise we can add edges to G to triangulate it. Let us label the vertices of G by integers $1, \ldots, n$ in order along C so that $e_0 = 1n$. Note that the drawing of the cycle C splits the plane into two regions; let A and B be the subgraphs of G drawn in the closure of these regions. Hence, A and B can be viewed as outerplanar graphs with the outer face bounded by the cycle C. For $2 \le v \le n$, let $a_{\min}(v)$ be the minimum positive integer such that the vertices labeled v and $a_{\min}(v)$ are adjacent in A; note that $a_{\min}(v) \le v-1$, since $v-1$ is adjacent to v in C. Let us also define $a_{\min}(1) = 0$. For $1 \le v \le n-1$, let $a_{\max}(v)$ be maximum such that the vertices labeled v and $a_{\max}(v)$ are adjacent in A, and let $a_{\max}(n) = n+1$. Let $b_{\min}$ and $b_{\max}$ be defined for the subgraph B analogously.

Recall that the vertices of the $n \times n$ grid are the pairs (i, j) with $1 \le i, j \le n$, with two vertices adjacent if they differ in exactly one coordinate and the difference is ± 1. For a vertex $v \in V(G)$, let

$$R_v = \{(v, i) : a_{\min}(v) < i < b_{\max}(v)\}$$
$$C_v = \{(j, v) : b_{\min}(v) < j < a_{\max}(v)\}.$$

Since $a_{\min}(v) < v < a_{\max}(v)$ and $b_{\min}(v) < v < b_{\max}(v)$, we have $(v, v) \in R_v \cap C_v$, and thus the subgraph $\mu(v)$ of the $n \times n$ grid induced by $R_v \cup C_v$ is connected. We claim that for any vertices $u \ne v$, the subgraphs $\mu(u)$ and $\mu(v)$ are vertex-disjoint. Note that $R_u \cap R_v = \emptyset$ and $C_u \cap C_v = \emptyset$ trivially. Suppose that say R_u intersects C_v, i.e., that $(u, v) \in R_u \cap C_v$, and thus

$$a_{\min}(u) < v < b_{\max}(u)$$
$$b_{\min}(v) < u < a_{\max}(v).$$

If $u < v$, then this would imply that B has edges $ub_{\max}(u)$ and $b_{\min}(v)v$ such that $b_{\min}(v) < u < v < b_{\max}(u)$, and since C bounds the outer face of B, these edges would cross. Similarly, if $v < u$, then the edges $a_{\min}(u)u$ and $va_{\max}(v)$ would cross in A. This is a contradiction, implying that indeed $V(\mu(u) \cap \mu(v)) = \emptyset$.

Consider now an edge $uv \in E(G)$ distinct from e_0, where say $uv \in E(A)$ and $u < v$. Then $a_{\min}(v) \le u$ and $v \le a_{\max}(u)$. Observe that since G is a triangulation, $uv \ne e_0$, and A is outerplanar, there exists a triangular face uvx of A such that $x < u$ or $x > v$. By symmetry, we can assume $x < u$, and since $xv \in E(G)$, we have $a_{\min}(v) \le x < u$. Therefore, $(v, u) \in R_v$ since $a_{\min}(v) < u < v < b_{\max}(v)$, and $(v-1, u) \in C_u$ since $b_{\min}(u) < u \le v-1 < a_{\max}(u)$. It follows that the $n \times n$ grid has an edge between $\mu(u)$ and $\mu(v)$. Since this holds for every edge $uv \in E(G - e_0)$, we conclude that μ gives a model of $G - e_0$ in the $n \times n$ grid. □

To get the result for planar graphs in general, we need the following observation.

Observation 2.32 *Every simple planar graph G with $n \ge 3$ vertices is a minor of a simple Hamiltonian planar graph with at most $2n-3$ vertices.*

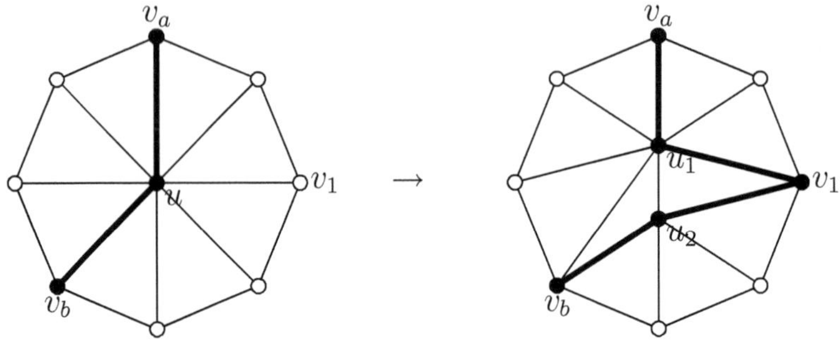

Fig. 2.1 Prolonging a cycle

Proof Without loss of generality, we can assume that G is a triangulation. Let C be an arbitrary cycle in G. We gradually modify the graph G and the cycle C, in each step decreasing the number of vertices in $V(G) \setminus V(C)$ at the expense of increasing the number of vertices by one, see Fig. 2.1: Choose a vertex $v_1 \in V(G) \setminus V(C)$ with a neighbor u in C, and let $v_1, \ldots, v_m$ be the neighbors of u in the counter-clockwise order in the drawing. Let $a < b$ be the indices such that $v_a u, uv_b \in E(C)$. Let us split u into two vertices u_1 and u_2, with u_1 having neighbors $v_1, \ldots, v_b, u_2$ and u_2 having neighbors $u_1, v_b, \ldots, v_m, v_1$. The resulting graph is planar and contains G as a minor. Moreover, we can turn C into a cycle in the modified graph containing v_1 by replacing its subpath $v_a u v_b$ by the path $v_a u_1 v_1 u_2 v_b$.

After repeating this transformation at most $n - 3$ times, the cycle C contains all vertices of the graph, while the number of vertices have not increased to more than $2n - 3$. □

Given a simple planar graph G with n vertices, we apply Observation 2.32 to get a simple Hamiltonian planar graph G_1 with at most $2n - 3$ vertices containing G as a minor. Next, we add a vertex z of degree two adjacent to both ends of an edge of a Hamiltonian cycle of G_1 and let e_0 be one of the edges incident with z, obtaining a simple planar graph G_2 with a Hamiltonian cycle containing an edge e_0 such that G is a minor of $G_2 - e_0$ and $|G_2| < 2n$. Finally, we apply Lemma 2.31. Note also that every planar graph with n vertices and m edges is a minor of a simple planar graph with at most $n + 2m$ vertices, obtained by subdividing each loop twice and each parallel edge once. Putting all these observations together, we obtain the following conclusion.

Corollary 2.33 *Every planar graph G with n vertices and m edges is a minor of the $(2n + 4m) \times (2n + 4m)$ grid. The bound on the size of the grid can be improved to $2n \times 2n$ if G is simple.*

Together with Theorem 2.30, this has the following useful consequence.

Corollary 2.34 *For every planar graph H, there exists a constant t_H such that every H-minor-free graph has treewidth at most t_H.*

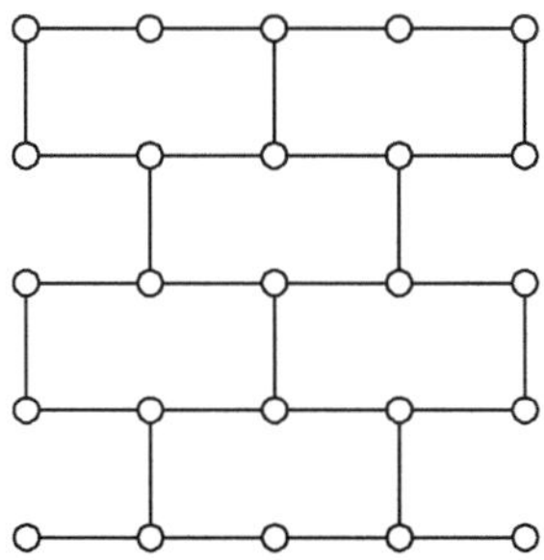

Fig. 2.2 A 5×5 wall

This fact can be nicely restated as follows.

Corollary 2.35 *A minor-closed class of graphs has unbounded treewidth if and only if it contains all planar graphs.*

A model of a grid minor is a quite explicit obstruction to small treewidth, but there is still is a little amount of uncertainty in how the connected subgraphs that model its vertices look like, which may cause difficulties in some applications. This can be avoided by using *walls* instead of grids. The $n \times n$ ***wall*** is the subgraph of the $n \times n$ grid of maximum degree at most three obtained by alternatingly deleting the upwards and downwards vertical edges incident with the vertices in each row. More precisely, the vertex set of the $n \times n$ wall is $\{(i, j) : i, j \in [n]\}$ and vertices (i_1, j_1) and (i_2, j_2) are adjacent if $|i_1 - i_2| + |j_1 - j_2| = 1$ and

- $i_1 = i_2$, or
- $j_1 = j_2$ and $\min(i_1, i_2) \equiv j_1 \pmod 2$.

See Fig. 2.2 for an illustration.

Note that contracting every second edge in each row of an $n \times n$ wall shows that it contains a $\lfloor \frac{n}{2} \rfloor \times \lfloor \frac{n}{2} \rfloor$ grid as a minor, and thus the treewidth of the $n \times n$ wall is at least $\lfloor \frac{n}{2} \rfloor$. Hence, a graph has large treewidth if and only if it contains a large wall as a minor. Moreover, a wall has maximum degree at most three, and thus by Observation 1.3, a wall is present as a minor if and only if it is present as a topological minor. Thus, we obtain a completely explicit subgraph certifying large treewidth, with the only uncertainty being the number of times each edge of the wall is subdivided. In summary:

Corollary 2.36 *Let $\mathcal{G}$ be a class of graphs. Then the following claims are equivalent:*

- *$\mathcal{G}$ contains graphs of arbitrarily large treewidth.*
- *Every planar graph is a minor of a graph from $\mathcal{G}$.*
- *For every n, the $n \times n$ grid is a minor of a graph from $\mathcal{G}$.*
- *For every n, the $n \times n$ wall is a topological minor of a graph from $\mathcal{G}$.*

Let us note that one might still want more. As we have seen, any graph G of large treewidth has a subgraph W which is a subdivision of a large wall. However,

G can contain additional edges between vertices of W not contained in $E(W)$. Can treewidth be characterized by a nice explicit set of obstructions in terms of *induced* subgraphs? After thinking about the problem for a bit, it becomes clear that one needs to forbid at least the following types of induced subgraphs for some n in order to force bounded treewidth:

- The clique K_n and biclique $K_{n,n}$,
- subdivisions of the $n \times n$ wall, and
- the graphs obtained from subdivisions of the $n \times n$ wall by replacing the vertices of degree three by triangles.

This is indeed enough in graphs of bounded maximum degree [16] (where the last two obstructions suffice), but not in general. Several intriguing families of graphs of large treewidth that do not contain any of the induced obstructions from the list above are known [10, 29], and it is unclear whether a nice characterization of treewidth in terms of forbidden induced subgraphs is possible.

2.8 Which Obstruction Is the Best One?

All the obstructions that we introduced (unbreakable sets, brambles, tangles, grid minors) are qualitatively equivalent—given any of them, we can construct the rest, though in general only certifying slightly smaller treewidth. Indeed, we have seen a transformation from unbreakable sets to brambles and tangles in our exposition, and we will see a transformation from tangles to a grid minor in Sect. 2.10. We leave the other transformations as an exercise to the reader.

Moreover, note that the transformations can be done so that the constructed objects correspond to the original ones, in the following sense. Suppose e.g. that we start with a vertex tangle $\mathcal{T}$ of order θ. Let $(A_0, B_0) \in \mathcal{T}$ be chosen so that B_0 is inclusionwise minimal, and subject to that A_0 is inclusionwise maximal, let $X = A_0 \cap B_0$, and let $k = \lfloor \theta/3 \rfloor$. Then X is a k-unbreakable set.[2] Let $\mathcal{T}_X$ be the corresponding tangle of order k constructed in Sect. 2.5. Then $\mathcal{T}_X$ is consistent with

[2] Note that $|X| = \theta - 1$, since otherwise (T1) applied to (A_0, B_0) and $(\{v\}, V(G))$ for an arbitrary vertex $v \in B_0 \setminus A_0$ shows that $(A_0 \cup \{v\}, B_0) \in \mathcal{T}$, contradicting the maximality of A_0. Suppose for a contradiction that X is not k-unbreakable, and thus there exists an X-balanced vertex separation (C, D) of G of order less than k. Consider the vertex separations $(A_0 \cup C, B_0 \cap D)$ and $(A_0 \cup D, B_0 \cap C)$. Note that $|(A_0 \cup C) \cap B_0 \cap D| \leq |X \cap D| + |C \cap D| < \frac{2}{3}|X| + k < \theta$. It follows that the vertex separation $(A_0 \cup C, B_0 \cap D)$ has order less than θ, and by a symmetric argument, so does the vertex separation $(A_0 \cup D, B_0 \cap C)$. Since $\mathcal{T}$ is an orientation of vertex $(<\theta)$-separations, the minimality of B_0 implies that $(B_0 \cap D, A_0 \cup C), (B_0 \cap C, A_0 \cup D) \in \mathcal{T}$. However, $G = G[A_0] \cup G[B_0 \cap D] \cup G[B_0 \cap C]$, contradicting (T1).

$\mathcal{T}$ in the sense that both orient the separations of order less than k in the same way, i.e., $\mathcal{T}_X \subset \mathcal{T}$.[3]

Hence, it is in principle possible to execute any argument with any of the obstructions and obtain qualitatively the same conclusions. Of course, the argument is not always equally easy, and there are upsides and downsides to each of the obstructions.

- Unbreakable sets are easy to represent explicitly, but are defined in terms of the connectivity in the wider graph and thus give relatively little in terms of the structure to work with. They are most often a technical tool that arises in the arguments, especially in connection to the recursive algorithm shown in Lemma 2.16.
- Tangles are the most canonical of the obstructions. This is not at all obvious, but as shown in [22], in each graph G there are at most $|G|$ maximal tangles such that every other tangle is the truncation of one of them to separations of certain size. In other words, there are at most $|G|$ distinct parts of G that force large treewidth, and each maximal tangle points to exactly one of them. In comparison, each such part might contain several different brambles or grid minors (of course, well connected to one another). On the other hand, since tangles operate on a rather abstract level, only speaking about orientations of separations rather than concrete objects in the graph, it is often difficult to obtain structural results based on them.
- Brambles have the unique advantage of precisely characterizing treewidth, and thus in comparison to other obstructions, we do not lose anything at all by using them. Moreover, they are substantially more explicit than unbreakable sets, though potentially harder to represent, since they can have exponentially many elements.
- Grid minors and especially wall topological minors are the most explicit of the obstructions, directly giving a structural reason for large treewidth. When we need to show properties of the underlying graph, they are usually the easiest obstructions to start from. On the negative side, we lose quite a bit on the treewidth when using them, unlike the other obstructions which approximate the treewidth at least up to a constant factor.

Throughout the book, we mostly work with (vertex) tangles, partly due to their useful canonicality property and partly because they are used in most of the papers in the field, including the graph minors series of Robertson and Seymour.

---

[3] Consider any vertex separation (C, D) of order less than k belonging to $\mathcal{T}_X$, i.e., satisfying $|X \setminus C| > \frac{2}{3}|X|$. Suppose for a contradiction that $(D, C) \in \mathcal{T}$. Note that $|(A_0 \cup D) \cap B_0 \cap C| < |X \cap C| + k < \theta$, and since $G[A_0] \cup G[D] \cup G[B_0 \cap C] = G$, (T1) implies that $(A_0 \cup D, B_0 \cap C) \in \mathcal{T}$. However, $B_0 \cap C$ does not contain all vertices of X, and thus $B_0 \cap C \subsetneq B_0$ and this vertex separation contradicts the minimality of B_0.

2.9 Operations on Tangles, Rank, and Freedom

As we are going to use tangles quite often, we need to study their properties in more detail. First, let us introduce several operations to obtain new tangles from existing ones. Let $\mathcal{T}$ be a vertex tangle of order θ in a graph H.

- For any positive integer $\theta' < \theta$, the set of vertex separations in $\mathcal{T}$ of order less than θ' clearly forms a vertex tangle of order θ', the *truncation* of $\mathcal{T}$ to order θ'.
- Consider any set $Z \subseteq V(H)$ of size less than θ. Let $\mathcal{T} - Z$ be the set of vertex separations (A, B) of $H - Z$ of order less than $\theta - |Z|$ such that $(A \cup Z, B \cup Z) \in \mathcal{T}$. It is easy to see that $\mathcal{T} - Z$ is a tangle of order $\theta - |Z|$ in $H - Z$.
- Suppose that H is a minor of another graph G, and let μ be a model of H in G. For a set C of vertices of G, let $\mu^{-1}(C)$ denote the set of vertices $v \in V(H)$ such that $V(\mu(v)) \cap C \neq \emptyset$. Let $\mathcal{T}_\mu$ be the set of vertex separations (A, B) of G of order less than θ such that $(\mu^{-1}(A), \mu^{-1}(B)) \in \mathcal{T}$. Note that $(\mu^{-1}(A), \mu^{-1}(B))$ is indeed a vertex separation of H, since if H had an edge $e = uv$ with $u \in \mu^{-1}(A) \setminus \mu^{-1}(B)$ and $v \in \mu^{-1}(B) \setminus \mu^{-1}(A)$, then $V(\mu(u)) \subseteq A \setminus B$ and $V(\mu(v)) \subseteq B \setminus A$, and thus $\mu(e)$ would be an edge of G with one end in $A \setminus B$ and the other end in $B \setminus A$.

  Moreover, if $v \in \mu^{-1}(A) \cap \mu^{-1}(B)$, then $V(\mu(v)) \cap A \neq \emptyset$ and $V(\mu(v)) \cap B \neq \emptyset$, and since the subgraph $\mu(v)$ is connected, we have $V(\mu(v)) \cap A \cap B \neq \emptyset$. Since the subgraphs that μ assigns to vertices of H are pairwise vertex-disjoint, it follows that the order of the vertex separation $(\mu^{-1}(A), \mu^{-1}(B))$ is at most as large as the order of (A, B). Hence, if (A, B) has order less than θ, then so does $(\mu^{-1}(A), \mu^{-1}(B))$, and thus $(\mu^{-1}(A), \mu^{-1}(B)) \in \mathcal{T}$ or $(\mu^{-1}(B), \mu^{-1}(A)) \in \mathcal{T}$, and $(A, B) \in \mathcal{T}_\mu$ or $(B, A) \in \mathcal{T}_\mu$. Consequently, $\mathcal{T}_\mu$ is an orientation of vertex $(<\theta)$-separations of G.

  And $\mathcal{T}_\mu$ is in fact a vertex tangle in G, since $G = G[A_1] \cup G[A_2] \cup G[A_3]$ implies $H = H[\mu^{-1}(A_1)] \cup H[\mu^{-1}(A_2)] \cup H[\mu^{-1}(A_3)]$. We say that $\mathcal{T}_\mu$ is the vertex tangle in G *μ-induced* by $\mathcal{T}$.

The last construction shows that a tangle in a minor H of G gives rise to a tangle in G. We are often going to need to view the relationship in the opposite direction: Suppose that $\mathcal{T}_0$ is a tangle in G of order $\theta_0 \geq \theta$, $\mathcal{T}$ is a tangle in H of order θ, and μ is a model of H in G. When the μ-induced tangle $\mathcal{T}_\mu$ is equal to the truncation of $\mathcal{T}_0$ to order θ, we say that $\mathcal{T}$ is ***μ-conformal*** with $\mathcal{T}_0$. In case that H is a subgraph of G and μ is the trivial model of H in G (such that for each $v \in V(H)$, $\mu(v)$ is the subgraph of G consisting of v, and $\mu(e) = e$ for each edge e of H), we simply say that $\mathcal{T}$ is *conformal* with $\mathcal{T}_0$.

In particular, the following strengthening of the Grid Theorem holds.

Theorem 2.37 *There exists a function $f_{2.37} : \mathbb{N} \to \mathbb{N}$ such that the following claim holds for every positive integer n. Let G_n be the $n \times n$ grid and let $\mathcal{T}_n$ be the canonical vertex tangle of order n in G_n. If $\mathcal{T}$ is a vertex tangle of order at least $f_{2.37}(n)$ in a graph G, then there exists a model μ of G_n in G such that $\mathcal{T}_n$ is μ-conformal with $\mathcal{T}$.*

Thus, in one direction, every model of a large grid in G induces a tangle of large order in G, and conversely, any tangle of large order is derived from some grid minor. It is easy to see that this implies the analogous result for wall subdivisions. Suppose W is a subdivision of the $n \times n$ wall, and let $\mathcal{T}$ be the set of vertex separations (A, B) of W of order less than $\lfloor \frac{n}{2} \rfloor$ such that A does not contain any row of the wall; it is easy to see that $\mathcal{T}$ is a vertex tangle (induced by the canonical vertex tangle in an $\lfloor \frac{n}{2} \rfloor \times \lfloor \frac{n}{2} \rfloor$ grid which it contains as a minor); we say that $\mathcal{T}$ is a *canonical vertex tangle* of the wall subdivision.

Corollary 2.38 *There exists a function $f_{2.38} : \mathbb{N} \to \mathbb{N}$ such that the following claim holds for every positive integer n. If $\mathcal{T}$ is a vertex tangle of order at least $f_{2.38}(n)$ in a graph G, then there exists a subgraph W of G isomorphic to a subdivision of the $n \times n$ wall such that the canonical vertex tangle of W is conformal with $\mathcal{T}$.*

Let us now introduce an important notion of the tangle rank function, which is in particular useful in showing that two tangles are conformal. Given a vertex tangle $\mathcal{T}$ of order θ in a graph G, the *rank* $\mathrm{rk}_{\mathcal{T}}(Z)$ of a set $Z \subseteq V(G)$ is defined as follows:

- If there exists a vertex separation $(A, B) \in \mathcal{T}$ such that $Z \subseteq A$, then let $\mathrm{rk}_{\mathcal{T}}(Z)$ be the minimum order of such a vertex separation.
- Otherwise, let $\mathrm{rk}_{\mathcal{T}}(Z) = \theta$.

Note that $\mathrm{rk}_{\mathcal{T}}(\emptyset) = 0$ by (T2), and clearly the rank function is monotone, i.e., if $Z' \subseteq Z$, then $\mathrm{rk}_{\mathcal{T}}(Z') \leq \mathrm{rk}_{\mathcal{T}}(Z)$. Furthermore, the rank function is *submodular*:

Lemma 2.39 *Let $\mathcal{T}$ be a vertex tangle of order θ in a graph G. All sets $Z_1, Z_2 \subseteq V(G)$ satisfy*

$$\mathrm{rk}_{\mathcal{T}}(Z_1) + \mathrm{rk}_{\mathcal{T}}(Z_2) \geq \mathrm{rk}_{\mathcal{T}}(Z_1 \cup Z_2) + \mathrm{rk}_{\mathcal{T}}(Z_1 \cap Z_2).$$

Proof The claim follows from the monotonicity if say $\mathrm{rk}_{\mathcal{T}}(Z_1) = \theta$, since then $\mathrm{rk}_{\mathcal{T}}(Z_1) = \theta \geq \mathrm{rk}_{\mathcal{T}}(Z_1 \cup Z_2)$ and $\mathrm{rk}_{\mathcal{T}}(Z_2) \geq \mathrm{rk}_{\mathcal{T}}(Z_1 \cap Z_2)$. Hence, suppose that there exist vertex separations $(A_1, B_1), (A_2, B_2) \in \mathcal{T}$ such that $Z_i \subseteq A_i$ and the order of (A_i, B_i) is $\mathrm{rk}_{\mathcal{T}}(Z_i) < \theta$ for $i \in [2]$.

Let o_1 be the order of the vertex separation $(A_1 \cup A_2, B_1 \cap B_2)$. If $o_1 < \theta$, then the vertex separation belongs to $\mathcal{T}$ by (T1), and since $Z_1 \cup Z_2 \subseteq A_1 \cup A_2$, we have $\mathrm{rk}_{\mathcal{T}}(Z_1 \cup Z_2) \leq o_1$. If $o_1 \geq \theta$, then this inequality holds trivially.

Let o_2 be the order of the vertex separation $(A_1 \cap A_2, B_1 \cup B_2)$. If $o_2 < \theta$, then the vertex separation belongs to $\mathcal{T}$ by (T1), and since $Z_1 \cap Z_2 \subseteq A_1 \cap A_2$, we have $\mathrm{rk}_{\mathcal{T}}(Z_1 \cap Z_2) \leq o_2$. If $o_2 \geq \theta$, then this inequality holds trivially. Therefore, by (1.1), we have

$$\begin{aligned}\mathrm{rk}_{\mathcal{T}}(Z_1) + \mathrm{rk}_{\mathcal{T}}(Z_2) &= |A_1 \cap B_1| + |A_2 \cap B_2| \\ &= |(A_1 \cup A_2) \cap B_1 \cap B_2| + |A_1 \cap A_2 \cap (B_1 \cup B_2)| \\ &= o_1 + o_2 \geq \mathrm{rk}_{\mathcal{T}}(Z_1 \cup Z_2) + \mathrm{rk}_{\mathcal{T}}(Z_1 \cap Z_2).\end{aligned}$$

□

Note that submodularity implies subadditivity, i.e., if Z_1 and Z_2 are disjoint, then $\mathrm{rk}_{\mathcal{T}}(Z_1) + \mathrm{rk}_{\mathcal{T}}(Z_2) \geq \mathrm{rk}_{\mathcal{T}}(Z_1 \cup Z_2) + \mathrm{rk}_{\mathcal{T}}(\emptyset) = \mathrm{rk}_{\mathcal{T}}(Z_1 \cup Z_2)$. Since $\mathrm{rk}_{\mathcal{T}}$ is monotone, submodular, and bounded by the size of the argument, it is a rank function of a matroid. We say that a set $Z \subseteq V(G)$ is *$\mathcal{T}$-free* if it is an independent set of this matroid, i.e., $|Z| = \mathrm{rk}_{\mathcal{T}}(Z)$, or in other words, $|Z| \leq \theta$ and there is no vertex separation $(A, B) \in \mathcal{T}$ of order less than $|Z|$ with $Z \subseteq A$. Clearly, any subset of a $\mathcal{T}$-free sets is also $\mathcal{T}$-free. Moreover, from the matroid connection, it is easy to see the following claim.

Observation 2.40 *If $\mathcal{T}$ is a vertex tangle in a graph G, then any set $Z_0 \subseteq V(G)$ contains a $\mathcal{T}$-free subset Z such that $|Z| = \mathrm{rk}_{\mathcal{T}}(Z_0)$.*

Proof It clearly suffices to argue that if Z_0 is not $\mathcal{T}$-free, then there exists $z \in Z_0$ such that $\mathrm{rk}_{\mathcal{T}}(Z_0 \setminus \{z\}) = \mathrm{rk}_{\mathcal{T}}(Z_0)$. Let Z' be a minimal subset of Z_0 that is not $\mathcal{T}$-free; we are going to show that every vertex $z \in Z'$ has this property.

Since Z' is not $\mathcal{T}$-free but $Z' \setminus \{z\}$ is and the rank function is monotone, we have $\mathrm{rk}_{\mathcal{T}}(Z') = |Z'| - 1 = \mathrm{rk}_{\mathcal{T}}(Z' \setminus \{z\})$. The submodularity then gives

$$\mathrm{rk}_{\mathcal{T}}(Z_0 \setminus \{z\}) + \mathrm{rk}_{\mathcal{T}}(Z') \geq \mathrm{rk}_{\mathcal{T}}(Z_0) + \mathrm{rk}_{\mathcal{T}}(Z' \setminus \{z\}),$$

and thus $\mathrm{rk}_{\mathcal{T}}(Z_0 \setminus \{z\}) \geq \mathrm{rk}_{\mathcal{T}}(Z_0)$. The equality follows by the monotonicity of the rank function. □

In particular, since $\mathrm{rk}_{\mathcal{T}}(V(G))$ is equal to the order θ of the vertex tangle $\mathcal{T}$ by (T2), there exists a $\mathcal{T}$-free set of order θ. Next, let us make a useful observation on vertex separations.

Lemma 2.41 *Let $\mathcal{T}$ be a vertex tangle of order θ in a graph G. If $(A, B) \in \mathcal{T}$, then $\mathrm{rk}_{\mathcal{T}}(A) = \mathrm{rk}_{\mathcal{T}}(A \cap B)$.*

Proof Since $(A, B) \in \mathcal{T}$, we have $\mathrm{rk}_{\mathcal{T}}(A \cap B) \leq |A \cap B| < \theta$. Hence, there exists $(C, D) \in \mathcal{T}$ such that $A \cap B \subseteq C$ and $|C \cap D| = \mathrm{rk}_{\mathcal{T}}(A \cap B)$. Note that since $A \cap B \setminus C = \emptyset$, we have

$$\begin{aligned} |(A \cup C) \cap B \cap D| &= |C \cap (B \cap D)| + |(A \setminus C) \cap B \cap D| \\ &= |C \cap (B \cap D)| \leq |C \cap D| < \theta, \end{aligned}$$

and by (T1), it follows that $(A \cup C, B \cap D) \in \mathcal{T}$. The order of this separation is $|(A \cup C) \cap B \cap D| \leq |C \cap D| = \mathrm{rk}_{\mathcal{T}}(A \cap B)$, and thus $\mathrm{rk}_{\mathcal{T}}(A) \leq \mathrm{rk}_{\mathcal{T}}(A \cap B)$. The equality follows by the monotonicity of the rank function. □

From this, it is easy to see that the tangle is determined by its rank function.

Observation 2.42 *If $\mathcal{T}_1 \neq \mathcal{T}_2$ are vertex tangles of the same order θ in a graph G, then $\mathrm{rk}_{\mathcal{T}_1} \neq \mathrm{rk}_{\mathcal{T}_2}$.*

Proof Suppose for a contradiction that $\mathrm{rk}_{\mathcal{T}_1} = \mathrm{rk}_{\mathcal{T}_2} = \mathrm{rk}$. Since $\mathcal{T}_1$ and $\mathcal{T}_2$ are different orientations of vertex $(< \theta)$-separations of G, there exists a vertex separation (A, B) of G of order less than θ such that $(A, B) \in \mathcal{T}_1$ and $(B, A) \in \mathcal{T}_2$. By Lemma 2.41 applied to $\mathcal{T}_1$ and $\mathcal{T}_2$, we have $\mathrm{rk}(A) = \mathrm{rk}(A \cap B) = \mathrm{rk}(B)$. However, then submodularity gives

$$2\,\mathrm{rk}(A \cap B) = \mathrm{rk}(A) + \mathrm{rk}(B) \geq \mathrm{rk}(V(G)) + \mathrm{rk}(A \cap B) = \theta + \mathrm{rk}(A \cap B),$$

which is a contradiction since $\mathrm{rk}(A \cap B) \leq |A \cap B| < \theta$. □

It is tempting to suggest that a tangle is determined by a single basis of its rank matroid, i.e., by a single $\mathcal{T}$-free set of size θ, but this is false. Consider e.g. a graph G consisting of two cliques L and R of size $3k$ intersecting in k vertices. Then G has two natural vertex tangles $\mathcal{T}_L$ and $\mathcal{T}_R$ of order $2k$ induced by the cliques: Consider any vertex separation (A, B) of order less than $2k$. Note that each of the two cliques is fully contained in exactly one part of the separation. We define $(A, B) \in \mathcal{T}_L$ if and only if $L \subseteq B$ and $(A, B) \in \mathcal{T}_R$ if and only if $R \subseteq B$; see the discussion at the end of Sect. 2.5 to see why the defined objects are vertex tangles. Note that these are indeed different tangles, since $(R, L) \in \mathcal{T}_L$ but $(L, R) \in \mathcal{T}_R$; however, the set Z consisting of k vertices of $L \setminus R$ and k vertices of $R \setminus L$ is free in both of them. Nevertheless, a bit weaker claim holds.

Observation 2.43 *Let $\mathcal{T}$ be a vertex tangle of order θ in a graph G and let $Z \subseteq V(G)$ be a $\mathcal{T}$-free set of size θ. Then a vertex separation (A, B) of order less than $\theta/2$ belongs to $\mathcal{T}$ if and only if $|A \cap Z| \leq |A \cap B|$.*

Proof Since Z is $\mathcal{T}$-free, the set $A \cap Z$ is also $\mathcal{T}$-free. If $(A, B) \in \mathcal{T}$, it follows that $|A \cap Z| \leq |A \cap B|$.

Conversely, suppose that $|A \cap Z| \leq |A \cap B|$. If $(A, B) \notin \mathcal{T}$, then $(B, A) \in \mathcal{T}$, and since Z is $\mathcal{T}$-free, we have $|B \cap Z| \leq |A \cap B|$. However, then $|Z| \leq |A \cap Z| + |B \cap Z| \leq 2|A \cap B| < \theta$, which is a contradiction. □

Observation 2.43 can be quite useful in algorithmic design; it shows that if we are willing to sacrifice a bit from the order of the tangle, then we can represent it concisely just by storing a single set of vertices which enables us to easily determine whether a separation belongs to the tangle.

Another important property of free sets is that they are hard to separate.

Lemma 2.44 *Let $\mathcal{T}$ be a vertex tangle of order θ in a graph G and let Z_1 and Z_2 be subsets of $V(G)$ of the same size $k \leq \theta$. If Z_1 and Z_2 are $\mathcal{T}$-free, then G contains k pairwise vertex-disjoint paths from Z_1 to Z_2.*

Proof Consider any vertex separation (A, B) of G with $Z_1 \subseteq A$ and $Z_2 \subseteq B$. If $(A, B) \in \mathcal{T}$ or $(B, A) \in \mathcal{T}$, then since $Z_1 \subseteq A$ and $Z_2 \subseteq B$ are $\mathcal{T}$-free, the order of the vertex separation must be at least k. If neither of the vertex separations belongs to $\mathcal{T}$, then their order is at least $\theta \geq k$. Therefore, any cut separating A from B in G has size at least k, and the conclusion follows from Menger's theorem. □

On a related note, any set that is well-linked to a free set must have large rank.

Lemma 2.45 *Let $\mathcal{T}$ be a vertex tangle of order θ in a graph G and let Z_1 and Z_2 be subsets of $V(G)$. If Z_1 is $\mathcal{T}$-free and G contains k pairwise vertex-disjoint paths from Z_1 to Z_2, then $\mathrm{rk}_{\mathcal{T}}(Z_2) \geq k/2$.*

Proof Consider $(A, B) \in \mathcal{T}$ with $Z_2 \subseteq A$. The graph G contains k pairwise vertex-disjoint paths from Z_1 to Z_2 and at most $|A \cap B|$ of them intersect $|A \cap B|$, and thus $|Z_1 \cap A| \geq k - |A \cap B|$. Since Z_1 is $\mathcal{T}$-free, the set $Z_1 \cap A$ is also $\mathcal{T}$-free, and thus $|Z_1 \cap A| \leq |A \cap B|$. A combination of these inequalities gives $|A \cap B| \geq k/2$. We conclude that $\mathrm{rk}_{\mathcal{T}}(Z_2) \geq k/2$. □

As we have mentioned, rank can be useful in showing that two tangles are conformal. The example given before Observation 2.43 shows that unfortunately the condition cannot be entirely straightforward; however, the usefulness of the following technical statement can be seen in the following section.

Lemma 2.46 (♀→) *Let $\mathcal{T}_H$ be a vertex tangle in a graph H of order $\theta_H \geq 2$ and let $\mathcal{T}_G$ be a vertex tangle in a graph G of order $\theta_G \geq 2\theta_H$. Let μ be a model of H in G and let Z be a $\mathcal{T}_G$-free set. If there exist pairwise vertex-disjoint trees $R_1, \ldots, R_{\theta_H} \subseteq G$ such that for $i \in [\theta_H]$, all leaves of R_i belong to Z and $\mathrm{rk}_{\mathcal{T}_H}(\mu^{-1}(V(R_i))) = \theta_H$, then $\mathcal{T}_H$ is μ-conformal with $\mathcal{T}_G$.*

Proof We need to show that the vertex tangle $\mathcal{T}_\mu$ μ-induced by $\mathcal{T}_H$ in G is the truncation of $\mathcal{T}_G$ to order θ_H. That is, for every vertex separation $(A, B) \in \mathcal{T}_G$ of order less than θ_H, we need to show that $(A, B) \in \mathcal{T}_\mu$.

Let $L = A \cap (Z \cup B) = (A \cap B) \cup (Z \setminus B)$. Since Z is $\mathcal{T}_G$-free, its subset $Z \setminus B \subseteq A$ is also $\mathcal{T}_G$-free, and thus $|Z \setminus B| \leq |A \cap B| < \theta_H$ and $|L| < 2\theta_H$.

Consider now any $i \in [\theta_H]$ such that $V(R_i) \not\subseteq B$, and let u be a vertex in $V(R_i) \setminus B$. Since $\mu^{-1}(V(R_i))$ has rank $\theta_H \geq 2$, the tree R_i has at least two vertices, and thus u lies on a path $P \subseteq R_i$ joining distinct leaves of R_i. Let $k \in \{0, 1, 2\}$ be the number of ends of P in B. Since $u \notin B$, observe that P contains at least k vertices in $A \cap B$. Moreover, since all leaves of R_i belong to Z, the path P also contains at least $2 - k$ vertices of $Z \setminus B$. We conclude that $P \subseteq R_i$ contains at least two vertices in L.

Since $R_1, \ldots, R_{\theta_H}$ are vertex-disjoint, it follows that there are at most $|L|/2 < \theta_H$ indices i such that $V(R_i) \not\subseteq B$. Hence, there exists $i \in [\theta_H]$ such that $V(R_i) \subseteq B$. Then $\mathrm{rk}_{\mathcal{T}_H}(\mu^{-1}(B)) \geq \mathrm{rk}_{\mathcal{T}_H}(\mu^{-1}(V(R_i))) = \theta_H$, and thus $(\mu^{-1}(B), \mu^{-1}(A)) \notin \mathcal{T}_H$. Consequently, $(\mu^{-1}(A), \mu^{-1}(B)) \in \mathcal{T}_H$ and $(A, B) \in \mathcal{T}_\mu$, as required. □

Let us note that if we are willing to sacrifice a bit on the order of $\mathcal{T}_H$, a simpler and more natural condition that the reader might expect suffices.

Lemma 2.47 (♀→) *Let $\mathcal{T}_H$ be a vertex tangle in a graph H of order $2\theta_H$, let Z_H be a $\mathcal{T}_H$-free subset of $V(H)$ of size $2\theta_H$ and let $\mathcal{T}_G$ be a vertex tangle in a graph G of order $\theta_G \geq 2\theta_H$. Let μ be a model of H in G and let Z_G be a subset of $V(G)$ of size $|Z_H|$ such that $|V(\mu(z)) \cap Z_G| = 1$ for every $z \in Z_H$. If Z_G is $\mathcal{T}_G$-free, then the truncation $\mathcal{T}'_H$ of $\mathcal{T}_H$ to order θ_H is μ-conformal with $\mathcal{T}_G$.*

Proof We need to show that the vertex tangle $\mathcal{T}_\mu$ μ-induced by $\mathcal{T}'_H$ in G is the truncation of $\mathcal{T}_G$ to order θ_H. That is, for every vertex separation $(A, B) \in \mathcal{T}_G$ of order less than θ_H, we need to show that $(A, B) \in \mathcal{T}_\mu$.

Since Z_G is $\mathcal{T}_G$-free, the set $A \cap Z_G$ is also $\mathcal{T}_G$-free and we have $|A \cap Z_G| \leq |A \cap B| < \theta_H$. Since $|Z_G| = |Z_H| = 2\theta_H$, it follows that $|B \cap Z_G| > \theta_H$. By the assumptions, for each $v \in Z_G$, there exists a distinct vertex $z \in Z_H$ such that $z \in \mu^{-1}(\{v\})$. Hence,

$$|\mu^{-1}(B) \cap Z_H| \geq |B \cap Z_G| > \theta_H > |A \cap B| \geq |\mu^{-1}(A) \cap \mu^{-1}(B)|.$$

Since the set Z_H is $\mathcal{T}_H$-free, we have $(\mu^{-1}(B), \mu^{-1}(A)) \notin \mathcal{T}_H$, and consequently $(\mu^{-1}(B), \mu^{-1}(A)) \notin \mathcal{T}'_H$. We conclude that $(\mu^{-1}(A), \mu^{-1}(B)) \in \mathcal{T}'_H$ and $(A, B) \in \mathcal{T}_\mu$, as required. □

Let us make a final simple observation on the freeness in conformal tangles.

Lemma 2.48 *Let H and G be graphs, let ρ be an injective function from a set* $\mathrm{dom}(\rho) \subseteq V(H)$ *to* $V(G)$, *and let μ be a ρ-rooted model of H in G. Let $\mathcal{T}_H$ be a vertex tangle in H which is μ-conformal with a vertex tangle $\mathcal{T}_G$ in G. If* $\mathrm{dom}(\rho)$ *is $\mathcal{T}_H$-free, then* $\mathrm{img}(\rho)$ *is $\mathcal{T}_G$-free.*

Proof Suppose for a contradiction that $(A, B) \in \mathcal{T}_G$ for a vertex separation (A, B) of G of order less than $|\mathrm{img}(\rho)|$ such that $\mathrm{img}(\rho) \subseteq A$. Note that since $\mathrm{dom}(\rho)$ is $\mathcal{T}_H$-free, the order of $\mathcal{T}_H$ is at least $|\mathrm{dom}(\rho)| = |\mathrm{img}(\rho)|$. Since $\mathcal{T}_H$ is μ-conformal with $\mathcal{T}_G$, we have $(\mu^{-1}(A), \mu^{-1}(B)) \in \mathcal{T}_H$. However, $(\mu^{-1}(A), \mu^{-1}(B))$ is a vertex separation of H of order at most $|A \cap B| < |\mathrm{img}(\rho)| = |\mathrm{dom}(\rho)|$ and $\mathrm{dom}(\rho) \subseteq \mu^{-1}(\mathrm{img}(\rho)) \subseteq \mu^{-1}(A)$, contradicting the assumption that $\mathrm{dom}(\rho)$ is $\mathcal{T}_H$-free. □

2.10 Proof of the Grid Theorem ($\looparrowright$)

We have now gathered all the ingredients needed for the proof of the Grid Theorem. We make no effort to optimize the bounds in the proof, though if we did, we would obtain a fairly reasonable exponential bound on the treewidth of graphs avoiding a fixed grid minor.

2.10.1 Pregrids

First, let us show how to get a grid minor from a slightly simpler object; see Fig. 2.3 for an illustration. For integers r, c, and m, an *m-thick $r \times c$ pregrid* in a graph G consists of a set $\mathcal{R}$ of r pairwise vertex-disjoint paths (the *rows* of the pregrid) with

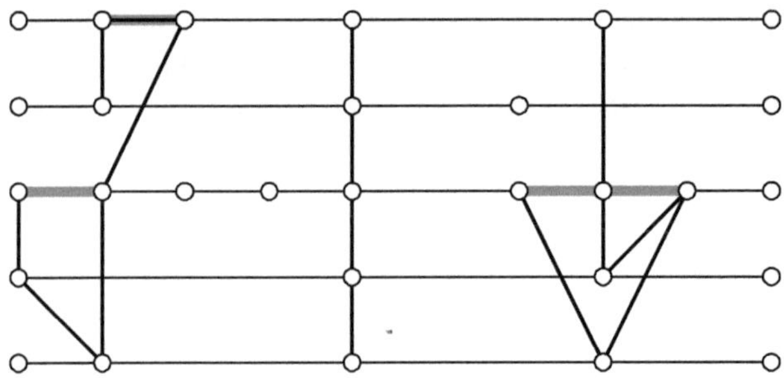

Fig. 2.3 A 4-thick 5×3 pregrid. The spines of the columns are emphasized

specified beginning vertices and a sequence $\mathcal{C} = C_1, \ldots, C_c$ of pairwise vertex-disjoint connected subgraphs of G (the *columns* of the pregrid) such that

- each column intersects at least m rows, and
- for $1 \le i < j \le c$, if C_i and C_j both intersect the row P, then all vertices of $C_i \cap P$ are closer to the beginning of P than all vertices of $C_j \cap P$.

For each column C and row P, let the *C-rib* of P be the minimal subpath of P containing all intersections of C with P; the C-rib is empty if C is disjoint from P. By the second condition, for any $i < j$ such that C_i and C_j both intersect P, the C_i-rib of P is disjoint from and appears closer to the beginning of P than the C_j-rib of P. Let the *C-spine* be the union of C with the C-ribs of all rows of the pregrid; for distinct columns C and C', the C-spine is vertex-disjoint from the C'-spine.

For a column C, a sequence $R_1, \ldots, R_n$ of rows of the pregrid is *C-path-like* if for every $i \in [n-1]$, there exists a path Q in the C-spine such that Q starts on R_i, ends on R_{i+1}, and Q is vertex-disjoint from $R_1, \ldots, R_{i-1}, R_{i+2}, \ldots, R_n$.

Observation 2.49 *If $(\mathcal{R}, \mathcal{C})$ is an m-thick pregrid and $n \ll m$, then for each column $C \in \mathcal{C}$, there exists a C-path-like sequence of n rows of the pregrid.*

Proof Let F be the auxiliary graph with vertex set consisting of the rows of $\mathcal{R}$ intersected by C and with two rows adjacent if and only if there exists a path in C between them that does not intersect any other row. If F has diameter at least $n-1$, then we can choose $R_1, \ldots, R_n$ along a path of length $n-1$ in F. Otherwise, since $|F| \ge m \gg n$ and F has diameter less than $n-1$, there exists a vertex $R \in V(F)$ of degree at least n. Then any two rows adjacent to R in F are connected by a path in the C-spine that intersects no other rows except possibly for R, and thus n neighbors of R taken in any order form a C-path-like sequence. □

A model μ of the $n \times n$ grid in G is *strung* on a system $\mathcal{R}$ of paths if there exists an injective function s from the rows of the grid to $\mathcal{R}$ such that each row R of the grid satisfies $R \subseteq \mu^{-1}(V(s(R)))$.

Lemma 2.50 *If a graph G contains an m-thick $r \times c$ pregrid $(\mathcal{R}, \mathcal{C})$ and $n \ll m \ll r \ll c$, then G contains a model of the $n \times n$ grid strung on the rows of the pregrid.*

Proof For each column $C \in \mathcal{C}$, Observation 2.49 gives a C-path-like sequence $\mathcal{R}_C$ of n rows of $\mathcal{R}$. Since there are only r^n sequences of n rows of $\mathcal{R}$ and $r, n \ll c$, there exist columns $C_1, \ldots, C_{n(n-1)}$ of the pregrid in order (but not necessarily consecutive) and distinct rows $R_1, \ldots, R_n \in \mathcal{R}$ such that $\mathcal{R}_{C_1} = \ldots = \mathcal{R}_{C_{n(n-1)}} = R_1, \ldots, R_n$.

The $n \times n$ grid minor in G is obtained as follows: For each $i \in [n]$, we partition R_i into n pairwise vertex-disjoint subpaths, the j-th one of them containing the $C_{(n-1)(j-1)+1}$-rib, the $C_{(n-1)(j-1)+2}$-rib, …, and the $C_{(n-1)(j-1)+n-1}$-rib of R_i, and we contract each of the subpaths to a single vertex. This turns R_i into a path with n vertices, which will form the i-th row of the grid minor.

For $j \in [n]$, we then form the j-th column by, for $i \in [n-1]$, choosing a path $Q_{i,j}$ in the spine of $C_{(n-1)(j-1)+i}$ between R_i and R_{i+1} and vertex-disjoint from $R_1, \ldots, R_{i-1}, R_{i+2}, \ldots, R_n$, and contracting $Q_{i,j}$ to a single edge. Observe that the model of the $n \times n$ grid corresponding to this construction is strung on the rows of the pregrid via the injection mapping the i-th row of the grid to R_i for $i \in [n]$. □

2.10.2 Cleaning Up Path Systems

For two sets A and B of vertices of a graph G, an ***(A, B)-linkage*** is a system of pairwise vertex-disjoint paths with one end in A and the other end in B. For a subgraph F of G, an (A, B)-linkage $\mathcal{L}$ in G is *F-minimal* if for each edge $e \in E\left(\bigcup \mathcal{L}\right) \setminus E(F)$, the graph $F \cup \bigcup \mathcal{L} - e$ does not contain any (A, B)-linkage of size $|\mathcal{L}|$.

Observation 2.51 *Let A and B be disjoint sets of vertices in a graph G, let $\mathcal{L}$ be an (A, B)-linkage in G, and let F be a subgraph of G. There exists an F-minimal (A, B)-linkage $\mathcal{L}'$ in G of size $|\mathcal{L}|$ such that $\bigcup \mathcal{L}' \subseteq F \cup \bigcup \mathcal{L}$.*

Proof It suffices to choose $\mathcal{L}'$ as an (A, B)-linkage of size $|\mathcal{L}|$ in $F \cup \bigcup \mathcal{L}$ such that the set $E\left(\bigcup \mathcal{L}'\right) \setminus E(F)$ is smallest possible. □

Note that if $\mathcal{F}$ is the set of paths forming the columns of a grid, and the leftmost and the rightmost columns of the grid are A and B, then every (A, B)-linkage in the grid intersects all paths from $\mathcal{F}$. The following key lemma shows that if the graph does not contain a large grid minor, then such a situation cannot occur and we can "disentangle" sufficiently large linkages.

Lemma 2.52 *Let A and B be disjoint sets of vertices in a graph G, let $\mathcal{L}$ be an (A, B)-linkage in G, and let $\mathcal{F}$ be a set of pairwise vertex-disjoint paths in G. If $f, l, n \ll |\mathcal{L}| \ll |\mathcal{F}|$, then either*

- *there exists a model of the $n \times n$ grid strung on $\mathcal{L}$, or*
- *there exists a set $\mathcal{F}' \subseteq \mathcal{F}$ of size f and an (A, B)-linkage $\mathcal{L}'$ of size l in G such that $\bigcup \mathcal{F}'$ and $\bigcup \mathcal{L}'$ are vertex-disjoint and $\bigcup \mathcal{L}' \subseteq \bigcup \mathcal{L} \cup \bigcup \mathcal{F}$.*

Proof Let $F = \bigcup \mathcal{F}$; by Observation 2.51, we can assume that the linkage $\mathcal{L}$ is F-minimal. Choose integers m and c so that $f, l, n \ll m \ll |\mathcal{L}| \ll c \ll |\mathcal{F}|$. If $\mathcal{F}$ contains a set $\mathcal{F}'$ of f paths such that each of them intersects at most m of the paths from $\mathcal{L}$, then since $|\mathcal{L}| \geq fm + l$, there exists a set $\mathcal{L}' \subset \mathcal{L}$ of size l of paths vertex-disjoint from those in $\mathcal{F}'$, as required.

Hence, suppose that this is not the case, and thus there is a set $\mathcal{F}_0 \subseteq \mathcal{F}$ of size more than $|\mathcal{F}| - f$ such that each path in $\mathcal{F}_0$ intersects more than m of the paths from $\mathcal{L}$. Let $r = |\mathcal{L}| - 1$. Since $|\mathcal{F}| \gg f, r, c, |\mathcal{L}|$, the number of pairs (C, Q) such that $C \in \mathcal{F}_0$, $Q \in \mathcal{L}$, and C and Q intersect is more than

$$|\mathcal{F}_0|m > \frac{|\mathcal{F}|m}{2} = \frac{|\mathcal{F}|m}{2|\mathcal{L}|}|\mathcal{L}| \geq rc^2|\mathcal{L}|,$$

and thus we can fix a path $Q \in \mathcal{L}$ intersecting at least rc^2 paths from $\mathcal{F}_0$.

We now divide Q into segments such that each of them intersects many distinct paths from $\mathcal{F}_0$, and use this division to obtain cuts that similarly partition the rest of the paths of $\mathcal{L}$. More precisely, let us partition Q into vertex-disjoint paths $Q_1, \ldots, Q_c$ such that for $i \in [c-1]$, there is a set $\mathcal{F}_i \subset \mathcal{F}_0$ of size exactly rc of paths that intersect Q_i but not $Q_1 \cup \ldots \cup Q_{i-1}$, and Q_i cannot be extended towards the end of Q while preserving this property. Note that since at least rc^2 paths of $\mathcal{F}_0$ intersect Q, the set $\mathcal{F}_c$ of paths from $\mathcal{F}_0$ intersect Q_c but not $Q_1 \cup \ldots \cup Q_{c-1}$ has size at least rc. For each $i \in [c-1]$, since Q_i cannot be extended, the first vertex v_i of Q_{i+1} is contained in a path from $\mathcal{F}_0$ that does not intersect Q_i, and thus the edge e_i of Q from the last vertex of Q_i to v_i is not contained in $\bigcup \mathcal{F}$. By the F-minimality of $\mathcal{L}$, there exists no (A, B)-linkage of size $r + 1$ in $F \cup \bigcup \mathcal{L} - e$, and by Menger's theorem, there exists a set $S_i \subset V(G)$ of size at most r such that every path in $F \cup \bigcup \mathcal{L}$ from A to B intersects $S_i \cup \{e_i\}$. Because of the (A, B)-linkage $\mathcal{L} \setminus \{Q\}$ of size r, the set S_i necessarily contains exactly one vertex from each path in $\mathcal{L} \setminus \{Q\}$.

Let $S = \bigcup_{i=1}^{c-1} S_i$. Since $|S| < rc$, for each $i \in [c]$ there exists a path $C_i \in \mathcal{F}_i$ disjoint from S; recall that such a path C_i intersects Q_i but not $Q_1 \cup \ldots \cup Q_{i-1}$. We claim that $\mathcal{R} = \mathcal{L} \setminus \{Q\}$ and $C = C_1, \ldots, C_c$ forms an m-thick $r \times c$ pregrid; and since $n \ll m \ll r \ll c$, Lemma 2.50 then implies that G contains a model of the $n \times n$ grid strung on $\mathcal{R} \subset \mathcal{L}$, as desired.

Since each path from $\mathcal{F}_0$ intersects at least m paths from $\mathcal{R}$, it suffices to verify that for every $P \in \mathcal{R}$ and $1 \leq i < j \leq c$, if C_i and C_j both intersect P, then all vertices of $C_i \cap P$ are closer to the beginning of P than all vertices of $C_j \cap P$. Indeed, let s be the vertex of S_i on P. Since the set $S_i \cup \{e_i\}$ separates A from B and is disjoint from C_i, and additionally C_i has an intersection with Q in Q_i, that is, before e_i, all the intersections of C_i with P must precede s. On the other hand, since C_j has an intersection with Q in Q_j, that is, after e_i, all its intersections with P must be after s. □

We can now apply Lemma 2.52 repeatedly to disentangle several linkages.

Lemma 2.53 *Let $A_1, \ldots, A_k, B_1, \ldots, B_k$ be pairwise disjoint sets of vertices in a graph G, and let Z be a set of vertices of G such that $Z \supseteq \bigcup_{i=1}^{k} A_i \cup B_i$. Let s be* *an integer, and for $i \in [k]$, let $\mathcal{L}_i$ be an (A_i, B_i)-linkage in G of size s intersecting Z only in its endpoints. If $k, n \ll s$, then either*

- *there exists $i \in [k]$ and a model of the $n \times n$ grid in G strung on a linkage from A_i to B_i, or*
- *there exist pairwise vertex-disjoint paths $P_1, \ldots, P_k$ such that for $i \in [k]$, P_i starts in A_i, ends in B_i, and is otherwise disjoint from Z.*

Proof Let $l_k = 1$. For $i = k-1, k-2, \ldots, 1$, choose integers $l_i, f_{i,i+1}, \ldots, f_{i,k}$ so that

$$l_{i+1} \ll f_{i,k} \ll f_{i,k-1} \ll \cdots \ll f_{i,i+1} \ll l_i.$$

Since $k \ll s$, we can assume that $l_1 = s$. Let us also define $f_{i,i} = l_i$ for each $i \in [k-1]$.

We are going to alter the linkages in $k-1$ phases so that at the end of the i-phase, the linkages $\mathcal{L}_1, \ldots, \mathcal{L}_i$ will be pairwise vertex-disjoint as well as vertex-disjoint from the linkages $\mathcal{L}_{i+1}, \ldots, \mathcal{L}_k$. In the i-th phase, we are going to repeatedly apply Lemma 2.52 to disentangle $\mathcal{L}_i$ from $\mathcal{L}_{i+1}, \ldots, \mathcal{L}_k$ in turn, modifying these linkages in the process. In these applications, $\mathcal{L}_i$ will play the role of $\mathcal{F}$, i.e., we are only going to remove paths from $\mathcal{L}_i$. For $j \in \{i+1, \ldots, k\}$, the linkage $\mathcal{L}_j$ will play the role of $\mathcal{L}$, and so it can be rerouted, but only within $\bigcup \mathcal{L}_j \cup \bigcup \mathcal{L}_i$, and thus it will stay vertex-disjoint from the linkages $\mathcal{L}_1, \ldots, \mathcal{L}_{i-1}$.

We are going to preserve the invariant that at the beginning of the i-th phase, we are going to have $|\mathcal{L}_i| = |\mathcal{L}_{i+1}| = \ldots = |\mathcal{L}_k| = l_i$, and for $j = i+1, \ldots, k$, after disentangling $\mathcal{L}_i$ from $\mathcal{L}_j$, we are going to have $|\mathcal{L}_i| = f_{i,j}$. To do the disentangling, we first choose an integer λ so that $l_{i+1} \ll \lambda \ll f_{i,j-1}$ and delete all but λ paths from $\mathcal{L}_j$. Then we apply Lemma 2.52 with $\mathcal{L} = \mathcal{L}_j, \mathcal{F} = \mathcal{L}_i$, $f = f_{i,j}$ and $l = l_{i+1}$, either obtaining a model of the $n \times n$ grid strung on $\mathcal{L}_j$, or vertex-disjoint linkages $\mathcal{L}'$ of size l_{i+1} and $\mathcal{F}'$ of size $f_{i,j}$. In the latter case, we replace $\mathcal{L}_j$ by $\mathcal{L}'$ and $\mathcal{L}_i$ by $\mathcal{F}'$, and continue the process.

In the end, the paths in $\mathcal{L}_1, \ldots, \mathcal{L}_k$ are pairwise vertex-disjoint, and we can choose one path from each of these linkages. Note that the resulting paths only intersect Z in their endpoints: This was true initially for all paths in the linkages, and thus it suffices to argue that this condition is preserved whenever we reroute a linkage. However, the rerouting is done in the subgraph $H = \bigcup \mathcal{L}_j \cup \bigcup \mathcal{L}_i$ for some distinct $i, j \in [k]$, and since the linkages are between disjoint subsets of Z, every vertex in $Z \cap V(H)$ has degree one in H. Therefore, the vertices of Z cannot end up as internal vertices of the rerouted paths. □

2.10.3 Building the Base

We are going to use the disjoint paths obtained in Lemma 2.53 to connect fixed connected subgraphs in order to form the grid minor. We need to be careful and keep the rerouted paths disjoint from the fixed subgraphs. To ensure that, we are going to find a suitable separation (A, B) of our graph and look for the connected subgraphs in A and for the linkages in B. Let $\mathcal{T}$ be a vertex tangle in a graph G of order at least 2. A vertex separation $(A, B) \in \mathcal{T}$ is a *base* if the set $Z = A \cap B$ is $\mathcal{T}$-free and there exists a model μ of the $|Z|$-vertex path P in $G[A]$ such that for each $x \in V(P)$, $\mu(x)$ intersects Z in exactly one vertex; i.e., $\mu^{-1}(Z) = V(P)$. Note that there always exists a vertex v such that $\text{rk}_{\mathcal{T}}(\{v\}) = 1$, and thus $(\{v\}, V(G))$ is a base.

Lemma 2.54 *Let $\mathcal{T}$ be a vertex tangle of order $\theta \geq 2$ in a graph G, and let $(A, B) \in \mathcal{T}$ be a base of largest order with A inclusionwise-maximal. Then the set $Z = A \cap B$ has size $\theta - 1$ and the subgraph $G[B]$ has no vertex separation (C, D) such that $|C \cap Z| > |C \cap D|$ and $|D \cap Z| > |C \cap D|$.*

Proof Let μ be a model of the $|Z|$-vertex path P in $G[A]$ such that $\mu^{-1}(Z) = V(P)$. We are going to need the following observation.

($\star$) Every vertex separation $(A', B') \in \mathcal{T}$ such that $A \subsetneq A'$ has order more than $|Z|$.

Indeed, without loss of generality, we can assume that $B' \subseteq B$, as otherwise we can instead consider the vertex separation $(A \cup A', B \cap B') = (A', B \cap B')$ whose order is $|A' \cap B \cap B'| \leq |A' \cap B'|$ and which belongs to $\mathcal{T}$ by (T1). Let $Z' = A' \cap B'$. By Lemma 2.41 and since Z is $\mathcal{T}$-free, we have $\text{rk}_{\mathcal{T}}(Z') = \text{rk}_{\mathcal{T}}(A') \geq \text{rk}_{\mathcal{T}}(Z) = |Z|$, and thus $|Z'| \geq |Z|$. Moreover, if $|Z'| = |Z|$, then Z' would be $\mathcal{T}$-free, and by Lemma 2.44, there would exist a (Z, Z')-linkage $\mathcal{L}$ of size $|Z| = |Z'|$ in G, necessarily contained in $G[A'] - (A \setminus Z) - E(G[Z])$. Then μ would combine with $\mathcal{L}$ to a model μ_0 of P in $G[A']$ such that $\mu_0^{-1}(Z') = V(P)$, and (A', B') would be a base of the same order as (A, B). This contradicts the maximality of A, and thus $|Z'| > |Z|$.

Suppose now for a contradiction that $|Z| < \theta - 1$. Let x be an end of P and let v be the vertex in $Z \cap V(\mu(x))$. If v had no neighbor in $B \setminus Z$, then by (T1) the vertex separation $(A, B \setminus \{v\})$ of order less than $|Z|$ would belong to $\mathcal{T}$, and since $Z \subseteq A$, this would contradict the assumption that Z is $\mathcal{T}$-free. Hence, v has a neighbor $u \in B \setminus Z$. Let $A_1 = A \cup \{u\}$ and $Z_1 = Z \cup \{u\}$. Note that (A_1, B) is a vertex separation of G of order $|Z_1| > |Z|$. Moreover, adding u to μ gives a model μ_1 of the $|Z_1|$-vertex path P_1 with $\mu_1^{-1}(Z_1) = V(P_1)$ in $G[A_1]$. Since $|Z_1| = |Z| + 1 < \theta$, (T1) shows that $(A_1, B) \in \mathcal{T}$.

By ($\star$), every vertex separation $(A', B') \in \mathcal{T}$ with $A_1 \subseteq A'$ has size at least $|Z| + 1 = |Z_1|$, and thus $\text{rk}_{\mathcal{T}}(A_1) \geq |Z_1|$. Therefore, Lemma 2.41 gives $\text{rk}_{\mathcal{T}}(Z_1) = \text{rk}_{\mathcal{T}}(A_1) \geq |Z_1|$, and thus Z_1 is $\mathcal{T}$-free. It follows that (A_1, B) is a base of order greater than $|A \cap B|$, contradicting the choice of the base (A, B). We conclude that $|Z| = \theta - 1$.

Finally, suppose for a contradiction there exists a vertex separation (C, D) of $G[B]$ of order less than $\min(|C \cap Z|, |D \cap Z|)$. Note that $(A \cup C, D)$ and $(A \cup D, C)$ are vertex separations of G. We have

$$|(A \cup C) \cap D| = |C \cap D| + |(A \setminus C) \cap D| = |C \cap D| + |(A \cap B) \setminus C|$$
$$= |C \cap D| + |Z| - |C \cap Z| < |Z| \leq \theta - 1,$$

and similarly $|(A \cup D) \cap C| < \theta - 1$. Since $G = G[A] \cup G[C] \cup G[D]$, (T1) implies that $(D, A \cup C) \notin \mathcal{T}$ or $(C, A \cup D) \notin \mathcal{T}$. By symmetry, we can assume that $(D, A \cup C) \notin \mathcal{T}$, and thus $(A \cup C, D) \in \mathcal{T}$. However, this vertex separation has order less than $|Z|$, contradicting (⋆). Therefore, $G[B]$ has no vertex separation (C, D) such that $|C \cap Z| > |C \cap D|$ and $|D \cap Z| > |C \cap D|$. □

Let us remark that the proof method from Lemma 2.54—building up a structure in the small side A of a vertex separation $(A, B) \in \mathcal{T}$ while keeping $A \cap B$ free, so that the structure is still well-connected to everything in B—is quite useful and its variations appear in many arguments.

2.10.4 Constructing the Grid

We are now ready to prove the Grid Theorem, in the stronger form of finding a grid minor pointed to by a specified tangle; we restate the theorem for convenience.

Theorem 2.55 *There exists a function $f_{2.37} : \mathbb{N} \to \mathbb{N}$ such that the following claim holds for every positive integer n. Let G_n be the $n \times n$ grid and let $\mathcal{T}_n$ be the canonical vertex tangle of order n in G_n. If $\mathcal{T}$ is a vertex tangle of order at least $f_{2.37}(n)$ in a graph G, then there exists a model μ of G_n in G such that $\mathcal{T}_n$ is μ-conformal with $\mathcal{T}$.*

Proof Without loss of generality, we can assume that $n \geq 2$. Let $\theta \geq f_{2.37}(n) \gg n$ be the order of the vertex tangle $\mathcal{T}$. Let $(A, B) \in \mathcal{T}$ be a base of largest order with A inclusionwise-maximal, let $Z = A \cap B$, and let μ_0 be a model of the $|Z|$-vertex path P in $G[A]$ such that $\mu_0^{-1}(Z) = V(P)$. By Lemma 2.54, we have $|Z| = \theta - 1$. Let $k = 2n(n-1)$ be the number of edges of the $n \times n$ grid G_n. Let us choose an integer s so that $k, n \ll s \ll \theta$, and assign to each vertex $v \in V(G_n)$ a subpath P_v of P with $4s$ vertices so that the subpaths for distinct vertices are vertex-disjoint and no edge of P joins two of them. For each edge $e = uv \in E(G_n)$, choose subsets $A_e, B_e \subset Z$ of size s such that $\mu_0^{-1}(A_e) \subset P_u$, $\mu_0^{-1}(B_e) \subset P_v$, and the sets chosen for distinct edges e are pairwise disjoint.

Consider any edge $e = uv \in E(G_n)$ and let $R_e = Z \setminus (A_e \cup B_e)$. Lemma 2.54 implies that there is no vertex separation (C', D') of $G[B] - R_e$ such that $A_e \subseteq C'$, $B_e \subseteq D'$, and $|C' \cap D'| < s$; indeed, otherwise $(C, D) = (C' \cup R_e, D' \cup R_e)$ would be a separation of $G[B]$ of order $|C' \cap D'| + |R_e| < s + |R_e|$ with $|C \cap Z| \geq |A_e| + |R_e| = s + |R_e| > |C \cap D|$ and similarly $|D \cap Z| > |C \cap D|$. Therefore, Menger's theorem implies that there is an (A_e, B_e)-linkage $\mathcal{L}_e$ in $G[B]$ of size s intersecting Z only in endpoints. Moreover, since P does not have an edge between P_u and P_v, the linkage $\mathcal{L}_e$ is actually contained in $H = G[B] - \mu_0(E(P))$.

Let us now apply Lemma 2.53 to the linkages $\mathcal{L}_e$ for $e \in E(G_n)$, and let us discuss the two possible outcomes.

- Suppose first that there exists a model μ of the $n \times n$ grid G_n in G strung on a linkage $\mathcal{L}$ of size n from A_e to B_e for an edge $e \in E(G_n)$. Recall that $\mathcal{T}_n$ is the canonical vertex tangle of order n in G_n. Since the model is strung on $\mathcal{L}$ and $|\mathcal{L}| = n$, for each path $L \in \mathcal{L}$, there exists a row R of G_n such that $R \subseteq \mu^{-1}(V(L))$. Consequently, $\mathrm{rk}_{\mathcal{T}_n}(\mu^{-1}(V(L))) \geq \mathrm{rk}_{\mathcal{T}_n}(R) = n$. Moreover, both ends of L belong to Z. Since Z is $\mathcal{T}$-free, Lemma 2.46 implies that $\mathcal{T}_n$ is μ-conformal with $\mathcal{T}$.
- Next, suppose that there exists a system $\{P_e : e \in E(G_n)\}$ of pairwise vertex-disjoint paths in H, where for every edge $e \in E(G_n)$, the path P_e starts in A_e, ends in B_e, and intersects Z only in its endpoints. Let μ be the model of G_n in G corresponding to the following way of obtaining the $n \times n$ grid: We contract $G[A]$ according to the model μ_0 to form a path P' on Z. For each $v \in V(G_n)$, we contract the subpath of P' corresponding to P_v to a single vertex. Finally, for each $e \in E(G_n)$, we contract the path P_e to a single edge.

  Note that the model μ has the property that for each vertex $v \in V(G_n)$, the subgraph $\mu(v)$ contains a vertex of Z. For $i \in [n]$, let R_i' be the i-th row of G_n, let $M_i = \bigcup_{v \in R_i'} V(\mu(v))$, and let R_i be a minimal connected subgraph of $G[M_i]$ containing all vertices of $Z \cap M_i$. Then R_i is a tree with all leaves in Z, and $\mathrm{rk}_{\mathcal{T}_n}(\mu^{-1}(V(R_i))) = \mathrm{rk}_{\mathcal{T}_n}(R_i') = n$. Lemma 2.46 applied for $R_1, \ldots, R_n$ thus implies that $\mathcal{T}_n$ is μ-conformal with $\mathcal{T}$.

□

2.11 Erdős–Pósa for Planar Minors

How can one certify that a graph G does not contain more than k pairwise vertex-disjoint copies of another graph H in some prescribed containment relation (subgraph, induced subgraph, minor, ...)? An easy way to do so is to find a set X of at most k vertices of G such that $G - X$ does not contain any copy of H. However, does there always exist such a set X? Or at least a slightly larger set of vertices intersecting all copies of H? Erdős–Pósa theory (named after the seminal result of Erdős and Pósa [12] on disjoint cycles) is devoted to the study of this question. As we are going to discuss in this section, the Grid Theorem plays an important role in Erdős–Pósa theory.

Let us now define the notion more precisely. Let $\mathcal{H}$ be a set of graphs and for a graph G, let $C_{\mathcal{H}}(G)$ be the system of sets $C \subseteq V(G)$ such that the induced subgraph $G[C]$ is isomorphic to a graph in $\mathcal{H}$. An *$\mathcal{H}$-hitting set* in G is a set $X \subseteq V(G)$ intersecting all sets in $C_{\mathcal{H}}(G)$; we let $\gamma_{\mathcal{H}}(G)$ be the smallest size of an $\mathcal{H}$-hitting set in G. An *$\mathcal{H}$-packing* in G is a set of pairwise disjoint elements of $C_{\mathcal{H}}(G)$; we let $\alpha_{\mathcal{H}}(G)$ be the maximum size of an $\mathcal{H}$-packing in G. We say that $\mathcal{H}$ has ***Erdős–Pósa property*** (in a graph class $\mathcal{G}$) if there exists a non-decreasing function $f : \mathbb{N} \to \mathbb{N}$

such that for every integer $k \geq 0$, every graph G (or every graph $G \in \mathcal{G}$) contains either

- an $\mathcal{H}$-packing of size more than k, or
- an $\mathcal{H}$-hitting set of size at most $f(k)$.

Thus, we can either find more than k pairwise vertex-disjoint induced subgraphs of G belonging to $\mathcal{H}$, or a certificate that it is not possible to find more than $f(k)$ such pairwise vertex-disjoint induced subgraphs. Equivalently, $\gamma_{\mathcal{H}}(G) \leq f(\alpha_{\mathcal{H}}(G))$ for every $G \in \mathcal{G}$. We say that f is an *EP-function* for $\mathcal{H}$. Let us give a few examples:

- Erdős and Pósa [12] proved that the minimum size of a ***feedback vertex set*** (a set of vertices intersecting all cycles, i.e., whose removal turns the graph into a forest) in a graph G is $O(k \log k)$, where k is the maximum number of pairwise vertex-disjoint cycles in G. That is, if $\mathcal{H}$ consists of all cycles, then $\mathcal{H}$ has Erdős–Pósa property with EP-function $f(k) = O(k \log k)$.
- Let $\mathcal{H}$ be the set of all graphs that have K_5 as a minor, and let $\mathcal{G}$ be the class of graphs that can be drawn in the projective plane. Then $\mathcal{H}$ does not have Erdős–Pósa property on $\mathcal{G}$: The graph $2K_5$ consisting of two disjoint copies of K_5 cannot be drawn in the projective plane without crossings, and thus $\alpha_{\mathcal{H}}(G) \leq 1$ for every graph $G \in \mathcal{G}$. On the other hand, any graph drawn in the projective plane with sufficiently large representativity r_0 contains K_5 as a minor (as we are going to see in Chap. 4), and thus if the drawing of a graph G in the projective plane has representativity at least $r \geq r_0$, then it is not possible to eliminate all minors of K_5 in G by deleting at most $r - r_0$ vertices.
- A well-known theorem of Kőnig states that the maximum size of a matching in a bipartite graph is equal to the size of the vertex cover. Thus, letting $\mathcal{G}$ be the class of all bipartite graphs and letting $\mathcal{H} = \{K_2\}$, we conclude that $\mathcal{H}$ has Erdős–Pósa property on $\mathcal{G}$, with the identity function being an EP-function.
- This tight connection does not hold in general graphs. However, deletion of the set X of vertices incident with the edges of a maximal matching results in an edgeless graph, and the size of X is at most twice the size of the matching. Hence, $\{K_2\}$ actually has Erdős–Pósa property on all graphs, with EP-function $f(k) = 2k$.

The last example has an obvious generalization.

Observation 2.56 *Let $\mathcal{G}$ and $\mathcal{H}$ be classes of graphs. If there exists a constant c such that for every $G \in \mathcal{G}$, every set $C \subseteq V(G)$ such that $G[C]$ is isomorphic to a graph in $\mathcal{H}$ has size at most c, then $\mathcal{H}$ has Erdős–Pósa property on $\mathcal{G}$ with EP-function $f(k) = ck$.*

Proof Let $\mathcal{C}$ be a largest $\mathcal{H}$-packing in G, and let $X = \bigcup \mathcal{C}$. Then X is an $\mathcal{H}$-hitting set and if $|\mathcal{C}| \leq k$, then $|X| \leq c \cdot |\mathcal{C}| \leq ck$. □

Hence, Erdős–Pósa property is interesting only when the considered objects can be arbitrarily large. Importantly, Erdős–Pósa property holds in very general circumstances on graphs of bounded treewidth.

Lemma 2.57 *Let $\mathcal{H}$ be a class of connected graphs. For every positive integer t, the class $\mathcal{H}$ has Erdős–Pósa property on the class $\mathcal{G}$ of graphs of treewidth at most t.*

Let us give two proofs of this result: A direct one and one through a construction of a tangle. The latter proof is more complicated, but it is the starting point for many arguments about Erdős–Pósa property in more general graph classes. The first proof follows from the tight Erdős-Pósa property of trees:

Observation 2.58 *Let $\mathcal{T}$ be a set of non-empty subtrees of a tree T. The maximum number d of pairwise vertex-disjoint trees in $\mathcal{T}$ is equal to the minimum size of a subset of $V(T)$ intersecting all trees in $\mathcal{T}$.*

Proof We prove the claim by induction on $V(T)$. If $|T| = 1$, then the claim is trivial. Suppose that $|T| > 1$, and let v be a leaf of T. If $\mathcal{T}$ contains a single-vertex tree T_0 with vertex set $\{v\}$, then let $\mathcal{T}' = \{T' \in \mathcal{T} : v \notin V(T')\}$. Note that $\mathcal{T}'$ contains only $d - 1$ pairwise vertex-disjoint trees, since we can add T_0 to them to get d pairwise vertex-disjoint trees in $\mathcal{T}$. By the induction hypothesis for $\mathcal{T}'$ and $T - v$, we can intersect all trees in $\mathcal{T}'$ by $d - 1$ vertices, and adding v gives a set of d vertices intersecting all trees in $\mathcal{T}$.

Suppose now that $\mathcal{T}$ does not contain such a tree T_0, and let $\mathcal{T}'' = \{T' - v : T' \in \mathcal{T}\}$. Note that trees in $\mathcal{T}$ are vertex-disjoint if and only if the corresponding trees in $\mathcal{T}''$ are vertex-disjoint, since if a tree contains v, then it also contains the unique neighbor of v in T. Consequently, the maximum number of pairwise vertex-disjoint trees in $\mathcal{T}''$ is d, and by the induction hypothesis for $\mathcal{T}'$ and $T - v$, we can intersect all trees in $\mathcal{T}''$ by a set D of d vertices. Observe that D also intersects all trees in $\mathcal{T}$. □

Proof 1 of Lemma 2.57 Consider any graph $G \in \mathcal{G}$ with a tree decomposition (T, β) of width at most t, and let k be a non-negative integer. Since $G[C]$ is connected for every $C \in C_{\mathcal{H}}(G)$, Lemma 2.2 implies that $\beta^{-1}(C)$ induces a connected subtree of T. Let $\mathcal{T} = \{T[\beta^{-1}(C)] : C \in C_{\mathcal{H}}(G)\}$, and let d be the maximum number of pairwise vertex-disjoint trees in $\mathcal{T}$. If $d > k$, then observe that the corresponding subsets of $V(G)$ form an $\mathcal{H}$-packing of size $d > k$. If $d \leq k$, then by Observation 2.58, there exists a set $X \subseteq V(T)$ of size $d \leq k$ that intersects all trees in $\mathcal{T}$. Then the set $\bigcup_{x \in X} \beta(x)$ of size at most $(t + 1)k$ intersects all sets in $C_{\mathcal{H}}(G)$. Hence, $\mathcal{H}$ has Erdős-Pósa property on $\mathcal{G}$, with EP-function $f(k) = (t+1)k$. □

Let us now give the second proof based on tangles.

Proof 2 of Lemma 2.57 Let us define $f(0) = 0$ and $f(k) = (t + 1)(4k - 1)$ for $k \geq 1$. We are going to show that $\mathcal{H}$ has Erdős-Pósa property on $\mathcal{G}$ with EP-function f; i.e., that every graph $G \in \mathcal{G}$ has an $\mathcal{H}$-hitting set of size at most $f(k)$, where $k = \alpha_{\mathcal{H}}(G)$. We prove the claim by induction on k. The case $k = 0$ is trivial, and thus assume that $k \geq 1$.

Suppose first that there exists a set $X_0 \subseteq V(G)$ of size at most $t + 1$ such that $\alpha_{\mathcal{H}}(G - X_0) < k$. By the induction hypothesis, there exists an $\mathcal{H}$-hitting set X_1 in

$G - X_0$ of size at most $f(k-1)$. Then $X_0 \cup X_1$ is an $\mathcal{H}$-hitting set in $G - X_0$ of size at most $f(k-1)+t+1 < f(k)$. Hence, we can assume that $\alpha_{\mathcal{H}}(G - X_0) = k$ for every set $X_0 \subseteq V(G)$ of size at most $t+1$.

Next, suppose that there exists a vertex separation (A, B) of G of order at most $t+1$ such that $k_A = \alpha_{\mathcal{H}}(G[A \setminus B]) < k$ and $k_B = \alpha_{\mathcal{H}}(G[B \setminus A]) < k$. Since $\mathcal{H}$-packings in $G[A \setminus B]$ and $G[B \setminus A]$ combine to an $\mathcal{H}$-packing in G and all graphs in $\mathcal{H}$ are connected, we have $k_A + k_B = \alpha_{\mathcal{H}}(G - (A \cap B)) = k$, where the last equality follows by considering the set $X_0 = A \cap B$. Since $k_A, k_B < k$, we have $k_A, k_B \geq 1$. By the induction hypothesis, there exist $\mathcal{H}$-hitting sets X_A and X_B in $G[A \setminus B]$ and $G[B \setminus A]$ such that $|X_A| \leq f(k_A)$ and $|X_B| \leq f(k_B)$. Observe that since $\mathcal{H}$ only consists of connected graphs, $X_A \cup X_B \cup (A \cap B)$ is an $\mathcal{H}$-hitting set in G of size at most

$$\begin{aligned} f(k_A) + f(k_B) + t + 1 &= (t+1)(4k_A - 1) + (t+1)(4k_B - 1) + t + 1 \\ &= (t+1)(4k_A + 4k_B - 1) = f(k). \end{aligned}$$

Hence, we can assume that for every vertex separation (A, B) of G of order at most $t+1$, we have either $\alpha_{\mathcal{H}}(G[A \setminus B]) = k$ (and $\alpha_{\mathcal{H}}(G[B]) = 0$), or $\alpha_{\mathcal{H}}(G[B \setminus A]) = k$ (and $\alpha_{\mathcal{H}}(G[A]) = 0$).

Let $\mathcal{T}$ be the set of vertex separations (A, B) of G of order at most $t+1$ such that $\alpha_{\mathcal{H}}(G[A]) = 0$. Since $\mathrm{tw}(G) \leq t$, Lemma 2.23 implies that $\mathcal{T}$ is not a vertex tangle of order $t+2$, and thus (T1) is false. Let $(A_1, B_1), (A_2, B_2), (A_3, B_3) \in \mathcal{T}$ be vertex separations such that $G = G[A_1] \cup G[A_2] \cup G[A_3]$, and consider any $C \in \mathcal{C}_{\mathcal{H}}(G)$. There exists $i \in [3]$ such that $C \cap A_i \neq \emptyset$. However, since $\alpha_{\mathcal{H}}(G[A_i]) = 0$, we have $C \nsubseteq A_i$, and thus $C \cap B_i \neq \emptyset$. Since $G[C]$ is connected, this implies that $C \cap (A_i \cap B_i) \neq \emptyset$. We conclude that $\bigcup_{i=1}^{3} A_i \cup B_i$ is an $\mathcal{H}$-hitting set of size at most $3(t+1) = f(1) \leq f(k)$. □

The second proof is of course much less elegant and gives a worse EP-function. However, in case we apply it for a graph G of large treewidth, it gives us a vertex tangle $\mathcal{T}$ which points towards all appearances of the graphs from $\mathcal{H}$ in G. We can then study the structure of the part of G pointed towards by $\mathcal{T}$, giving us a common starting point for more involved arguments in Erdős-Pósa theory.

The assumption in Lemma 2.57 that the graphs in $\mathcal{H}$ are connected can be relaxed to having a bounded number of components subject to an additional technical condition on $\mathcal{H}$. To see that, let us start with the following observation on independent transversals in chordal graphs.

Lemma 2.59 *Let F be a chordal graph and let $Z_1, \ldots, Z_d$ be independent sets in F, each of size at least d. Then there exists an independent set $Z = \{z_1, \ldots, z_d\}$ in F such that $z_i \in Z_i$ for each $i \in [d]$.*

Proof We prove the claim by induction on d, with the basic case $d = 1$ being trivial. Suppose that $d > 1$. Without loss of generality, we can assume $V(F) = Z_1 \cup \ldots \cup Z_d$, as otherwise we can delete the vertices not belonging to $Z_1 \cup \ldots \cup Z_d$. Since F is chordal, it has a simplicial vertex v (i.e., a vertex whose neighborhood forms a clique in F), where say $v \in Z_d$. Since the closed neighborhood $N[v]$ of v is a clique, it intersects each of the sets $Z_1, \ldots, Z_{d-1}$ in at most one vertex, and thus the set $Z'_i = Z_i \setminus N[v]$ has size at least $d-1$ for each $i \in [d-1]$. By the induction hypothesis for $Z'_1, \ldots, Z'_{d-1}$, there exists an independent set $Z' = \{z_1, \ldots, z_{d-1}\}$ in $F - N[v]$ with $z_i \in Z'_i$ for $i \in [d-1]$. The claim of the lemma follows by letting $z_d = v$. □

This implies the following variation on Observation 2.58.

Lemma 2.60 *Let T be a tree, let d be a positive integer and for $i \in [d]$, let $\mathcal{T}_i$ be a set of non-empty subtrees of T. Then for every integer $k \geq 0$, either*

- *there exist a system $\{T_{ij} : i \in [d], j \in [k+1]\}$ of pairwise vertex-disjoint trees such that $T_{ij} \in \mathcal{T}_i$ for every $i \in [d]$ and $j \in [k+1]$, or*
- *there exists $i \in [d]$ and a set of at most $d(k+1) - 1$ vertices of T intersecting all trees in $\mathcal{T}_i$.*

Proof Let G be the intersection graph of subtrees of T, i.e., the vertices of G are the subtrees of T and distinct subtrees are adjacent in G if and only if they intersect. Observe that the graph G is chordal.

Suppose that for every $i \in [d]$, the trees in $\mathcal{T}_i$ cannot be intersected by less than $(d+1)k$ vertices. By Observation 2.58, there exists a set Z_i of $d(k+1)$ pairwise vertex-disjoint trees from $\mathcal{T}_i$; equivalently, Z_i is an independent set in G. For $i \in [d]$ and $j \in [k+1]$, let $Z_{i,j} = Z_i$. Lemma 2.59 applied for the system $\{Z_{i,j} : i \in [d], j \in [k+1]\}$ of $d(k+1)$ independent sets of size at least $d(k+1)$ gives a system $\{T_{ij} : i \in [d], j \in [k+1]\}$ of pairwise vertex-disjoint trees such that $T_{ij} \in Z_{i,j} = Z_i \subseteq \mathcal{T}_i$ for every $i \in [d]$ and $j \in [k+1]$. □

By using Lemma 2.60 instead of Observation 2.58 in the first proof of Lemma 2.57, we obtain the following strengthening of Lemma 2.57.

Lemma 2.61 *Let d and t be positive integers, and for $i \in [d]$, let $\mathcal{H}_i$ be a set of connected graphs. Let $\mathcal{H}$ be the set of all graphs H whose vertex set can be partitioned into parts $L_1, \ldots, L_d$ such that $H[L_i] \in \mathcal{H}_i$ for $i \in [d]$. Then $\mathcal{H}$ has Erdős-Pósa property on graphs of treewidth at most t, with EP-function $f(k) = (t+1)(d \cdot (k+1) - 1)$.*

Corollary 2.34 together with Lemma 2.61 show that minors of planar graphs have the Erdős-Pósa property.

Corollary 2.62 *Let H be a planar graph and let $\mathcal{H}$ be the set of all graphs that contain H as a minor. Then $\mathcal{H}$ has Erdős-Pósa property.*

Proof Let $H_1, \ldots, H_d$ be the components of H. For $i \in [d]$, let $\mathcal{H}_i$ be the set of connected graphs containing H_i as a minor. Let $\mathcal{H}'$ be the set of graphs whose vertex set can be partitioned into parts $L_1, \ldots, L_d$ such that $H[L_i] \in \mathcal{H}_i$ for $i \in [d]$. Note that every graph in $\mathcal{H}'$ contains H as a minor, and conversely, every graph containing H as a minor has an induced subgraph belonging to $\mathcal{H}'$. It follows that $\alpha_{\mathcal{H}'}(G) = \alpha_{\mathcal{H}}(G)$ and $\gamma_{\mathcal{H}'}(G) = \gamma_{\mathcal{H}}(G)$ for every graph G, and thus it suffices to prove that $\mathcal{H}'$ has the Erdős-Pósa property.

For every positive integer k, the graph kH consisting of the disjoint union of k copies of H is planar, and thus by Corollary 2.34, there exists an integer t_k such that every kH-minor-free graph has treewidth at most t_k. Let us define $f(k) = (t_{k+1} + 1)(d \cdot (k + 1) - 1)$. Consider any graph G with $\alpha_{\mathcal{H}'}(G) = k$. Since G does not contain $k + 1$ disjoint models of H, it is $(k + 1)H$-minor-free, and thus $\mathrm{tw}(G) \leq t_{k+1}$. Therefore, there exists an $\mathcal{H}'$-hitting set in G of size at most $(t_{k+1} + 1)(d \cdot (k + 1) - 1) = f(k)$ by Lemma 2.61. □

This is a qualitative strengthening of the aforementioned theorem of Erdős and Pósa [12] on disjoint cycles—Corollary 2.62 with $H = C_3$ implies that cycles have Erdős-Pósa property (though this argument gives a suboptimal EP-function). As we have seen on the example of K_5, the analogue of Corollary 2.62 does not necessarily hold for non-planar graphs H. In fact, a straightforward modification of the argument for K_5 shows that it does not hold for any non-planar graph.

Theorem 2.63 (Robertson and Seymour [21]) *Let H be a graph and let $\mathcal{H}$ be the set of all graphs that contain H as a minor. Then $\mathcal{H}$ has Erdős-Pósa property if and only if H is planar.*

2.12 Treewidth in Graphs on Surfaces

While the proof of the Grid Theorem in general (especially in its tighter forms) is involved, it is quite easy to prove a strong form of the Grid Theorem in planar graphs. A key ingredient is an argument to cleanup a system of intersecting paths, which is also useful in other proofs. Let us first present the argument in the cylindrical setting, which is not quite what we need at the moment but will come handy later.

Let G be a plane graph. For positive integers r and p, an $r \times p$ *sieve* in G consists of

- a sequence $\mathcal{R}$ of pairwise vertex-disjoint cycles $C_1, \ldots, C_r \subseteq G$ whose drawings are "nested", i.e., for $1 \leq i < j \leq r$, the disk in the plane bounded by C_j contains C_i, and
- a set $\mathcal{P}$ of pairwise vertex-disjoint paths $P_1, \ldots, P_p \subseteq G$, each intersecting all the cycles in $\mathcal{R}$.

We call the elements of $\mathcal{R}$ *rows* and the elements of $\mathcal{P}$ *pillars*. Note that because of the nested way the row cycles are drawn in the plane, it suffices to require that the pillars intersect C_1 and C_r, and this forces them to intersect all other rows.

We say that the sieve is *orderly* if for each $i \in [r]$ and $j \in [p]$, the intersection of C_i and P_j is connected (i.e., a common subpath of C_i and P_j). Note that in that case, contracting each of the intersections to a single vertex and contracting the subpaths of the rows and pillars between them to single edges gives us a cylindrical grid as a minor of G. The following lemma shows that rows and pillars of every sieve can be rerouted to make the sieve orderly; and moreover, that we have quite a bit of control over the resulting sieve.

Lemma 2.64 *Let G be a plane graph and let r and p be positive integers. If G contains an $r \times p$ sieve $(\mathcal{R}, \mathcal{P})$, then it also contains an orderly $r \times p$ sieve $(\mathcal{R}', \mathcal{P}')$ such that $\bigcup \mathcal{R}' \cup \bigcup \mathcal{P}' \subseteq \bigcup \mathcal{R} \cup \bigcup \mathcal{P}$. Moreover, the first and last rows of $\mathcal{R}$ and $\mathcal{R}'$ coincide, and for $j \in [r]$, the disk in the plane bounded by the j-th row of $\mathcal{R}'$ contains the j-th row of $\mathcal{R}$.*

Proof Let $\mathcal{R} = C_1, \ldots, C_r$ and $\mathcal{P} = \{P_1, \ldots, P_p\}$. Without loss of generality, we can assume that the pillars intersect C_1 and C_r exactly in their endpoints, as otherwise we can replace each pillar by its shortest subpath between C_1 and C_r. We are going to gradually modify the sieve, cleaning it up row by row starting from row r (the outermost one). To ensure that the process terminates, each transformation will decrease the *size* of the sieve, defined as the number of edges in $E\left(\bigcup \mathcal{R} \cup \bigcup \mathcal{P}\right)$; note that edges that belongs both to rows and pillars are counted only once. For an integer $a \in [r]$, we say that the sieve is *orderly up to row a* if for each $j \in [p]$, the pillar P_j intersects each of the rows $C_a, \ldots, C_r$ in a connected subpath. Since each pillar intersects C_r in a single vertex, the sieve is orderly up to row r initially. Moreover, since each pillar intersects C_1 in a single vertex, a sieve orderly up to row 2 is orderly.

Suppose we have already made the sieve orderly up to row $a + 1 \geq 3$, and let us describe how to make it orderly up to row a. Let Λ be the open region of the plane between C_a and C_{a+1}. Consider any pillar P, and suppose first that $P \cap \Lambda$ is disconnected (i.e., after P dives down to or below C_a, it comes back up above it at least once). The first component of $P \cap \Lambda$ joins C_{a+1} to C_a. Any other component has both ends in C_a, since the intersection of P with C_{a+1} is connected; let Q be the subpath of P whose drawing is the closure of such a component. Note that C_a is the only row intersected by Q, and $C_a \cap Q$ consists of the ends of the path Q. Let q be a point in the open disk bounded by C_1. The graph $C_a \cup Q$ contains exactly three cycles, and for two of them, the open disk bounded by the cycle contains q. One of the two cycles is C_a; let C'_a denote the other one. We modify the sieve by replacing C_a by C'_a (i.e., we redirect C_a over Q between the endpoints of Q). Note that the resulting sieve is still orderly up to row $a + 1$, and that the transformation decreases the size of the sieve, since the edges in $E(C_a) \setminus E(C'_a)$ are removed from the sieve.

Suppose now that $P_j \cap \Lambda$ is connected (i.e., consists only of a path from C_{a+1} to C_a) for every $j \in [p]$, but there is a pillar P such that $P \cap C_a$ is disconnected. Let Q be a subpath of P with both ends in C_a and otherwise disjoint from C_a, necessarily drawn in the disk bounded by C_a. Let Λ' be the unique face of the graph $C_a \cup Q$ not bounded by C_a and not containing the point q, and let Q' be the subpath of C_a contained in the boundary of Λ'. Since P intersects C_1 only in its last vertex, the path Q is disjoint from C_1, and since $q \notin \Lambda'$, we conclude that C_1 is disjoint from the closure of Λ'. Consequently, the closure of Λ' does not contain an endpoint of any pillar. It follows that every pillar $P' \neq P$ is disjoint from Q': Otherwise, since the endpoints of P' are outside of the closure of Λ', P' would have to enter and leave the closure of Λ' through Λ, implying that $P' \cap \Lambda$ is disconnected. Let us modify the sieve by replacing the subpath Q of the pillar P by Q'. Since $Q' \subset C_a$, this decreases the size of the sieve.

Since each of the transformations described above decreases the size of the sieve, after finitely many iterations the sieve becomes orderly up to row $a + 1$. Repeating this procedure, we eventually reach an orderly sieve. Since all transformations were performed within the sieve, the resulting sieve only uses vertices and edges contained in the original one. Moreover, observe that the transformations do not alter the first and the last row, and that we only increase the disks bounded by the other rows. □

We are also going to need a normal grid version of this statement. Let G be a plane graph and let Δ be a closed disk in the plane with G-normal boundary. Let A and B be disjoint sets of vertices of G drawn in order in the boundary of Δ, with $|A| = |B| = r$. Recall that an (A, B)-linkage is a system of pairwise vertex-disjoint paths with one end in A and the other end in B. An *$r \times p$ grill across Δ with ends A and B* consists of an (A, B)-linkage $\mathcal{R}$ of size r in G drawn in Δ and of a system $\mathcal{P}$ of p pairwise vertex-disjoint paths in G drawn in Δ such that every path in $\mathcal{P}$ intersects every path in $\mathcal{R}$. The grill is *orderly* if the intersection of every path from $\mathcal{R}$ with every path in $\mathcal{P}$ is connected.

Lemma 2.65 *Let G be a plane graph and let Δ be a closed disk in the plane with G-normal boundary. Let r and p be positive integers. If there exists an $r \times p$ grill $(\mathcal{R}, \mathcal{P})$ across Δ with ends A and B, then there also exists an orderly one.*

The proof is exactly the same as the one of Lemma 2.64—it suffices to turn the grill into a sieve by drawing auxiliary edges outside of Δ between the corresponding vertices of A and B, then verifying that the argument from the proof never causes these auxiliary edges to be added to pillars or removed from rows. Note also that an orderly $r \times p$ grill can be contracted to an $r \times p$ grid minor.

To apply this result, we need a planar version of Menger's theorem.

Lemma 2.66 *Let G be a plane graph and let u and v be non-adjacent vertices of G. For every integer k, exactly one of the following claims holds:*

- *There exist k paths in G from u to v intersecting only in their endpoints.*
- *There exists a G-normal simple closed curve in the plane separating u from v and intersecting G in less than k vertices.*

This statement is easiest to prove in terms of an important auxiliary graph, the ***radial graph*** of G, which is the natural drawing of the vertex-face incidence graph of G. More precisely, let G be a connected plane graph, and let $F(G)$ denote the set of faces of G. The radial graph R_G of G is the bipartite graph with vertex set $V(G) \cup \{x_f : f \in F(G)\}$, where for a face $f \in F(G)$ bounded by the closed walk $v_1, e_1, v_2, \ldots, v_m, e_m$ in G, the vertex x_f is drawn inside f and joined to vertices $v_1, \ldots, v_m$ by edges drawn inside f. See Fig. 2.4 for an illustration. Note that if a vertex appears multiple times in the boundary of f (i.e., the vertex forms a 1-cut), it is joined to x_f by multiple edges. This way, every edge $e \in E(G)$ corresponds to

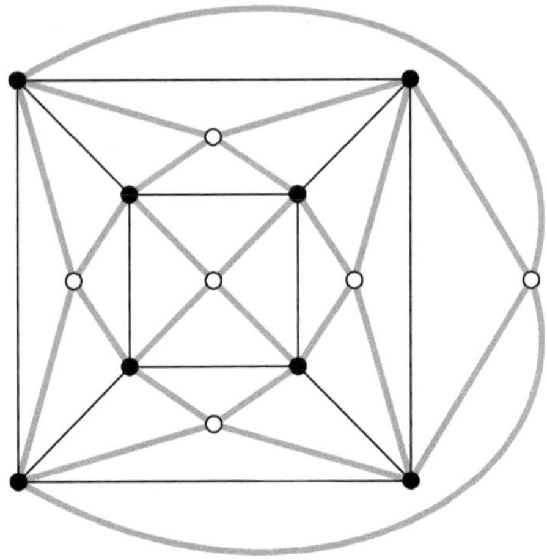

Fig. 2.4 A plane graph and its radial graph (in gray)

a unique face f_e of R_G of length four in which e is drawn. Hence, the radial graph R_G is accompanied by the function ρ_G such that

- for $v \in V(G)$, $\rho_G(v) = v$ is the same vertex in R_G,
- for $f \in F(G)$, $\rho_G(f) = x_f$ is the corresponding vertex in R_G, and
- for $e \in E(G)$, $\rho_G(e) = f_e$ is the corresponding face of R_G.

Observe that a cycle of length $2t$ in R_G corresponds to a G-normal simple closed curve intersecting G in t vertices. Hence, Lemma 2.66 is implied by the following slightly stronger claim.

Lemma 2.67 *Let G be a connected plane graph and let u and v be non-adjacent vertices of G. For every integer k, exactly one of the following claims holds:*

(i) There exist k paths in G from u to v intersecting only in their endpoints.
(ii) There exists a cycle of length less than $2k$ in the radial graph R_G separating u from v.

Proof Suppose that (i) is false. By Menger's theorem, there exists a separation (A, B) of G of order less than k such that $u \in V(A) \setminus V(B)$ and $v \in V(B) \setminus V(A)$. Note that $\{\rho_G(E(A)), \rho_G(E(B))\}$ is a partition of the faces of R_G. Let S be the subgraph of R_G consisting of the edges separating a face in $\rho_G(E(A))$ from a face in $\rho_G(E(B))$, and of the vertices incident with such edges. Each vertex in $V(S) \cap V(G)$ is incident both with an edge of A and an edge of B, and thus $V(S) \cap V(G) \subseteq V(A \cap B)$. In particular, $u, v \notin V(S)$. Moreover, observe that the definition of S ensures that if f_1 and f_2 are faces of R_G contained inside the same face of S, then either $\rho_G^{-1}(f_1), \rho_G^{-1}(f_2) \in E(A)$ or $\rho_G^{-1}(f_1), \rho_G^{-1}(f_2) \in E(B)$. Since all edges incident with u belong to A and all edges incident with v belong to B, u and v are drawn inside different faces of S. Therefore, there exists a cycle $K \subseteq S$ separating u from v, and since $|K \cap G| \leq |S \cap G| \leq |A \cap B| < k$ and every second vertex of K belongs to G, the cycle K has length less than $2k$.

Conversely, if (i) is true, then (ii) clearly cannot hold, since each of the paths from u to v must intersect every cycle in the radial graph R_G separating u from v. □

We can now combine Lemmas 2.65 and 2.67 to get a sufficient condition for the existence of a grid minor in a plane graph.

Lemma 2.68 *Let G be a graph drawn without crossings in a closed disk Δ in the plane, such that G intersects the boundary of Δ exactly in a set Z of vertices, and let n be an integer such that $4n \leq |Z|$. If G does not contain an $n \times n$ grid as a minor, then there exists a simple G-normal curve γ with ends in* bd(Δ) *and otherwise contained in the interior of Δ dividing Δ into two closed disks Δ_1 and Δ_2 such that $|Z \cap \Delta_1| > |G \cap \gamma|$ and $|Z \cap \Delta_2| > |G \cap \gamma|$.*

Proof Let A_1, A_2, B_1, and B_2 be pairwise disjoint subsets of Z of size n, appearing in the boundary of Δ in order. If G contains n pairwise vertex-disjoint paths from A_i to B_i for both $i \in [2]$, then the two systems of paths form an $n \times n$ grill across Δ. By Lemma 2.65, it follows that G contains an orderly $n \times n$ grill, which can be contracted into an $n \times n$ grid minor, and the second outcome holds.

Otherwise, we can assume that G does not contain n pairwise vertex-disjoint paths from say A_1 to B_1. To simplify the presentation, we assume that G contains at least one path from A_1 to B_1; we leave the details of the case that there are no paths between A_1 and B_1 for the reader to work out.

Let G' be a plane graph obtained from G as follows. If there exists a component K disjoint from $A_1 \cup B_1$, we connect K to another component incident with a common face by a single edge drawn inside that face (and inside Δ); note that this does not change the number of pairwise vertex-disjoint paths from A_1 to B_1. We repeat this step as long as such a component K exists. Finally, we add a vertex u with neighborhood A_1 and a vertex v with neighborhood B_1, both drawn outside of Δ so that they are both incident with the outer face f of G'. Note that G' is connected and does not contain n paths from u to v intersecting only in endpoints.

Next, consider the radial graph $R_{G'}$, and observe that its drawing can be chosen so that

- the vertex $\rho_{G'}(f)$ is drawn outside of Δ and all other vertices corresponding to faces of G' are drawn inside Δ, and
- the drawing of any edge e of $R_{G'}$ intersects bd(Δ) in a point different from the ends of e only if exactly one of the ends of e is in the interior of Δ and the other end of e is outside of Δ, and in that case the drawing of e intersects bd(Δ) in a single point.

Since G' does not contain n paths from u to v intersecting only in endpoints, Lemma 2.67 implies that there exists a cycle K in $R_{G'}$ of length less than $2n$ separating u from v.

Since u and v are both incident with the outer face f, the cycle K necessarily contains the vertex $\rho_{G'}(f)$. Let z_1 and z_2 be the neighbors of $\rho_{G'}(f)$ in K. Without loss of generality, we can assume that K is an induced cycle, since if e were a chord of K, then one of the shorter cycles in $K + e$ would also separate u from v.

Consequently, $V(K) \cap \mathrm{bd}(\Delta) \subseteq \{z_1, z_2\}$. It follows that the drawing of K intersects $\mathrm{bd}(\Delta)$ in exactly two points. We can choose γ as the intersection of the drawing of K with Δ. Then $|G \cap \gamma| = |G' \cap K| < n$, but the two closed disks to which γ splits Δ contain the sets $A_1, B_1 \subset Z$, respectively, of size n. □

To finish the proof of the Grid Theorem in a plane graph, we use tangles to obtain a suitable disk.

Lemma 2.69 *Let G be a plane graph and let n be a positive integer. If there exists a tangle $\mathcal{T}$ in G of order $\theta > 4n$, then G contains the $n \times n$ grid as a minor.*

Proof For a closed disk Δ in the plane with G-normal boundary, let B_Δ be the subgraph of G consisting of vertices and edges drawn in Δ and let A_Δ be the subgraph of G consisting of vertices and edges of G drawn in the closure of the complement of Δ. Hence, (A_Δ, B_Δ) is a separation of G whose order is equal to $|G \cap \mathrm{bd}(\Delta)|$. Let us fix a disk Δ with G-normal boundary such that $(A_\Delta, B_\Delta) \in \mathcal{T}$, A_Δ is maximal among all disks Δ with this property, and subject to that B_Δ is minimal. Note that such a disk Δ exists, since for a disk containing the whole graph G, the corresponding separation belongs to $\mathcal{T}$ by (T2).

Observe that since $\mathcal{T}$ is a tangle of order more than two, $(A_\Delta, B_\Delta) \in \mathcal{T}$ implies that $E(B_\Delta) \neq \emptyset$. Let e be an arbitrary edge incident with the outer face of B_Δ. If both ends of e are in $\mathrm{bd}(\Delta)$, then let Δ' be obtained from Δ by shifting its boundary across e, so that $B_{\Delta'} = B_\Delta - e$ and $A_{\Delta'} = A_\Delta + e$. Letting K_e be the subgraph of G with edge set e and the vertex set consisting of the ends of e, (T2) implies that $(K_e, G - e) \in \mathcal{T}$, and thus by (T1), we have $(A_{\Delta'}, B_{\Delta'}) \in \mathcal{T}$. This would contradict the maximality of A_Δ. Hence, the edge e is incident with a vertex $z \notin \mathrm{bd}(\Delta)$.

If the order of the separation (A_Δ, B_Δ) is less than $4n$, then consider the disk Δ' obtained from Δ by shifting a part of its boundary to touch z. Then $A_{\Delta'} = A_\Delta + z$ and $B_{\Delta'} = B_\Delta$. Note that (T1) and (T2) imply that $(A_{\Delta'}, B_{\Delta'}) \in \mathcal{T}$, contradicting the maximality of A_Δ. Hence, $Z = V(G) \cap \mathrm{bd}(\Delta)$ has size at least $4n$.

Consider any simple G-normal curve γ with ends in $\mathrm{bd}(\Delta)$ and otherwise contained in the interior of Δ. Let Δ_1 and Δ_2 be the closed disks to which γ divides Δ, and suppose that $|Z \cap \Delta_1| > |G \cap \gamma|$ and $|Z \cap \Delta_2| > |G \cap \gamma|$. Then $(A_{\Delta_1}, B_{\Delta_1})$ and $(A_{\Delta_2}, B_{\Delta_2})$ are separations of G of order less than $|Z| < \theta$. Since $G = A_\Delta \cup B_{\Delta_1} \cup B_{\Delta_2}$, by (T1) and symmetry we can assume that $(B_{\Delta_1}, A_{\Delta_1}) \notin \mathcal{T}$, and thus $(A_{\Delta_1}, B_{\Delta_1}) \in \mathcal{T}$. Note that $A_{\Delta_1} \supseteq A_\Delta$ and $B_{\Delta_1} \subsetneq B_\Delta$, since B_{Δ_1} does not contain all vertices of $|Z \cap \Delta_2|$. This contradicts the minimality of A_Δ or, when $A_{\Delta_1} = A_\Delta$, the maximality of B_Δ.

We conclude that $|Z \cap \Delta_1| \leq |G \cap \gamma|$ or $|Z \cap \Delta_2| \leq |G \cap \gamma|$ for any such curve γ. By Lemma 2.68 applied to $G \cap \Delta$, it follows that G contains the $n \times n$ grid as a minor. □

By Theorem 2.25, we obtain the following strong form of the Grid Theorem for planar graphs.

Corollary 2.70 *Let n be a positive integer. If G is a planar graph of treewidth at least $6n$, then G contains the $n \times n$ grid as a minor.*

Hence, in contrast to the general graphs, the grid minor size in planar graphs approximates treewidth up to a constant multiplicative factor. Let us remark that using a slight variation of the idea, an analogous linear bound can be proved for graphs drawn on any fixed surface; see Lemma 8.1 for details.

Corollary 2.70 has several interesting consequences. Clearly, if $|G| < n^2$, then G does not contain the $n \times n$ grid minor; hence, we obtain a sublinear bound on the maximum treewidth of a planar graph.

Corollary 2.71 *Every planar graph G has treewidth less than $6\sqrt{|G|+1}$.*

Together with Lemma 2.15, this gives a proof of famous planar separator theorem of Lipton and Tarjan [18].

Corollary 2.72 *Every planar graph G has a balanced cut of size at most $\lceil 6\sqrt{|G|+1}\rceil$.*

Observe also that if the $n \times n$ grid is a minor of a planar graph G, then every path in G from the boundary of the grid to the center of the grid has at least $n/2$ vertices. If G has radius r, this implies that $n/2 \le 2r + 1$. By Corollary 2.70, we conclude that every planar graph of radius r has treewidth $O(r)$. This important result has an even simpler proof, based on the useful concept of ***interdigitating trees***. Let us give the argument in the setting of graphs on surfaces.

Consider a triangulation G of a surface Σ, and let T be a spanning tree of G. Let us cut the surface along the edges of T, resulting in a graph G_1 drawn on a surface Σ_1 obtained from Σ by drilling a hole (the cycle in G_1 bounding this hole corresponds to a closed walk in T passing through every edge twice). Let $G_1^\star$ be the dual graph to this drawing; i.e., the vertices of $G_1^\star$ are faces of G and they are adjacent if and only if the faces share an edge not belonging to T. Since Σ_1 is a connected surface, the graph $G_1^\star$ is connected, and thus it has a spanning tree T_1. We say that T_1 is an *interdigitating tree* for T in G. See Fig. 2.5 for an illustration. If Σ is the sphere, then it is easy to see that $E(T_1)$ consists exactly of the duals of the edges in $E(G) \setminus E(T)$, and thus the interdigitating tree is uniquely determined by T; this is not the case for other surfaces.

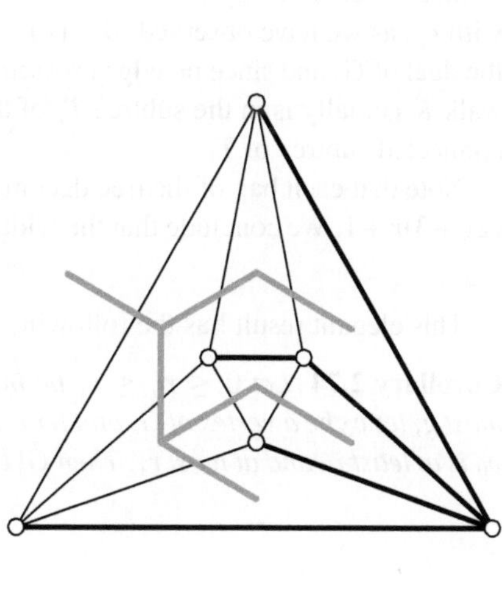

Fig. 2.5 A plane triangulation and the interdigitating tree (in gray) to its spanning tree (in bold)

Theorem 2.73 *Let G be a graph with a drawing on a surface Σ of Euler genus g. If G has radius r, then* $\mathrm{tw}(G) \le (2g+3)r$.

Proof Without loss of generality, we can assume that G is a triangulation of Σ, since adding edges to G neither increases its radius nor decreases its treewidth. Let v_0 be a vertex of G such that any other vertex of G is at distance at most r from v_0, and let T be a breadth-first search spanning tree from v_0. For each $v \in V(G)$, let $P(v)$ be the set of vertices on the path from v to v_0 in T, including v but excluding v_0; we have $|P(v)| \le r$.

Let T_1 be an interdigitating tree for T, and let E_1 be the set of edges of G whose duals are the edges of T_1. Note that $|T|$ is the number n of vertices of G, $\|T\| = n-1$, $|T_1|$ is the number s of faces of G, and $|E_1| = \|T_1\| = s-1$. Using the generalized Euler's formula, there are

$$\|G\| - \|T\| - |E_1| = (n+s+g-2) - (n-1) - (s-1) = g$$

edges of G that belong neither to $E(T)$ nor to E_1. Let $B_0 \subseteq V(G)$ consist of the vertices of G incident with an edge in $X = E(G) \setminus (E(T) \cup E_1)$, and note that $|B_0| \le 2|X| = 2g$. Let $B = \{v_0\} \cup \bigcup_{v \in B_0} P(v)$; we have $|B| \le 2gr + 1$.

For each face $f \in V(T_1)$, bounded by the triangle $v_1v_2v_3$ in G, we define $\beta(f) = B \cup P(v_1) \cup P(v_2) \cup P(v_3)$. We claim that (T_1, β) is a tree decomposition of G. To see that (D1) holds, note that each edge $e = uv$ of G is incident with a face f, and $\{u, v\} \subseteq \{v_0\} \cup P(u) \cup P(v) \subseteq \beta(f)$.

Hence, we only need to prove that (T_1, β) satisfies (D2). For a vertex $v \in B$, $\beta^{-1}(v)$ is the vertex set of the whole tree T_1. Suppose now that $v \in V(G) \setminus B$. We view T as rooted in v_0 and let T_v be the subtree of T rooted in v. Observe that $\beta^{-1}(v)$ consists of faces f incident with a vertex $v' \in V(G)$ such that $v \in P(v')$, i.e., exactly with the vertices of T_v. Moreover, since $v \notin B$, no vertex of T_v is incident with an edge of X. Consider a simple closed curve γ going around T_v infinitesimally close to it, so that the faces intersected by γ are exactly those belonging to $\beta^{-1}(v)$. Let e_v be the edge of T from v to its parent. Let $K = f_1, \ldots, f_m$ be the sequence of faces intersected by γ in order when γ is traversed starting from its intersection with e_v; as we have observed, $\beta^{-1}(v) = \{f_1, \ldots, f_m\}$. Note that K forms a walk in the dual of G, and since no edge crossed by γ except for e_v belongs to $E(T) \cup X$, the walk K actually is in the subtree T_1 of the dual. We conclude that $\beta^{-1}(v)$ induces a connected subtree of T_1.

Note that each bag of the tree decomposition (T_1, β) has size at most $|B| + 3r = (2g+3)r + 1$. We conclude that the width of this decomposition is at most $(2g+3)r$. □

This elegant result has the following simple but useful strengthening.

Corollary 2.74 *Let $0 \le r_0 \le r_1$ be integers. Let G be a graph of Euler genus at most g, let v_0 be a vertex of G, and let L be a set of vertices of G whose distance from v_0 is at least r_0 and at most r_1. Then $G[L]$ has treewidth at most* $(2g+3)(r_1 - r_0 + 1)$.

Proof Let G' be obtained from G by deleting all vertices at distance greater than r_1 from v_0 and then contracting to v_0 all vertices at distance less than r_0 from v_0. The contracted vertices induce a connected subgraph of G, since they form a subtree of a breadth-first search tree from v_0, and thus G' is a minor of G. Consequently, the Euler genus of G' is at most g. Note that $G[L]$ is a subgraph of G', and that every vertex in G' is at distance at most $r_1 - r_0 + 1$ from v_0. Hence, by Theorem 2.73, we have

$$\operatorname{tw}(G[L]) \leq \operatorname{tw}(G') \leq (2g+3)(r_1 - r_0 + 1).$$

□

This observation forms the basis for Baker's technique [2], an argument used especially in design of approximation algorithms: A graph on a surface is decomposed into layers based on the distance from an arbitrary vertex v_0, i.e., so that the distances from v_0 to distinct vertices of a single layer differ at most by a fixed constant ℓ. By Corollary 2.74, each layer induces a subgraph of treewidth $O(\ell)$, which makes it possible to solve the problem exactly in this subgraph. Finally, the individual solutions for the layers are stitched together with a small additional loss. We are going to discuss this technique in more detail in Sects. 7.2 and 16.2, but for now, let us demonstrate it by giving another proof of Lipton-Tarjan separator theorem. The following lemma is based on a choice of narrow layers covering almost the whole graph.

Lemma 2.75 *Let G be a graph of Euler genus at most g and let ℓ be a positive integer. For any assignment w of non-negative weights to vertices of G, there exists a set $S \subseteq V(G)$ such that $w(S) \leq \frac{1}{\ell} w(G)$ and $\operatorname{tw}(G - S) \leq (2g+3)(\ell - 1)$.*

Proof Without loss of generality, we can assume that G is connected, as otherwise we can consider each component of G separately. Let v_0 be an arbitrary vertex of G, and for any non-negative integer i, let L_i be the set of vertices of G at distance exactly i from v_0. For $m \in \{0, \ldots, \ell - 1\}$, let

$$S_m = \bigcup_{i \equiv m \pmod{\ell}} L_i.$$

Note that each component of $G - S_m$ consists of vertices at distance at least $c\ell + m + 1$ but at most $(c+1)\ell + m - 1$ from v_0 for some integer c, and thus its treewidth is at most $(2g+3)(\ell - 1)$ by Corollary 2.74. Hence, $\operatorname{tw}(G - S_m) \leq (2g+3)(\ell - 1)$ for each m.

Moreover, the sets $S_0, \ldots, S_{\ell-1}$ are pairwise disjoint, and thus

$$\sum_{m=0}^{\ell-1} w(S_m) \leq w(G).$$

Therefore, there exists $m \in \{0, \ldots, \ell - 1\}$ such that $w(S_m) \leq \frac{1}{\ell} w(G)$, and we can choose $S = S_m$.

□

Balancing the size of the removed part with the treewidth of the rest of the graph gives us the following strengthening of Corollary 2.71. By Lemma 2.15, this strengthening implies that every n-vertex graph of Euler genus g has a balanced cut of size $O(\sqrt{(g+1)n})$.

Corollary 2.76 *Every graph G of Euler genus at most g has treewidth at most $2\sqrt{(2g+3)|G|}$.*

Proof Let $\ell = \left\lceil \sqrt{\frac{|G|}{2g+3}} \right\rceil$. We give each vertex weight 1 and apply Lemma 2.75, obtaining a set $S \subseteq V(G)$ of size at most $\frac{1}{\ell}|G|$ such that $\mathrm{tw}(G-S) \le (2g+3)(\ell-1)$. Adding S to all bags of an optimal tree decomposition of $G-S$ gives a tree decomposition of G of width at most

$$\mathrm{tw}(G-S) + |S| \le (2g+3)(\ell-1) + \tfrac{1}{\ell}|G| \le 2\sqrt{(2g+3)|G|}.$$ □

For future use, let us also note the following variation on Theorem 2.73.

Corollary 2.77 *Let G be a graph with a drawing on a surface Σ of Euler genus g, let $f_1, \ldots, f_m$ be faces of G (where $m \ge 1$), and let U be the set of vertices of G incident with these faces. If every vertex of G is at distance at most r from U, then* $\mathrm{tw}(G) \le (2g+4m-1)(r+1)$.

Proof By adding $m-1$ handles to Σ, we can join $f_1, \ldots, f_m$ into a single face of a drawing of G on a surface of Euler genus $g+2m-2$. Let G' be obtained from G by adding a vertex u drawn inside this face and adjacent to all vertices of U. Then G' has radius at most $r+1$, and thus $\mathrm{tw}(G) \le \mathrm{tw}(G') \le (2g+4m-1)(r+1)$ by Theorem 2.73. □

2.13 Respectful Tangles

Let us now introduce the notion of respectful tangles in graphs on surfaces, which turns out to be extremely important in understanding the minors of these graphs. To motivate this notion, consider a graph G drawn on a surface Σ other than the sphere and a G-normal contractible simple closed curve γ. This curve naturally defines a separation of G into the subgraph A drawn in the closed disk Δ in Σ bounded by γ and the subgraph B drawn in the closure of $\Sigma \setminus \Delta$. Recall that a tangle $\mathcal{T}$ can be viewed as orienting all small separations towards the part of the graph that is more "substantial", and in this case it would be pleasing if the tangle pointed towards the part B which is drawn in the part of $\Sigma - \gamma$ of non-zero genus. Naturally, not every tangle has this property; it is of course possible that $\mathcal{T}$ is derived from a grid minor contained in A. Nevertheless, it is easy to imagine that a tangle with this pleasing "surface respecting" property could be useful in study of treewidth in graphs on surfaces.

Can we actually use this property to *construct* a tangle in a drawing? An issue with this in general is that not all separations in a graph drawn on a surface can be defined in terms of such contractible curves. E.g., in a graph on the torus, another natural separation is obtained by taking two disjoint homotopically equivalent non-contractible simple closed curves, which split the torus into two cylinders and thus do not offer a natural choice for the more substantial part. To avoid this issue, we restrict our attention to graphs of large representativity, so that all small separations correspond to combinations of contractible G-normal closed curves. And indeed, as one of the results that we are going to present in this section, a connected graph drawn on a surface other than the sphere with representativity θ admits a unique respectful tangle of order θ defined by the discussed property.

For another source of motivation, consider a graph G drawn on the sphere and a tangle $\mathcal{T}$ in G of order θ. A G-normal simple closed curve γ intersecting G in less than θ vertices again naturally defines a separation (A, B) of G. Now both parts are drawn in disks and thus cannot be distinguished based by the properties of a drawing. Nevertheless, $\mathcal{T}$ gives an orientation to each such separation, and we can use this orientation to select one of the disks as the interior of γ. This puts us on the same footing as in the non-sphere case: The (respectful) tangle shows a disk bounded by γ that contains the "small" part of the graph. Respectful tangles thus give a way to treat the spherical and non-spherical cases uniformly, avoiding the need to overcome the ambiguity in the choice of the disks bounded by simple closed curves in the spherical case.

Let us now give the necessary definitions. To avoid dealing with general simple closed curves, it is more convenient to work in the radial graph setting. For a graph G with a cellular drawing on any surface, the ***radial graph*** R_G and the associated function ρ_G is defined in the same way as we did in the plane case—to form R_G, a new vertex is put in each face f of G and joined to all vertices incident with f by edges (if a vertex is incident with f multiple times, an edge is added for each incidence), and the original edges of G are deleted. The function ρ_G maps vertices and faces of G to corresponding vertices of R_G, and edges of G to the faces of R_G that contain them. Paths (and cycles) in R_G naturally correspond to simple (closed) G-normal curves. Conversely, any closed G-normal curve can be shifted to a homotopically equivalent closed walk in R_G. Observe that consequently, the edgewidth of R_G is twice the representativity of G.

Let $\mathcal{T}$ be an orientation of $(<\theta)$-separations in a graph G with cellular drawing on a surface Σ. We say that $\mathcal{T}$ is ***respectful*** if for every cycle C in R_G of length less than 2θ, there exists a closed disk $\Delta \subset \Sigma$ bounded by C such that

$$(G \cap \Delta, G \cap \overline{\Sigma \setminus \Delta}) \in \mathcal{T}.$$

In that case, we define $\mathrm{ins}_{\mathcal{T}}(C) = \Delta$. Let us remark that if Σ is not the sphere, respectfulness implies that every such cycle C is contractible, and thus G has representativity at least θ; and in this case, the disk Δ is uniquely determined by C.

In case that Σ is the sphere, the respectfulness condition is automatically satisfied (i.e., every orientation of separations in a connected graph drawn on the sphere is respectful), but it serves as a definition of the interior $\mathrm{ins}_{\mathcal{T}}(C)$ of the cycle C relative to $\mathcal{T}$.

Let us remark that although we generally prefer to work with the cycles in the radial graph, a general simple closed curve perspective is sometimes necessary (e.g., when we compare tangles in two different graphs drawn on the same surface), and equivalent according to the following lemma.

Lemma 2.78 *Let $\mathcal{T}$ be a pretangle of order θ in a graph G with cellular drawing on a surface Σ. The pretangle $\mathcal{T}$ is respectful if and only if for every simple closed G-normal curve γ in Σ intersecting G in less than θ vertices, there exists a closed disk $\Delta \subset \Sigma$ bounded by γ such that*

$$(G \cap \Delta, G \cap \overline{\Sigma \setminus \Delta}) \in \mathcal{T}.$$

Proof Since cycles in R_G define simple closed G-normal curves, only the forward implication is non-trivial. We prove that the conclusion of the lemma holds by induction on $|\gamma \cap G|$. Suppose first that there exists a face f of G such that $\gamma \cap f$ is not connected. Let γ' be a simple curve inside f joining two components of $\gamma \cap f$ and disjoint from γ except for its ends, and let γ_1 and γ_2 be the two simple closed curves formed by γ' and the parts of γ between the ends of γ'. Clearly $|G \cap \gamma_1| + |G \cap \gamma_2| = |G \cap \gamma|$ and $G \cap \gamma_i \neq \emptyset$ for $i \in [2]$. Hence, for $i \in [2]$, we have $|G \cap \gamma_i| < |G \cap \gamma|$, and by the induction hypothesis, there exists a disk $\Delta_i \subset \Sigma$ bounded by γ_i such that $(G \cap \Delta_i, G \cap \overline{\Sigma \setminus \Delta_i}) \in \mathcal{T}$. If $\Delta_1 \not\subseteq \Delta_2$ and $\Delta_2 \not\subseteq \Delta_1$, then note that $\Delta = \Delta_1 \cup \Delta_2$ is a closed disk bounded by γ, and $(G \cap \Delta, G \cap \overline{\Sigma \setminus \Delta}) \in \mathcal{T}$ by (T1). Otherwise, if say $\Delta_1 \subseteq \Delta_2$, then γ bounds the closed disk $\Delta = \overline{\Delta_2 \setminus \Delta_1}$ and $(G \cap \Delta, G \cap \overline{\Sigma \setminus \Delta}) \in \mathcal{T}$ by Observation 2.18.

Hence, suppose that $\gamma \cap f$ is connected for every face f of G. Then there exists a cycle C in R_G obtained by continuously deforming γ inside each face. Since the pretangle $\mathcal{T}$ is respectful, there exists a closed disk $\Delta_C \subset \Sigma$ bounded by C such that $(G \cap \Delta_C, G \cap \overline{\Sigma \setminus \Delta_C}) \in \mathcal{T}$. We can deform Δ_C inside faces of G back into a closed disk Δ bounded by γ so that $(G \cap \Delta, G \cap \overline{\Sigma \setminus \Delta}) = (G \cap \Delta_C, G \cap \overline{\Sigma \setminus \Delta_C})$, and thus the conclusion of the lemma holds. □

As the first step in the study of respectful tangles, we aim to show that prescribing the interior for each short cycle in the radial graph is enough to uniquely determine the respectful tangle (this is not completely obvious, since not all small separations are defined by a single cycle—e.g., they may arise from several vertex-disjoint cycles in the radial graph). More precisely, we show this for prescriptions called *slopes* satisfying additional consistency conditions which are easily seen to hold when the prescription corresponds to a tangle.

2.13.1 Slopes

Consider a graph H with a cellular drawing on a surface Σ (you should imagine that H is the radial graph of another graph drawn on Σ). A ***slope*** of order θ in H is a function ins that to every cycle $C \subseteq H$ of length less than 2θ in H assigns a closed disk $\text{ins}(C) \subset \Sigma$ bounded by C, such that

(S1) if C_1 and C_2 are cycles of length less than 2θ in H and $C_1 \subseteq \text{ins}(C_2)$, then $\text{ins}(C_1) \subseteq \text{ins}(C_2)$; and,
(S2) if $F \subseteq H$ is a theta graph and all three cycles in F have length less than 2θ, then there exists a cycle $C \subset F$ such that every other cycle $C' \subset F$ satisfies $\text{ins}(C') \subseteq \text{ins}(C)$.

Here, a *theta graph* is a graph consisting of the union of three paths intersecting exactly in their common endpoints.

Observation 2.79 *Let H be a graph with a cellular drawing on a surface Σ other than the sphere and let θ be a positive integer. There exists a slope in H of order θ if and only if H has edgewidth at least 2θ, and such a slope is unique.*

Proof The existence of the disk $\text{ins}(C)$ for every cycle C of length less than 2θ implies that each such cycle is contractible, and thus H has edgewidth at least 2θ.

Conversely, if H has edgewidth at least 2θ, then every cycle C of length less than 2θ bounds a unique disk Δ in Σ, and we necessarily have to define $\text{ins}(C) = \Delta$. It is easy to see that this defines a slope: For (S1), if $C_1 \subseteq \text{ins}(C_2)$, then C_1 bounds a subdisk of $\text{ins}(C_2)$, and since it only bounds one disk, we have $\text{ins}(C_1) \subseteq \text{ins}(C_2)$. For (S2), since all cycles in F are contractible, there exists a disk $\Lambda \subset \Sigma$ containing F. Let C be the cycle in F bounding the face that contains the boundary of Λ; then F is contained in the subdisk $\text{ins}(C)$ of Λ, and (S2) follows from (S1). □

In case that Σ is the sphere, there are more ways how to choose ins. One is to drill a hole into one of the faces, transforming the sphere into an open disk (or, up to a homeomorphism, the plane), in which case each cycle encloses a unique closed disk. However, this construction gives slopes which are "degenerate" and clearly do not relate to any tangle, and we are going to exclude them from our consideration later. Before that, let us establish basic properties of slopes.

We can naturally extend a slope from cycles to more general subgraphs. Let ins be a slope of order θ in a graph H. We say that a subgraph F of H is ***confined*** if every cycle in F has length less than 2θ. For a confined subgraph, we define $\text{ins}(F)$ as the union of F and the disks $\text{ins}(C)$ for every cycle C in F. Thus, (S2) can be restated as saying that every confined theta-subgraph F satisfies $\text{ins}(F) = \text{ins}(C)$ for a cycle C in F.

As we are going to see in Lemma 2.81, $\text{ins}(F)$ is always the complement of a face f of F, and coincides with $\text{ins}(F')$ for the subgraph F' of F consisting of vertices and edges incident with f. To prove that, we need a description of such graphs with all vertices and edges incident with a single face. More precisely, we say that a graph H drawn on a surface Σ is ***strictly outerplanar*** if there exists an open disk $\Lambda \subset \Sigma$ containing H and the face f of H containing the boundary of Λ is incident with all vertices and edges of H; we say that f is an *outer face* of H (if Σ is the sphere and H is the cycle, then both faces of H are outer). A ***cactus*** is a graph in which any two cycles intersect in at most one vertex, or equivalently, every 2-connected block is a vertex, an edge or a cycle.

Observation 2.80 *If H is a strictly outerplanar graph drawn on a surface Σ with outer face f, then H is a cactus and every cycle in H bounds a cellular face different from f.*

Proof Since H is strictly outerplanar with outer face f, there exists an open disk $\Lambda \subset \Sigma$ containing H such that $\text{bd}(\Lambda) \subset f$. Consider any cycle $K \subseteq H$. Note that the open disk $f' \subset \Lambda$ bounded by K cannot contain any vertex or edge of G, since such a vertex or edge would not be incident with the outer face f. It follows that $f' \neq f$ is a cellular face of G bounded by K.

Consider now a 2-connected block B of H such that B is neither a vertex nor an edge, and thus there exists a cycle K' in B. If $B \neq K'$, then B would contain a path P intersecting K' exactly in its ends; but then a cycle in $K' \cup P$ separates an edge of K from f, contradicting the assumption that H is outerplanar with outer face f. Therefore, $B = K'$ is a cycle. We conclude that H is indeed a cactus. □

Note in particular that the outer face of a strictly outerplanar graph H drawn on Σ is uniquely determined by the drawing of H, unless H is a cycle and Σ is the sphere. We are now ready to give a description of $\text{ins}(F)$ for a general subgraph F.

Lemma 2.81 *Let H be a graph with a cellular drawing on a surface Σ and let θ be a positive integer. Let* ins *be a slope of order θ in G. For every confined subgraph F of H, there exists a face f of F and a strictly outerplanar graph $F' \subseteq F$ with outer face f such that* $\text{ins}(F) = \text{ins}(F') = \Sigma \setminus f$.

Proof Let $\mathcal{C}$ be the set of cycles $C \subseteq F$ such that $\text{ins}(C)$ is inclusionwise-maximal, i.e., $\text{ins}(C) \nsubseteq \text{ins}(C')$ for any other cycle $C' \subseteq F$. Consider any cycle $C' \subseteq F$; since $C' \subset \text{ins}(C')$, there exists a cycle C with $C' \subset \text{ins}(C)$ and with $\text{ins}(C)$ inclusionwise-maximal, i.e., with $C \in \mathcal{C}$. By (S1), it follows that $\text{ins}(C') \subseteq \text{ins}(C)$. Therefore, $\text{ins}(F)$ is the union of F and $\bigcup_{C \in \mathcal{C}} \text{ins}(C)$.

We claim that for distinct cycles $C_1, C_2 \in \mathcal{C}$, the disks $\text{ins}(C_1)$ and $\text{ins}(C_2)$ intersect in at most one point. Indeed, consider a point $p \in \text{ins}(C_1) \cap \text{ins}(C_2)$. We can move p inside the intersection until it hits C_1 or C_2; we can by symmetry assume that it hits C_1, and thus C_1 intersects $\text{ins}(C_2)$. By (S1) and the maximality of $\text{ins}(C_1)$, we have $C_1 \nsubseteq \text{ins}(C_2)$, and thus C_1 and C_2 must intersect. Observe that C_1 and C_2 intersect only in a single vertex, and thus the intersection $\text{ins}(C_1) \cap \text{ins}(C_2)$ consists only of this single vertex. Indeed, otherwise there would exist a path P in C_1 with both ends in C_2 and otherwise drawn outside of $\text{ins}(C_2)$, and by (S2), a cycle in the theta graph $C_2 \cup P$ would contradict the inclusionwise-maximality of $\text{ins}(C_2)$.

Let F' be the subgraph of F obtained by deleting the vertices and edges contained in the interiors of the disks $\text{ins}(C)$ for $C \in \mathcal{C}$. By the previous paragraph, each cycle in C is a subgraph of F'. Moreover, since every cycle in F is contained in $\mathrm{ins}(C)$ for some $C \in \mathcal{C}$, the subgraph F' only contains the cycles belonging to $\mathcal{C}$, and thus F' is a cactus. Furthermore, we have

$$\mathrm{ins}(F') = F' \cup \bigcup_{C \in \mathcal{C}} \mathrm{ins}(C) = F \cup \bigcup_{C \in \mathcal{C}} \mathrm{ins}(C) = \mathrm{ins}(F).$$

The cactus F' has exactly one face f different from the interiors of all disks $\mathrm{ins}(C)$ for $C \in \mathcal{C}$, and the drawing of F' is strictly outerplanar with outer face f. Since F' is obtained from F by deleting the interiors of the disks $\mathrm{ins}(C)$ for $C \in \mathcal{C}$, f is also a face of F. Moreover, $\mathrm{ins}(F) = \mathrm{ins}(F') = \Sigma \setminus f$. □

This has the following consequence, generalizing (S1).

Corollary 2.82 *Let H be a graph with a cellular drawing on a surface Σ and let θ be a positive integer. Let* ins *be a slope of order θ in H and let F_1 and F_2 be confined subgraphs of H. If $F_1 \subseteq \mathrm{ins}(F_2)$, then $\mathrm{ins}(F_1) \subseteq \mathrm{ins}(F_2)$.*

Proof It suffices to prove this in the case that F_1 is a cycle. By Lemma 2.81, we can assume that F_2 is a strictly outerplanar cactus with outer face f such that $\mathrm{ins}(F_2) = \Sigma \setminus f$. Since F_1 is 2-connected, it follows that there exists a cycle C in the cactus F_2 such that $F_1 \subseteq \mathrm{ins}(C)$. The claim then follows by (S1). □

Let G be a graph with a cellular drawing on a surface and let ins be a slope of order θ in the radial graph R_G of G. We would like to use ins to define an orientation of $(< \theta)$-separations of G. Note that for any separation (A, B) of G, $\{\rho_G(E(A)), \rho_G(E(B))\}$ is a partition of the set of faces of R_G. This motivates the following definition, which we use for $H = R_G$ and $X = V(G)$.

Suppose that H is a bipartite graph with a cellular drawing on a surface, and let X be one of the parts of its bipartition. For a set Z of faces of H, let $\overline{Z}$ be the set of faces of H not in Z and let $N(Z)$ denote the subgraph of H consisting of vertices and edges incident both with a face in Z and in $\overline{Z}$. For a slope ins of order θ in H, we say that Z is *X-small* if $|V(N(Z)) \cap X| < \theta$ and $Z \subset \mathrm{ins}(N(Z))$. Let us remark that the former condition implies that $N(Z)$ is confined, since every second vertex of each cycle in H belongs to X. Since $N(Z) = N(\overline{Z})$, Lemma 2.81 has the following consequence.

Observation 2.83 *Let H be a bipartite graph with a cellular drawing on a surface, and let X be one of the parts of its bipartition. Let* ins *be a slope of order θ in H. If Z is a set of faces of H with $|V(N(Z)) \cap X| < \theta$, then exactly one of the sets Z and $\overline{Z}$ is X-small.*

We are going to need a property of slopes analogous to the condition (T1) from the definition of a tangle; the proof of this property is rather technical and we omit it here.

Lemma 2.84 (Roberson and Seymour [23, (5.5)]) *Let H be a bipartite graph with a cellular drawing on a surface, and let X be one of the parts of its bipartition.*

Let ins *be a slope of order θ in H. If sets Z_1, Z_2, and Z_3 of faces of H are X-small, then there exists a face of H not contained in $Z_1 \cup Z_2 \cup Z_3$.*

This has the following consequence.

Corollary 2.85 *Let G be a graph with a cellular drawing on a surface Σ and let* ins *be a slope of order θ in R_G. If F_1, F_2, and F_3 are subgraphs of R_G with $|F_i \cap G| < \theta$ for $i \in [3]$, then there exists a face of R_G not contained in* $\text{ins}(F_1) \cup \text{ins}(F_2) \cup \text{ins}(F_3)$.

Proof For $i \in [3]$, let Z_i be the set of faces of R_G contained in $\text{ins}(F_i)$. By Lemma 2.81, there exists a face f_i of F_i and a strictly outerplanar graph $F_i' \subseteq F_i$ with outer face f_i such that $\text{ins}(F_i) = \text{ins}(F_i') = \Sigma \setminus f_i$. Observe that $N(Z_i)$ is exactly the union of cycles of F_i', and thus $\text{ins}(F_i) = \text{ins}(F_i') = F_i' \cup \text{ins}(N(Z_i))$. We conclude that Z_i is exactly the set of faces of R_G contained in $\text{ins}(N(Z_i))$, and thus the set Z_i is $V(G)$-small. The claim then follows from Lemma 2.84. □

2.13.2 Pretangles and Slopes

As we have discussed, there exist "degenerate" slopes that cannot correspond to any tangle. However, as we show next, slopes correspond to respectful *pretangles*, i.e., orientations of separations that satisfy the condition (T1) from the definition of a tangle. Let us start with the direction from pretangles to slopes.

Lemma 2.86 *Let G be a graph with a cellular drawing on a surface Σ. If $\mathcal{T}$ is a respectful pretangle of order θ in G, then $\text{ins}_{\mathcal{T}}$ is a slope of order θ in R_G.*

Proof If Σ is not the sphere, then the existence of the respectful pretangle $\mathcal{T}$ implies that G has representativity at least θ, and thus R_G has edgewidth at least 2θ. In that case, $\text{ins}_{\mathcal{T}}$ clearly matches the unique slope in R_G that exists by Observation 2.79.

Therefore, suppose that Σ is the sphere. We need to verify that $\text{ins}_{\mathcal{T}}$ satisfies the slope axioms (S1) and (S2). For any cycle C_i in R_G of length less than 2θ, let $(A_i, B_i) = (G \cap \text{ins}_{\mathcal{T}}(C_i), G \cap \Sigma \setminus \text{ins}_{\mathcal{T}}(C_i))$ denote the corresponding separation belonging to $\mathcal{T}$.

(S1) Let C_1 and C_2 be cycles of length less than 2θ in R_G such that C_1 is drawn in the disk $\text{ins}_{\mathcal{T}}(C_2)$. If $\text{ins}_{\mathcal{T}}(C_1) \not\subseteq \text{ins}_{\mathcal{T}}(C_2)$, then the union of the disks $\text{ins}_{\mathcal{T}}(C_1)$ and $\text{ins}_{\mathcal{T}}(C_2)$ is the whole sphere, and thus $A_1 \cup A_2 = G$. But this contradicts (T1).

(S2) Let F be a confined theta-subgraph of R_G and let C_1, C_2, and C_3 be the distinct cycles in F. If there exists $i \in [3]$ such that $F \subseteq \text{ins}_{\mathcal{T}}(C_i)$, then (S2) holds by (S1). Otherwise, observe that $\text{ins}_{\mathcal{T}}(C_1)$, $\text{ins}_{\mathcal{T}}(C_2)$, and $\text{ins}_{\mathcal{T}}(C_3)$ are the closures of the faces of F bounded by C_1, C_2, and C_3, respectively. Consequently, $\text{ins}_{\mathcal{T}}(C_1) \cup \text{ins}_{\mathcal{T}}(C_2) \cup \text{ins}_{\mathcal{T}}(C_3)$ is the whole sphere and $A_1 \cup A_2 \cup A_3 = G$, contradicting (T1).

□

We say that the slope $\text{ins}_{\mathcal{T}}$ is *derived* from the pretangle $\mathcal{T}$. Next, let us describe the converse construction of a pretangle from a slope ins. Let G be a graph with a cellular drawing on a surface. For a separation (A, B) of G of order less than θ, let $Z_A = \rho_G(E(A))$ be the set of faces of R_G corresponding to the edges of A. Note that a vertex v of G belongs to $N(Z_A)$ only if v is incident with edges of both A and B, and thus $v \in V(A \cap B)$; hence, $|V(N(Z_A)) \cap V(G)| \leq |A \cap B| < \theta$. We define $\mathcal{T}_{\text{ins}}$ as the set of all separations (A, B) of G of order less than θ such that Z_A is $V(G)$-small, i.e., $Z_A \subset \text{ins}(N(Z_A))$.

Lemma 2.87 *Let G be a graph with a cellular drawing on a surface* Σ*, and let* ins *be a slope in* R_G *of order* θ*. Then* $\mathcal{T}_{\text{ins}}$ *is a respectful pretangle in G of order* θ*, and* ins *is the slope derived from* $\mathcal{T}_{\text{ins}}$.

Proof By Observation 2.83, $\mathcal{T}_{\text{ins}}$ is an orientation of $(< \theta)$-separations of G. Consider any separations $(A_1, B_1), (A_2, B_2), (A_3, B_3) \in \mathcal{T}_{\text{ins}}$, meaning that Z_{A_1}, Z_{A_2}, Z_{A_3} are $V(G)$-small. By Lemma 2.84, R_G has a face f not belonging to $Z_{A_1} \cup Z_{A_2} \cup Z_{A_3}$, and thus the edge $\rho_G^{-1}(f)$ of G does not belong to $A_1 \cup A_2 \cup A_3$. Therefore, $\mathcal{T}_{\text{ins}}$ satisfies (T1), and thus it is a pretangle of order θ.

Next, let us argue that $\mathcal{T}_{\text{ins}}$ is respectful and ins is derived from it. Consider any cycle C in R_G of length less than 2θ, and let $(A, B) = (G \cap \text{ins}(C), G \cap \overline{\Sigma \setminus \text{ins}(C)})$ be the corresponding separation of G. It suffices to argue that $(A, B) \in \mathcal{T}_{\text{ins}}$. This is clearly the case, since $N(Z_A) = C$ and $Z_A \subset \text{ins}(C)$, implying that Z_A is $V(G)$-small. □

We say that $\mathcal{T}_{\text{ins}}$ is the pretangle *induced* by the slope. So far, we have shown that if $\mathcal{T}$ is the respectful pretangle induced by the slope ins, then ins is derived from $\mathcal{T}$. We want to establish a $(1:1)$-correspondence between respectful pretangles and slopes in the radial graph, and thus we also need the converse.

Lemma 2.88 *Let G be a graph with a cellular drawing on a surface* Σ*, and let* $\mathcal{T}$ *be a respectful pretangle in G of order* θ*. Let* $\text{ins} = \text{ins}_{\mathcal{T}}$ *be the slope derived from* $\mathcal{T}$*. Then* $\mathcal{T}$ *is the pretangle induced by* ins.

Proof Consider any separation $(A, B) \in \mathcal{T}_{\text{ins}}$, of order less than θ. We need to argue that $\mathcal{T}$ agrees with $\mathcal{T}_{\text{ins}}$ on this separation, i.e., that $(A, B) \in \mathcal{T}$. Since $(A, B) \in \mathcal{T}_{\text{ins}}$, Z_A is $V(G)$-small, and thus $Z_A \subset \text{ins}(N(Z_A))$. Therefore, for each edge $e \in E(A)$, there exists a cycle $C_e \subseteq N(Z_A)$ such that the face $\rho_G(e)$ is a subset of $\text{ins}(C_e)$, and thus e is drawn in $\text{ins}(C_e)$. Let $(A_e, B_e) = (G \cap \text{ins}(C_e), G \cap \overline{\Sigma \setminus \text{ins}(C_e)})$. We have $e \in E(A_e)$, and since $\text{ins} = \text{ins}_{\mathcal{T}}$ is derived from $\mathcal{T}$, we have $(A_e, B_e) \in \mathcal{T}$. Moreover, $V(A_e \cap B_e) = V(C_e \cap G) \subseteq V(N(Z_A)) \subseteq V(A \cap B)$. Let (A_0, G) be the separation of G with $V(A_0) = V(A \cap B)$ and $E(A_0) = \emptyset$; by (T1), we have $(A_0, G) \in \mathcal{T}$. By Lemma 2.22, we have

$$(A, B) = \left(A_0 \cup \bigcup_{e \in E(A)} A_e, G \cap \bigcap_{e \in E(A)} B_e\right) \in \mathcal{T},$$

as required. □

As a consequence, we get the following description of separations forming respectful pretangles.

Lemma 2.89 *Let $\mathcal{T}$ be a respectful pretangle of order θ in a graph G with a cellular drawing on a surface Σ, and let (A, B) be a separation of G of order less than θ. Then the following claims are equivalent:*

(i) $(A, B) \in \mathcal{T}$.
(ii) There exists a strictly outerplanar cactus $F \subseteq R_G$ with outer face f such that $V(F) \cap V(G) \subseteq V(A \cap B)$ and $\rho_G(V(A) \cup E(A)) \subseteq \mathrm{ins}_{\mathcal{T}}(F) = \Sigma \setminus f$.
(iii) There exists a confined subgraph $F \subseteq R_G$ such that $\rho_G(E(A)) \subseteq \mathrm{ins}_{\mathcal{T}}(F)$.

Proof Suppose first that $(A, B) \in \mathcal{T}$. Let $Z_A = \rho_G(E(A))$, let $F_0 = N(Z_A)$ and note that $V(F_0) \cap V(G) \subseteq V(A \cap B)$. Since $\mathcal{T}$ is induced by $\mathrm{ins}_{\mathcal{T}}$, we have $Z_A \subset \mathrm{ins}_{\mathcal{T}}(F_0)$. Since the drawing of G is cellular and $A \neq G$, each vertex $v \in V(A) \setminus V(B)$ is incident with an edge of A, and thus $\rho_G(v) \in \mathrm{ins}_{\mathcal{T}}(F_0)$. Let $F_1 = F_0 \cup V(A \cap B)$, so that $\rho_G(V(A) \cup E(A)) \subseteq \mathrm{ins}_{\mathcal{T}}(F_1)$. The existence of the cactus F then follows from Lemma 2.81. Therefore, (i) implies (ii).

Note that (ii) implies (iii), since the cactus F from (ii) is clearly confined.

Finally, suppose that there exists a confined subgraph $F \subseteq R_G$ such that $Z_A = \rho_G(E(A)) \subseteq \mathrm{ins}_{\mathcal{T}}(F)$. This implies that $N(Z_A) \subseteq \mathrm{ins}_{\mathcal{T}}(F)$, and by Corollary 2.82, $\mathrm{ins}_{\mathcal{T}}(N(Z_A)) \subseteq \mathrm{ins}_{\mathcal{T}}(F)$. By Lemma 2.81, $\mathrm{ins}_{\mathcal{T}}(F) = \Sigma \setminus f$ for a face f of F, and thus there exists a face $f' \subseteq f$ of R_G such that $f' \not\subseteq \mathrm{ins}_{\mathcal{T}}(F)$. Since $Z_A \subseteq \mathrm{ins}_{\mathcal{T}}(F)$, the face f' belongs to $Z_B = \rho_G(E(B))$, and thus $Z_B \not\subseteq \mathrm{ins}_{\mathcal{T}}(F)$, which implies that $Z_B \not\subseteq \mathrm{ins}_{\mathcal{T}}(N(Z_A))$. Since $Z_B = \overline{Z_A}$, Observation 2.83 implies that Z_A is $V(G)$-small, i.e., $Z_A \subseteq \mathrm{ins}_{\mathcal{T}}(N(Z_A))$. Since the pretangle $\mathcal{T}$ is induced by the slope $\mathrm{ins}_{\mathcal{T}}$, we conclude that $(A, B) \in \mathcal{T}$. Consequently, (iii) implies (i). □

2.13.3 Tangles and Slopes

As we have seen, on the sphere there are slopes which have nothing to do with tangles, namely those where we choose any face f of R_G and for every cycle C in R_G define $\mathrm{ins}(C)$ to be the disk bounded by C that does not contain f. These are actually the only "degenerate" slopes irrelevant with respect to tangles: The following lemma shows that if every face of R_G is contained in $\mathrm{ins}(C)$ for some cycle C, then $\mathcal{T}_{\mathrm{ins}}$ is a tangle.

Lemma 2.90 *Let G be a graph with a cellular drawing on a surface Σ, and let* ins *be a slope in R_G of order $\theta \geq 3$. For an edge $e \in E(G)$, let C_e be the subgraph of R_G consisting of the vertices and edges drawn in the boundary of the face $\rho_G(e)$. The pretangle $\mathcal{T}_{\mathrm{ins}}$ induced by the slope* ins *is a tangle if and only if $\rho_G(e) \subset \mathrm{ins}(C_e)$ for every $e \in E(G)$.*

Proof Suppose first that there exists an edge $e \in E(G)$ such that $\rho_G(e) \not\subset \text{ins}(C_e)$. By Lemma 2.81, we have $\text{ins}(C_e) = \Sigma \setminus \rho_G(e)$, and thus $(G - e, e) \in \mathcal{T}_{\text{ins}}$ by the definition of $\mathcal{T}_{\text{ins}}$. Since $V(G - e) = V(G)$, this means that $\mathcal{T}_{\text{ins}}$ does not satisfy (T2), and thus it is not a tangle.

Conversely, suppose that $\rho_G(e) \subset \text{ins}(C_e)$ for every edge $e \in E(G)$. It suffices to argue that $\mathcal{T}_{\text{ins}}$ satisfies (T2). Suppose for a contradiction there exists a separation $(A, B) \in \mathcal{T}_{\text{ins}}$ such that $V(A) = V(G)$, and thus $V(B) = V(A \cap B)$. For any edge $e \in E(B)$, we have $\rho_G(e) \subset \text{ins}(C_e)$, and thus there exists a separation $(A_e, B_e) \in \mathcal{T}_{\text{ins}}$ with $e \in E(A_e)$ and $V(A_e \cap B_e) \subseteq V(C_e) \cap V(G) \subseteq V(B)$. By Lemma 2.22, we have

$$(G, V(B)) = \Big(A \cup \bigcup_{e \in E(A)} A_e, B \cap \bigcap_{e \in E(B)} B_e\Big) \in \mathcal{T}_{\text{ins}},$$

which contradicts (T1) for the pretangle $\mathcal{T}_{\text{ins}}$. □

As a consequence, we obtain the promised result concerning the existence and uniqueness of respectful tangles in graphs on surfaces other than the sphere.

Theorem 2.91 *Let G be a graph with a cellular drawing on a surface Σ other than the sphere. Then G contains a respectful tangle of order $\theta \geq 3$ if and only if the representativity of G is at least θ. Moreover, this respectful tangle is unique.*

Proof As we argued before, representativity at least θ is a necessary condition for the existence of a respectful tangle of order θ. Conversely, suppose that the drawing of G has representativity at least θ. Then there exists a unique slope ins of order θ in R_G. We have established a $(1:1)$-correspondence between slopes and respectful pretangles, and thus G contains a unique respectful pretangle $\mathcal{T}$, specifically the one induced by ins. Consider any edge $e \in E(G)$, and let C_e be the subgraph of R_G consisting of the vertices and edges drawn in the boundary of the face $\rho_G(e)$. By Lemma 2.81, C_e has a non-cellular face f such that $\text{ins}(C_e) = \Sigma \setminus f$, and since the face $\rho_G(e)$ of C_e is cellular, we have $\rho_G(e) \subset \text{ins}(C_e)$. Therefore, Lemma 2.90 implies that $\mathcal{T}$ is a tangle. □

On the other hand, recall that for graphs drawn on the sphere, every tangle is respectful. This has the following interesting corollary.

Theorem 2.92 *Let G be a connected plane graph and let $G^\star$ be its dual. Then there is a $(1:1)$-correspondence between tangles in G and $G^\star$, and in particular the largest orders of tangles in G and $G^\star$ are the same.*

Proof Tangles in G are in $(1:1)$-correspondence with slopes in R_G that satisfy the condition of Lemma 2.90, and the same is true for the tangles in $G^\star$. The claim of the theorem thus follows from the fact that $R_G = R_{G^\star}$. □

In particular, together with Lemma 2.23 and Theorem 2.25, this implies that for a connected plane graph G, the treewidth of its dual is at most $\lceil \frac{3}{2}\,\text{tw}(G) \rceil$. In fact, a much stronger connection holds, see [5, 17]: The treewidth of a connected plane graph and its dual differs by at most one.

2.14 Related Notions (⤼)

The notions of treewidth and tree decompositions inspired many related graph parameters. While we certainly cannot speak about all of them, much less explore their theory in depth, let us go on a brief detour surveying at least the most important ones. Even then, we can only scratch the surface and mention just a few interesting results, generally focusing on relationships between the parameters and on approximate structural characterizations.

2.14.1 Branch Decompositions and Related Parameters

Branch decompositions are an important alternative to tree decompositions. A ***branch decomposition*** of a set U is a pair (T, σ), where T is a subcubic tree (i.e., $\Delta(T) \leq 3$) and $\sigma : U \to V(T)$ injectively maps U to leaves of T. Each edge e of T then naturally defines a partition $\{A, B\}$ of U, where $A = \sigma^{-1}(V(T_1))$ and $B = \sigma^{-1}(V(T_2))$ for the two components T_1 and T_2 of $T - e$; let this partition be denoted by $\sigma^{-1}(T - e)$. Let ρ be a function from partitions of U to non-negative integers, called a ***connectivity function***, with the property that $\rho(\{\emptyset, U\}) = 0$. Then the *$\rho$-width* of the branch decomposition is defined to be $\max_{e \in E(T)} \rho(\sigma^{-1}(T - e))$ if $|U| \geq 2$ and 0 otherwise. The ρ-width of U is the minimum of the ρ-widths of its branch decompositions.

The specific graph parameters are obtained by choosing U as either the set of vertices or the set of edges of the graph and by choosing a particular connectivity function. Before we give specific examples, let us note that we can without loss of generality assume that the tree T of the branch decomposition is cubic (i.e., all non-leaf vertices have degree exactly three) and an element of U is assigned by σ to every leaf of T, since we can delete leaves to which no element is assigned and suppress vertices of degree two without affecting the ρ-width.

The most important graph parameter defined in terms of branch decompositions is the ***branchwidth***. The branchwidth $\mathrm{bw}(G)$ of a graph G is defined as the ρ_{branch}-width of $E(G)$, with the connectivity function $\rho_{\mathrm{branch}}(\{A, B\})$ defined as the number of vertices $v \in V(G)$ such that v is incident both with an edge in A and in an edge in B; that is, $\rho_{\mathrm{branch}}(\{A, B\})$ is the minimum order of a separation (A', B') of G with $E(A') = A$ and $E(B') = B$.

Branchwidth is closely related (but not equal) to treewidth. On one hand, consider a tree decomposition (T, β) of G of width $k \geq 1$. We can easily transform the decomposition without increasing its width so that for each edge $g = uv \in E(G)$, there is a distinct leaf $\sigma(g)$ of T with $\beta(\sigma(g)) = \{u, v\}$, by adding such a leaf adjacent to a bag containing $\{u, v\}$ which exists by (D1). Furthermore, T can be made subcubic by splitting vertices of larger degree, with all the resulting vertices having the same bag. We end up with (T, σ) being a branch decomposition of $E(G)$. Consider an edge $e \in E(T)$, let T_1 and T_2 be the components of $T - e$, and let $\{E_1, E_2\} = \sigma^{-1}(T - e)$. By Observation 2.1, $(\beta(T_1), \beta(T_2))$ is a vertex separation of G of order at most $k + 1$. If $g_1 \in E_1$ and $g_2 \in E_2$ are incident with the same vertex $v \in V(G)$, then $v \in \beta(\sigma(g_1)) \cap \beta(\sigma(g_2))$, and thus $v \in \beta(T_1) \cap \beta(T_2)$. It follows that $\rho_{\text{branch}}(\sigma^{-1}(T - e)) \leq |\beta(T_1) \cap \beta(T_2)| \leq k + 1$, and thus G has branchwidth at most $k + 1$.

Conversely, let G be a graph without isolated vertices (deleting isolated vertices affects neither treewidth nor branchwidth) and consider a branch decomposition (T, σ) of $E(G)$ of ρ_{branch}-width b, where without loss of generality, T is cubic and an edge is assigned by σ to every leaf of T. For a leaf $x \in V(T)$, let $\beta(x)$ consists of the ends of the edge $\sigma^{-1}(x)$. For a non-leaf vertex $x \in V(T)$, let $\beta(x)$ consist of vertices $v \in V(G)$ such that there exist edges $g_1, g_2 \in E(G)$ incident with v and mapped by σ to leaves of different components of $T - v$. Then (T, β) is a tree decomposition of G. Consider a non-leaf node $x \in V(T)$ incident with edges e_1, e_2, and e_3. Observe that every vertex $v \in \beta(x)$ contributes 1 to $\rho_{\text{branch}}(\sigma^{-1}(T - e_i))$ for at least two indices $i \in [3]$, and thus

$$|\beta(x)| \leq \frac{1}{2}\sum_{i=1}^{3} \rho_{\text{branch}}(\sigma^{-1}(T - e_i)) \leq \tfrac{3}{2}b.$$

If $b \geq 2$, then $|\beta(x)| \leq 2 \leq \lfloor\tfrac{3}{2}b\rfloor$ holds for each leaf x of T as well.

In summary, we get the following relationship between the treewidth and the branchwidth.

Lemma 2.93 *If G is a graph of branchwidth $b \geq 2$, then*

$$b - 1 \leq \text{tw}(G) \leq \left\lfloor \tfrac{3}{2}b \right\rfloor - 1.$$

The importance of branchwidth stems from the fact that tangles are the exact obstructions for it. Let the ***tangle number*** $\theta(G)$ of a graph G be the maximum positive integer θ such that G has a tangle of order θ.

Theorem 2.94 (Robertson and Seymour [22]) *If G is a graph of branchwidth $b \geq 2$, then $\theta(G) = b$.*

Let us remark that Theorem 2.25 follows by combining Theorem 2.94 with Lemma 2.93. Moreover, by Theorem 2.92, the branchwidth of a connected plane graph is equal to the branchwidth of its dual.

Branch decompositions are in general less convenient in applications than tree decompositions. However, an advantage of branchwidth is that it is defined in terms of a decomposition on the edge set of the graph (with vertices only playing a role in the connectivity function). This makes it possible to generalize branchwidth to "vertex-free" settings, such as to matroids.

Let us now briefly discuss two graph parameters defined in terms of branch decompositions on the vertex set of a graph G:

- ***Carvingwidth*** $\mathrm{carw}(G)$ is defined as the ρ_{carving}-width of $V(G)$, where $\rho_{\mathrm{carving}}(\{V_1, V_2\})$ is the number of edges of G between V_1 and V_2,
- ***Rankwidth*** $\mathrm{rw}(G)$ is defined by the connectivity function $\rho_{\mathrm{rank}}(\{V_1, V_2\})$ equal to the linear-algebraic rank of the bipartite adjacency matrix of V_1 and V_2 over $\mathbb{Z}_2$, i.e., of the matrix A with rows indexed by V_1 and columns by V_2 and with $A_{v_1,v_2} = 1$ if $v_1v_2 \in E(G)$ and $A_{v_1,v_2} = 0$ otherwise.

One of the motivations for the notion of carvingwidth is that for a connected plane graph G, the branchwidth of G is exactly equal to half the carvingwidth of the medial graph of G (the ***medial graph*** of a plane graph is obtained by replacing each vertex by a cycle on the midpoints of the incident edges). Moreover, the carvingwidth of a plane graph can be computed in polynomial time. This gives a polynomial-time algorithm to exactly determine the branchwidth of a planar graph; see [28]. Let us remark that it is open whether it is possible to determine the treewidth of a planar graph exactly in polynomial time (though of course, by Lemma 2.93, we can approximate it within the factor of $3/2$).

Unlike treewidth and branchwidth, carvingwidth can only be bounded on graphs of bounded maximum degree. Indeed, in any branch decomposition (T, σ) of the vertex set of a graph G and any vertex $v \in V(G)$, the ρ_{carving}-width of the edge of T incident with the leaf $\sigma(v)$ is $\deg v$, and thus $\mathrm{carw}(G) \ge \Delta(G)$.

Moreover, carvingwidth is lower-bounded by branchwidth. More precisely, a branch decomposition (T, σ) of $V(G)$ can be turned into a branch decomposition of $E(G)$ as follows. We choose an arbitrary map $\alpha : E(G) \to V(G)$ assigning to each edge one of its incident vertices. For each $v \in V(G)$, we attach a binary tree with $|\alpha^{-1}(v)|$ leaves below the leaf $\sigma(v)$ of T, and we assign each edge of $\alpha^{-1}(v)$ to one of these leaves. Let (T_0, σ_0) be the resulting branch decomposition of $E(G)$. Let us analyze the branchwidth of the decomposition (T_0, σ_0). Consider any edge $e \in E(T_0)$:

- If $e \in E(T_0) \setminus E(T)$, then all edges in one part of $\sigma_0^{-1}(T_0 - e)$ are incident with a single vertex $v \in V(G)$. Therefore, only v and its at most $\Delta(G)$ neighbors can be incident with edges in both parts of $\sigma_0^{-1}(T_0 - e)$. It follows that $\rho_{\mathrm{branch}}(\sigma_0^{-1}(T_0 - e)) \le \Delta(G) + 1 \le \mathrm{carw}(G) + 1$.
- If $e \in E(T)$, consider any vertex $v \in V(G)$ incident with edges $g_1 = vu_1$ and $g_2 = vu_2$ contained in different parts of the partition $\sigma_0^{-1}(T_0 - e)$. Then $\alpha(g_1) \in \{v, u_1\}$ and $\alpha(g_2) \in \{v, u_2\}$ are in different parts of the partition $\sigma^{-1}(T - e)$. Hence, there exists $i \in [2]$ such that u_i is in different part of $\sigma^{-1}(T - e)$ from v and $\alpha(g_i) = u_i$, and thus g_i contributes 1 towards $\rho_{\mathrm{carving}}\sigma^{-1}(T - e)$. Therefore, $\rho_{\mathrm{branch}}(\sigma_0^{-1}(T_0 - e)) \le \rho_{\mathrm{carving}}\sigma^{-1}(T - e)$.

We conclude that

$$\mathrm{bw}(G) \leq \mathrm{carw}(G) + 1$$

for every graph G.

On the other hand, rankwidth is at most as large as branchwidth (and indeed, unlike branchwidth and treewidth, it can be small even on very dense graphs–e.g., the rankwidth of a clique is 1). To see that this is the case, consider any branch decompositon (T, σ) of the edge set $E(G)$ of a graph G without isolated vertices. We can turn (T, σ) into a branch decomposition (T_0, σ_0) of $V(G)$ as follows. For each vertex $v \in V(G)$, choose an incident edge $\alpha(v)$. For each edge $e \in E(G)$, we attach $|\alpha^{-1}(e)|$ leaves below the leaf $\sigma(e)$ of T, thus obtaining the tree T_0. Finally, for each vertex $v \in V(G)$, we let $\sigma_0(v)$ be a distinct leaf attached below $\sigma(\alpha(e))$.

Let us analyze the rankwidth of (T_0, σ_0). The ρ_{rank}-width of each edge of $E(T_0) \setminus E(T)$ is one. For any edge $e \in E(T)$, consider the partition $\{V_1, V_2\} = \sigma_0^{-1}(T_0 - e)$, and let M be the set of $\rho_{\text{branch}}(\sigma^{-1}(T-e))$ vertices incident with edges in both parts of $\sigma^{-1}(T - e)$. Suppose $g = v_1v_2$ is an edge of G with $v_1 \in V_1$ and $v_2 \in V_2$. Since $\alpha(v_1)$ and $\alpha(v_2)$ are in different parts of $\sigma^{-1}(T - e)$, there is $i \in [2]$ such that g and $\alpha(v_i)$ are in different parts of $\sigma^{-1}(T - e)$, and thus $v_i \in M$. Consequently, all edges of G between V_1 and V_2 are incident with M. Let A be the bipartite adjacency matrix of V_1 and V_2; we conclude that for every $v \in V_1 \setminus M$, all non-zero entries of the row of A indexed by v are in the columns indexed by $V_2 \cap M$, and thus the submatrix of A formed by these rows has rank at most $|V_2 \cap M|$. Since A has exactly $|V_1 \cap M|$ additional rows, we have

$$\rho_{\text{rank}}(\sigma^{-1}(T_0 - e)) = \mathrm{rk}(A) \leq |V_1 \cap M| + |V_2 \cap M| = |M| \leq \rho_{\text{branch}}(\sigma^{-1}(T - e)).$$

Therefore,

$$\mathrm{rw}(G) \leq \mathrm{bw}(G).$$

The notion of rankwidth may be easier to grasp intuitively in terms of the *neighborhood complexity*. The ***neighborhood complexity*** $\mathrm{nc}_G(V_1, V_2)$ of the vertex partition (V_1, V_2), is the number of vertices in V_1 with pairwise different sets of neighbors in V_2 (note that $\mathrm{nc}_G(V_2, V_1)$ may be different from $\mathrm{nc}_G(V_1, V_2)$). In other words, the neighborhood complexity of (V_1, V_2) is the number of different rows of the bipartite adjacency matrix A of V_1 and V_2, and thus the rank of A is at most $\mathrm{nc}(V_1, V_2)$. On the other hand, since there are only 2^k different linear combinations of k vectors over $\mathbb{Z}_2$, if A has rank k, then it has at most 2^k different rows. Therefore, a branch decomposition of $V(G)$ has small ρ_{rank}-width if and only if each of the corresponding partitions of the vertex set has small neighborhood complexity.

Rankwidth is especially important in the theory of *vertex-minors* (graphs obtained by a sequence of vertex deletions and neighborhood complementations, i.e., exchanging edges with non-edges on the neighborhood of a vertex), since the neighborhood complementation does not affect the value of ρ_{rank}. The following analogue of the Grid Theorem (see Corollary 2.35 for a comparison) was shown recently.

Theorem 2.95 (Geelenet al. [13]) *A graph class closed under vertex-minors has unbounded rankwidth if and only if it contains all circle graphs.*

Rankwidth is related to another well-known width parameter, ***cliquewidth***, which is defined in terms of a construction using a bounded number of colors. We say that a graph G with vertices colored (not necessarily properly) by k colors is k-*constructible* if G can be obtained from k-colored single-vertex graphs by finite number of the following operations:

- Disjoint union of two smaller k-constructible graphs.
- For distinct $i, j \in [k]$, changing the color of all vertices of color i to j.
- For distinct $i, j \in [k]$, adding all edges between vertices colored i and j.

The cliquewidth $\mathrm{cw}(G)$ is the minimum k such that G with some k-coloring is k-constructible. For example, cliques have cliquewidth two: To obtain a clique with all vertices of color 1, we can repeatedly do the disjoint union with a single vertex of color 2, add all edges between colors 1 and 2, and recolor 2 to 1. A similar argument shows that all ***cographs***, i.e., graphs obtained from single-vertex graphs by disjoint unions and complete joins (or equivalently, complementations), have cliquewidth two. As shown in [20], cliquewidth and rankwidth are qualitatively equivalent; every graph G satisfies

$$\mathrm{rw}(G) \leq \mathrm{cw}(G) \leq 2^{\mathrm{rw}(G)+1}.$$

The notion of cliquewidth is often easier to use in algorithmic applications than the notion of rankwidth. On the other hand, rankwidth can be approximated up to a constant ratio in polynomial time, and for any fixed k, it is possible to determine in polynomial time whether the graph has rankwidth at most k in polynomial time [20]; it is open whether either of these is possible for cliquewidth.

Let us mention one more result on relationship between branchwidth (or treewidth) and rankwidth (or cliquewidth): They qualitatively coincide on graphs without 4-cycles. Indeed, more is true, showing that in a sense, the extra power of bounded cliquewidth compared to treewidth is only in allowing complete bipartite subgraphs:

Theorem 2.96 (Gurski and Wanke [14]) *Let G be a graph and let $k \geq 2$ be an integer. If $K_{k,k}$ is not a subgraph of G, then*

$$\mathrm{tw}(G) \leq 3(k-1)\,\mathrm{cw}(G) - 1.$$

Let us note that it is possible to define parameters stronger than cliquewidth (yet still useful). One interesting example is the notion of ***twinwidth*** developed recently [4], which is bounded on all graph classes with bounded cliquewidth, but also on all proper minor-closed classes. This generality of course comes with some drawbacks. For example, the meta-algorithm of Theorem 2.8 on testing the properties expressible in the Monadic Second Order Logic in graphs with bounded treewidth extends to graphs with bounded cliquewidth [9], up to the restriction that it is not possible to quantify over sets of edges. However, since many problems expressible in the Monadic Second Order Logic (e.g., 3-colorability) are NP-hard on planar graphs, such a meta-algorithm is unlikely for graphs of bounded twinwidth; instead, only a much weaker result on testing the properties expressible in the First Order Logic holds in this setting [4].

2.14.2 Pathwidth

Recall that a path decomposition (P, β) is a special case of a tree decomposition where P is required to be a path, and the ***pathwidth*** $\mathrm{pw}(G)$ of a graph G is the minimum possible width of a path decomposition of G. Hence, we trivially have $\mathrm{tw}(G) \le \mathrm{pw}(G)$. Initially, one might suspect that conversely, pathwidth might be bounded by a function of treewidth. Perhaps supporting this idea, for the $n \times n$ grid G_n, the tree decomposition described in Sect. 2.6 is actually a path decomposition, showing that $\mathrm{pw}(G_n) = \mathrm{tw}(G_n) = n$. However, this intuition is false.

Lemma 2.97 *The complete rooted ternary tree T_n of depth n has pathwidth at least $n - 1$.*

Proof Let us prove this by induction on n. The claim is trivial for $n = 1$, and thus suppose that $n \ge 2$. Let r be the root of T_n and let G_1, G_2, and G_3 be the components of $T_n - r$; each of them is a complete rooted ternary tree of depth $n-1$, and thus $\mathrm{pw}(G_i) \ge n-2$ for $i \in [3]$.

Consider any path decomposition (P, β) of G, and let x and y be the ends of the path P. Clearly, we can assume that $\beta(x) \ne \emptyset \ne \beta(y)$, as otherwise we can shorten P. Let $u \in \beta(x)$ and $v \in \beta(y)$ be chosen arbitrarily, and let Q be the path between u and v in T_n. By symmetry, we can assume that $u \in \{r\} \cup V(G_1)$ and $v \in \{r\} \cup V(G_1) \cup V(G_2)$, and thus $G_3 \subseteq T_n - V(Q)$. By Lemma 2.2, we have $\beta^{-1}(V(Q)) = V(P)$; i.e., letting $\beta'(z) = \beta(z) \setminus V(Q)$ for each $z \in V(P)$, we have $|\beta'(z)| < |\beta(z)|$. Note that (P, β') is a path decomposition of $T_n - V(Q)$. Consequently,

$$\mathrm{width}(P, \beta) \ge \mathrm{width}(P, \beta') + 1 \ge \mathrm{pw}(T_n - V(Q)) + 1 \ge \mathrm{pw}(G_3) + 1 \ge n - 1.$$

Therefore, T_n indeed has pathwidth at least $n - 1$. □

Hence, there exist graphs of treewidth one and arbitrarily large pathwidth. Let us remark that the fact that we did the argument for complete ternary trees is just for convenience; a complete rooted binary tree of depth $2n$ contains a complete rooted ternary (even quaternary) tree of depth n as a minor, obtained by contracting the edges between every other level, and thus its pathwidth is also at least $n - 1$. Conversely, Bienstock et al. [3] gave the following elegant result.

Theorem 2.98 *If F is a tree with k vertices, then every F-minor-free graph has pathwidth at most $k - 2$.*

To prove this theorem, let us start with an auxiliary result. Let G be a graph and let X be a subset of its vertices. We define $\partial_G X$ to be the subset X formed by the vertices that have neighbors outside of X. We say that X an *n-tail* of G if there exists a path decomposition (P, β) of $G[X]$ of width less than n such that $\partial_G X \subseteq \beta(x)$ for an end x of the path P. Note that if Y is a subset of X, then Y also has a path decomposition of width less than n, but it is not necessarily an n-tail, since we do not have control over $\partial_G Y$. However, the following technical statement holds.

Lemma 2.99 *Let G be a graph, n a positive integer and let $X \subseteq V(G)$ be an n-tail of G. Let Y be a subset of X such that G contains a system of pairwise vertex-disjoint paths $\{Q_v : v \in \partial_G Y\}$, where for $v \in \partial_G Y$, Q_v joins v to a vertex of $\partial_G X$. Then Y is an n-tail of G.*

Proof By taking the paths Q_v for $v \in \partial_G Y$ as short as possible, we can assume that they intersect $\partial_G X$ exactly in their endpoints, and thus Q_v is a path in $G[X]$. Note also that since the paths are pairwise vertex-disjoint and cover $\partial_G Y$, they cannot pass through any vertex in $Y \setminus \partial_G Y$. Let μ be the model of $G[Y]$ in $G[X]$ such that $\mu(v) = Q_v$ for $v \in \partial_G Y$, $\mu(v)$ is the single-vertex graph consisting of v for $v \in Y \setminus \partial_G Y$, and $\mu(e) = e$ for $e \in E(G[Y])$.

Let (P, β) be a path decomposition of $G[X]$ of width less than n such that $\partial_G X \subseteq \beta(x)$ for an end x of the path P. Then the path decomposition (P, β_μ) of $G[Y]$ induced by (P, β) and μ has width less than n, and $v \in \beta_\mu(x)$ for every $v \in \partial_G Y$, since Q_v intersects $\partial_G X \subseteq \beta(x)$. Therefore, Y is an n-tail. □

We are now ready to prove the forbidden tree theorem.

Proof of Theorem 2.98 Let $n = k - 1$. It suffices to prove that every connected graph G of pathwidth at least n has a minor isomorphic to F. This is trivial for $n = 0$, since then F has only one vertex, and thus assume that $n \geq 1$. Let v_1, $v_2, \ldots, v_{n+1}$ be the vertices of F listed in an order such that the induced subgraph $F_i = F[\{v_1, \ldots, v_i\}]$ is connected for every $i \in [n + 1]$; equivalently, for $i \in [n]$, v_{i+1} is a vertex with exactly one neighbor in $V(F_i)$. For $i = 1, \ldots, n$ in order, we are going to construct a model μ_i of F_i in an induced subgraph $G[X_i]$ of G such that

(i) $|\partial_G X_i| = i$, and for each $v \in V(F_i)$, the subgraph $\mu_i(v)$ intersects $\partial_G X_i$ in a single vertex,
(ii) X_i is an n-tail of G, and
(iii) every n-tail $X \subseteq V(G)$ of G such that $X_i \subsetneq X$ satisfies $|\partial_G X_i| > i$.

Note that since G has pathwidth at least n, the set $V(G)$ is not an n-tail. Since G is connected, it follows that every non-empty n-tail X satisfies $|\partial_G X| \geq 1$. Moreover, since $n \geq 1$, the single-vertex subsets of $V(G)$ are n-tails of G. Therefore, for

$i = 1$, we can let $X_1 \subseteq V(G)$ be a maximal n-tail of G with $|\partial_G X_1| = 1$, with $\mu_1(v_1)$ consisting of the single-vertex subgraph with the vertex set $\partial_G X_1$.

Suppose now that $2 \le i \le n+1$ and we have already constructed X_{i-1} and μ_{i-1}. For $j \in [i-1]$, let z_j be the unique vertex in $\partial_G X_{i-1} \cap V(\mu_{i-1}(v_j))$. Let $i' \in [i-1]$ be the index such that $v_i v_{i'} \in E(F)$, and let $z_i \in V(G) \setminus X_{i-1}$ be a neighbor of $z_{i'}$, which exists since $z_{i'} \in \partial_G X_{i-1}$. Let $X'_i = X_{i-1} \cup \{z_i\}$ and let μ'_i be the model of F_i in $G[X'_i]$ obtained from μ_{i-1} by letting $\mu'_i(v_i)$ be the single-vertex subgraph formed by z_i and letting $\mu'_i(v_i v_{i'}) = z_i z_{i'}$. If $i = n+1$, then μ'_{n+1} is a model of F in G, and the proof is finished.

If $i \le n$, we need to show how to construct the model μ_i and the set X_i satisfying (i), (ii), and (iii). Note that the set X'_i an n-tail, since X_{i-1} is an n-tail and we can extend the corresponding path decomposition to a path decomposition of $G[X'_i]$ of width less than n by adding the bag $\{z_i\} \cup \partial_G X_{i-1} \supseteq \partial_G X'_i$ of size $i \le n$ at its end. By the condition (iii) for $i-1$, we have $|\partial_G X'_i| > i-1$, and thus $\partial_G X'_i = \{z_i\} \cup \partial_G X_{i-1}$ has size exactly i. Also, μ'_i satisfies the condition (i). However, (iii) could be violated for X'_i.

Let $X_i \subseteq V(G)$ be a maximal n-tail of G such that $X'_i \subseteq X_i$ and $|\partial_G X_i| = i$. Then X_i clearly satisfies (ii) and (iii). By Menger's theorem, the number of pairwise vertex-disjoint paths from X'_i to $\partial_G X_i$ is equal to the size of the smallest cut separating these sets; i.e., there exists a vertex separation (A, B) of $G[X_i]$ such that $X'_i \subseteq A$ and $\partial_G X_i \subseteq B$, and a system $\{Q_v : v \in A \cap B\}$ of pairwise vertex-disjoint paths from X'_i to $\partial_G X_i$. By Lemma 2.99 applied with $\{Q_v \cap G[B] : v \in A \cap B\}$, $X = X_i$ and $Y = A$, we conclude that A is an n-tail of G. Since $\partial_G X_i \subseteq B$ and (A, B) is a vertex separation of $G[X_i]$, we have $\partial_G A \subseteq A \cap B$. Since $X_{i-1} \subset X'_i \subseteq A$, the condition (iii) for $i-1$ implies $|A \cap B| \ge |\partial_G A| > i-1$. Consequently, $|A \cap B| = i = |\partial_G X'_i|$, and the system $\{Q_v : v \in A \cap B\}$ joins each vertex of $\partial_G X'_i$ to a vertex in $\partial_G X_i$. Let μ_i be obtained from μ'_i by, for $j \in [i]$, letting $\mu_i(v_j)$ consist of $\mu'_i(v_j)$ and the path of the system that contains z_j. The resulting model of F_i in $G[X_i]$ clearly satisfies (i), finishing the construction. □

Thus, we obtain the following analogue of Corollary 2.35 for pathwidth.

Corollary 2.100 *A minor-closed class of graphs has unbounded pathwidth if and only if it contains all trees.*

2.14.3 More Restrictive Parameters

Let us now discuss several graph parameters that are more restrictive than treewidth. The first of these parameters is *treedepth*. A *rooted forest* F is a forest where each component has one special vertex, a *root*. We can naturally extend the terminology for rooted trees to rooted forests: The *depth* of F is the maximum number of vertices of a path in F starting in a root. A vertex u is an *ancestor* of a vertex v if u lies on the path from u to the root of the component containing u. We say that two vertices are *comparable* if one is an ancestor of the other one. An ***elimination forest*** of a graph G is a rooted forest F with $V(F) = V(G)$ such that adjacent vertices of G are comparable in F. Equivalently, G is a spanning subgraph of the *closure* of F, i.e., the graph obtained from F by adding edges between all comparable vertices. The ***treedepth*** $\mathrm{td}(G)$ is then defined as the minimum depth of an elimination forest of G. For example, graphs of treedepth one are exactly those without edges, while graphs of treedepth two are *star forests*, i.e., forests where every component is a star.

Observe that if G is connected, then every elimination forest of G is connected, i.e., a tree. Moreover, if T is a connected elimination forest of G of depth $\mathrm{td}(G)$ and r is the root of T, then $T - r$ is an elimination forest of $G - r$ of depth $\mathrm{td}(G) - 1$. Based on this, it is easy to see that the following observation holds.

Observation 2.101 *Let G be a graph with more than one vertex. If G is disconnected, then* $\mathrm{td}(G)$ *is the maximum of treedepths of its components. If G is connected, then* $\mathrm{td}(G) = 1 + \max_{v \in V(G)} \mathrm{td}(G - v)$.

For a fixed integer k, this gives a recursive algorithm to decide whether G has treedepth at most k in time $O(|G|^k)$.

Another characterization of treedepth is in terms of *centered colorings*. A coloring φ of a graph G is ***centered*** if for every connected subgraph H of G, there exists a color assigned by φ to exactly one vertex of H. If F is an elimination forest of G of depth $\mathrm{td}(G)$, for each $v \in V(G)$ we can define $\psi(v)$ as the distance in F from v to the root of the component containing v. Then ψ is a centered coloring of G using $\mathrm{td}(G)$ colors. Conversely, it is easy to turn a centered coloring of G using d colors into an elimination forest of G of depth d.

Note that treedepth is minor-monotone, i.e., if H is a minor of G, then $\mathrm{td}(H) \leq \mathrm{td}(G)$. Indeed:

- For any edge $e \in E(G)$, an elimination forest of G is also an elimination forest of $G - e$.
- For a vertex $v \in V(G)$, an elimination forest F of G can be turned into an elimination forest of $G - v$ by deleting v if v is a root and by contracting the edge between v and its parent in F otherwise.
- For an edge $e = uv \in E(G)$, an elimination forest F of G where say v is an ancestor of u can be turned into an elimination forest of G/e by contracting the edge between u and its parent in F.

Hence, treedepth can be characterized by forbidden minors. For example, a graph is a star forest (i.e., has treedepth at most two) if and only if it avoids C_3 and P_4 as minors. Interestingly, for every positive integer k, even the set of forbidden *subgraphs* for having treedepth at most k is finite [19].

For an approximate characterization by forbidden subgraphs, note that a path with 2^n vertices has treedepth more than n. This can be easily seen by induction using Observation 2.101, since deleting a vertex from a path with 2^n vertices results in a graph containing a path with 2^{n-1} vertices. Conversely, note that a depth-first search spanning tree F of a graph G only contains tree edges (contained in F) and back edges (joining a vertex to its predecessor in F), and thus F is an elimination forest of G. This gives us the following nice result.

Observation 2.102 *If every path in a graph G has at most n vertices, then there exists an elimination forest F of G of depth at most n such that $F \subseteq G$.*

Combining these two argument, we obtain an analogue of Corollary 2.35 for treedepth.

Corollary 2.103 *A subgraph-closed class of graphs has unbounded treedepth if and only if it contains all paths.*

Note also that $\mathrm{pw}(G) \leq \mathrm{td}(G) - 1$ for every graph G. Clearly, it suffices to prove this for connected graphs. Let T be an optimal elimination forest of G. There exists a closed walk $W = w_1 w_2 \ldots w_m$ in T starting and ending in the root and traversing each edge exactly twice (an Euler tour in the graph obtained from T by doubling each edge). Let $P = x_1 \ldots x_m$ be a path and for $i \in [m]$, let $\beta(x_i)$ be the set of ancestors of w_i in T (including w_i itself). Consider a vertex $v \in V(G)$, and let a be the minimum and b the maximum index such that $w_a = v = w_b$. Observe that $v \in \beta(x_i)$ if and only if $a \leq i \leq b$. Moreover, if uv is an edge of G, where u is an ancestor of v in T, then $\{u, v\} \subseteq \beta(x_a)$. Hence, (P, β) is a path decomposition of G of width $\mathrm{td}(G) - 1$.

Even more restrictive parameter is the ***vertex cover size*** $\mathrm{vcs}(G)$, i.e., the minimum size of a set $S \subseteq V(G)$ intersecting all edges of G. Note that a tree consisting of a path on S and with all vertices in $V(G) \setminus S$ attached as leaves below the last vertex of S is an elimination forest of G, and thus $\mathrm{td}(G) \leq \mathrm{vcs}(G) + 1$. Vertex cover size is closely related to the maximum size $\mu(G)$ of a matching in G. Indeed, a vertex cover must intersect all edges of such a matching, and thus $\mathrm{vcs}(G) \geq \mu(G)$. By Kőnig's theorem, if G is bipartite, then actually $\mathrm{vcs}(G) = \mu(G)$. This is not true in general graphs; e.g., $\mathrm{vcs}(K_{2n+1}) = 2n$ and $\mu(K_{2n+1}) = n$. However, the set of vertices incident with an inclusionwise maximal matching is a vertex cover, implying that $\mathrm{vcs}(G) \leq 2\mu(G)$ in general.

Corollary 2.104 *A subgraph-closed class of graphs has unbounded vertex cover size if and only if it contains all matchings.*

Finally, let us mention another parameter that is less restrictive than vertex cover size but more restrictive than treewidth, namely the minimum size $\mathrm{fvs}(G)$ of a ***feedback vertex set*** in G, i.e., a set $S \subseteq V(G)$ intersecting all cycles (or equivalently, such that $G - S$ is a forest). A vertex cover is a feedback vertex set, implying $\mathrm{fvs}(G) \leq \mathrm{vcs}(G)$. On the other hand, the forest $G - S$ has a tree decomposition of width one, and adding S to all its bags gives a tree decomposition of G showing that $\mathrm{tw}(G) \leq \mathrm{fvs}(G) + 1$. For the relation to treedepth and pathwidth, note that there are trees (graphs with feedback vertex set of size 0) of arbitrarily large pathwidth, and that the disjoint union T_k of k triangles (a graph of treedepth 3) has $\mathrm{fvs}(T_k) = k$; i.e., these parameters are incomparable with $\mathrm{fvs}(G)$. Erdős-Pósa theorem on disjoint cycles [12] whose proof we saw in Sect. 2.11 can be restated as follows.

Corollary 2.105 *A minor-closed class of simple graphs has unbounded feedback vertex set size if and only if it contains all disjoint unions of triangles.*

References

1. S. Arnborg, A. Proskurowski, D.G. Corneil, Forbidden minors characterization of partial 3-trees. Discrete Math. **80**, 1–19 (1990)
2. B.S. Baker, Approximation algorithms for NP-complete problems on planar graphs. J. ACM **41**, 153–180 (1994)
3. D. Bienstock, N. Robertson, P.D. Seymour, R. Thomas, Quickly excluding a forest. J. Combin. Theory Ser. B **52**, 274–283 (1991)
4. É. Bonnet, E.J. Kim, S. Thomassé, R. Watrigant, Twin-width I: tractable FO model checking. J. ACM **69**, 1–46 (2021)
5. V. Bouchitté, F. Mazoit, I. Todinca, Treewidth of planar graphs: connections with duality. Electron. Notes Discrete Math. **10**, 34–38 (2001)
6. C. Chekuri, J. Chuzhoy, Polynomial bounds for the Grid-minor Theorem. J. ACM **63**, 40:1–40:65 (2016)
7. J. Chuzhoy, Z. Tan, Towards tight(er) bounds for the excluded Grid Theorem. J. Combin. Theory Ser. B **146**, 219–265 (2021)
8. B. Courcelle, The monadic second-order logic of graphs. I. Recognizable sets of finite graphs. Inf. Comput. **85**, 12–75 (1990)
9. B. Courcelle, J.A. Makowsky, U. Rotics, Linear time solvable optimization problems on graphs of bounded clique-width. Theory Comput. Syst. **33**, 125–150 (2000)
10. J. Davies, *Graph Theory*, ed. by J. Geelen, D. Král', A. Scott. Oberwolfach Technical Report, vol. 19 (2022)
11. Z. Dvořák, S. Norin, Treewidth of graphs with balanced separations. J. Combin. Theory Ser. B **137**, 137–144 (2019)
12. P. Erdős, L. Pósa, On independent circuits contained in a graph. Can. J. Math. **17**, 347–352 (1965)
13. J. Geelen, O. Kwon, R. McCarty, P. Wollan, The grid theorem for vertex-minors. J. Combin. Theory Ser. B **158**, 93–116 (2023)
14. F. Gurski, E. Wanke, The tree-width of clique-width bounded graphs without $K_{n,n}$, in *International Workshop on Graph-Theoretic Concepts in Computer Science* (Springer, Berlin, 2000), pp. 196–205
15. R. Karp, Reducibility among combinatorial problems, in *Complexity of Computer Computations*, ed. by R. Miller, J. Thatcher (eds.) (Plenum Press, New York, 1972), pp. 85–103
16. T. Korhonen, Grid induced minor theorem for graphs of small degree. J. Combin. Theory Ser. B **160**, 206–214 (2023)
17. D. Lapoire, Treewidth and duality for planar hypergraphs (preprint, 1996)
18. R. Lipton, R. Tarjan, A separator theorem for planar graphs. SIAM J. Appl. Math. **36**, 177–189 (1979)
19. J. Nešetřil, P. Ossona de Mendez, Tree-depth, subgraph coloring and homomorphism bounds. Eur. J. Combin. **27**, 1022–1041 (2006)
20. S. Oum, P.D. Seymour, Approximating clique-width and branch-width. J. Combin. Theory Ser. B **96**, 514–528 (2006)
21. N. Robertson, P.D. Seymour, Graph minors. V. Excluding a planar graph. J. Combin. Theory Ser. B **41**, 92–114 (1986)
22. N. Robertson, P. D. Seymour, Graph minors. X. Obstructions to tree-decomposition. J. Combin. Theory Ser. B **52**, 153–190 (1991)
23. N. Robertson, P.D. Seymour, Graph minors. XI. Circuits on a surface. J. Combin. Theory Ser. B **60**, 72–106 (1994)
24. N. Robertson, P.D. Seymour, Graph minors. XVI. Excluding a non-planar graph. J. Combin. Theory Ser. B **89**, 43–76 (2003)
25. N. Robertson, P.D. Seymour, R. Thomas, Quickly excluding a planar graph. J. Combin. Theory Ser. B **62**, 323–348 (1994)
26. A. Satyanarayana, L. Tung, A characterization of partial 3-trees. Networks **20**, 299–322 (1990)

27. P.D. Seymour, R. Thomas, Graph searching and a min-max theorem for tree-width. J. Combin. Theory Ser. B **58**, 22–33 (1993)
28. P.D. Seymour, R. Thomas, Call routing and the ratcatcher. Combinatorica **14**, 217–241 (1994)
29. N.D.L. Sintiari, N. Trotignon, (Theta, triangle)-free and (even hole, K_4)-free graphs—Part 1: layered wheels. J. Graph Theory **97**, 475–509 (2021)

Chapter 3
Linkedness

In this chapter, we study several topics vaguely linked by the fact that they concern the existence of disjoint paths with prescribed ends. It is clear that this issue often arises when we are trying to construct minors by connecting their small fragments, and this chapter will provide us with several useful tools for such situations.

The basic problem that we are about to consider is as follows: Suppose that $(s_1, t_1), \ldots, (s_m, t_m)$ are pairs of vertices of a graph G. An $\{(s_i, t_i) : i \in [m]\}$-*linkage* is a system $\{P_i : i \in [m]\}$ of pairwise vertex-disjoint paths in G such that for $i \in [m]$, the path P_i has ends s_i and t_i. What can we say about the existence of an $\{(s_i, t_i) : i \in [m]\}$-linkage in G, in terms of sufficient conditions, structural constraints, and algorithms?

If $|G| \geq 2m$ and an $\{(s_i, t_i) : i \in [m]\}$-linkage exists for every choice of distinct vertices $s_1, \ldots, s_m, t_1, \ldots, t_m$, we say that the graph G is ***m-linked***. Let us remark that although the definition requires the vertices to be distinct, we can at no extra cost allow $s_i = t_i$ for some indices $i \in [m]$: In this situation, we can replace t_i by an arbitrary vertex in $V(G) \setminus \{s_j, t_j : j \in [m]\}$, find the linkage with these modified ends, and truncate the i-th path to consist just of s_i.

Finding an $\{(s_i, t_i) : i \in [m]\}$-linkage is a substantially more difficult problem than deciding whether there are m pairwise vertex-disjoint paths from $\{s_i : i \in [m]\}$ to $\{t_i : i \in [m]\}$, and not one that can be answered purely in terms of connectivity. E.g., consider a plane graph G drawn with vertices s_1, s_2, t_1, t_2 in the boundary of the outer face in order. By planarity, any path from s_1 to t_1 intersects every path from s_2 to t_2, and thus G is not 2-linked. Yet, G can be up to 5-connected. This is in contrast to Menger's theorem, which shows that 2-connectivity is sufficient to ensure the existence of an $(\{s_1, s_2\}, \{t_1, t_2\})$-linkage of size two. This example also indicates that there are topological aspects to the linkage problem; we postpone their more detailed discussion till Chap. 4.

Note that the linkage problem is a special case of finding a rooted minor. Indeed, suppose that H is a matching with m edges and for $i \in [m]$, the function ρ maps the ends of the i-th edge to s_i and t_i. Then H is a ρ-rooted minor of G if and only if G

© The Author(s), under exclusive license to Springer Nature Switzerland AG 2025

Z. Dvořák, *Graph Minors*, Springer Monographs in Mathematics,
https://doi.org/10.1007/978-3-031-87469-7_3

contains an $\{(s_i, t_i) : i \in [m]\}$-linkage. Some of the tools we develop in this chapter will also be useful in the context of this more general problem.

3.1 Exploiting a Clique Minor

For any fixed graph H, if a graph G contains a sufficiently large clique minor, then it clearly also contains H as a minor. What about rooted minors? Of course, we need to avoid the situation where the clique minor is separated from the roots by a small cut, but other than that, it seems intuitively clear that we will be able to route the model of H through the clique as desired.

Before giving the details of an argument confirming this intuition, let us first discuss its general idea, which is quite useful in similar circumstances. Suppose we have k roots R; to ensure that the clique minor is useful, we are going to assume that there is no vertex separation (A, B) of order less than k separating the roots from the minor. If there is no such separation of order exactly k (except for the trivial one with A consisting exactly of the roots and $B = V(G)$), this gives us freedom to perform various reductions (deleting a vertex, contracting an edge) without violating the connectivity assumption. If there is such a vertex separation (A, B) of order k, then by Menger's theorem, we can connect the roots to $A \cap B$ by an $(R, A \cap B)$-linkage $\mathcal{L}$. We can now apply induction to $G[B]$ with $A \cap B$ playing the role of the roots, and connect the resulting minor of H to the original roots through the linkage $\mathcal{L}$.

A technical detail with the last step is that we cannot guarantee that the whole clique minor is contained in $G[B]$. More precisely, consider a connected subgraph $\nu(v)$ of G representing a vertex v of the clique. If $\nu(v)$ extends into A, then the intersection $\nu(v) \cap G[B]$ is not necessarily connected. However, each component of $\nu(v) \cap G[B]$ must then intersect the root set $A \cap B$. Similarly, if $\nu(u)$ represents another vertex u of the clique, and the edge $\nu(e)$ of G representing the edge $e = uv$ of the clique is in $G[A]$, then $G[B]$ does not necessarily have to contain an edge between $\nu(u) \cap G[B]$ and $\nu(v) \cap G[B]$. However, this is again only possible if both $\nu(u) \cap G[B]$ and $\nu(v) \cap G[B]$ intersect the root set $A \cap B$. This motivates the following definition.

Let R be a set of vertices of a graph G and let K be a complete graph. An *R-partial model of K in G* is a function ν assigning pairwise vertex-disjoint non-empty subgraphs of G to vertices of K such that

(P1) For every vertex $v \in V(K)$, if the subgraph $\nu(v)$ is not connected, then every component of $\nu(v)$ intersects R.
(P2) For distinct $u, v \in V(K)$, if G does not have an edge with one end in $\nu(u)$ and the other end in $\nu(v)$, then $\nu(u)$ and $\nu(v)$ both intersect R.

A vertex separation (A, B) of G *separates ν from the roots* if $R \subseteq A$ and there exists $v \in V(K)$ such that $V(\nu(v)) \subseteq B \setminus A$. We say that ν is *linked to the roots* if every vertex separation that separates ν from the roots has order at least $|R|$.

Theorem 3.1 *Let H be a simple graph and let ρ be a bijection between a subset $\mathrm{dom}(\rho)$ of $V(H)$ and a set R of vertices of a graph G. If G contains an R-partial model ν of a complete graph K with at least $|R| + |H|$ vertices linked to the roots, then H is a ρ-rooted minor of G.*

Proof We prove the claim by induction on $|G| + \|G\|$. Suppose first that a vertex separation (A, B) of order $|R|$ with $B \neq V(G)$ separates ν from the roots, and let u_0 be a vertex of K such that $V(\nu(u_0)) \subseteq B \setminus A$. Let $R' = A \cap B$ and let us define $\nu'(u) = \nu(u) \cap G[B]$ for each $u \in V(K)$.

We claim that ν' is an R'-partial model of K in $G[B]$. First, note that since $\nu(u_0) \subseteq G[B] - A$ is disjoint from R, the condition (P2) for ν implies that G contains an edge between $\nu(u_0)$ and $\nu(u)$ for any other vertex $u \in V(K)$, and thus so does $G[B]$. Consequently, $\nu'(u)$ is non-empty. If $\nu'(u)$ has a component C disjoint from R', then C is also a component of $\nu(u)$ disjoint from R, and by (P1) for ν, we have $\nu(u) = C$. Consequently, $\nu'(u) = C$ is connected, and thus (P1) holds for ν'. Similarly, consider distinct $u_1, u_2 \in V(K)$ such that say $\nu'(u_1)$ is disjoint from R'. As we have just argued, this implies that $\nu(u_1) = \nu'(u_1)$, and thus $\nu(u_1)$ is disjoint from R. By (P2) for ν, G has an edge e between $\nu(u_1)$ and $\nu(u_2)$, and since $\nu'(u_1) = \nu(u_1)$ is contained in $G[B] - A$, we have $e \in E(G[B])$. Therefore, e is also an edge between $\nu'(u_1)$ and $\nu'(u_2)$. This implies that (P2) holds for ν' as well.

Furthermore, the R'-partial model ν' is linked to the roots in $G[B]$. Indeed, consider a vertex separation (A', B') of $G[B]$ separating ν' from the roots, i.e., there exists $u_1 \in V(K)$ such that $\nu'(u_1) \subseteq G[B'] - A'$. In particular, $\nu'(u_1)$ is disjoint from R', and thus $\nu(u_1) = \nu'(u_1)$. Hence, the vertex separation $(A \cup A', B')$ of G separates the R-partial model ν from the roots. Since ν is linked to the roots, we have $|A' \cap B'| = |(A \cup A') \cap B'| \geq |R| = |R'|$, as required.

Note also that $G[A]$ does not have a vertex separation (A', B') of order less than $|R|$ with $R \subset A'$ and $R' \subseteq B'$, since otherwise the vertex separation $(A', B' \cup B)$ of G of order less than $|R|$ would contradict the assumption that ν is linked to the roots. By Menger's theorem, there exists an (R, R')-linkage $\mathcal{P} = \{P_v : v \in R\}$ in $G[A]$, where P_v starts in v for each $v \in R$. Let $\rho' : \mathrm{dom}(\rho) \to R'$ be the bijection defined by letting $\rho'(x)$ be the end of the path $P_{\rho(x)}$ in R' for every $x \in \mathrm{dom}(\rho)$. Since the R'-partial model ν' of K is linked to the roots in $G[B]$ and $B \neq V(G)$, by the induction hypothesis it follows that H has a ρ'-rooted model μ' in $G[B]$. The combination of this model μ' with the linkage $\mathcal{P}$ gives us a ρ-rooted model of H in G, as desired.

Therefore, we can assume that the following condition holds:

($\star$) Every vertex separation (A, B) of G with $B \neq V(G)$ separating ν from the roots has order greater than $|R|$.

Let $K_0 \subseteq K$ be the clique of size at least $|K| - |R| \geq |H|$ consisting of the vertices $u \in V(K)$ such that $V(\nu(u)) \cap R = \emptyset$. Suppose next that there exists $u \in V(K_0)$ such that $\nu(u)$ is not a single vertex, and let e be any edge of $\nu(u)$. Consider the graph G/e obtained by contracting the ends of the edge e to a vertex z, with the R-

partial model ν/e of K obtained by replacing $\nu(u)$ by $\nu(u)/e$. Note that if (A', B') is a vertex separation of G/e separating ν/e from the roots, then the vertex separation (A, B) obtained by replacing z by x and y both in A' and in B' separates ν from the roots. By ($\star$), we have $|A' \cap B'| \geq |A \cap B| - 1 \geq |R|$ if $B' \neq V(G/e)$, and $|A' \cap B'| \geq |A'| \geq |R|$ otherwise. Consequently, ν/e is linked to the roots in G/e, and by the induction hypothesis, there exists a ρ-rooted model of H in G/e; and decontracting the edge e turns it to a ρ-rooted model of H in G.

Hence, we can assume that $\nu(u)$ is a single vertex of every $u \in V(K_0)$. By (P2), the set $M = \bigcup_{u \in V(K_0)} V(\nu(u))$ induces a clique in G. Moreover, since ν is linked to the roots, Menger's theorem imples that there exists an (R, M)-linkage $\mathcal{P}$ in G of size $|R|$. We can use the paths of $\mathcal{P}$ as the subgraphs representing the vertices of dom(ρ) in the model of H in G, and vertices of the clique $G[M] \setminus V(\bigcup \mathcal{P})$ to represent the rest of the vertices of H. □

Theorem 3.1 is a key ingredient in the polynomial-time algorithm of Robertson and Seymour [6] to test the presence of an $\{(s_i, t_i) : i \in [m]\}$-linkage (with m fixed, not part of an input) or more generally, of a rooted minor of a fixed graph H in an input graph G. The main idea of the algorithm is that either G has bounded treewidth, or we can efficiently locate an ***irrelevant vertex*** in G whose removal does not change the outcome. We can then delete such irrelevant vertices until the treewidth becomes bounded, and then conclude testing the presence of a rooted minor of H using Theorem 2.8.

More precisely, let ρ be an injective function from dom$(\rho) \subseteq V(H)$ to $V(G)$. We say that a vertex $v \in V(G) \setminus \text{img}(H)$ is *(H, ρ)-irrelevant* if G contains a ρ-rooted model of H if and only if $G - v$ does. That is, either G does not contain a ρ-rooted model of H, or G contains one avoiding v. Assuming we are given a model of a large clique minor in G, locating such an irrelevant vertex is easy.

Corollary 3.2 *Let H be a simple graph and let ρ be a bijection between a subset dom(ρ) of $V(H)$ and a set R of vertices of a graph G. Let ν be a model of a complete graph K with at least $2|R| + |H| + 1$ vertices in G. Let (A, B) be a vertex separation of G that separates ν (viewed as an R-partial model) from the roots, chosen so that $|A \cap B|$ is smallest possible and subject to that $|B|$ is smallest possible. Then every vertex $v \in B \setminus A$ is (H, ρ)-irrelevant.*

Proof The claim is trivial if H is not a ρ-rooted minor of G. Hence, suppose that μ is a ρ-rooted model of H in G; we need to show that $G - v$ also contains a ρ-rooted model of H.

Let H' be the complete graph whose vertex set consists of $R' = A \cap B$ and the vertices of H such that $V(\mu(x)) \subseteq B \setminus A$; clearly, we have $|H'| \leq |H| + |R'| \leq |H| + |R|$. Let ρ' be the identity function on R'. Observe that it suffices to show that H' has a ρ'-rooted model μ' in $G[B] - v$. Indeed, let μ_1 be defined as follows: For $x \in V(H)$, we let

$$\mu_1(x) = \begin{cases} \mu(x) & \text{if } V(\mu(x)) \cap B = \emptyset \\ \mu'(x) & \text{if } V(\mu(x)) \cap A = \emptyset \\ (\mu(x) \cap G[A]) \cup G\big[\bigcup_{v \in V(\mu(x)) \cap R'} V(\mu'(v))\big] & \text{otherwise.} \end{cases}$$

For an edge $e = x_1x_2 \in E(H)$, we define $\mu_1(e)$ as follows: If $\mu(e) \in E(G[A])$, then let $\mu_1(e) = \mu(e)$. Otherwise, for $i \in [2]$, the subgraph $\mu(x_i)$ intersects $G[B]$; we let $x_i' = x_i$ if $V(\mu(x_i)) \subseteq B \setminus A$ and we let x_i' be a vertex of R' contained in $\mu(x_i)$ otherwise, and we define $\mu_1(e) = \mu'(x_1'x_2')$. Observe that μ_1 is a model of H in $G - v$.

The existence of μ' can be shown using Theorem 3.1 as follows. There is at most one vertex $u_0 \in V(K)$ such that $v \in \nu(u_0)$. For every $u \in V(K - u_0)$, let $\nu'(u) = \nu(u) \cap G[B]$. Since (A, B) separates ν from the roots, there exists a vertex $u_1 \in V(K)$ such that $V(\nu(u_1)) \subseteq B \setminus A$. Observe that for each edge $e = u_1u \in E(K)$, $\nu(e)$ has both ends in $G[B]$, and thus if $u \neq u_0$, then $\nu'(u)$ is a non-empty subgraph of $G[B] - v$. It is easy to see that ν' is an R'-partial model of $K - u_0$ in $G[B] - v$.

Consider any vertex separation (A', B') of $G[B]-v$ separating ν' from the roots. Then $(A'', B'') = (A \cup A' \cup \{v\}, B' \cup \{v\})$ is a vertex separation of G of order $|A' \cap B'| + 1$ separating ν from the roots. Recall that the vertex separation (A, B) was chosen with $|A \cap B|$ minimum and subject to that with $|B|$ minimum. If $B'' \neq B$, this implies that $|A'' \cap B''| \geq |A \cap B| + 1 = |R'| + 1$. If $B'' = B$, then $|A'' \cap B''| \geq |(A \cap B) \cup \{v\}| \geq |R'| + 1$. In either case, we conclude that (A', B') has order at least $|R'|$. Therefore, the R'-partial model ν' is linked to the roots.

Since $|K - u_0| \geq 2|R| + |H| \geq |R'| + |H'|$, Theorem 3.1 implies that H' is a ρ'-rooted minor of $G[B] - v$, as desired. □

3.2 Linkedness in Graphs of High Connectivity

Note that a k-linked graph G must be $(2k-1)$-connected: Indeed, otherwise it has a vertex separation (A, B) of order $2k - 2$ with $A \setminus B \neq \emptyset \neq B \setminus A$, and letting $A \cap B = \{s_1, \ldots, s_{k-1}, t_1, \ldots, t_{k-1}\}$ and choosing $s_k \in A \setminus B$ and $t_k \in B \setminus A$, it is clear that G does not have an $\{(s_i, t_i) : i \in [k]\}$-linkage. The converse is false; as we have seen at the beginning of the chapter, even 5-connectivity does not imply 2-linkedness. However, a combination of Theorem 3.1 with Theorem 1.2 shows that a near-converse holds in graphs with high maximum average degree. For every positive integer k, let $d_{\text{link}}(k)$ be the minimum integer such that every $2k$-connected simple graph of maximum average degree at least $d_{\text{link}}(k)$ is k-linked.

Corollary 3.3 *The function $d_{\text{link}}(k)$ satisfies $d_{\text{link}}(k) = O(k\sqrt{\log k})$.*

Proof Let $f(k) = O(k\sqrt{\log k})$ be the function from the statement of Theorem 1.2. We claim that $d_{\text{link}}(k) \leq f(4k)$. Indeed, consider any $2k$-connected simple graph G

of maximum average degree at least $f(4k)$. By Theorem 1.2, there exists a model ν of K_{4k} in G.

We need to show that G has an $\{(s_i, t_i) : i \in [k]\}$-linkage for any set $R = \{s_1, \ldots, s_k, t_1, \ldots, t_k\}$ of $2k$ distinct vertices of G. Let H be the matching with k edges and let $\rho : V(H) \to R$ be a function mapping the ends of the i-th edge of H to s_i and t_i for each $i \in [k]$. Since G is $2k$-connected, ν is an R-partial model of K_{4k} linked to the roots, and since $4k = |R| + |H|$, Theorem 3.1 implies that H is a ρ-rooted minor of G. This minor clearly gives the desired $\{(s_i, t_i) : i \in [k]\}$-linkage in G. □

Let us remark that since the minimum degree is greater or equal to the connectivity of the graph, $d_{\text{link}}(k)$-connectivity implies k-linkedness. In Sect. 5.1, we are going to see that $d_{\text{link}}(2) = 6$; this is tight, as the example of 5-connected non-2-linked planar graphs with average degree arbitrarily close to 6 shows. Thomas and Wollan [10] proved that $d_{\text{link}}(3) = 10$. Furthermore, the bound on $d_{\text{link}}(k)$ can be improved to $O(k)$; the basic idea of the improvement is that instead of linking through a clique minor, it suffices to link through a minor of a sufficiently dense graph.

Theorem 3.4 (Thomas and Wollan [9]) *For every positive integer k, the function* $d_{\text{link}}(k)$ *satisfies* $d_{\text{link}}(k) \le 10k$.

Sometimes, we only need to find a subgraph that is highly linked. By the results of this chapter, to do so, it suffices to find a highly connected subgraph. And, by the following observation of Mader, this can always be done in graphs of large average degree.

Lemma 3.5 *For any positive integer c and real number* $b \ge c$, *if a simple graph G has at least* $(b+c)|G|$ *edges, then G contains a* $(c+1)$*-connected induced subgraph H with* $\|H\| > b|H|$ *and of minimum degree more than* $b + c$.

Proof Since G has average degree at least $2(b + c)$, it has more than $2(b + c)$ vertices. Let H be an induced subgraph of G such that $n = |H| \ge 2c$ and $\|H\| > (b + c)(n - c)$, chosen with n minimum. If $n \le b + c$, then

$$\|H\| > (b+c)(n-c) \ge n \cdot \frac{n}{2} > \binom{n}{2},$$

which is a contradiction. Therefore, $n > b + c$. By the minimality of n, for every vertex $v \in V(H)$ we have

$$\|H - v\| \le (b+c)(|H-v| - c) = (b+c)(n-c) - (b+c) < \|H\| - (b+c),$$

and thus H has minimum degree more than $b + c$. Moreover,

$$\|H\| > (b+c)(n-c) = bn + c(n - (b+c)) > bn.$$

Consider any vertex separation (A, B) of H with $A \setminus B \neq \emptyset \neq B \setminus A$. Any vertex $v \in A \setminus B$ has all its neighbors in A, and thus $|A| \geq \deg_H v + 1 > b + c + 1 > 2c$, and similarly $|B| > 2c$. By the minimality of H, we have $\|H[A]\| \leq (b+c)(|A|-c)$ and $\|H[B]\| \leq (b+c)(|B|-c)$. Consequently,

$$\begin{aligned}\|H\| &\leq \|H[A]\| + \|H[B]\| \leq (b+c)(|A|+|B|-2c) \\ &= (b+c)(n+|A \cap B|-2c).\end{aligned}$$

Since $\|H\| > (b+c)(n-c)$, it follows that $|A \cap B| > c$. Therefore, H has no cut of size at most c, and so it is $(c+1)$-connected. □

Hence, we get the following consequence.

Corollary 3.6 *For any positive integer k, any simple graph G of average degree $d \geq \max(8k, d_{\text{link}}(k) + 4k)$ has a k-linked induced subgraph of minimum degree more than $d/2$.*

Proof Note that $\|G\| = \frac{d}{2}|G|$, and thus by Lemma 3.5 with $c = 2k$ and $b = \frac{d}{2} - c$, G has a $(2k+1)$-connected induced subgraph H of minimum degree more than $d/2$ and with average degree

$$\frac{2\|H\|}{|H|} > 2b = d - 2c \geq d_{\text{link}}(k),$$

which is k-linked by the definition of $d_{\text{link}}(k)$. □

In particular, combining these results, we can get a quite good average degree bound forcing the existence of a clique as a topological minor.

Lemma 3.7 *For every positive integer k, if a simple graph G has average degree at least $14\binom{k}{2}$, then it contains K_k as a topological minor.*

Proof We can assume $k \geq 2$. By Corollary 3.6 and Theorem 3.4, G has a $\binom{k}{2}$-linked subgraph H of minimum degree more than k^2. For each $i \in [k]$, choose c_i as an arbitrary element of $[k] \setminus \{i\}$. For each $i \in [k]$ and $j \in [k] \setminus \{i\}$, choose a distinct vertex $v_{i,j}$ of H, so that the vertex v_{i,c_i} is adjacent to all vertices $v_{i,j}$ for $j \in [k] \setminus \{i, c_i\}$. Such a choice is possible, since we are selecting $k(k-1) < k^2 < \delta(H)$ vertices in total. Since H is $\binom{k}{2}$-linked, it contains an $\{(v_{i,j}, v_{j,i}) : 1 \leq i < j \leq k\}$-linkage $\mathcal{L}$. The paths of $\mathcal{L}$ together with the edges $v_{i,c_i}v_{i,j}$ for $i \in [k]$ and $j \in [k] \setminus \{i, c_i\}$ form a subdivision of K_k in G, with branch vertices v_{i,c_i} for $i \in [k]$. □

The quadratic dependence on k is optimal, as can be seen by considering the biclique $K_{n,n}$ of minimum degree n: Suppose K_k is a topological minor of $K_{n,n}$, with $a \geq k/2$ of the branch vertices contained in one of the parts of $K_{n,n}$. Then the paths representing the edges between these branch vertices intersect the other part, and thus

$$n \geq \binom{a}{2} + k - a = \frac{a^2 - 3a + 2k}{2} \geq \frac{k^2}{8}.$$

3.3 Unique Linkage Theorem

Let $\mathcal{S} = \{(s_i, t_i) : i \in [k]\}$ be a set of pairs of distinct vertices of a graph G. As we have seen in Sect. 3.1, one approach to deciding whether G has an $\mathcal{S}$-linkage is to locate an *$\mathcal{S}$-linkage irrelevant vertex*, i.e., a vertex $v \in V(G) \setminus \{s_1, \ldots, s_k, t_1, \ldots, t_k\}$ such that if G has an $\mathcal{S}$-linkage, then so does $G - v$. We can then delete v to decrease the size of the instance. It is natural to ask when this process stops; what can we say about a graph G with an $\mathcal{S}$-linkage such that deletion of any vertex disrupts all $\mathcal{S}$-linkages?

Let us first consider a simpler problem, the analogous question for linkages between sets. Let A and B be sets of vertices of a graph G of size k. We say that G is ***(A, B)-linkage critical*** if G has an (A, B)-linkage of size k, but for every $v \in V(G) \setminus (A \cup B)$, the graph $G - v$ does not have such an (A, B)-linkage.

Lemma 3.8 *Let A and B be sets of vertices of a graph G of size k. The graph G is (A, B)-linkage critical if and only if it has a path decomposition (P, β) of adhesion k such that, letting $P = x_0 x_1 \ldots x_m$, we have $\beta(x_0) = A$, $\beta(x_m) = B$, and $\beta(x_i) = \beta(x_i x_{i-1}) \cup \beta(x_i x_{i+1})$ for $i \in [m-1]$.*

Proof Suppose first that G has such a path decomposition, and consider any vertex $v \in V(G) \setminus (A \cup B)$. Let $i \in [m-1]$ be an index such that $v \in \beta(x_i)$. Since $\beta(x_i) = \beta(x_i x_{i-1}) \cup \beta(x_i x_{i+1})$, by symmetry we can assume that $v \in \beta(x_i x_{i-1})$. By Observation 2.1, we conclude that $\beta(x_i x_{i-1}) \setminus \{v\}$ is a cut of size $k-1$ separating A from B in $G - v$, and thus there is no (A, B)-linkage of size k in $G - v$. Therefore, G is (A, B)-linkage critical.

Let us now prove the converse, by induction on the number of vertices of the (A, B)-linkage critical graph G. If $V(G) = A \cup B$, then we can let $m = 2$ and $\beta(x_1) = A \cup B$. Hence, suppose that there exists a vertex $v \in V(G) \setminus (A \cup B)$.

Since G is (A, B)-linkage critical, $G - v$ does not have an (A, B)-linkage of size k, and by Menger's theorem, there exists a vertex separation (C, D) of size at most k with $A \subseteq C$, $B \subseteq D$, and $v \in C \cap D$. Note that since G has an (A, B)-linkage of size k, we have $C \cap D = k$. Since $|A| = k$ and $v \notin A$, there exists a vertex in $A \setminus (C \cap D) \subseteq C \setminus D$, and thus $D \neq V(G)$; and similarly, $C \neq V(G)$. Since G is (A, B)-linkage-critical, clearly $G[C]$ is $(A, C \cap D)$-linkage-critical and $G[D]$ is $(C \cap D, B)$-linkage-critical. Let (P_1, β_1) and (P_2, β_2) be the path decompositions of $G[C]$ and $G[D]$, respectively, obtained by the induction hypothesis. Let $P_1 = a_0 a_1 \ldots a_{m_1}$ and $P_2 = b_0 b_1 \ldots b_{m_2}$. Let $m = m_1 + m_2$, and for a path $P = x_0 x_1 \ldots x_m$, define $\beta(x_i) = \beta_1(a_i)$ for $0 \leq i \leq m_1$ and $\beta(x_i) = \beta_2(b_i - m_1)$ for $m_1 \leq i \leq m$. Note that $\beta_1(a_{m_1}) = \beta_2(b_0) = C \cap D$, and thus (P, β) is a path decomposition of G. Moreover, we have $\beta(x_0) = A$ and $\beta(x_m) = B$.

We need to show that $\beta(x_i) = \beta(x_i x_{i-1}) \cup \beta(x_i x_{i+1})$ holds for every $i \in [m-1]$. This follows from the same property of (P_1, β_1) and (P_2, β_2) for every $i \neq m_1$. Hence, suppose that $i = m_1$, and thus $\beta(x_i) = C \cap D$. Consider any vertex $z \in C \cap D$, and suppose for a contradiction that $z \notin \beta(x_{i-1}) \cup \beta(x_{i+1})$. By (D2), $\beta(x_i)$ is the only bag containing z, and in particular $z \notin A \cup B$. Moreover, by (D1), it follows that all neighbors of z belong to $C \cap D$. However, then $C \cap D \setminus \{z\}$ is a cut of size $k - 1$ separating A from B, contradicting the assumption that G contains an (A, B)-linkage of size k. □

Lemma 3.8 implies that if $|A| = |B| = k$ and G is (A, B)-linkage critical, then $\mathrm{pw}(G) \leq 2k - 1$.

Let us now consider the analogous problem for linkages where each path has prescribed ends. For a system $\mathcal{S} = \{(s_i, t_i) : i \in [k]\}$ of pairs of distinct vertices of a graph G, we say that G is *$\mathcal{S}$-linkage critical* if G has an $\mathcal{S}$-linkage and no vertex of G is $\mathcal{S}$-linkage irrelevant. The situation is much more complicated, since there is no characterization of existence of an $\mathcal{S}$-linkage as convenient as Menger's theorem. Mayhew et al. [4] gave an exact description for the case $k = 2$, and it might be possible to give such a description for other small values of k. However, there likely does not exist a succinct exact characterization in general, since the $\mathcal{S}$-linkage problem for unbounded k (i.e., with k being a part of the input) is NP-complete [2]. On the other hand, Robertson and Seymour [7] proved that at least qualitatively, Lemma 3.8 extends to this setting.

Theorem 3.9 *There exists a function $f_{3.9} : \mathbb{N} \to \mathbb{N}$ such that if $\mathcal{S} = \{(s_i, t_i) : i \in [k]\}$ is a system of pairs of distinct vertices of a graph G and the graph G is $\mathcal{S}$-linkage critical, then* $\mathrm{pw}(G) \leq f_{3.9}(k)$.

They show that this result is a fairly simple consequence of a weaker one, only bounding the treewidth.

Theorem 3.10 (Unique Linkage Theorem) *There exists a function $f_{3.10} : \mathbb{N} \to \mathbb{N}$ such that if $\mathcal{S} = \{(s_i, t_i) : i \in [k]\}$ is a system of pairs of distinct vertices of a graph G and the graph G is $\mathcal{S}$-linkage critical, then* $\mathrm{tw}(G) \leq f_{3.10}(k)$.

Although Theorem 3.10 is weaker than Theorem 3.9, it is sufficient for most applications and generally easier to use. The proof of Theorem 3.10 given in [7] is surprisingly involved, using the Minor Structure Theorem (by Corollary 3.2, it suffices to prove it for K_{4k+1}-minor-free graphs). A much simpler (though still complicated) direct proof was given by Kawarabayashi and Wollan [3].

By Theorem 3.10, every graph of large treewidth has a vertex that is irrelevant for the existence of an $\mathcal{S}$-linkage. However, to use this result algorithmically, we also need to be able to locate such a vertex, and it is not at all obvious how to do that. We are going to discuss a polynomial-time algorithm in Sect. 5.3. For now, let us showcase how the Unique Linkage Theorem is used in this algorithm (and other applications) on a simpler example.

In this example application, we are going to need a lemma dealing with a special case. Consider a plane graph G. A *nest of depth d* is a sequence $C_1, \ldots, C_d$ of pairwise vertex-disjoint cycles in G such that, denoting by Λ_i for $i \in [d]$ the open disk in the plane bounded by C_i, the cycle C_j is drawn in Λ_i for every $j < i$. The *egg* of the nest is the set of vertices of G drawn in Λ_1.

Lemma 3.11 *Let G be a plane graph, let $\mathcal{S} = \{(s_i, t_i) : i \in [k]\}$ be a system of pairs of distinct vertices of G incident with the outer face of G, and let $\mathcal{C} = C_1, \ldots, C_k$ be a nest of depth k in G. If G has an $\mathcal{S}$-linkage, then it also has one disjoint from the egg of $\mathcal{C}$. In particular, all vertices of the egg of $\mathcal{C}$ are $\mathcal{S}$-linkage irrelevant.*

Proof We prove the claim by induction on k, with the basic case $k = 0$ being trivial. Consider any $\mathcal{S}$-linkage $\mathcal{P}$ in G and a path $P \in \mathcal{P}$. We can draw a simple closed curve γ_P along P and then through the outer face of G between the ends of P; this curve divides G into two subgraphs G_1 and G_2 intersecting in P. Choose the path $P \in \mathcal{P}$ and the labels of G_1 and G_2 so that G_1 is inclusionwise-minimal. Observe that G_1 then intersects $\mathcal{P}$ only in P: If a path $P' \in \mathcal{P}$ different from P intersected G_1, then since the paths P and P' are vertex-disjoint, the planarity would imply $P' \subset G_1$. But then the simple closed curve $\gamma_{P'}$ could be chosen to be disjoint from γ_P and one of the subgraphs to which $\gamma_{P'}$ splits G would be a proper subgraph of G_1, contradicting the choice of P.

We can assume that the ends of P are s_k and t_k. Let Λ_k be the open disk bounded by C_k, and let Q be the subgraph of G_1 consisting of $G_1 \cap C_k$ and the part of P drawn in the complement of Λ_k. Let us double the edges of Q belonging both to C and P and observe that with this change, the vertices s_k and t_k have odd degree in Q and all other vertices of Q have even degree (e.g., if $v \in V(P) \setminus \{s_k, t_k\}$ is incident with exactly one edge of P drawn in the complement of Λ_k, then the curve γ_P at v enters Λ_k and exactly one of the incident edges of C belongs to G_1). Consequently, there exists a path P_k between s_k and t_k in $Q \subseteq G_1$, and in particular, this path P_k is disjoint from Λ_k.

The claim then follows by the induction hypothesis applied to $\mathcal{S}' = \{(s_i, t_i) : i \in [k-1]\}$ in the graph $G - V(P_k)$ with the nest $C_1, \ldots, C_{k-1}$, by combining an $\mathcal{S}'$-linkage disjoint from the egg with the path P_k. □

Thus, if G is a plane graph, all vertices forming $\mathcal{S}$ are incident with its outer face, and G is $\mathcal{S}$-linkage critical, then G does not have any nest of depth d. This implies that the auxiliary graph G' obtained from G by adding a vertex into each face and joining it by edges to all incident vertices has radius less than $2d$, and by Theorem 2.73, $\mathrm{tw}(G) \le \mathrm{tw}(G') \le 6d$. Let us remark that this gives a proof of Theorem 3.10 in this very special case. We can combine Lemma 3.11 with Theorem 3.10 to obtain the following useful result, which may seem as just a small modification of Lemma 3.11, but it is actually much deeper.

Lemma 3.12 *Let H be a graph with separation (H', G) and let $\mathcal{S} = \{(s_i, t_i) : i \in [k]\}$ be a system of pairs of distinct vertices of H'. Suppose that G is a plane graph and all vertices of $V(G \cap H')$ are drawn in the boundary of the outer face of G. Let $d = 2f_{3.10}(k) + 2$ and let $\mathcal{C} = C_1, \ldots, C_d$ be a nest of depth d in G. If the graph*

H has an $\mathcal{S}$-linkage, then it also has one disjoint from the egg of $\mathcal{C}$. In particular, all vertices of the egg of $\mathcal{C}$ are $\mathcal{S}$-linkage irrelevant.

Proof We prove the claim by induction on the number of vertices of H. Suppose first that H has an $\mathcal{S}$-linkage irrelevant vertex v. If v is not contained in any of the cycles of the nest $\mathcal{C}$, then the claim follows by the induction hypothesis applied to $H - v$. If $v \in V(C_i)$ for some $i \in [d]$, then the claim follows by the induction hypothesis applied to H/e for an edge $e \in E(C_i)$ incident with v (unless v is the only vertex of C_i, i.e., C_i is a loop, in which case v separates the vertices of $\mathcal{S}$ from the egg of $\mathcal{C}$ and the conclusion of the lemma holds trivially). Therefore, we can assume that H is $\mathcal{S}$-linkage critical, and thus $\mathrm{tw}(H) \leq f_{3.10}(k)$.

Let Δ be the closed disk in the plane bounded by C_d. Let G_0 be the subgraph of G obtained from the one drawn in Δ by deleting the edges of C_d. Let $\mathcal{P}$ be an $\mathcal{S}$-linkage in H, and let $\mathcal{P}_0$ be the set of components of $G_0 \cap \bigcup \mathcal{P}$. Then $\mathcal{P}_0$ is a set of pairwise vertex-disjoint paths in G_0 with ends in $V(C_d)$. Note that it may be the case that $|\mathcal{P}_0|$ is much larger than k, since each path of $\mathcal{P}$ may enter and leave Δ multiple times; thus, at this point, we cannot simply apply Lemma 3.11 to G_0.

Let $m = f_{3.10}(k)+2$ and let $\mathcal{P}_1$ be the set of paths from $\mathcal{P}_0$ that intersect C_{d-m+1} (and thus also all the cycles $C_{d-m+2}, \ldots, C_d$). Observe that $\mathcal{B} = \{P \cup C_{d-m+i} : P \in \mathcal{P}_1, i \in [m]\}$ is a bramble. A hitting set for $\mathcal{B}$ must intersect all paths in $\mathcal{P}_1$ or all the cycles $C_{d-m+1}, \ldots, C_d$, and thus the order of $\mathcal{B}$ is $\min(m, |\mathcal{P}_1|)$. Corollary 2.27 implies that

$$\min(m, |\mathcal{P}_1|) - 1 \leq \mathrm{tw}(H) \leq f_{3.10}(k),$$

and thus $|\mathcal{P}_1| \leq f_{3.10}(k) + 1$.

Let $\mathcal{C}_1$ be the nest $C_1, \ldots, C_{d-m+1}$ of depth $d - m + 1 \geq |\mathcal{P}_1|$, let $G_1 = \bigcup \mathcal{P}_1 \cup \bigcup \mathcal{C}_1$, and let $\mathcal{S}_1$ be the set of the pairs of ends of the paths in $\mathcal{P}_1$. Then by Lemma 3.11, G_1 contains an $\mathcal{S}_1$-linkage $\mathcal{P}_1'$ disjoint from the egg of $\mathcal{C}_1$, which is the same as the egg of $\mathcal{C}$. An $\mathcal{S}$-linkage in H disjoint from the egg of $\mathcal{C}$ is then obtained by replacing the subpaths of $\mathcal{P}$ belonging to $\mathcal{P}_1$ by those in $\mathcal{P}_1'$. □

Hence, in case that we can locate a planar part of the graph containing a deep nest, we can also locate an irrelevant vertex. Lemma 3.12 and its variants are also useful in constructions of minors in graphs (partially) drawn on a surface when we need to show that pieces of the minor obtained in different parts of the drawing do not interfere.

3.4 Linked Tree Decompositions

Let us now turn our attention to a quite different notion of linkedness in the context of tree decompositions. Let G be a graph and consider a tree decomposition (T, β) of G of optimal width $\mathrm{tw}(G)$. Observation 2.1 states that edges of T correspond to vertex separations of G. It would be convenient if all small separations were displayed in the tree decomposition, at least to the extent that we can say whether two bags of the tree decomposition are joined by many pairwise vertex-disjoint paths in G just by looking on the size of the bags that separate them in T. This is not necessarily true for every optimal tree decomposition; e.g., suppose that G is a disjoint union of a clique K of size n and an n-vertex tree H. Then $\mathrm{tw}(G) = n - 1$ and we have many options for the part of the tree decomposition containing H, but very few of them will correctly reflect the lack of connectivity in H. However, it turns out it is possible to always rearrange an optimal tree decomposition so that it has the desired property.

We say that a tree decomposition (T, β) of a graph G is ***lean*** if it has the following additional property.

(D3) For all (not necessarily distinct) nodes $x_1, x_2 \in V(T)$ and sets $X_1 \subseteq \beta(x_1)$ and $X_2 \subseteq \beta(x_2)$ of the same size k, either

- G contains an (X_1, X_2)-linkage of size k, or
- there exists a node x on the path between x_1 and x_2 in T such that $|\beta(x)| < k$.

Note that this condition is quite interesting even in the case $x_1 = x_2$, as then it shows that the bags of the decomposition must be formed by sets that are very well connected inside G.

Theorem 3.13 (Thomas [8]) *Every graph G has a lean tree decomposition of width* $\mathrm{tw}(G)$.

The proof of Theorem 3.13 is based on splitting and rearranging the bags of the tree decomposition based on the cuts between sets violating the condition (D3). The main issue is to ensure that the process actually terminates, which is achieved by showing that a technical measure of the complexity of the tree decomposition decreases with each rearrangement.

Instead of going into the details of the (rather technical) proof, let us consider a related concept in the setting of path decompositions and let us show its typical application. Let (P, β) be a path decomposition of a graph G, not necessarily of bounded width. For a positive integer k, we say that the path decomposition is ***k*-*linked*** if $|\beta(e)| = k$ for every $e \in E(P)$ and for every node $x \in V(P)$ incident with distinct edges e_1 and e_2, the graph $G[\beta(x)]$ contains a $(\beta(e_1), \beta(e_2))$-linkage of size k, which we call a *boundary linkage of x*. Note that this condition does not say anything about the end nodes y_1 and y_2 of P, which are only incident with one edge. Let f_1 and f_2 be the first and the last edge of P. Observe that the boundary linkages for the nodes in $V(P) \setminus \{y_1, y_2\}$ can be combined to a $(\beta(f_1), \beta(f_2))$-linkage $\mathcal{L}$ of size k in G, such that for every node $x \in V(P) \setminus \{y_1, y_2\}$, the intersection of $\mathcal{L}$ with $G[\beta(x)]$ is a boundary linkage of x. We say that $\mathcal{L}$ is a ***boundary linkage*** of the path decomposition.

Of course, not all path decompositions are linked. Making a path decomposition linked involves splitting and combining its bags, and it is important to ensure that we do not end up collapsing it into a trivial one. So, if we start with a decomposition with many bags, we would like to end up with one which still has many bag. However, we need to avoid the triviality of a single bag being repeated many times. We say that a general decomposition (B, β) is *proper* if for every edge $xy \in E(B)$, we have $\beta(x) \not\subseteq \beta(y)$ and $\beta(y) \not\subseteq (x)$. A path decomposition (P', β') is a *coarsening* of a path decomposition (P, β) if it is obtained from (P, β) by combining some of the consecutive bags; that is, there exists a model μ of P' in P with $V(P) = \bigcup_{x \in V(P')} V(\mu(x))$ such that $\beta'(x) = \beta(\mu(x))$ for every $x \in V(P')$. Note that a coarsening of a proper path decomposition is again proper, since if $\beta(x) \not\subseteq \beta(y)$ for an edge $xy \in E(P)$, then (D2) implies that $\beta(x) \not\subseteq \beta(Q)$ for any subpath Q of P containing y but not x.

Lemma 3.14 *If a graph G has a proper path decomposition (P, β) of adhesion at most k and with at least n^{k+1} nodes, then G also has a proper path decomposition with at least n nodes that is k'-linked for some $k' \leq k$.*

Proof We prove the claim by the induction on k. If $k = 0$, then (P, β) is already 0-linked and we are done; hence, suppose that $k \geq 1$.

If there exist at least $n^k - 1$ edges $e \in E(P)$ such that $|\beta(e)| \leq k - 1$, then combining the bags between these edges gives us a coarsening (P', β') of (P, β) with adhesion at most $k - 1$ and with at least n^k nodes. The claim then follows by the induction hypothesis. Hence, assume that there are at most $n^k - 2$ such edges. Combining the bags joined by these edges gives us a coarsening (P_1, β_1) of (P, β) with at least $n^{k+1} - n^k + 2$ nodes such that $|\beta_1(e)| = k$ for every $e \in E(P_1)$.

Let $R \subset V(P_1)$ consist of the nodes $x \in V(P_1)$ such that x is not an end of P_1 and, letting e_1 and e_2 be the edges of P_1 incident with x, the graph $G[\beta_1(x)]$ does not contain a $(\beta_1(e_1), \beta_1(e_2))$-linkage of size k. By Menger's theorem, for each such node $x \in R$ there exists a vertex separation (A_x, B_x) of $G[\beta_1(x)]$ of order at most $k - 1$ with $\beta_1(e_1) \subset A_x$ and $\beta_1(e_2) \subset B_x$. For the first vertex x_1 of P_1, let us define $B_{x_1} = \beta_1(x_1)$, and for the last vertex x_2, let $A_{x_2} = \beta_1(x_2)$. Let us partition P_1 into $|R| + 1$ pairwise edge-disjoint subpaths $H_0, \ldots, H_{|R|}$ intersecting exactly in the vertices of R. For $i \in [m]$, let s_i and t_i be the first and the last node of the subpath H_i.

Suppose next that $|R| \geq n^k - 1$. Let (P_2, β_2) be the path decomposition of G where $P_2 = y_0 y_1 \ldots y_{|R|}$ and $\beta_2(y_i) = B_{s_i} \cup A_{t_i} \cup \beta_1(H_i - \{s_i, t_i\})$ for each $i \in \{0, \ldots, |R|\}$. This path decomposition has $|R| + 1 \geq n^k$ nodes and adhesion at most $k - 1$. Moreover, note that for $i \in [|R|]$, we have $A_{s_i} \not\subseteq B_{s_i}$ and $B_{s_i} \not\subseteq A_{s_i}$, and thus $\beta_2(y_{i-1}) \not\subseteq \beta_2(y_i)$ and $\beta_2(y_i) \not\subseteq \beta_2(y_{i-1})$. Consequently, the path decomposition (P_2, β_2) is proper. The conclusion of the lemma then follows from the induction hypothesis.

Therefore, suppose that $|R| \leq n^k - 2$, and thus there exists $i \in \{0, \ldots, |R|\}$ such that the path H_i has at least

$$\frac{\|P_1\|}{|R| + 1} + 1 \geq \frac{n^{k+1} - n^k + 1}{n^k - 1} + 1 \geq n$$

nodes. Combining the bags not in H_i with the bags of the first and last node of H_i gives us a coarsening (H_i, β_3) of (P_1, β_1) with at least n nodes. Moreover, for each internal node x of H_i, we have $\beta_3(x) = \beta_1(x)$ and $x \notin R$, and thus if e_1 and e_2 are the incident edges of H_i, the graph $G[\beta_3(x)]$ contains a $(\beta_3(e_1), \beta_3(e_2))$-linkage of size k. Therefore, (H_i, β_3) is a proper k-linked path decomposition of G with at least n nodes. □

Just having a proper path decomposition is typically not good enough for applications; instead, we need a decomposition where each bag contains some "private" vertices not belonging to any other bag. We can achieve this by further coarsening the path decomposition (observe that this preserves k-linkedness). For a node x of a general decomposition (B, β) of a graph G, let $\beta^\circ(x)$ denote the set of vertices of G that belong to $\beta(x)$ and no other bag of the decomposition; note that

$$\beta^\circ(x) = \beta(x) \setminus \bigcup_{xy \in E(B)} \beta(y).$$

We say that a path decomposition (P, β) of a graph G has *non-empty interiors* if $\beta^\circ(x) \neq \emptyset$ for every node $x \in V(P)$ different from the ends of P.

Lemma 3.15 *Let $n \geq 2$ and $k \geq 0$ be integers, let G be a graph, and let (P, β) be a path decomposition of adhesion at most k of G with at least $(k+1)n$ nodes. If the path decomposition (P, β) is proper, then it has a coarsening with non-empty interiors and at least n nodes.*

Proof For each subpath Q of P that does not contain either of the ends of P, let us define $e_1(Q)$ and $e_2(Q)$ to be the edges of $E(P) \setminus E(Q)$ incident with the ends of Q. Let us partition P into pairwise vertex-disjoint subpaths $Q_1, Q_2, \ldots, Q_t$ so that

- Q_1 consists of the first vertex of P,
- for $i = 2, \ldots, t-1$, the path Q_i is the shortest initial segment of $P - V(Q_1 \cup \ldots \cup Q_{i-1})$ such that $\beta(Q_i) \not\subseteq \beta(e_1(Q_i)) \cup \beta(e_2(Q_i))$, and
- Q_t does not have such an initial segment.

Consider $i \in \{2, \ldots, t-1\}$ and let $Q_i = x_1 \ldots x_m$. Since the path decomposition (P, β) is proper, there exists a vertex $v_1 \in \beta(x_1) \setminus \beta(e_1(Q_i))$ and for $j \in \{2, \ldots, m\}$, there exists a vertex $v_j \in \beta(x_j) \setminus \beta(x_{j-1})$. The vertices $v_1, \ldots, v_m$ are pairwise different and do not belong to $\beta(e_1(Q_i))$ by the property (D2). Recall that Q_i is the shortest initial segment of $P - V(Q_1 \cup \ldots \cup Q_{i-1})$ such that $\beta(Q_i) \not\subseteq \beta(e_1(Q)) \cup \beta(e_2(Q))$, and thus $\beta(Q_i - x_m) \subseteq \beta(e_1(Q_i)) \cup \beta(x_{m-1}x_m)$. Hence, the vertices $v_1, \ldots, v_{m-1}$ belong to $\beta(Q_i - x_m) \setminus \beta(e_1(Q_i)) \subseteq \beta(x_{m-1}x_m)$. Since the path decomposition (P, β) has adhesion at most k, it follows that $m \leq k+1$. Similarly, $|Q_t| \leq k+1$. Therefore, $t \geq 1 + \frac{|P|-1}{k+1} \geq n$. Contracting each of the paths $Q_1, \ldots, Q_t$ results in a coarsening of (P, β) with non-empty interiors. □

We now have all the tools we need to demonstrate how linkedness can be used to construct minors. For a motivation, note that examples based on random graphs prove that there are K_t-minor-free graphs of minimum degree $\Omega(t\sqrt{\log t})$. However, these graphs are themselves relatively small, with $\Theta(t\sqrt{\log t})$ vertices. Can there be arbitrarily large K_t-minor-free graphs with superlinear minimum degree? Of course, we can cheat and take a disjoint union of small dense K_t-minor-free graphs; but are there such graphs that additionally are say t-connected? Norin and Thomas [5] proved that this is not the case: Every sufficiently large t-connected graph of minimum degree at least $t+1$ contains K_t as a minor. In fact, they proved an even stronger statement, which we discuss in more detail in Chap. 10.

The proof of this result of Norin and Thomas has not yet been fully published and in any case is too complicated for us to present. Instead, we prove a related claim restricted to graphs of bounded treewidth, postulating the existence of a $K_{t,n}$ minor for any fixed n, but assuming $2t$-connectivity and large minimum degree. This special case will form a basis for the proof of a similar result for graphs without the treewidth restriction, which we give in Chap. 10. To facilitate this strengthening, we need to prove a slightly more technical version of the following lemma than what we would strictly need for the application to graphs of bounded treewidth. On the first reading, I suggest to only consider the following lemma with the additional restrictions that $X=\emptyset$ and the set U consists of all nodes of the path P except for its ends.

Lemma 3.16 *For any integers $t, n \geq 1$ and $k \geq t$, there exists an integer n_0 such that the following claim holds. Let G be a graph, let (P, β) be a k-linked path decomposition of G, and let X be a set of vertices of G of degree at least $2t+1$. Let U be the set of nodes $x \in V(P)$ different from the ends of P such that*

- *$\beta^\circ(x) \neq \emptyset$ and $|\beta^\circ(x) \cap X| \leq 1$,*
- *all vertices in $\beta^\circ(x) \setminus X$ have degree at least $30t+1$, and*
- *there is no vertex separation (R, S) of $G[\beta(x)]$ of order less than $2t$ with $(\beta(x) \cap X) \cup (\beta(x) \setminus \beta^\circ(x)) \subseteq R$ and $S \not\subseteq R$.*

If $|U| \geq n_0$, then $K_{t,n}$ is a minor of G.

Proof We choose $n_0 = \binom{k}{t}n$.

Let $\mathcal{L} = \{L_1, \ldots, L_k\}$ be a boundary linkage of (P, β), Consider any node $x \in U$, let e_1 and e_2 be the incident edges of P, and let $B = \beta(e_1) \cup \beta(e_2)$ and $K = G[\beta(x) \setminus B] = G[\beta^\circ(x)]$. For $i \in [k]$, let s_i and t_i be the vertices of L_i in $\beta(e_1)$ and $\beta(e_2)$, respectively. We are going to define

(i) the *type* D of x, a subset of t indices in $[k]$,
(ii) the *focus* f of x, a vertex of $\beta^\circ(x)$,
(iii) the *realization* of x, a subgraph of $G[\beta(x)]$ containing the focus and the vertices s_i and t_i for $i \in D$ and otherwise disjoint from B, and
(iv) the *routing* of x, an $\{(s_i, t_i) : i \in D\}$-linkage $\mathcal{L}_x$ in the realization disjoint from the focus such that for each $P \in \mathcal{L}_x$, the realization contains a path from the focus f to P disjoint from all other paths of $\mathcal{L}_x$.

Suppose we succeed in this for every node $x \in U$. Since there are only $\binom{k}{t}$ possible types and $|U| \geq n_0$, there exists a set $A \subseteq U$ of n nodes of the same type D. Let $\mathcal{L}'$ be the union of the routings of the nodes in A and of the paths in $\{L_i \cap G[\beta(x)] : i \in D\}$ for $x \in V(P) \setminus A$. Then $\mathcal{L}'$ is formed by t pairwise vertex-disjoint paths. Let T be the set of vertices obtained by contracting each of the paths of $\mathcal{L}'$ to a single vertex. For each $x \in A$, the condition (iv) implies that the realization of x contains paths from the focus f of x to all vertices of T intersecting only in f. Contracting all but the last edge of each of the paths gives us the desired minor of $K_{t,n}$ in G.

It remains to show how to obtain the type, focus, realization, and routing for each node $x \in U$. Let us first consider the simpler case that there exists a vertex $f \in V(K)$ with at least $2t + 1$ neighbors in B. Since each path of $\mathcal{L}$ intersects B in at most two vertices, there exists a set $D \subset [k]$ of size t such that for every $i \in D$, the vertex f does not lie on the path L_i, but has a neighbor in $V(L_i) \cap B$. In this case, let us define the *type* of x to be D and the *focus* of x to be the vertex f. Let the *routing* of x consist of the paths $L_i \cap G[\beta(x)]$ for $i \in D$, and let the *realization* of x consist of the union of the paths of the routing, the vertex of f, and the edges of G from f to these paths.

Let us now consider the more involved case that no vertex of K has more than $2t$ neighbors in B. In particular, if there exists a vertex $z \in V(K) \cap X$, then since $\deg z \geq 2t + 1$, it follows that z has a neighbor in K. Since $|\beta^{\circ}(x) \cap X| \leq 1$, this neighbor does not belong to X, and thus $K - X$ is non-empty. Since each vertex of $V(K) \setminus X$ has degree at least $30t+1$ in G and at most $2t$ neighbors in B, we conclude that $K - X$ has minimum degree at least $28t$. By Theorem 3.4 and Corollary 3.6, $K - X$ has a $2t$-linked induced subgraph M of minimum degree more than $14t$. By the assumptions, $G[\beta(x)]$ has no vertex separation (R, S) of order less than $2t$ with $B \cup (\beta(x) \cap X) \subseteq R$ and $V(M) \subseteq S$, and thus by Menger's theorem, there exists an $(V(M), B \cup (\beta(x) \cap X))$-linkage $\mathcal{Q}$ of size $2t$ in $G[\beta(x)]$. We can assume that the paths of $\mathcal{Q}$ intersect M only in their ends.

Choose $\mathcal{Q}$ so that none of its paths end in $V(K) \cap X$ if possible, and subject to that with as few edges outside of $\mathcal{L}$ as possible. Consider a path $L_i \in \mathcal{L}$. If L_i is intersected by more than one path of $\mathcal{Q}$, then we claim that both s_i and t_i are ends of paths of $\mathcal{Q}$: If say s_i is not, we could choose the path $Q \in \mathcal{Q}$ with the nearest intersection with L_i to s_i and reroute Q along L_i to s_i, thus decreasing the number of edges of Q outside of $\mathcal{L}$ or making Q not end in $V(K) \cap X$. Similarly, if L_i is intersected by exactly one path $Q \in \mathcal{Q}$, then Q ends in s_i or t_i.

Let D_1 be the set of indices i such that L_i is intersected by exactly one path $Q \in \mathcal{Q}$ (ending in s_i or t_i, where possibly $s_i = t_i$), and let D_2 be the set of indices i such that L_i is intersected by more than one path from $\mathcal{Q}$ (and thus $s_i \neq t_i$ and both s_i and t_i is an end of a path from $\mathcal{Q}$). Since at most one of the paths of $\mathcal{Q}$ ends in $V(K) \cap X$ and all other intersect $\mathcal{L}$ at least in their ends, we have $|D_1| + 2|D_2| \geq 2t - 1$, and thus $|D_1| + |D_2| \geq t$.

Let D be a subset of $D_1 \cup D_2$ of size t; we define the *type* of x to be D. For $i \in D$, choose vertices $a_i, b_i \in V(M)$ as follows:

- If $i \in D_2$, then let a_i and b_i be the first vertices of the paths of $\mathcal{Q}$ ending in s_i and t_i.
- If $i \in D_1$ and L_i intersects M, then let a_i and b_i be the first and the last vertex of M on L_i.
- If $i \in D_1$ and L_i does not intersect M, then let $a_i = b_i$ be the first vertex of the path of $\mathcal{Q}$ ending on L_i.

For each $i \in D$, choose a neighbor c_i of a_i in M different from all vertices chosen so far. Choose $i_0 \in D$ arbitrarily and a vertex d_{i_0} in M different from already chosen vertices, and for $i \in D \setminus \{i_0\}$ select a neighbor d_i of d_{i_0} in M different from all vertices chosen so far. Note that this is possible, since M has minimum degree more than $14t$. Since M is $2t$-linked, it contains an $\{(a_i, b_i), (c_i, d_i) : i \in D\}$-linkage $\mathcal{R}$. We define the *focus* of x to be the vertex d_{i_0}, and the *realization* of x to consist of

- the paths of $\mathcal{R}$, the edges $a_i c_i$ for $i \in D$ and $d_i d_{i_0}$ for $i \in D \setminus \{i_0\}$,
- the paths of $\mathcal{Q}$ from a_i to s_i and b_i to t_i for each $i \in D \cap D_2$,
- $L_i \cap G[\beta(x)]$ and the path of $\mathcal{Q}$ from $a_i = b_i$ to L_i for each $i \in D \cap D_1$ such that L_i is disjoint from M, and
- the subpaths of L_i from s_i to a_i and from b_i to t_i for each $i \in D \cap D_1$ such that L_i intersects M.

Finally, the *routing* of x is defined to consist of

- the path of $\mathcal{R}$ from a_i to b_i and the paths of $\mathcal{Q}$ from a_i to s_i and b_i to t_i for each $i \in D \cap D_2$,
- $L_i \cap G[\beta(x)]$ for each $i \in D \cap D_1$ such that L_i is disjoint from M, and
- the path of $\mathcal{R}$ from a_i to b_i and the subpaths of L_i from s_i to a_i and from b_i to t_i for each $i \in D \cap D_1$ such that L_i intersects M.

Observe that these choices satisfy the conditions (i)–(iv), concluding the argument. □

For now, we are only going to need the following consequence.

Corollary 3.17 *For any integers $t, n \geq 1$ and $k \geq t$, there exists an integer n_1 such that the following claim holds. Let G be a $2t$-connected graph of minimum degree at least $30t + 1$ and let (P, β) be a proper path decomposition of G of adhesion at most k. If $|P| \geq n_1$, then $K_{t,n}$ is a minor of G.*

Proof Let n_0 be the maximum of the constants from Lemma 3.16 for t, n, and $k' \leq k$, and choose $n_1 \gg n_0, k$. By Lemmas 3.14 and 3.15, for some $k' \leq k$ we obtain a k'-linked path decomposition (P', β') of G with non-empty interiors such that $|P'| \geq n_0 + 2$. The claim now follows by Lemma 3.16 with $X = \emptyset$ and U consisting of all nodes of the path P' except for its ends. □

Using this corollary, it is now quite easy to prove the promised result for graphs of bounded treewidth. The proof showcases a useful trick for turning a tree decomposition into a long path decomposition.

Theorem 3.18 *For any integers $t, n \geq 1$ and $k \geq t$, there exists an integer n_2 such that the following claim holds. If a $2t$-connected graph G of minimum degree at least $30t + 1$ and treewidth at most k has at least n_2 vertices, then it contains $K_{t,n}$ as a minor.*

Proof Let n_1 be the constant from Corollary 3.17 for t, n, and $k + 1$, and choose $n_2 \gg n_1, k$. Let $n \geq n_2$ be the number of vertices of G. Let (T, β_0) be a tree decomposition of G of width at most k. Without loss of generality, we can assume that the tree decomposition is proper, as otherwise we can contract edges xy of T such that $\beta_0(x) \subseteq \beta_0(y)$, preserving only the larger bag $\beta_0(y)$. Note that $|T| \geq n/(k+1) \gg n_1$. Since a tree of maximum degree $\Delta \geq 2$ and radius r has at most $2\Delta^r$ vertices, we conclude that T either contains a path P with at least n_1 nodes, or has a node x_0 of degree at least n_1,

In the former case, for each $y \in P$, let T_y be the component of $T - E(P)$ containing y, and let $\beta(y) = \beta_0(T_y)$; observe that $|\beta(e)| = |\beta_0(e)| \leq k$ for every edge $e \in E(P)$. In the latter case, let P a path with $\deg x_0 \geq n_1$ nodes, let each node $y \in V(P)$ correspond to a component T_y of $T - x_0$, and let $\beta(y) = \beta_0(T_y) \cup \beta_0(y)$; we have $|\beta(e)| = |\beta_0(x_0)| \leq k + 1$ for every edge $e \in E(P)$.

In either case, (P, β) is a proper path decomposition of G of adhesion at most $k + 1$, and thus the claim follows by Corollary 3.17. □

Let us remark that the large minimum degree assumption in this result is necessary (though the bound $30t + 1$ clearly is not the best possible): As proved in [1], the strong product of K_t with an arbitrarily long cycle (which is $2t$-connected and $(3t - 1)$-regular) does not contain $K_{t,n}$ as a minor for any $n \geq 2t + 1$.

References

1. T. Böhme, J. Maharry, B. Mohar, $K_{a,k}$ minors in graphs of bounded tree-width. J. Combin. Theory Ser. B **86**, 133–147 (2002)
2. R. Karp, Reducibility among combinatorial problems, in *Complexity of Computer Computations*, ed. by R. Miller, J. Thatcher (Plenum Press, New York, 1972), pp. 85–103
3. K. Kawarabayashi, P. Wollan, A shorter proof of the graph minor algorithm: the unique linkage theorem, in *Proceedings of the Forty-Second ACM Symposium on Theory of Computing* (2010), pp. 687–694
4. D. Mayhew, G. Whittle, S.H.M. van Zwam, The structure of graphs with a vital linkage of order 2. J. Graph Theory **71**, 176–181 (2012)
5. S. Norin, R. Thomas, K_t-minors in large t-connected graphs (in preparation, 2009)
6. N. Robertson, P.D. Seymour, Graph minors. XIII. The disjoint paths problem. J. Combin. Theory Ser. B **63**, 65–110 (1995)
7. N. Robertson, P.D. Seymour, Graph minors. XXI. Graphs with unique linkages. J. Combin. Theory Ser. B **99**, 583–616 (2009)
8. R. Thomas, A Menger-like property of tree-width: the finite case. J. Combin. Theory Ser. B **48**, 67–76 (1990)
9. R. Thomas, P. Wollan, An improved linear edge bound for graph linkages. Eur. J. Combin. **26**, 309–324 (2005)
10. R. Thomas, P. Wollan, The extremal function for 3-linked graphs. J. Combin. Theory Ser. B **98**, 939–971 (2008)

Chapter 4
Graphs on Surfaces

Both in the proof of the Minor Structure Theorem and in its applications, we often need to to show that a fixed graph H appears in a given graph G drawn on a surface Σ as a (possibly rooted) minor. There are two kinds of obstructions that can prevent this:

- *Topological*: If H cannot be drawn on Σ, then it clearly cannot be a minor of G.
- *Connectivity*: If there is a linkage of size k between subsets of roots in H and the corresponding subsets of vertices are separated in G by a cut of size less than k, then H cannot be a rooted minor of G.

Furthermore, we can "mix" the two kinds of obstructions; e.g., if G and H have the same genus and the genus of G can be decreased by deleting a few vertices, but this is not the case for H, then H cannot be a minor of G. The goal of this chapter is to present results showing that if H satisfies the topological constraints and G satisfies the connectivity constraints with a wide margin, then H is a minor of G.

Rather than directly considering rooted minors, it turns out to be convenient to start with a simpler problem (which nevertheless generalizes the problem of finding a linkage between prescribed pairs of vertices). Let $\mathcal{S} = \{S_1, \ldots, S_k\}$ be a system of non-empty pairwise disjoint sets of vertices of a graph G; we call such a system a ***root partition***, and the elements of $\bigcup \mathcal{S}$ are *roots*. An *$\mathcal{S}$-linkage* in G is a system $\{T_1, \ldots, T_k\}$ of pairwise vertex-disjoint connected subgraphs of G such that $S_i \subseteq V(T_i)$ for each $i \in [k]$. The *incidence graph* $I_{\mathcal{S}}$ of $\mathcal{S}$ is the bipartite graph with vertex set $\{v_S : S \in \mathcal{S}\} \cup \bigcup \mathcal{S}$, where for $S \in \mathcal{S}$, the vertex v_S is adjacent exactly to the elements of S. Note that $I_{\mathcal{S}}$ is a star forest. In case that G is drawn in a subset Π of a surface, we say that $\mathcal{S}$ is ***topologically feasible in*** Π if $I_{\mathcal{S}}$ has a drawing in Π with the vertices of $\bigcup \mathcal{S}$ represented by the same points as in the drawing of G. We say that such a drawing is a ***topological realization*** of $\mathcal{S}$, and for each $S \in \mathcal{S}$, we say that the component of $I_{\mathcal{S}}$ containing v_S *represents* S and that v_S is its *center*.

© The Author(s), under exclusive license to Springer Nature Switzerland AG 2025

Z. Dvořák, *Graph Minors*, Springer Monographs in Mathematics,
https://doi.org/10.1007/978-3-031-87469-7_4

Observation 4.1 *Let G be a graph drawn on a surface Π (possibly with boundary), and let $\mathcal{S}$ be a root partition of a subset of $V(G)$. If G contains an $\mathcal{S}$-linkage, then $\mathcal{S}$ is topologically feasible in Π.*

Proof Without loss of generality, we can assume that each vertex $v \in \bigcup \mathcal{S}$ that does not form a part of $\mathcal{S}$ by itself has degree exactly one in G and its neighbor in G does not belong to $\bigcup \mathcal{S}$: Otherwise, we can shift the drawing of G slightly away from the point p representing v, add a new vertex v' drawn at p and adjacent only to v, and replace v by v' in $\mathcal{S}$ to form a root partition $\mathcal{S}'$ of vertices of the modified graph G'. An $\mathcal{S}$-linkage in G clearly can be turned into an $\mathcal{S}'$-linkage in G'. Conversely, since v does not form a part of $\mathcal{S}$ by itself, any $\mathcal{S}'$-linkage in G' contains v and v' in the same connected subgraph, and thus corresponds to an $\mathcal{S}$-linkage in G.

Let $\mathcal{S} = \{S_1, \ldots, S_k\}$, and consider now an $\mathcal{S}$-linkage $\{T_1, \ldots, T_k\}$ in G. For each $i \in [k]$ with $|S_i| > 1$, contract $T_i - S_i$ to a single vertex, and let T_i' be the resulting star. For $i \in [k]$ with $|S_i| = 1$, let T_i' be the single-vertex graph with vertex set S_i. The resulting star forest $T = \bigcup_{i\in[k]} T_i'$ is a minor of G with the roots of $\mathcal{S}$ fixed, and thus drawn on the surface Π with vertices of $\bigcup \mathcal{S}$ represented by the same points as in G. We can turn T to a topological realization of $\mathcal{S}$ by, for each part $S_i = \{v_i\}$ of size one, adding a neighbor v_{S_i} of v_i representing S_i. □

Therefore, topological feasibility is a necessary condition for the existence of an $\mathcal{S}$-linkage. Let us now look for sufficient conditions subject to various additional restrictions on G and $\mathcal{S}$.

4.1 Linkage Across a Disk or Cylinder

Many arguments about graphs on surfaces start by first solving the special case of the restriction of the problem to the disk. Hence, let us consider the $\mathcal{S}$-linkage problem in the following setting: Let G be a graph drawn in the closed disk Δ with G-normal boundary $\mathrm{bd}(\Delta)$, and let $\mathcal{S}$ be a root partition of the vertices of G drawn in $\mathrm{bd}(\Delta)$. Note that in this case, $\mathcal{S}$ is topologically feasible in Δ exactly if no two parts of $\mathcal{S}$ "cross"; i.e., $\mathcal{S}$ is *not* topologically feasible exactly if there exist roots s_1, s_2, t_1, t_2 drawn in $\mathrm{bd}(\Delta)$ in order such that $s_1, t_1 \in S_1$ and $s_2, t_2 \in S_2$ for distinct parts $S_1, S_2 \in \mathcal{S}$.

The main result of this section is that in this special case, the following simple connectivity condition together with the topological feasibility is sufficient to ensure the existence of an $\mathcal{S}$-linkage: For a separation (A, B) of G, we say that $S \in \mathcal{S}$ is *split by* (A, B) if $S \cap V(A) \neq \emptyset \neq S \cap V(B)$. Note that this implies that every connected subgraph T with $S \subseteq V(T)$ necessarily intersects $A \cap B$. Let $\tau_{\mathcal{S}}(A, B)$ denote the number of elements of $\mathcal{S}$ that are split by (A, B). We say that G satisfies the ***basic $\mathcal{S}$-connectivity condition*** if the order of every separation (A, B) of G is at least $\tau_{\mathcal{S}}(A, B)$. Clearly, if G has an $\mathcal{S}$-linkage, then it has this property.

Actually, in the disk case it even suffices to consider only a special kind of separations. An ***I-arc*** in Δ is a simple curve with ends in $\mathrm{bd}(\Delta)$ and otherwise drawn in the interior of Δ. A G-normal I-arc γ splits Δ into two closed disks Δ_A and Δ_B (intersecting exactly in γ). Let A and B be the subgraphs of G drawn in Δ_A and Δ_B; then (A, B) is a separation of G, whose order is equal to the number of intersections of γ with G. We call any separation of G that can be obtained in this way an *I-arc separation.* It turns out that it suffices to enforce the $\mathcal{S}$-connectivity condition for I-arc separations. Note that (A, B) splits a set $S \in \mathcal{S}$ if and only if S intersects both Δ_A and Δ_B, i.e., S has vertices in both closed arcs of $\mathrm{bd}(\Delta)$ between the ends of γ.

Theorem 4.2 *Let G be a graph drawn in the closed disk Δ with G-normal boundary* $\mathrm{bd}(\Delta)$, *and let $\mathcal{S}$ be a root partition of the vertices of G drawn in* $\mathrm{bd}(\Delta)$. *If $\mathcal{S}$ is topologically feasible in Δ, then the following claims are equivalent:*

(i) G satisfies the basic $\mathcal{S}$-connectivity condition.
(ii) Every I-arc separation (A, B) of G has order at least $\tau_{\mathcal{S}}(A, B)$.
(iii) G contains an $\mathcal{S}$-linkage.

Proof The implications (iii) $\Rightarrow$ (i) $\Rightarrow$ (ii) are clear, and thus it suffices to prove the implication (ii) $\Rightarrow$ (iii). We prove this implication by induction on $|G| + \|G\|$.

Suppose first that there exists an I-arc separation (A, B) of G such that $|A \cap B| = \tau_{\mathcal{S}}(A, B)$ and $A \neq G \neq B$. Let γ be the corresponding I-arc splitting the disk Δ into subdisks Δ_A and Δ_B intersecting G in A and B, respectively. Let s and t be the ends of γ and let σ_A and σ_B be the closed arcs of $\mathrm{bd}(\Delta)$ from s to t. Let $S_1, \ldots, S_m \in \mathcal{S}$ be the sets split by (A, B), listed in order in which σ_A first encounters them. Note that σ_B first encounters all vertices of $S_1 \cap \sigma_B$, then all vertices of $S_2 \cap \sigma_B, \ldots,$ and finally all vertices of $S_m \cap \sigma_B$, as otherwise $\mathcal{S}$ would not be topologically feasible in Δ. In particular, if s is a vertex of G, then $s \in S_1$, and if t is a vertex of G, then $t \in S_m$. Let $v_1, \ldots, v_m$ be the vertices of G encountered by the I-arc γ in order. For $i \in [m]$, let $S_{i,A} = (S_i \cap \sigma_A) \cup \{v_i\}$ and $S_{i,B} = (S_i \cap \sigma_B) \cup \{v_i\}$. Let $\mathcal{S}_A$ consist of the sets $S_{i,A}$ for $i \in [m]$ and of the sets $S \in \mathcal{S}$ not split by (A, B) and contained in σ_A. Let $\mathcal{S}_B$ consist of the sets $S_{i,B}$ for $i \in [m]$ and of the sets $S \in \mathcal{S}$ not split by (A, B) and contained in σ_B. If A contains an $\mathcal{S}_A$-linkage and B contains an $\mathcal{S}_B$-linkage, then their combination gives an $\mathcal{S}$-linkage in G.

Suppose for a contradiction that say A does not contain an $\mathcal{S}_A$-linkage. Note that $\mathcal{S}_A$ is topologically feasible in Δ_A, since the sets of $\mathcal{S}_A$ intersect γ in the same order as the corresponding sets of $\mathcal{S}$ intersect σ_B. By the induction hypothesis, it follows that there exists an I-arc separation (C, D) of A of order less than $\tau_{\mathcal{S}_A}(C, D)$. Let γ' be the corresponding I-arc of Δ_A splitting Δ_A into disks Δ_C and Δ_D containing C and D, respectively. We can assume that the labels of C and D are chosen so that $\Delta_C \cap \sigma_A$ is non-empty and connected. If both ends of γ' are in σ_A, then $(C, D \cup B)$ is an I-arc separation of G of order $|C \cap D| < \tau_{\mathcal{S}_A}(C, D) = \tau_{\mathcal{S}}(C, D \cup B)$, which is a contradiction. If both ends of γ' are in γ, then rerouting γ along γ' gives us an I-arc separation $(C, D \cup B)$ of G which splits the sets $S_1, \ldots, S_m$ but has order less than m, again a contradiction.

Finally, consider the case that γ' has one end in the interior of σ_A and the other end in the interior of γ. Let $c = |V(C) \cap \{v_1, \ldots, v_m\}|$, $d = |V(D) \cap \{v_1, \ldots, v_m\}|$, let c' be the number of sets in $\{S_i : i \in [m]\}$ intersecting C, and let d' be the number of such sets intersecting D. Note that $m \leq c+d \leq m+1$ and $m \leq c'+d' \leq m+1$, since at most one vertex may belong to $C \cap D \cap B$ and at most one set S_i may intersect both C and D. By symmetry between C and D, we can assume that

($\star$) $c' \geq c$, and if $c+d = m$ and $c'+d' = m+1$, then $c' > c$.

We can choose the labels so that $V(C) \cap \{v_1, \ldots, v_m\} = \{v_1, \ldots, v_c\}$ and the sets $S_1, \ldots, S_{c'}$ intersect C. The I-arc formed by γ' and the part of γ between the point $\gamma \cap \gamma'$ and s shows that $(C, D \cup B)$ is an I-arc separation of G. Since $c' \geq c$, the sets $S_1, \ldots, S_{c-1}$ are split by $(C, D \cup B)$, but the sets $S_{1,A}, \ldots, S_{c-1,A}$ are not split by (C, D). Moreover, consider any set $S \in \mathcal{S}_A$ split by (C, D).

- If $S = S_{i,A}$ for $i \in [m] \setminus [c]$, then since $v_i \notin V(C)$, we have $S_i \cap V(C) \neq \emptyset$, and thus S_i is split by $(C, D \cup B)$.
- If $S = S_{c,A}$, then $S_c \cap V(C) \neq \emptyset$ since $c' \geq c$, and again, it follows that S_c is split by $(C, D \cup B)$.
- If $S \in \mathcal{S}_A \setminus \{S_{i,A} : i \in [m]\}$, then S belongs to $\mathcal{S}$ and intersects both C and D, and thus it is split by $(C, D \cup B)$ as well.

We conclude that the number of sets of $\mathcal{S}$ split by $(C, D \cup B)$ is by at least $c-1$ larger than the number of sets of $\mathcal{S}_A$ split by (C, D), and thus $\tau_{\mathcal{S}}(C, D \cup B) \geq c - 1 + \tau_{\mathcal{S}_A}(C, D)$. Recall that $\tau_{\mathcal{S}_A}(C, D) > |C \cap D|$; it follows that

$$\tau_{\mathcal{S}}(C, D \cup B) \geq c - 1 + \tau_{\mathcal{S}_A}(C, D) \geq c + |C \cap D| \geq |C \cap (D \cup B)|. \quad (4.1)$$

By the assumptions, the order of $(C, D \cup B)$ is at least $\tau_{\mathcal{S}}(C, D \cup B)$, and thus all the inequalities in (4.1) have to hold with equality. In particular $|C \cap (D \cup B)| = c + |C \cap D|$, and thus $v_c \notin V(D)$; this implies that $c+d = m$. Moreover, $\tau_{\mathcal{S}}(C, D \cup B) = c - 1 + \tau_{\mathcal{S}_A}(C, D)$, and thus $S_{c,A}$ must be split by (C, D), as otherwise S_c contributes to $\tau_{\mathcal{S}}(C, D \cup B)$ but not to $\tau_{\mathcal{S}_A}(C, D)$. Since $v_c \notin V(D)$, it follows that S_c intersects D in addition to C, and thus $c' = c$ and $c'+d' = m+1$. However, this contradicts ($\star$).

Therefore, we can assume that

($\star\star$) $\tau_{\mathcal{S}}(A, B) < |A \cap B|$ holds for every I-arc separation (A, B) of G such that $A \neq G \neq B$.

Suppose now that G has a vertex v distinct from the roots. We claim that $G - v$ satisfies the condition (ii). Indeed, consider an I-arc separation (A', B') of $G - v$, given by a $(G - v)$-normal I-arc γ'. In case that $A' \neq G - v \neq B'$, we modify γ' as follows: Whenever γ' enters the face of $G - v$ containing v, we redirect it to pass through v without intersecting any edges of G. Afterwards, we simplify the resulting curve by eliminating the parts between its self-intersections, and denote the resulting G-normal I-arc by γ. Note that γ intersects G only in v and vertices in which γ' intersects $G - v$, and thus for the I-arc separation (A, B) of G defined by γ, the condition ($\star\star$) gives

$$\tau_{\mathcal{S}}(A', B') = \tau_{\mathcal{S}}(A, B) \leq |A \cap B| - 1 \leq |A' \cap B'|.$$

In case that $A' = G - v$, we have $\tau_{\mathcal{S}}(A', B') \leq |B'| = |A' \cap B'|$. Similarly $\tau_{\mathcal{S}}(A', B') \leq |A' \cap B'|$ holds if $B' = G - v$. We conclude that every I-arc separation (A', B') of $G - v$ has order at least $\tau_{\mathcal{S}}(A', B')$, and thus the induction hypothesis implies that there exists an $\mathcal{S}$-linkage in $G - v$, and also in G. Hence, we can assume that all vertices of G are roots.

If G is not connected, then there exists an I-arc γ disjoint from G such that each part of $\Delta - \gamma$ contains a non-empty subgraph of G. Then the corresponding I-arc separation (A, B) of G has order 0 and $\tau_{\mathcal{S}}(A, B) \geq 0$, contradicting ($\star\star$). Therefore, G is connected. If $|\mathcal{S}| = 1$, this implies that G itself forms an $\mathcal{S}$-linkage, and thus suppose that $|\mathcal{S}| \geq 2$. Then there exists an edge $e = uv$ with $u \in V(S_1)$ and $v \in V(S_2)$ for distinct $S_1, S_2 \in \mathcal{S}$. In this case, an I-arc intersecting $\mathrm{bd}(\Delta)$ in u and v and drawn next to e gives an I-arc separation (A, B) with $|A \cap B| = |\{u, v\}| = 2 \leq \tau_{\mathcal{S}}(A, B)$, since both S_1 and S_2 are split by (A, B). Since we can freely decide which part of the separation contains the edge e, this contradicts ($\star\star$) unless $V(G) = \{u, v\}$ and $E(G) = \{e\}$, in which case we cannot choose A and B so that $A \neq G \neq B$. However, in that case $\mathcal{S} = \{S_1, S_2\}$, $S_1 = \{u\}$, $S_2 = \{v\}$, and an $\mathcal{S}$-linkage trivially exists. □

Note that it is possible to obtain a polynomial-time algorithm for the $\mathcal{S}$-linkage problem across a disk based on the presented proof of Theorem 4.2; e.g., to verify the condition ($\star\star$) or to find an I-arc separation violating it, we just need to find I-arcs with prescribed ends (with $O(k^2)$ choices of the ends, where k is the number of roots) and intersecting G in as few vertices as possible, which easily translates to finding shortest paths in the radial graph of G. However, there is a simpler and more efficient algorithm for the problem, see [5] for details.

The basic $\mathcal{S}$-connectivity condition is no longer sufficient when we consider graphs drawn in the cylinder. To see that, consider e.g. a cylindrical grid with boundary cycles $s_1 \ldots s_k$ and $t_1 \ldots t_k$, where for $i \in [k]$, s_i and t_i are in the same column, and let $\mathcal{S} = \{\{s_i, t_{i+1}\} : i \in [k]\}$, where $t_{k+1} = t_1$. The graph does not contain an $\mathcal{S}$-linkage, even though it is topologically feasible and basic $\mathcal{S}$-connectivity holds. However, a somewhat more complicated necessary-and-sufficient condition combining connectivity and topological considerations was found in [5]. Rather than formulating it, let us just note that it gives a way to test the existence of an $\mathcal{S}$-linkage in this case.

Theorem 4.3 (Robertson and Seymour [5]) *There exists a polynomial-time algorithm that, given a graph G drawn in the disk or the cylinder and a root partition $\mathcal{S}$ of the vertices of G drawn in the boundary of the surface, finds an $\mathcal{S}$-linkage in G or correctly decides that no $\mathcal{S}$-linkage in G exists.*

For other surfaces, the situation is less clear, partly because the cuts in them have more complicated structure. It seems plausible that a more elaborate condition similar to the one of Theorem 4.2 is also necessary and sufficient on any fixed surface; e.g., Schrijver [9] got a quite strong result of similar flavor. Nevertheless, for our purposes, we do not need a precise answer; rather, we prefer a sufficient (but not necessary) condition that is convenient to use. The condition can be most concisely stated in terms of a metric deriving from respectful tangles, defined in the next section.

4.2 Metric from Respectful Tangles

To give a bit of a motivation, let us first formulate a version of Theorem 4.2 in terms of tangles.

Corollary 4.4 *Let G be a graph with cellular drawing on the sphere Σ, let f be a face of G, and let $\mathcal{S}$ be a root partition of a subset Z of vertices of G incident with f. Let $\mathcal{T}$ be a tangle in G of order at least $|\mathcal{S}|$. If $\mathcal{S}$ is topologically feasible in $\Sigma \setminus f$ and Z is $\mathcal{T}$-free, then G contains an $\mathcal{S}$-linkage.*

Proof Consider any separation (A, B) of G of order less than $|\mathcal{S}|$. Without loss of generality, we can assume $(A, B) \in \mathcal{T}$. Since Z is $\mathcal{T}$-free, every subset of Z is also $\mathcal{T}$-free, and thus $\tau_{\mathcal{S}}(A, B) \le |V(A) \cap Z| \le |A \cap B|$. Therefore, G satisfies the basic $\mathcal{S}$-connectivity condition, and the claim follows from Theorem 4.2. □

Although Corollary 4.4 is not an exact characterization, its assumptions are compact and convenient to verify. When applying Corollary 4.4 and its generalizations, the following description of free sets for respectful tangles is often useful.

Lemma 4.5 *Let $\mathcal{T}$ be a respectful tangle of order θ in a graph G with a cellular drawing on a surface Σ, and let Z be a set of vertices of G of size at most θ. Then the following claims are equivalent:*

(i) The set Z is not $\mathcal{T}$-free.
(ii) There exists a confined subgraph $F \subseteq R_G$ with $Z \subseteq \mathrm{ins}_{\mathcal{T}}(F)$ and $|Z| > |F \cap G|$.
(iii) There exists a 2-edge-connected confined strictly outerplanar cactus $F \subseteq R_G$ with outer face f such that $\mathrm{ins}_{\mathcal{T}}(F) = \Sigma \setminus f$ and $|Z \cap \mathrm{ins}_{\mathcal{T}}(F)| > |F \cap G|$, and moreover, for every cycle K in F, the interior of the disk $\mathrm{ins}_{\mathcal{T}}(K)$ contains a vertex of Z,

Proof If Z is not $\mathcal{T}$-free, then there exists a separation $(A, B) \in \mathcal{T}$ of order less than $|Z|$ such that $Z \subseteq V(A)$. By Lemma 2.89, there exists a confined strictly outerplanar cactus F_0 with outer face f such that $|F_0 \cap G| \le |A \cap B| < |Z|$ and $Z \subseteq V(A) \subset \mathrm{ins}_{\mathcal{T}}(F_0) = \Sigma \setminus f$. Hence, (i) implies (ii). Moreover, without loss of generality, we can assume that for every cycle K in F_0, the interior of $\mathrm{ins}_{\mathcal{T}}(K)$ contains a vertex of Z, as otherwise we can delete an edge of K from F_0 while preserving the described conditions. Observe that for distinct 2-edge-connectivity components F and F' of F_0, the sets $\mathrm{ins}_{\mathcal{T}}(F)$ and $\mathrm{ins}_{\mathcal{T}}(F')$ are disjoint. Since $|Z| > |F_0 \cap G|$ and $Z \subset \mathrm{ins}_{\mathcal{T}}(F_0)$, there exists a 2-edge-connectivity component F of F_0 such that $|Z \cap \mathrm{ins}_{\mathcal{T}}(F)| > |F \cap G|$. Hence, (i) also implies (iii).

Let us now argue that the conditions (ii) and (iii) imply (i). To show that the set Z is not $\mathcal{T}$-free, it suffices to show that a subset of Z is not $\mathcal{T}$-free. Hence, in the situation of (iii), we can consider the set $Z \cap \text{ins}_{\mathcal{T}}(F)$ instead of Z, and thus it suffices to show that (ii) implies (i). Let (A, B) be the separation of G with $V(A) = \{v \in V(G) : v \in \text{ins}_{\mathcal{T}}(F)\}$, $E(A) = \{e \in E(G) : \rho_G(e) \subset \text{ins}_{\mathcal{T}}(F)\}$, and $V(A \cap B) = V(F) \cap V(G)$. By Lemma 2.89, we have $(A, B) \in \mathcal{T}$, and thus Z is not $\mathcal{T}$-free. □

Let us remark that in (iii), F is not necessarily a cycle; e.g., consider the case that F is the union of two cycles K_1 and K_2 intersecting in a single vertex of G, Z is disjoint from F, and $|Z \cap \text{ins}_{\mathcal{T}}(K_i)| = |K_i \cap G|$ for $i \in [2]$.

Let us now consider the general case of the $\mathcal{S}$-linkage problem in a graph G drawn on a surface (and without the restriction that the roots are incident with only one face). It is easy to see that assuming the freeness of the set of roots is no longer sufficient. It is natural to additionally assume that the tangle is respectful and has large order. This forces large representativity in the non-sphere case, and the existence of a large grid minor in the sphere case, which should give us enough freedom to route the linkage paths according to their topological realization. However, this is still not sufficient—the topological realization may force us to route many disjoint paths so that they pass between two faces of G incident with distinct roots, and this may not be possible if the faces are not "far apart". In this section, we develop the theory necessary to state this distance condition precisely.

4.2.1 *The Definition of the Metric*

For a graph with a cellular drawing on a surface, let its ***atoms*** be its vertices, edges, and faces. Let G be a graph with a cellular drawing on a surface and let $\mathcal{T}$ be a respectful tangle in G of order θ. Consider a closed walk W in R_G and let U be the subgraph of R_G formed by vertices and edges traversed by W. Note that U may have fewer edges than the length $|W|$ of the walk W in case that W traverses an edge several times. If the subgraph U is confined, we define $\text{ins}_{\mathcal{T}}(W) = \text{ins}_{\mathcal{T}}(U)$. Note that if $|W| < 2\theta$, then $|U \cap G| < \theta$ and U is confined. For atoms a and b of R_G, let us define the distance $d_{\mathcal{T}}(a, b)$ between a and b as follows:

- If $a = b$, then $d_{\mathcal{T}}(a, b) = 0$; otherwise,
- if there exists a closed walk W in R_G of length less than 2θ such that $a \cup b \subseteq \text{ins}_{\mathcal{T}}(W)$ and ℓ is the minimum length of such a walk, then $d_{\mathcal{T}}(a, b) = \ell/2$;
- otherwise, $d_{\mathcal{T}}(a, b) = \theta$.

For atoms a' and b' of G, we define $d_{\mathcal{T}}(a', b') = d_{\mathcal{T}}(\rho_G(a), \rho_G(b))$.

Lemma 4.6 *Let G be a graph with a cellular drawing on a surface and let $\mathcal{T}$ be a respectful tangle in G of order θ. The function $d_{\mathcal{T}}$ is a metric on the atoms of R_G, and thus also on the atoms of G.*

Proof Clearly $d_{\mathcal{T}}$ is symmetric and $d_{\mathcal{T}}(a,b) = 0$ if and only if $a = b$, and thus it suffices to show that $d_{\mathcal{T}}$ satisfies the triangle inequality. Consider atoms a, b, and c of R_G; we aim to show that

$$d_{\mathcal{T}}(a,c) \le d_{\mathcal{T}}(a,b) + d_{\mathcal{T}}(b,c).$$

This is clearly true if a, b, and c are not pairwise different, or if $d_{\mathcal{T}}(a,b) + d_{\mathcal{T}}(b,c) \ge \theta$. Hence, we can assume that there exist closed walks W_1 and W_2 in R_G of lengths $\ell_1 = 2d_{\mathcal{T}}(a,b)$ and $\ell_2 = 2d_{\mathcal{T}}(b,c)$ such that $a \cup b \subseteq \mathrm{ins}_{\mathcal{T}}(W_1)$, $b \cup c \subseteq \mathrm{ins}_{\mathcal{T}}(W_2)$, and $\ell_1 + \ell_2 < 2\theta$.

The intersection $\mathrm{ins}_{\mathcal{T}}(W_1) \cap \mathrm{ins}_{\mathcal{T}}(W_2)$ is non-empty, since it contains b. We can move continuously from a point in this intersection to a point in its complement until we hit a point in $W_1 \cup W_2$. Hence, by symmetry, we can assume that there exists a vertex $v \in V(W_1)$ with $v \in \mathrm{ins}_{\mathcal{T}}(W_2)$. If $W_1 \subseteq \mathrm{ins}_{\mathcal{T}}(W_2)$, then Corollary 2.82 implies $a \subseteq \mathrm{ins}_{\mathcal{T}}(W_1) \subseteq \mathrm{ins}_{\mathcal{T}}(W_2)$, and thus $d_{\mathcal{T}}(a,c) \le \ell_2/2 < (\ell_1 + \ell_2)/2$. Otherwise, we can walk around W_1 from v until we would leave W_2, i.e., till we encounter a vertex $z \in V(W_1) \cap V(W_2)$. Let W be the closed walk in G of length $\ell_1 + \ell_2$ obtained by concatenating W_1 and W_2 in z. Then $a \cup c \subseteq \mathrm{ins}_{\mathcal{T}}(W_1) \cup \mathrm{ins}_{\mathcal{T}}(W_2) \subseteq \mathrm{ins}_{\mathcal{T}}(W)$, and $d_{\mathcal{T}}(a,c) \le (\ell_1 + \ell_2)/2 = d_{\mathcal{T}}(a,b) + d_{\mathcal{T}}(b,c)$. □

Let us remark that if v is a vertex incident with a face f, then $d_{\mathcal{T}}(v,f) = 1$. Moreover, if a and b are vertices or faces incident with the same edge e, then $d_{\mathcal{T}}(a,b) \le 2$, as seen by considering $\mathrm{ins}_{\mathcal{T}}(W)$ for the closed walk W bounding the 4-face $\rho_G(e)$. By triangle inequality, it follows that the tangle metric is (up to a constant factor) bounded by the graph distance; i.e., if u and v are vertices of G joined by a path of length ℓ, then $d_{\mathcal{T}}(u,v) \le 2\ell$. On the other hand, vertices may be far in the graph distance yet close in the tangle metric, e.g., if they are contained in the disk bounded by a short cycle. Let us also note the following simple observation.

Observation 4.7 *Let G be a graph with a cellular drawing on a surface and let $\mathcal{T}$ be a respectful tangle in G of order θ. If $\mathcal{T}'$ is the truncation of $\mathcal{T}$ to order $\theta' \le \theta$, then*

$$d_{\mathcal{T}'}(a,b) = \min(d_{\mathcal{T}}(a,b), \theta')$$

for all atoms a and b of G.

Lemma 2.81 implies that the distance is actually determined by quite special types of walks. Let $\mathcal{T}$ be a respectful tangle in a graph G with a cellular drawing on a surface Σ. For distinct atoms a and b of G or R_G, we say that a and b are *nearby* if $d_{\mathcal{T}}(a,b) < \theta$. Suppose now that a and b are atoms of R_G. An (a,b)-*restraint* is a connected confined strictly outerplanar cactus $F \subseteq R_G$ with outer face f such that $a \cup b \subseteq \mathrm{ins}_{\mathcal{T}}(F) = \Sigma \setminus f$. The *perimeter* of F is the boundary walk W of f; note that $\mathrm{ins}_{\mathcal{T}}(W) = \mathrm{ins}_{\mathcal{T}}(F)$. An (a,b)-restraint is *optimal* if its perimeter has length $d_{\mathcal{T}}(a,b)$. We say that an (a,b)-restraint F is an (a,b)-***tie*** if

- a and b are vertices and F is a path with ends a and b, or
- F is a cycle with $|\{a, b\} \cap V(F)| \leq 1$, or
- F is the union of a path P with distinct ends x and y and a cycle C intersecting P only in y, with $\{a, b\} \cap V(F) = \{x\}$, or
- F is the union of two cycles intersecting in a single vertex, or of two vertex-disjoint cycles and a path connecting them, and $a, b \notin V(F)$.

Lemma 4.8 *Let G be a graph with a cellular drawing on a surface and let $\mathcal{T}$ be a respectful tangle in G of order θ. If a and b are distinct nearby atoms of G, then there exists an optimal $(\rho_G(a), \rho_G(b))$-restraint which is a $(\rho_G(a), \rho_G(b))$-tie.*

Proof Let W_0 be a closed walk in R_G of length $2d_{\mathcal{T}}(a, b)$ such that $\rho_G(a) \cup \rho_G(b) \subseteq \text{ins}_{\mathcal{T}}(W_0)$. By Lemma 2.81, there exists a $(\rho_G(a), \rho_G(b))$-restraint F_0 with $\text{ins}_{\mathcal{T}}(W_0) = \text{ins}_{\mathcal{T}}(F_0)$. Note that W_0 traverses each edge of F_0, and traverses each bridge of F_0 at least twice, and thus W_0 is at least as long as the perimeter of F_0; hence, without loss of generality, we can assume that W_0 is the perimeter of F_0.

If $\rho_G(a) \in V(F_0)$, then let $C_a = \rho_G(a)$, otherwise let C_a be the cycle in F_0 such that $\rho_G(a) \subset \text{ins}_{\mathcal{T}}(C_a)$. Let C_b be defined analogously. If $C_a = C_b$, then let $F = C_a$, otherwise let F consist of $C_a \cup C_b$ together with a shortest path P in F_0 between C_a and C_b. Then F is a $(\rho_G(a), \rho_G(b))$-tie. Let W be the perimeter of F. Note that W_0 traverses all edges of C_a and C_b, and that it contains two subwalks disjoint from $E(C_a \cup C_b)$ between the ends of P. We conclude that $|W| = \|C_a\| + \|C_b\| + 2\|P\| \leq |W_0| = 2d_{\mathcal{T}}(a, b)$. Since $\rho_G(a) \cup \rho_G(b) \subseteq \text{ins}_{\mathcal{T}}(C_a \cup C_b) \subseteq \text{ins}_{\mathcal{T}}(F)$, it follows that F is an optimal $(\rho_G(a), \rho_G(b))$-restraint. □

4.2.2 The Structure of Balls

An important insight into the behavior of the metric defined by a respectful tangle is that every ball of radius $k < \theta$ (the union of the atoms at distance at most k from a fixed atom) forms a simply connected subset of the surface. We give a proof of this result (Theorem 4.13), as the ideas used in the argument (Lemma 4.9 and Corollaries 4.11 and 4.12) perhaps provide a better understanding of the metric; nevertheless, the proof is rather technical and the reader should feel free to skip it.

An optimal (a, b)-restraint F is ***extremal*** if $\text{ins}_{\mathcal{T}}(F)$ is inclusionwise-maximal among all optimal (a, b)-restraints. Note that optimal restraints are not necessarily ties; e.g., in the construction described in the proof of Lemma 4.8, P can be replaced by two paths of length $\|P\|$ between the ends of P. Extremal restraints have the following no-shortcut property.

Lemma 4.9 (↬) *Let G be a graph with a cellular drawing on a surface Σ, let $\mathcal{T}$ be a respectful tangle in G of order θ, and let a and b be distinct nearby atoms of R_G. Let F be an extremal (a, b)-restraint with perimeter walk W, let P be a path in R_G intersecting $\text{ins}_{\mathcal{T}}(F)$ exactly in its endpoints u and v, and let W_1 and W_2 be* *subwalks of W between u and v such that W is the concatenation of W_1 and W_2. Then at least one of W_1 and W_2 is strictly shorter than P.*

Proof Suppose for a contradiction that $|W_1|, |W_2| \geq |P|$. For $i \in [2]$, let W_i' be the closed walk formed by the concatenation of W_i with P, and note that $|W_i'| \leq |W| < 2\theta$. Let f be the face of F such that $\mathrm{ins}_{\mathcal{T}}(F) = \Sigma \setminus f$. Note that the path P is drawn in f (except for its endpoints), and it splits f into two parts f_1 and f_2, where W_i' forms the boundary of f_i for $i \in [2]$. By Corollary 2.85, there exists a face f_0 of R_G such that f_0 is not contained in $\mathrm{ins}_{\mathcal{T}}(W_1') \cup \mathrm{ins}_{\mathcal{T}}(W_2') \cup \mathrm{ins}_{\mathcal{T}}(F)$. By symmetry, we can assume $f_0 \subseteq f_1$. Lemma 2.81 implies that $\mathrm{ins}_{\mathcal{T}}(W_1') = \Sigma \setminus f_1$, and that there exists a strictly outerplanar cactus F_1 with outer face f_1 such that $\mathrm{ins}_{\mathcal{T}}(F_1) = \mathrm{ins}_{\mathcal{T}}(W_1')$. Note that the perimeter of F_1 has length at most $|W_1'| \leq |W|$, and that $\mathrm{ins}_{\mathcal{T}}(F_1) = \Sigma \setminus f_1 \supsetneq \Sigma \setminus f = \mathrm{ins}_{\mathcal{T}}(F)$. Therefore, F_1 is an (a, b)-restraint contradicting the extremality of F. □

Given a graph H drawn on a surface Σ, a set X is *H-nice* if it is a union of vertices, edges, and faces of H. We need the following topological facts (which are less obvious than they might seem, and rely on the niceness of the sets; see [2, 3]). Recall that a space X is *simply connected* if it is arcwise-connected and every closed curve in X is contractible.

Lemma 4.10 (↪) *Let H be a graph with a cellular drawing on a surface Σ, and let X and Y be closed H-nice simply connected subsets of Σ such that $X \cap Y$ is arcwise-connected. Then $X \cup Y$ is simply connected. Moreover, if $X \cup Y \neq \Sigma$, then $X \cap Y$ is simply connected.*

We get the following consequence for extremal restraints.

Corollary 4.11 (↪) *Let G be a graph with a cellular drawing on a surface Σ, let $\mathcal{T}$ be a respectful tangle in G of order θ, and for $i \in [2]$, let a_i and b_i be distinct nearby atoms of R_G and let F_i be an extremal (a_i, b_i)-restraint. If $\mathrm{ins}_{\mathcal{T}}(F_1) \cap \mathrm{ins}_{\mathcal{T}}(F_2) \neq \emptyset$, then both the union and the intersection of $\mathrm{ins}_{\mathcal{T}}(F_1)$ and $\mathrm{ins}_{\mathcal{T}}(F_2)$ is simply connected.*

Proof We have $\mathrm{ins}_{\mathcal{T}}(F_1) \cup \mathrm{ins}_{\mathcal{T}}(F_2) \neq \Sigma$ by Corollary 2.85. By Lemma 4.10, it suffices to argue that $Z = \mathrm{ins}_{\mathcal{T}}(F_1) \cap \mathrm{ins}_{\mathcal{T}}(F_2)$ is arcwise-connected.

Consider any component X of arcwise-connectivity of Z. We can assume that $\mathrm{ins}_{\mathcal{T}}(F_1) \neq X \neq \mathrm{ins}_{\mathcal{T}}(F_2)$, as otherwise $X = Z$ and the claim clearly holds. For $i \in [2]$, consider a curve in $\mathrm{ins}_{\mathcal{T}}(F_i)$ from a point in X to a point in $\mathrm{ins}_{\mathcal{T}}(F_i) \setminus X$; this curve necessarily intersects the boundary of X in a point of $\mathrm{ins}_{\mathcal{T}}(F_{3-i})$. We conclude that every component of Z is incident with a vertex of F_1 and a vertex of F_2.

Suppose for a contradiction that Z is not arcwise-connected, and for $i \in [2]$, let P_i be a shortest path in F_i joining vertices u_i and v_i incident with different components of Z. Clearly P_i has no internal vertices in $\mathrm{ins}_{\mathcal{T}}(F_{3-i})$, as otherwise these vertices would belong to $F_i \cap \mathrm{ins}_{\mathcal{T}}(F_{3-i}) \subseteq Z$ and P_i would have a proper subpath joining different components of Z. Since $u_i \in Z$ but the first edge of P is not contained in Z, it follows that $u_i \in V(F_{3-i})$, and analogously $v_i \in V(F_{3-i})$.

Therefore, P_i is a path joining distinct vertices of $V(F_{3-i})$ and otherwise disjoint from $\text{ins}_{\mathcal{T}}(F_{3-i})$, and Lemma 4.9 implies that there is a path between u_i and v_i in F_{3-i} shorter than P_i. Since P_{3-i} was chosen to be a shortest path in F_{3-i} with ends in different components of Z, this implies $|P_{3-i}| < |P_i|$. However, this cannot hold for both $i \in [2]$, giving a contradiction. □

This gives us the following laminarity property for extremal restraints.

Corollary 4.12 (⇸) *Let G be a graph with a cellular drawing on a surface Σ, let $\mathcal{T}$ be a respectful tangle in G of order θ, and for $i \in [2]$, let a_i and b_i be distinct nearby atoms of R_G and let F_i be an extremal (a_i, b_i)-restraint. If $a_2 \cup b_2 \subseteq \text{ins}_{\mathcal{T}}(F_1)$, then $\text{ins}_{\mathcal{T}}(F_2) \subseteq \text{ins}_{\mathcal{T}}(F_1)$.*

Proof Let Q_1 and Q_2 be the perimeters of F_1 and F_2. Let $Z_1 = \text{ins}_{\mathcal{T}}(F_1) \cap \text{ins}_{\mathcal{T}}(F_2)$ and $Z_2 = \text{ins}_{\mathcal{T}}(F_1) \cup \text{ins}_{\mathcal{T}}(F_2)$. Since $a_2 \cup b_2 \subseteq Z_1$, Z_1 is non-empty, and thus by Corollary 4.11, both Z_1 and Z_2 are simply connected. Moreover, $Z_2 \neq \Sigma$ by Corollary 2.85. For $i \in [2]$, let H_i be the subgraph of R_G formed by vertices and edges contained in the boundary of Z_i, and observe that H_i is a connected strictly outerplanar graph with outer face $f_i = \Sigma \setminus Z_i$. Let W_i be the boundary walk of f_i.

We claim that $|W_1| \geq |Q_2|$. This is clearly true if $|W_1| \geq 2\theta$, since a_2 and b_2 are nearby. Hence, suppose that $|W_1| < 2\theta$. Since $H_1 \subseteq Z_1 \subseteq \text{ins}_{\mathcal{T}}(F_1)$, Corollary 2.82 implies $\text{ins}_{\mathcal{T}}(W_1) = \text{ins}_{\mathcal{T}}(H_1) \subseteq \text{ins}_{\mathcal{T}}(F_1)$, and since $Z_1 \cup f_1 = \Sigma$ and $\text{ins}_{\mathcal{T}}(F_1) \neq \Sigma$, it follows that $f_1 \not\subseteq \text{ins}_{\mathcal{T}}(W_1)$. By Lemma 2.81, we conclude that $\text{ins}_{\mathcal{T}}(W_1) = Z_1$. Since $a_2 \cup b_2 \subseteq Z_1$, it follows that W_1 is an (a_2, b_2)-restraint. Since F_2 is an optimal (a_2, b_2)-restraint, we obtain the inequality $|W_1| \geq |Q_2|$, as desired.

Consider now any edge $e \in E(R_G)$ and let k be the number of times e is traversed by W_1 and W_2. We claim that Q_1 and Q_2 traverses e at least k times:

- If $k = 4$, then e is bridge in H_1 as well as H_2, and thus neither of the faces incident with e belongs to $\text{ins}_{\mathcal{T}}(F_1) \cup \text{ins}_{\mathcal{T}}(F_2)$. Consequently, e is a bridge of F_1 and F_2, and thus Q_1 and Q_2 also traverse e four times.
- If $k = 3$, then e a bridge in H_1 and exactly one face g of R_G incident with e belongs to Z_2. Then for some $i \in [2]$, e is a bridge of F_i and an edge of F_{3-i}, and $g \in \text{ins}_{\mathcal{T}}(F_{3-i})$. Then Q_i traverses e twice and Q_{3-i} once.
- If $k = 2$ and e is a bridge in H_1, but not an edge of H_2, then one face incident with e belongs to $\text{ins}_{\mathcal{T}}(F_1) \setminus \text{ins}_{\mathcal{T}}(F_2)$ and the other one to $\text{ins}_{\mathcal{T}}(F_2) \setminus \text{ins}_{\mathcal{T}}(F_1)$ and each of Q_1 and Q_2 traverses e once.
- If $k = 2$ and e is a bridge in H_2, but not an edge of H_1, then neither of the incident faces belongs to $\text{ins}_{\mathcal{T}}(F_1) \cup \text{ins}_{\mathcal{T}}(F_2)$ and e is a bridge in one of F_1 and F_2, and thus Q_1 or Q_2 traverses e twice.
- If $k = 2$ and e is a non-bridge edge of both H_1 and H_2, then one face incident with e belongs to both $\text{ins}_{\mathcal{T}}(F_1)$ and $\text{ins}_{\mathcal{T}}(F_2)$ and the other one to neither, and thus each of Q_1 and Q_2 traverses e once.
- If $k = 1$, then observe that e is necessarily an edge of F_1 or F_2, and thus Q_1 or Q_2 traverses e.

Consequently, $|Q_2| + |W_2| \le |W_1| + |W_2| \le |Q_1| + |Q_2|$, and thus $|W_2| \le |Q_1|$. Since $\mathrm{ins}_{\mathcal{T}}(F_1) \cup \mathrm{ins}_{\mathcal{T}}(F_2) \cup f_2 = Z_2 \cup f_2 = \Sigma$, Corollary 2.85 implies that $f_2 \not\subseteq \mathrm{ins}_{\mathcal{T}}(W_2)$; by Lemma 2.81, it follows that $\mathrm{ins}_{\mathcal{T}}(W_2) = Z_2 \supseteq a_1 \cup b_1$. Therefore, H_2 is an (a_1, b_1)-restraint, and since F_1 is an extremal (a_1, b_1)-restraint and $\mathrm{ins}_{\mathcal{T}}(F_1) \subseteq Z_2 = \mathrm{ins}_{\mathcal{T}}(W_2)$, we conclude that $\mathrm{ins}_{\mathcal{T}}(F_1) = Z_2$. This implies that $\mathrm{ins}_{\mathcal{T}}(F_2) \subseteq \mathrm{ins}_{\mathcal{T}}(F_1)$. □

And finally, this gives us the aforementioned key result.

Theorem 4.13 *Let G be a graph with a cellular drawing on a surface Σ, let $\mathcal{T}$ be a respectful tangle in G of order θ, let a be an atom of R_G, and let $k < \theta$ be a positive integer. Let Z be the union of the atoms b of R_G such that $d_{\mathcal{T}}(a, b) \le k$. Then Z is simply connected.*

Proof (⇝) For every atom b distinct from a such that $d_{\mathcal{T}}(a, b) \le k$, let F_b be an extremal (a, b)-restraint, which exists by Lemma 2.81. Note that $d_{\mathcal{T}}(a, b') \le d_{\mathcal{T}}(a, b) \le k$ for every atom $b' \subseteq \mathrm{ins}_{\mathcal{T}}(F_b)$, and thus $\mathrm{ins}_{\mathcal{T}}(F_b) \subseteq Z$, and moreover, either $Z = a$ (in which case the conclusion of the theorem holds), or

$$Z = \bigcup_{b \ne a : d_{\mathcal{T}}(a,b) \le k} \mathrm{ins}_{\mathcal{T}}(F_b).$$

This implies that Z is arcwise-connected. Hence, it suffices to show that every cycle C of $R_G \cap Z$ is contractible inside Z. Let $v_1, e_1, v_2, e_2, \ldots, v_m, e_m$ be the vertices of C in order, and let $v_{m+1} = v_1$. Fix a point p of a, and for $i \in [m]$, let γ_i be a curve from p to v_i contained in $\mathrm{ins}_{\mathcal{T}}(F_{v_i})$, with $\gamma_{m+1} = \gamma_1$. Let γ_i' be the concatenation of the reversal of γ_i with γ_{i+1}. Since $a, v_i, v_{i+1} \in \mathrm{ins}_{\mathcal{T}}(F_{e_i})$, Corollary 4.12 implies that $\mathrm{ins}_{\mathcal{T}}(F_{v_i}) \cup \mathrm{ins}_{\mathcal{T}}(F_{v_{i+1}}) \subseteq \mathrm{ins}_{\mathcal{T}}(F_{e_i})$, and thus $\gamma_i' \subseteq \mathrm{ins}_{\mathcal{T}}(F_{e_i})$. Since $\mathrm{ins}_{\mathcal{T}}(F_{e_i})$ is simply connected, the closed curve consisting of γ_i' and e_i is contractible in $\mathrm{ins}_{\mathcal{T}}(F_{e_i}) \subseteq Z$, and thus the curve tracing e_i is homotopically equivalent to γ_i' in Z. It follows that the concatenation γ of γ_1', …, γ_m' is homotopically equivalent to the cycle C (viewed as the concatenation of curves tracing the edges of C). The closed curve γ is trivially contractible in Z, and thus C is contractible in Z as well. □

In case that Σ is not the sphere, Theorem 4.13 implies that there is a face of R_G not contained in Z. This is also true if Σ is the sphere, but the proof is rather lengthy and we omit it in our exposition. Let us state this result in an equivalent form speaking about the atoms of G, which will be more useful for us.

Theorem 4.14 (Robertson and Seymour [7, 8]) *Let G be a graph with a cellular drawing on a surface Σ and let $\mathcal{T}$ be a respectful tangle in G of order $\theta \ge 2$. For every atom a of G, there exists an edge e of G such that $d_{\mathcal{T}}(a, e) = \theta$.*

We are actually going to need a more refined version of Theorem 4.13, showing that for every atom a and every positive integer $k < \theta$, there exists a disk in the surface containing exactly the part of R_G at distance at most k from a, except for a bit of fuzziness concerning the atoms at distance exactly k.

Corollary 4.15 *Let G be a graph with a cellular drawing on a surface Σ and let $\mathcal{T}$ be a respectful tangle in G of order θ. For every atom a of G and an integer k such that $2 \le k < \theta$, there exists a cycle C in R_G bounding a disk $\Delta \subset \Sigma$ such that $\rho_G(a) \subset \Delta$,*

(i) every atom b of R_G such that $b \subseteq \Delta$ satisfies $d_{\mathcal{T}}(\rho_G(a), b) \le k$,
(ii) every atom b of R_G with $d_{\mathcal{T}}(\rho_G(a), b) \le k-1$ satisfies $b \subseteq \Delta$, and
(iii) for every vertex $u \in V(C)$, there exists a face q of R_G incident with u such that $d_{\mathcal{T}}(\rho_G(a), q) \ge k+1$, and in particular, $d_{\mathcal{T}}(\rho_G(a), u) \ge k-1$.

Proof Let Z be the union of the atoms b of R_G such that $d_{\mathcal{T}}(\rho_G(a), b) \le k$. By Theorems 4.13 and 4.14, Z is simply connected and does not contain all faces of R_G. Let H be the subgraph of R_G contained in the boundary of Z; then H has a face f such that $Z = \Sigma \setminus f$, and H is a strictly outerplanar cactus with the outer face f. Since $k \ge 2$ and all faces of R_G have length four, if a is a vertex or a face of G, then Z contains all faces of R_G incident with $\rho_G(a)$; and if a is an edge of G, then the face $\rho_G(a)$ is contained in Z. We conclude that there exists a cellular face $f_0 \ne f$ of H bounded by a cycle C such that $\rho_G(a) \subseteq f_0$. Let Δ be the closure of f_0. Then $\Delta \subseteq Z$, and thus (i) holds.

Note that if $u \in V(H)$, then u is incident with a face q of R_G not in Z, and thus $d_{\mathcal{T}}(\rho_G(a), u) \ge d_{\mathcal{T}}(\rho_G(u), q) - d_{\mathcal{T}}(q, u) \ge (k+1) - 2 = k-1$, implying that (iii) holds.

Consider now any atom b of Z not contained in Δ, let F be an optimal $(\rho_G(a), b)$-restraint, and let W be the perimeter of F. Since b is not contained in Δ and $\mathrm{ins}_{\mathcal{T}}(F) \subseteq Z$, observe that there exists a cutvertex u of H contained in C such that u is also a cutvertex of F. Let W' be the closed walk obtained from W by removing the part of W outside of Δ between two appearances of u in W, and let $\ell \ge 2$ be the length of the removed part. Since $\rho_G(a) \cup u \subset \mathrm{ins}_{\mathcal{T}}(W')$, we have $|W'| \ge 2d_{\mathcal{T}}(\rho_G(a), u)$. Consequently,

$$k \ge d_{\mathcal{T}}(\rho_G(a), b) = |W|/2 = (|W'| + \ell)/2 \ge d_{\mathcal{T}}(\rho_G(a), u) + \ell/2 \ge k - 1 + \ell/2.$$

It follows that $\ell = 2$ and $d_{\mathcal{T}}(\rho_G(a), b) = k$, and thus (ii) holds. □

As we will see in the next section, it is convenient to get a similar disk bounded by a cycle in G rather than R_G. We say that a disk Δ in the surface is a ***zone*** if Δ is bounded by a cycle in G. It is a zone *around* an atom a of G if $a \subset \Delta$, and a *k-zone around a* if additionally $d_{\mathcal{T}}(a, b) \le k$ for every atom b of G contained in Δ.

Corollary 4.16 *Let G be a graph with a cellular drawing on a surface Σ and let $\mathcal{T}$ be a respectful tangle in G of order θ. For every atom a of G and an integer k such that $4 \le k < \theta$, there exists a cycle C in G bounding a k-zone Δ around a such that*

- *every atom b of G with $d_{\mathcal{T}}(a, b) \le k-3$ satisfies $b \subset \Delta$, and*
- *for every vertex $v \in V(C)$, there exists a face g incident with v and an edge e incident with g such that $d_{\mathcal{T}}(a, e) \ge k+1$, and in particular $d_{\mathcal{T}}(a, v) \ge k-2$.*

Proof Let C_0 be the cycle in R_G and Δ_0 the disk bounded by C_0 with $\rho_G(a) \subset \Delta_0$ obtained in Corollary 4.15. Consider any face f of G with $\rho_G(f) \in V(C_0)$. Let $v_1, g_1, v_2, g_2, \ldots, g_{m-1}, v_m$ be the neighbors of $\rho_G(f)$ and the faces incident with $\rho_G(f)$ contained in $R_G \cap \Delta_0$, in order around $\rho_G(f)$ according to the drawing of R_G; in particular, v_1 and v_m are the neighbors of $\rho_G(f)$ in C_0. Let $e_i = \rho_G^{-1}(g_i)$ for $i \in [m-1]$ and note that $W_f = v_1, e_1, v_2, \ldots, e_{m-1}, v_m$ is a walk in G. Let γ_f be the curve tracing the subpath $v_1 \rho_G(f) v_m$ of C_0. Cutting the face f of G along the curve γ_f splits f into two parts homeomorphic to open disks, and W_f together with γ_f forms the boundary of the one contained in Δ; let this part be denoted by Λ_f. Corollary 4.15(iii) implies that $d_{\mathcal{T}}(\rho_G(a), \rho_G(f)) \geq k-1$. Therefore, for every $i \in [m]$ we have $d_{\mathcal{T}}(a, v_i) \geq d_{\mathcal{T}}(\rho_G(a), \rho_G(f)) - 1 \geq k-2$. Moreover, $d_{\mathcal{T}}(\rho_G(a), g_i) \geq d_{\mathcal{T}}(a, v_i) - 1 \geq k-3 > 0$ for $i \in [m-1]$, and thus $\rho_G(a)$ is not contained in the closure of Λ_f. Let Δ' be the space obtained from the disk Δ_0 by deleting a point in $\rho_G(a)$. Then W_f is homotopically equivalent to γ_f in Δ'.

Let W be the closed walk in G formed by the concatenation of the walks W_f for all faces f of G with $\rho_G(f) \in V(C_0)$. We conclude that W is homotopically equivalent to the cycle C_0 in the space Δ', and consequently W is non-contractible in Δ'. Therefore, there also exists a cycle C consisting of vertices and edges traversed by W that is non-contractible in Δ'; such a cycle C clearly bounds a disk $\Delta \subset \Delta_0$ containing $\rho_G(a)$. By the property (i) of Corollary 4.15, the disk Δ is a k-zone around a.

Consider now any atom b of G such that $d_{\mathcal{T}}(a, b) \leq k-3$, and let F be an optimal $(\rho_G(a), \rho_G(b))$-restraint. Since $d_{\mathcal{T}}(a, v) \geq k-2$ for every vertex $v \in V(C)$, the set $\text{ins}_{\mathcal{T}}(F)$ is disjoint from C, and since $\rho_G(a) \subseteq \text{ins}_{\mathcal{T}}(F) \cap \Delta$, we conclude that $\text{ins}_{\mathcal{T}}(F) \subset \Delta$. It follows that $\rho_G(b) \in \Delta$, and since b is an atom of G and Δ is bounded by a cycle in G, it follows that $b \subseteq \Delta$.

Finally, for every $v \in V(C)$, there exists a face g with $\rho_G(g) \in V(C_0)$ such that $v \in W_g$, and thus v is incident with g in G. By the property (iii) of Corollary 4.15, there exists a face q of R_G incident with $\rho_G(g)$ such that $d_{\mathcal{T}}(\rho_G(a), q) \geq k+1$. The edge $e = \rho_G^{-1}(q)$ is incident with g and $d_{\mathcal{T}}(a, e) = d_{\mathcal{T}}(\rho_G(a), q) \geq k+1$. □

Next, let us show that the slope derived from the tangle behaves as expected with respect to zones.

Lemma 4.17 *Let G be a graph with a cellular drawing on a surface Σ, let $\mathcal{T}$ be a respectful tangle in G of order θ, and let Δ be a k-zone around an atom a of G. If W is a closed walk in R_G of length less than $2(\theta - k)$ contained in Δ, then* $\text{ins}_{\mathcal{T}}(W) \subseteq \Delta$.

Proof By Lemma 2.81, $\text{ins}_{\mathcal{T}}(W) = \Sigma \setminus f$ for a face f of the subgraph F of R_G formed by the vertices and edges traversed by W. Let e be an edge of G such that $d_{\mathcal{T}}(a, e) = \theta$, which exists by Theorem 4.14. Note that $e \not\subseteq \text{ins}_{\mathcal{T}}(F)$, since otherwise for any vertex v of F, we would have $d_{\mathcal{T}}(a, v) \geq d_{\mathcal{T}}(a, e) - d_{\mathcal{T}}(v, e) > \theta - (\theta - k) \geq k$, contradicting the assumption that W is contained in the k-zone Δ around a. Consequently, $e \subset f$. Since $\Sigma \setminus \Delta$ is arcwise-connected, disjoint from F, and contains e, we conclude that $\Sigma \setminus \Delta \subseteq f$, and thus $\text{ins}_{\mathcal{T}}(W) = \Sigma \setminus f \subseteq \Delta$. □

From this, we can easily conclude that boundaries of sufficiently large zones around the same atom are well connected to each other.

Lemma 4.18 *Let G be a graph with a cellular drawing on a surface, let $\mathcal{T}$ be a respectful tangle in G of order θ, let a be an atom of G, and let k_1 and k_2 be positive integers such that $k_1 + k_2 \leq \theta$. Let $\Delta_1 \subset \Delta_2$ be zones around a, where Δ_2 is a k_2-zone, such that the cycles C_1 and C_2 bounding Δ_1 and Δ_2 are vertex-disjoint. If there exists a vertex $z \in V(C_1)$ such that $d_{\mathcal{T}}(a, z) \geq k_1$, then G contains a $(V(C_1), V(C_2))$-linkage of size k_1.*

Proof Let G_1 be the subgraph of G drawn in the annulus between C_1 and C_2, and consider the auxiliary plane graph G_1' obtained from $G_1 - E(C_1 \cup C_2)$ by adding vertices u and v adjacent to all vertices of C_1 and C_2, respectively. If G_1 contains k_1 paths from u to v intersecting only in their endpoints, then the restriction of these paths to G_1 forms the desired linkage. Otherwise, by Lemma 2.67, there exists a cycle Q of length less than $2k_1$ in the radial graph of G_1' separating u from v. We can naturally interpret Q as a cycle in R_G separating C_1 from C_2. Since $Q \subseteq \Delta_2$ and $|Q| < 2k_1 \leq 2(\theta - k_2)$, Lemma 4.17 implies that $\mathrm{ins}_{\mathcal{T}}(Q) \subseteq \Delta_2$, and thus $\Delta_1 \subset \mathrm{ins}_{\mathcal{T}}(Q)$. However, this implies that $d_{\mathcal{T}}(a, z) < k_1$ for every vertex $z \in V(C_1)$, which is a contradiction. □

4.2.3 *Clearing a Zone*

In many constructions, we "use up" a part of a graph G contained inside a k-zone Δ around an atom, and to ensure that the rest of the construction is independent from this part, we delete all vertices and edges of G drawn in the interior of Δ. We say that the resulting subgraph G' is obtained by ***clearing*** the zone Δ. The drawing of G' on Σ is clearly cellular, and since we obtained it from G by deleting a region of bounded radius k, we would intuitively expect the distances in G' to be smaller by at most $O(k)$ than the distances in G. Of course, this implicitly assumes that we can actually define distances in G' somewhat consistently to the distances in G, which is not completely obvious—there is no general procedure to turn a tangle $\mathcal{T}$ in a graph G into a consistent tangle $\mathcal{T}'$ in a subgraph G' of G. The following lemma formalizes the described intuition; its proof is straightforward but technical and the reader should feel free to skip it.

Lemma 4.19 *Let G be a graph with a cellular drawing on a surface Σ and let $\mathcal{T}$ be a respectful tangle in G of order θ. Let a_0 be an atom of G and let Δ be a k-zone around a_0 for a positive integer k such that $2k < \theta$. Let $G' \subseteq G$ be obtained from G by clearing Δ and let f_0 be the resulting face of G'. There exists a respectful tangle $\mathcal{T}'$ in G' of order $\theta' = \theta - 2k$ conformal with $\mathcal{T}$, such that*

(i) $d_{\mathcal{T}}(a, b) - 4k \leq d_{\mathcal{T}'}(a, b) \leq d_{\mathcal{T}}(a, b)$ for all atoms a and b of G' different from f_0, and

(ii) every atom b of G' different from f_0 satisfies $d_{\mathcal{T}}(a_0, b) - 2k \leq d_{\mathcal{T}'}(f_0, b) \leq d_{\mathcal{T}}(a_0, b)$.

Moreover, every $\mathcal{T}$-free set Z consisting of vertices of G at distance at least $|Z|+2k$ from a_0 is also $\mathcal{T}'$-free.

Proof (⊶) Let C be the cycle in G bounding Δ. Let e_0 be an edge of G such that $d_{\mathcal{T}}(a_0, e_0) = \theta$, which exists by Theorem 4.14, and let g_0 be the corresponding face of R_G; clearly, g_0 is also a face of $R_{G'}$. Let $\Sigma' = \Sigma \setminus g_0$. Let $v_0 = \rho_{G'}(f_0)$ and without loss of generality assume that v_0 is drawn at a point belonging to the drawing of $\rho_G(a_0)$. Note that by Corollary 4.15 applied for $k + 1$, there exists a disk $\Delta_0 \supseteq \Delta$ in Σ' bounded by a cycle in R_G, such that Δ_0 contains all atoms of R_G at distance at most k from $\rho_G(a_0)$, but no atoms at distance more than $k + 1$. Since every vertex $v \in V(C)$ satisfies $d_{\mathcal{T}}(a_0, v) \leq k$, there exists a $(\rho_G(a_0), v)$-restraint F_v with perimeter of length at most $2k$. Note that F_v intersects $V(G)$ in at most k vertices. Let γ_v be a simple curve in $\mathrm{ins}_{\mathcal{T}}(F_v) \subseteq \Delta_0$ joining the points representing v_0 and v; since γ_v and the edge v_0v of $R_{G'}$ are both drawn inside Δ_0, they are homotopically equivalent in Σ'.

For a cycle K in $R_{G'}$ of length less than $2\theta'$, let us define $\mathrm{ins}'(K)$ as follows.

- If $v_0 \notin V(K)$, then let $\mathrm{ins}'(K) = \mathrm{ins}_{\mathcal{T}}(K)$.
- Otherwise, let $v_1, v_2 \in V(K)$ be the neighbors of v_0 in K. Note that the subgraph $F = F_{v_1} \cup F_{v_2} \cup (K - v_0)$ of R_G has less than $\theta' + 2k \leq \theta$ vertices in $V(G)$, and thus it is confined. By Lemma 2.81, $\mathrm{ins}_{\mathcal{T}}(F) = \Sigma \setminus f$ for a face f of F. The boundary of f traverses each edge of $K - v_0$ at most once, and otherwise consists of subwalks of the perimeters of $\mathrm{ins}_{\mathcal{T}}(F_{v_1})$ and $\mathrm{ins}_{\mathcal{T}}(F_{v_2})$, and thus the closed walk W bounding f has length less than $2\theta'+4k \leq 2\theta$; it follows that $g_0 \subset f$. Note that the closed curve γ formed by $K - v_0$, γ_{v_1}, and γ_{v_2} is contained in $\mathrm{ins}_{\mathcal{T}}(F)$, and thus it is contractible in Σ'. It follows that the homotopically equivalent cycle K is also contractible in Σ', and thus it bounds a unique disk $\Delta_K \subseteq \Sigma'$. Let us define $\mathrm{ins}'(K) = \Delta_K$. Note that equivalently, $\mathrm{ins}'(K)$ is the face of K in Σ that does not contain g_0.

We claim that ins' is a slope of order θ' in G'. The verification of the properties (S1) and (S2) is mostly straightforward; let us explain the argument for the most complicated subcase of (S2): Let P_1, P_2, and P_3 be paths in $R_{G'}$ forming a confined theta graph T, where v_0 is an internal vertex of P_3. Let $K_1 = P_1 \cup P_3$, $K_2 = P_2 \cup P_3$, and $K = P_1 \cup P_2$. Suppose for a contradiction that (S2) is false, and thus T has three 2-cell faces whose closures $\mathrm{ins}'(K_1)$, $\mathrm{ins}'(K_2)$, and $\mathrm{ins}'(K)$ cover the whole surface Σ. For $i \in [2]$, the vertex v_0 belongs to K_i, and thus by the definition of $\mathrm{ins}'(K_i)$, it follows that $g_0 \not\subseteq \mathrm{ins}'(K_i)$. Therefore, $g_0 \subseteq \mathrm{ins}'(K) = \mathrm{ins}_{\mathcal{T}}(K)$, and thus $d_{\mathcal{T}}(g_0, u) < \theta'$ for every vertex $u \in V(K)$. Consequently, $d_{\mathcal{T}}(\rho_G(a_0), u) \geq d_{\mathcal{T}}(\rho_G(a_0), g_0) - d_{\mathcal{T}}(g_0, u) > \theta - \theta' = 2k$. For the neighbors v_1 and v_2 of v_0 in P_3, this implies that $\mathrm{ins}_{\mathcal{T}}(F_{v_1})$ and $\mathrm{ins}_{\mathcal{T}}(F_{v_2})$ are disjoint from K. Let P_3' and P_3'' be the shortest subpaths of P_3 from its ends to $\mathrm{ins}_{\mathcal{T}}(F_{v_1} \cup F_{v_2})$, let Q' be a path joining the ends of P_3' and P_3'' in $R_G \cap \mathrm{ins}_{\mathcal{T}}(F_{v_1} \cup F_{v_2})$, and let Q be the concatenation of P_3', Q', and P_3''; then P_1, P_2, and Q form a theta graph in R_G. For $i \in [2]$, let $F_i = F_{v_1} \cup F_{v_2} \cup P_3' \cup P_3'' \cup P_i$. Note that each cycle in F_i has length at most $|P_3 \cup P_i| + |F_{v_1}| + |F_{v_2}| < 2\theta' + 4k = 2\theta$, and thus F_i is confined. Since $\rho_G(a_0) \subseteq \text{ins}_{\mathcal{T}}(F_{v_1}) \subseteq \text{ins}_{\mathcal{T}}(Q_i)$, every atom contained in $\text{ins}_{\mathcal{T}}(F_i)$ is at $d_{\mathcal{T}}$-distance less than θ from $\rho_G(a_0)$, and thus we have $g_0 \not\subseteq \text{ins}_{\mathcal{T}}(F_i)$. By Lemma 2.81, it follows that $\text{ins}_{\mathcal{T}}(F_i) = \Sigma \setminus g_i$ for the face g_i of F_i that contains g_0. Let f_i be the face of the theta graph $P_1 \cup P_2 \cup Q$ bounded by the cycle $P_i \cup Q$. Since g_0 is contained in the face bounded by $P_1 \cup P_2 = K$, we have $g_0 \not\subseteq f_i$. Since the cycle $P_i \cup Q$ is contained in $\text{ins}_{\mathcal{T}}(F_i) = \Sigma \setminus g_i$, we have $\overline{f_i} \subseteq \text{ins}_{\mathcal{T}}(F_i)$. But then $\Sigma = \overline{f_1} \cup \overline{f_2} \cup \text{ins}_{\mathcal{T}}(K) = \text{ins}_{\mathcal{T}}(F_1) \cup \text{ins}_{\mathcal{T}}(F_2) \cup \text{ins}_{\mathcal{T}}(K)$, contradicting Corollary 2.85.

It is also easy to check that the assumptions of Lemma 2.90 hold, and thus the respectful pretangle $\mathcal{T}'$ of order θ' induced by ins' is a tangle in G'. We now need to show that this tangle and the metric derived from it satisfy the properties described in the statement of the lemma.

First, let us show that $\mathcal{T}'$ is conformal with $\mathcal{T}$. Let id be the trivial model of G' in G and let $\mathcal{T}'_{\text{id}}$ be the tangle of order θ' in G id-induced by $\mathcal{T}'$. Consider any cycle K in R_G of length less than $2\theta'$, let $\Delta_K = \text{ins}_{\mathcal{T}}(K)$, and let $(A, B) \in \mathcal{T}$ be the separation of G with $A = G \cap \Delta_K$ and $B = G \cap \overline{\Sigma \setminus \Delta_K}$.

- If K does not contain any vertex or edge drawn in the interior of Δ, then K is also a cycle in $R_{G'}$, and the definition of the slope ins' implies that $\text{ins}'(K) = \text{ins}_{\mathcal{T}}(K)$. Since the tangle $\mathcal{T}'$ is induced by the slope ins', we have $(A \cap G', B \cap G') = (G' \cap \Delta_K, G' \cap \overline{\Sigma \setminus \Delta_K}) \in \mathcal{T}'$.
- Suppose that K intersects the interior of Δ. Then note that $d_{\mathcal{T}}(a_0, b) \le k + |K|/2 < \theta$ for every atom b of R_G with $\rho_G(b) \subset \Delta_K$, and thus $e_0 \not\subseteq \Delta_K$.
  - If $K \subseteq \Delta$, then this implies that $\Delta_K \subseteq \Delta$, and thus $(A \cap G', B \cap G') = (A \cap C, G') \in \mathcal{T}'$ by (T2).
  - Otherwise, let W be the closed walk in $R_{G'}$ obtained from K by replacing each subpath P of K intersecting the interior of Δ by the 2-edge path P' in $R_{G'}$ with the same ends passing through v_0. Consider any edge $e \in E(A \cap G')$, and let Δ_e be the space obtained from Δ' by drilling a hole in e. Since e is disjoint from the interior of Δ, the paths P and P' are homotopically equivalent in Σ_e, and thus the closed walk W is homotopically equivalent to K in Σ_e. Since $e \in E(A) \subseteq \Delta_K$, the cycle K is non-contractible in Σ_e, and thus so is W. The definition of ins' for cycles passing through v_0 consequently implies that $\rho_G(e) \subset \text{ins}'(W)$. Since this holds for every edge $e \in E(A \cap G')$, we conclude using Lemma 2.89 that $(A \cap G', B \cap G') \in \mathcal{T}'$.

In all cases, we concluded that $(A \cap G', B \cap G') \in \mathcal{T}'$, and thus $(A, B) \in \mathcal{T}'_{\text{id}}$. Therefore, $\mathcal{T}'_{\text{id}}$ is a respectful tangle whose derived slope is equal to $\text{ins}_{\mathcal{T}_1}$, where $\mathcal{T}_1$ is the truncation of $\mathcal{T}$ to order θ'. By Lemma 2.88, we conclude that $\mathcal{T}'_{\text{id}} = \mathcal{T}_1$, and thus $\mathcal{T}'$ is conformal with $\mathcal{T}$.

Next, let us consider the condition (ii) on the $d_{\mathcal{T}'}$-distance from f_0. Consider any atom b of G' different from f_0, and let us first give a lower bound on the $d_{\mathcal{T}'}$-distance of b from f_0.

- If $d_{\mathcal{T}'}(f_0, b) = \theta'$, then clearly

$$d_{\mathcal{T}}(a_0, b) - 2k \leq \theta - 2k = \theta' = d_{\mathcal{T}'}(f_0, b).$$

- Hence, suppose that $d_{\mathcal{T}'}(f_0, b) < \theta'$, and let F' be an optimal $(v_0, \rho_{G'}(b))$-restraint in $R_{G'}$. By Lemma 4.8, we can assume that F' is a $(v_0, \rho_{G'}(b))$-tie.
  - If $v_0 \notin V(F')$, then $\rho_G(a_0) \subseteq \Delta \subseteq \text{ins}'(F') = \text{ins}_{\mathcal{T}}(F')$, and thus F' is also a $(\rho_G(a_0), \rho_G(b))$-restraint in G. This implies that $d_{\mathcal{T}}(a_0, b) \leq d_{\mathcal{T}'}(f_0, b)$.
  - If $v_0 \in V(F')$ has degree one in F', then let v_1 be the neighbor of v_0 in F' and note that $F' - v_0$ is $(v_1, \rho_G(b))$-restraint in G, implying that

$$d_{\mathcal{T}}(a_0, b) \leq d_{\mathcal{T}}(a_0, v_1) + d_{\mathcal{T}'}(f_0, b) - 1 \leq d_{\mathcal{T}'}(f_0, b) + k - 1.$$

  - If $v_0 \in V(F')$ has degree more than one in F', then since F' is a $(v_0, \rho_{G'}(b))$-tie, we conclude that F' is a cycle. If $\rho_{G'}(b) \subseteq \Delta_0$, then $d_{\mathcal{T}}(a_0, b) \leq k + 1 \leq d_{\mathcal{T}'}(f_0, b) + k + 1$. Otherwise, letting v_1 and v_2 be the neighbors of v_0 in F', the definition of ins$'$ implies $\rho_G(a_0) \cup \rho_G(b) \subseteq \text{ins}_{\mathcal{T}}(F_{v_1} \cup F_{v_2} \cup (F' - v_0))$, and thus

$$d_{\mathcal{T}}(a_0, b) \leq d_{\mathcal{T}'}(f_0, b) + 2k.$$

In all cases, we conclude that $d_{\mathcal{T}}(a_0, b) - 2k \leq d_{\mathcal{T}'}(f_0, b)$.

Let us now upper-bound the $d_{\mathcal{T}'}$-distance of b from f_0. If $d_{\mathcal{T}}(a_0, b) \geq \theta'$, then clearly $d_{\mathcal{T}}(a_0, b) \geq d_{\mathcal{T}'}(f_0, b)$. Hence, suppose that $d_{\mathcal{T}}(a_0, b) < \theta'$, and let W be a closed walk in R_G of length $2d_{\mathcal{T}}(a_0, b)$ with $\rho_G(a_0) \cup \rho_G(b) \subseteq \text{ins}_{\mathcal{T}}(W)$. Note that in particular $e_0 \notin \text{ins}_{\mathcal{T}}(W)$. Let W' be the closed walk in R'_G obtained from W by replacing all vertices drawn in the interior of Δ by v_0 and eliminating loops at v_0. Note that $v_0 \in \text{ins}'(W')$; this is clear if $v_0 \in V(W')$, and otherwise we have $W = W'$ and $\text{ins}'(W') = \text{ins}_{\mathcal{T}}(W)$ contains $\rho_G(a_0) \supseteq v_0$. Moreover, e_0 is disjoint from $\text{ins}'(W')$ by the definition of ins$'$, and W' is obtained from W by replacing subwalks inside Δ by homotopically equivalent subwalks. If $\rho_G(b) \in V(W)$, then $\rho_{G'}(b) \in V(W')$ as well. Otherwise, W is contractible in Σ' but non-contractible in the space obtained from Σ' by drilling a hole in $\rho_G(b)$, and thus the same is the case for W'. Since $\text{ins}'(W')$ is by Lemma 2.81 equal to the complement of the face containing e_0, we conclude that $\rho_{G'}(b) \in \text{ins}'(W')$. Clearly $|W'| \leq |W|$, and thus $d_{\mathcal{T}'}(f_0, b) \leq d_{\mathcal{T}}(a_0, b)$. Therefore, (ii) holds.

Let us turn our attention to the condition (i). Consider distinct atoms a and b of G' different from f_0, and let us first give the lower bound on their $d_{\mathcal{T}'}$-distance.

- If $d_{\mathcal{T}'}(a, b) = \theta'$, then

$$d_{\mathcal{T}}(a, b) - 2k \leq \theta - 2k = \theta' = d_{\mathcal{T}'}(a, b).$$

- Hence, we can assume that $d_{\mathcal{T}'}(a,b) < \theta'$. In this case, let F' be an optimal $(\rho_{G'}(a), \rho_{G'}(b))$-restraint in $R_{G'}$. By Lemma 4.8, we can assume that F' is a $(\rho_{G'}(a), \rho_{G'}(b))$-tie.
  - If $v_0 \notin V(F')$, then $\text{ins}_{\mathcal{T}}(F') = \text{ins}'(F')$, and thus F' is also a $(\rho_G(a), \rho_G(b))$-restraint in R_G, implying that $d_{\mathcal{T}'}(a,b) \ge d_{\mathcal{T}}(a,b)$.
  - If v_0 is a cutvertex of F', then F' decomposes into a $(\rho_{G'}(a), v_0)$-restraint and a $(\rho_{G'}(b), v_0)$-restraint, showing by (ii) that

$$\begin{aligned} d_{\mathcal{T}}(a,b) &\le d_{\mathcal{T}}(a,a_0) + d_{\mathcal{T}}(a_0,b) \le d_{\mathcal{T}'}(a,f_0) + d_{\mathcal{T}'}(b,f_0) + 4k \\ &\le d_{\mathcal{T}'}(a,b) + 4k. \end{aligned}$$

  - If $d_{\mathcal{T}}(a_0,b) \le k+1$, then by (ii),

$$\begin{aligned} d_{\mathcal{T}}(a,b) &\le d_{\mathcal{T}}(a,a_0) + d_{\mathcal{T}}(a_0,b) \le d_{\mathcal{T}'}(a,f_0) + 3k + 1 \\ &\le d_{\mathcal{T}'}(a,b) + 3k + 1. \end{aligned}$$

  - Hence, we can assume that $d_{\mathcal{T}}(a_0,b) > k+1$ and $\rho_{G'}(b) \not\subseteq \Delta_0$, and symmetrically, that $\rho_{G'}(a) \not\subseteq \Delta_0$. Moreover, we can assume that $v_0 \in V(F')$ is not a cutvertex of F', and thus it is a vertex of degree two of a cycle of F'. Letting v_1 and v_2 be the neighbors of v_0 in F', the definition of ins$'$ implies $\rho_G(a) \cup \rho_G(b) \subseteq \text{ins}_{\mathcal{T}}(F_{v_1} \cup F_{v_2} \cup (F' - v_0))$, and thus $d_{\mathcal{T}}(a,b) \le d_{\mathcal{T}'}(a,b) + 2k$.

In all cases, we conclude that $d_{\mathcal{T}}(a,b) - 4k \le d_{\mathcal{T}'}(a,b)$.

It remains to upper-bound $d_{\mathcal{T}'}(a,b)$. If $d_{\mathcal{T}}(a,b) \ge \theta'$, then clearly $d_{\mathcal{T}}(a,b) \ge d_{\mathcal{T}'}(a,b)$. Hence, suppose that $d_{\mathcal{T}}(a,b) < \theta'$, and let W be closed walk in R_G of length $2d_{\mathcal{T}}(a,b)$ with $\rho_G(a) \cup \rho_G(b) \subseteq \text{ins}_{\mathcal{T}}(W)$. Let W' be the closed walk in R'_G obtained from W by replacing all vertices drawn in the interior of Δ by v_0 and eliminating loops at v_0. If $v_0 \notin V(W')$, then $\text{ins}'(W') = \text{ins}_{\mathcal{T}}(W)$, showing that $d_{\mathcal{T}'}(a,b) \le d_{\mathcal{T}}(a,b)$. Otherwise, the same argument as in the case (ii) shows that $\rho_{G'}(a), \rho_{G'}(b) \subseteq \text{ins}'(W')$, and thus $d_{\mathcal{T}'}(a,b) \le |W'|/2 \le |W|/2 = d_{\mathcal{T}}(a,b)$. Therefore, (i) holds.

Finally, let us relate the $\mathcal{T}$-freeness to $\mathcal{T}\mathcal{V}'$-freeness. Consider a $\mathcal{T}$-free set Z of vertices of G at distance at least $|Z| + 2k$ from a_0. If Z were not $\mathcal{T}'$-free, then by Lemma 4.5(iii), there would exist a 2-edge-connected confined cactus $F \subseteq R_{G'}$ such that

- the set $Z' = Z \cap \text{ins}'(F)$ has size greater than $|F \cap G'|$, and
- for every cycle $K \subseteq F$, the interior of $\text{ins}'(K)$ contains a vertex of Z'.

Since F is 2-edge-connected and has more than one vertex, every vertex $v \in V(F)$ is incident with a cycle K of F; letting $z \in Z'$ be a vertex in $\text{ins}'(K)$, we have $d_{\mathcal{T}'}(v,z) \le |K|/2 \le |F \cap G'| < |Z'|$, and thus using (ii),

$$d_{\mathcal{T}'}(v_0,v) \ge d_{\mathcal{T}'}(v_0,z) - d_{\mathcal{T}'}(v,z) > (d_{\mathcal{T}}(a_0,z) - 2k) - |Z'| \ge 0.$$

Therefore, $v_0 \notin V(F)$, and thus $\mathrm{ins}_{\mathcal{T}}(F) = \mathrm{ins}'(F) \supseteq Z'$ and $|F \cap G| = |F \cap G'| < |Z'|$. By Lemma 4.5(ii), this implies that Z' is not $\mathcal{T}$-free, and thus its superset Z also is not $\mathcal{T}$-free. This is a contradiction, and thus Z is $\mathcal{T}'$-free. □

We say that a tangle $\mathcal{T}'$ satisfying the conclusions of Lemma 4.19 is *obtained from $\mathcal{T}$ by clearing the zone* Δ. Let us remark that such a tangle $\mathcal{T}'$ is actually uniquely determined, see [8] for more details. Later, we are also going to need a statement complementary to Lemma 4.19: If the zone Δ is sufficiently broad, then we can also obtain a respectful tangle inside Δ consistent with $\mathcal{T}$.

Lemma 4.20 (⇸) *Let G be a graph with a cellular drawing on a surface Σ and let $\mathcal{T}$ be a respectful tangle in G of order θ. Let k and θ' be positive integers such that $3\theta' + k \le \theta$. Let v_0 be a vertex of G and let $\Delta \subseteq \Sigma$ be a k-zone around v_0 bounded by a cycle $K \subseteq G$ such that $d_{\mathcal{T}}(v_0, V(K)) \ge \frac{9}{2}\theta'$. Let G' be the subgraph of G drawn in Δ. Let Σ_0 be the sphere obtained from Δ by patching the hole with the open disk f_0, and let us view G' as drawn on Σ_0 (in particular, f_0 is a face of G' bounded by K). There exists a respectful tangle $\mathcal{T}'$ in G' of order θ' such that $d_{\mathcal{T}'}(v_0, f_0) = \theta'$,*

- *each atom a of G' distinct from f_0 satisfies $d_{\mathcal{T}'}(v_0, a) = \min(d_{\mathcal{T}}(v_0, a), \theta')$, and*
- *for all atoms a and a' of G', if $d_{\mathcal{T}'}(v_0, a), d_{\mathcal{T}'}(v_0, a') < \theta'$, then $d_{\mathcal{T}'}(a, a') = d_{\mathcal{T}}(a, a')$.*

Suppose moreover that for a positive integer $m \le \theta'$, there exist m pairwise vertex-disjoint cycles $K_1, \dots, K_m \subset G'$ such that for each $i \in [m]$, v_0 is contained in the open subdisk of Δ bounded by K_i, and that there exist m vertex-disjoint paths $Q_1, \dots, Q_m \subset G'$, where for $i \in [m]$, the path Q_i starts in a vertex $z_i \in V(K)$ and intersects all cycles $K_1, \dots, K_m$. Then the set $Z = \{z_1, \dots, z_m\}$ is $\mathcal{T}'$-free.

Proof Consider a cycle C in $R_{G'}$ of length less than $6\theta'$. Since $d_{\mathcal{T}}(v_0, V(K)) > 3\theta'$, note that C cannot contain both $\rho_{G'}(f_0)$ and v_0.

- If $\rho_{G'}(f_0) \notin V(C)$, then let $\mathrm{ins}(C)$ be the disk in Σ_0 not containing $\rho_{G'}(f_0)$.
- If $\rho_{G'}(f_0) \in V(C)$, then let $\mathrm{ins}(C)$ be the disk in Σ_0 not containing v_0.

We claim that ins is a slope of order $3\theta'$ in $R_{G'}$. Indeed, (S1) is easy to verify, and for (S2), the only problematic case is the one when the theta-subgraph of $R_{G'}$ consist of paths P_1, P_2, and P_3 and $\rho_{G'}(f_0)$ is an internal vertex of one of them, say of P_1. Suppose for a contradiction that (S2) is false, and thus $\mathrm{ins}(P_1 \cup P_2) \cup \mathrm{ins}(P_1 \cup P_3) \cup \mathrm{ins}(P_2 \cup P_3) = \Sigma_0$. By the definition of ins, we have $v_0 \notin \mathrm{ins}(P_1 \cup P_2) \cup \mathrm{ins}(P_1 \cup P_3)$, and thus $v_0 \in \mathrm{ins}(P_2 \cup P_3)$. Moreover, the definition of ins implies that $\rho_{G'}(f_0) \notin \mathrm{ins}(P_2 \cup P_3)$, and thus $\mathrm{ins}(P_2 \cup P_3) \subseteq \Delta$. We have $\|P_2 \cup P_3\| < 6\theta' \le 2(\theta - k)$, and thus by Lemma 4.17, we conclude that $\mathrm{ins}_{\mathcal{T}}(P_2 \cup P_3) = \mathrm{ins}(P_2 \cup P_3)$. Therefore, there exists a closed walk W in $(P_1 - \rho_{G'}(f_0)) \cup P_2 \cup P_3$ of length at most $\|P_1 \cup P_2 \cup P_3\|$ such that $\mathrm{ins}_{\mathcal{T}}(W)$ contains both v_0 and a vertex of K. Consequently, $\|P_1 \cup P_2 \cup P_3\| \ge 2d_{\mathcal{T}}(v_0, V(K)) \ge 9\theta'$. This implies that $P_1 \cup P_2 \cup P_3$ contains a cycle of length at least $6\theta'$, contradicting the assumptions of (S2). Therefore, (S2) holds as well, and ins is indeed a slope.

It is easy to see that the slope ins satisfies the assumptions of Lemma 2.90, and thus it induces a respectful tangle $\mathcal{T}_0$ in G' of order $3\theta'$. Let us now argue that $d_{\mathcal{T}_0}(v_0, f_0) = 3\theta'$. Indeed, if $d_{\mathcal{T}_0}(v_0, f_0) < 3\theta'$, then by Lemma 4.8, there would exist a $(v_0, \rho_{G'}(f_0))$-tie F' with perimeter of length less than $6\theta'$ such that $v_0, \rho_{G'}(f_0) \in \text{ins}(F')$. The definition of ins implies that $\rho_{G'}(f_0) \in V(F')$, and by the definition of ins and Lemma 4.17, we have $v_0 \in \text{ins}(F' - \rho_{G'}(f_0)) = \text{ins}_{\mathcal{T}}(F' - \rho_{G'}(f_0))$. We conclude that the perimeter of F' has length at least $2d_{\mathcal{T}}(v_0, V(K)) > 3\theta'$, which is a contradiction.

Consider now an atom $a \notin \{v_0, f_0\}$ of G'; we aim to show that $d_{\mathcal{T}_0}(v_0, a) = \min(d_{\mathcal{T}}(v_0, a), 3\theta')$.

- The inequality $\min(d_{\mathcal{T}}(v_0, a), 3\theta') \leq d_{\mathcal{T}_0}(v_0, a)$ is trivial when $d_{\mathcal{T}_0}(v_0, a) = 3\theta'$, and thus suppose that $d_{\mathcal{T}_0}(v_0, a) < 3\theta'$. Let F' be a $(v_0, \rho_{G'}(a))$-tie with $v_0, \rho_{G'}(a) \subset \text{ins}(F')$ and perimeter of length $2d_{\mathcal{T}_0}(v_0, a)$, which exists by Lemma 4.8. By the previous paragraph, we have $\rho_{G'}(f_0) \notin V(F')$, and thus Lemma 4.17 and the definition of ins implies $\text{ins}_{\mathcal{T}}(F') = \text{ins}(F')$. Consequently, $d_{\mathcal{T}}(v_0, a) \leq d_{\mathcal{T}_0}(v_0, a)$.
- The inequality $d_{\mathcal{T}_0}(v_0, a) \leq \min(d_{\mathcal{T}}(v_0, a), 3\theta')$ is trivial when $d_{\mathcal{T}}(v_0, a) \geq 3\theta'$, and thus suppose that $d_{\mathcal{T}}(v_0, a) < 3\theta'$. Let F be a $(v_0, \rho_G(a))$-tie with $v_0, \rho_G(a) \subset \text{ins}_{\mathcal{T}}(F)$ and perimeter of length $2d_{\mathcal{T}}(v_0, a)$, which exists by Lemma 4.8. Since $d_{\mathcal{T}}(v_0, V(K)) > 3\theta'$, no vertex of K is contained in $\text{ins}_{\mathcal{T}}(F)$, and since $v_0 \in \text{ins}_{\mathcal{T}}(F)$, we conclude that F is contained in the interior of Δ. By Lemma 4.17 and the definition of ins, we have $\text{ins}(F) = \text{ins}_{\mathcal{T}}(F)$, and thus $d_{\mathcal{T}_0}(v_0, a) \leq d_{\mathcal{T}}(v_0, a)$.

Consider now distinct atoms $a \neq v_0 \neq a'$ of G' such that $d_{\mathcal{T}_0}(v_0, a) < \theta'$ and $d_{\mathcal{T}_0}(v_0, a') < \theta'$; we aim to show that $d_{\mathcal{T}_0}(a, a') = d_{\mathcal{T}}(a, a')$.

- By the triangle inequality, we have $d_{\mathcal{T}_0}(a, a') < 2\theta'$. By Lemma 4.8, there exists a $(\rho_{G'}(a), \rho_{G'}(a'))$-tie F' with $\rho_{G'}(a), \rho_{G'}(a') \subset \text{ins}(F')$ and perimeter of length $2d_{\mathcal{T}_0}(a, a')$. Note that $\rho_{G'}(f_0) \notin V(F')$, as otherwise the tie F' would show that $d_{\mathcal{T}_0}(v_0, f_0) \leq d_{\mathcal{T}_0}(v_0, a) + d_{\mathcal{T}_0}(a, f_0) < \theta' + 2\theta' = 3\theta'$. By Lemma 4.17 and the definition of ins, we have $\text{ins}_{\mathcal{T}}(F') = \text{ins}(F')$, and thus $d_{\mathcal{T}}(a, a') \leq d_{\mathcal{T}_0}(a, a')$.
- Let us now prove the converse inequality. We have $d_{\mathcal{T}}(v_0, a) = d_{\mathcal{T}_0}(v_0, a) < \theta'$ and $d_{\mathcal{T}}(v_0, a') = d_{\mathcal{T}_0}(v_0, a') < \theta'$. By the triangle inequality, we have $d_{\mathcal{T}}(a, a') < 2\theta'$; let F be a $(\rho_G(a), \rho_G(a'))$-tie with $\rho_G(a), \rho_G(a') \subset \text{ins}_{\mathcal{T}}(F)$ and perimeter of length $2d_{\mathcal{T}}(a, a')$, which exists by Lemma 4.8. Note that $d_{\mathcal{T}}(a, V(K)) \geq d_{\mathcal{T}}(v_0, V(K)) - d_{\mathcal{T}}(v_0, a) > 2\theta'$, and thus no vertex of K is contained in $\text{ins}_{\mathcal{T}}(F)$. Since $\rho_G(a), \rho_G(a') \subset \text{ins}_{\mathcal{T}}(F)$, we conclude that F is contained in the interior of Δ. By Lemma 4.17 and the definition of ins, we have $\text{ins}(F) = \text{ins}_{\mathcal{T}}(F)$, and thus $d_{\mathcal{T}_0}(a, a') \leq d_{\mathcal{T}}(a, a')$.

Finally, let $K_1, \ldots, K_m, Q_1, \ldots, Q_m$, and Z be as in the statement of the lemma, and suppose for a contradiction that Z is not $\mathcal{T}_0$-free. By Lemma 4.5, there exists a 2-edge-connected cactus $F \subseteq R_{G'}$ such that for every cycle C in F, the interior of $\text{ins}(C)$ contains a vertex of Z, and $|Z \cap \text{ins}(F)| > |F \cap G'|$. Since $Z \subseteq K$, the definition of ins implies that F is a union of cycles intersecting only in $\rho_{G'}(f_0)$, and that $v_0 \notin \text{ins}(F)$. In particular, there exists a cycle $C \subseteq F$ such that $|Z \cap \text{ins}(C)| > |C \cap G'|$. Since the paths $Q_1, \ldots, Q_m$ are pairwise vertex-disjoint, there exists $i \in [m]$ such that $z_i \in Z \cap \text{ins}(C)$ and $V(Q_i \cap C) = \emptyset$, and thus $Q_i \subset \text{ins}(C)$. Moreover, since $m \geq |Z \cap \text{ins}(C)| > |C \cap G'|$, there exists $j \in [m]$ such that $V(K_j \cap C) = \emptyset$, and since K_j intersects Q_i, we have $K_j \subset \text{ins}(C)$. However, by the definition of ins, the disk ins(C) does not contain v_0, while the subdisk of Δ bounded by K_j does, which is a contradiction. Therefore, the set Z is $\mathcal{T}_0$-free.

The conclusions of the lemma now follow by letting $\mathcal{T}'$ be the truncation of $\mathcal{T}_0$ to order θ'. □

4.3 Linking Through a Cylindrical Grid

Corollary 4.16 enables us to build a nest of cycles around a given atom a_0. As we have seen in Sect. 3.3, such nests are useful in redirecting linkages to avoid the atom a_0. Lemma 4.18 moreover gives us an additional property for this nest: There exists a large linkage between the innermost and the outermost cycle of the nest; we have seen an application of such a linkage at the end of the proof of Lemma 4.20. Using Lemma 2.64, we can furthermore assume that the nest and the linkage form a cylindrical grid minor around a_0. We are going to use such a cylindrical grid (which we call a *battlefield around* a_0) many times throughout the book to connect parts of constructions near a_0 to those far from a_0. In particular, after describing the construction of the battlefield in more detail (and with additional properties which will be useful in later applications), we are going use it to generalize Corollary 4.4 to graphs drawn on the cylinder.

Recall that an (A, B)-linkage is a system of pairwise vertex-disjoint paths with one end in A and the other end in B. The linkage is ***strict*** if only the first vertex of each of the paths belongs to A and only the last vertex belongs to B. Let G be a graph with a cellular drawing on a surface Σ, let $\mathcal{T}$ be a respectful tangle in G of order θ, and let a_0 be an atom of G. For integers $r \geq 2$ and $p \geq 1$, an $r \times p$ ***battlefield around*** a_0 in G consists of a sequence $\mathcal{R} = C_1, \ldots, C_r$ of pairwise vertex-disjoint cycles of G and of a strict $(V(C_1), V(C_r))$-linkage $\mathcal{P}$ of size p in G such that

(F1) for $i \in [r]$, the cycle C_i bounds a disk $\Delta_i \subset \Sigma$ such that $a_0 \subset \Delta_1 \subset \ldots \subset \Delta_r$, and

(F2) the linkage $\mathcal{P}$ is contained in Δ_r and $(\mathcal{R}, \mathcal{P})$ is an orderly sieve.

For a positive integer k, the battlefield is *k-local* if additionally

(F3) Δ_r is a k-zone around a_0,

and *graded* if

(F4) for $1 \leq i < j \leq r$, all atoms $a_1 \subset \Delta_i$ and $a_2 \not\subseteq \Delta_j$ satisfy $d_{\mathcal{T}}(a_1, a_2) \geq 2(j - i)$.

The battlefield is *free* if

(F5) the set Z of ends of $\mathcal{P}$ in C_r is $\mathcal{T}'$-free in the tangle $\mathcal{T}'$ obtained from $\mathcal{T}$ by clearing the zone Δ_r, as well as $\mathcal{T}$-free.

The *egg* of the battlefield is Δ_1.

Theorem 4.21 *Let G be a graph with a cellular drawing on a surface Σ, let $\mathcal{T}$ be a respectful tangle in G of order θ, and let a_0 be an atom of G. Let $r \geq 2$, $p \geq 1$, and $k \geq p^2 + 3p + 5r$ be integers. If $2k < \theta$, then there exists a free k-local graded $r \times p$ battlefield around a_0 whose egg contains all atoms a of G such that $d_{\mathcal{T}}(a_0, a) \leq k - 3(r + p)$.*

Proof Let $r_1 = r+p$ and $p_1 = p^2$. For $i \in [r_1]$, let $k_i = k-3(r_1-i)$. Corollary 4.16 implies that there exists a cycle C_i' bounding a k_i-zone Δ_i' around a_0 such that $k_i - 2 \leq d_{\mathcal{T}}(a_0, v) \leq k_i$ for every $v \in V(C_i')$. In particular, the cycles $C_1', \ldots, C_{r_1}'$ are pairwise vertex-disjoint, and $a_0 \subset \Delta_1' \subset \ldots \subset \Delta_{r_1}'$. Moreover, Corollary 4.16 also implies that every atom a of G such that $d_{\mathcal{T}}(a_0, a) \leq k_1 - 3 = k - 3(r + p)$ is contained in Δ_1'.

Each vertex $z \in V(C_1')$ satisfies $d_{\mathcal{T}}(a_0, z) \geq k_1 - 2 = k - 3(r + p) + 1 > p_1$, and thus by Lemma 4.18, there exists a $(V(C_1'), V(C_{r_1}'))$-linkage in G of size p_1. By Lemma 2.64, we conclude that there exists a sequence $\mathcal{R}_1 = C_1, \ldots, C_{r_1}$ of pairwise vertex-disjoint cycles and a strict $(V(C_1), V(C_{r_1}))$-linkage $\mathcal{P}_1$ of size p_1 in G such that $C_1 = C_1'$, $C_{r_1} = C_{r_1}'$, there exist disks $\Delta_i \subset \Sigma$ bounded by C_i for $i \in [r_1]$ such that $a_0 \subset \Delta_1 \subset \ldots \subset \Delta_{r_1}$, and $(\mathcal{R}_1, \mathcal{P}_1)$ forms an orderly sieve. Let $\mathcal{P}$ be the strict $(V(C_1), V(C_r))$-linkage of size p obtained from $\mathcal{P}_1$ by taking every p-th path in order of their ends along C_1 and truncating them at their first intersection with C_r. Let $\mathcal{R}$ be the sequence $C_1, \ldots, C_r$.

We claim that $(\mathcal{R}, \mathcal{P})$ is a free k-local graded $r \times p$ battlefield around a_0. The conditions (F1) and (F2) are clearly satisfied, and (F3) holds since Δ_r is a subset of the k-zone $\Delta_{r_1} = \Delta_{r_1}'$.

For (F4), consider atoms $a_1 \subset \Delta_i$ and $a_2 \not\subseteq \Delta_j$, where $1 \leq i < j \leq r$, and suppose for a contradiction that $d_{\mathcal{T}}(a_1, a_2) < 2(j - i)$. Let F be an optimal $(\rho_G(a_1), \rho_G(a_2))$-restraint. Since $a_2 \not\subseteq \Delta_j$, we have $\mathrm{ins}_{\mathcal{T}}(F) \not\subseteq \Delta_j$, and thus $F \not\subseteq \Delta_j$ by Lemma 4.17. Consequently, $F \cap \Delta_i = \emptyset$, since if $F \cap \Delta_i \neq \emptyset$, then F would have to intersect $C_i, C_{i+1}, \ldots, C_j$, and thus its perimeter would have length at least $4(j - i)$. Since $a_1 \subset \Delta_i$ is contained in $\mathrm{ins}_{\mathcal{T}}(F)$, it follows that $\Delta_i \subset \mathrm{ins}_{\mathcal{T}}(F)$. Since the disk $\Delta_1 \subseteq \Delta_i$ is bounded by C_1', the restraint F shows that $d_{\mathcal{T}}(a_0, z) < 2(j - i) < 2r < k_1 - 2$ for every $z \in V(C_1')$. This contradicts the choice of C_1'.

Finally, let Z be the set of ends of the paths of $\mathcal{P}$ in C_r, and let us argue that (F5) holds. Let G' be the graph obtained from G by clearing the zone Δ_r, let f_0 be the resulting face, and let $v_0 = \rho_{G'}(f_0)$. If Z is not $\mathcal{T}'$-free, then by Lemma 4.5, there exists a 2-edge-connected confined strictly outerplanar cactus $F \subseteq R_{G'}$ with outer face f such that $\mathrm{ins}_{\mathcal{T}'}(F) = \Sigma \setminus f$, for every cycle K in F, the interior of $\mathrm{ins}_{\mathcal{T}'}(K)$ contains a vertex of Z, and $|Z \cap \mathrm{ins}_{\mathcal{T}'}(F)| > |F \cap G'|$.

If a cycle K in F does not contain v_0, then since a vertex of Z is contained in the interior of $\text{ins}_{\mathcal{T}'}(K)$, we conclude that $\Delta_r \subseteq \text{ins}_{\mathcal{T}'}(K) = \text{ins}_{\mathcal{T}}(K)$. However, this implies that $d_{\mathcal{T}}(a_0, z) \leq |K|/2 \leq |F \cap G'| < |Z| = p < k_1 - 2$ for every $z \in V(C_1')$, contradicting the choice of C_1'.

Therefore, all cycles in F intersect in v_0. This implies that the sets $Z \cap \text{ins}_{\mathcal{T}'}(K)$ and $V(K) \cap V(G')$ are disjoint for distinct cycles K of F, and thus there exists a cycle K in F such that $p \geq |Z \cap \text{ins}_{\mathcal{T}'}(K)| > |K \cap G'| = |K|/2$. Note that $K \subset \Delta_{r_1}$, since K cannot intersect all cycles $C_r, C_{r+1}, \ldots, C_{r+p} = C_{r_1}$. But since $|Z \cap \text{ins}_{\mathcal{T}'}(K)| \geq 2$ and we have taken every p-th path of $\mathcal{P}_1$ to $\mathcal{P}$, we conclude that K must intersect at least $p > |Z|$ paths of $\mathcal{P}_1$, which is a contradiction. Therefore, Z is $\mathcal{T}'$-free, and since $\mathcal{T}'$ is conformal with $\mathcal{T}$, Lemma 2.48 implies that Z is also $\mathcal{T}$-free. We conclude that (F5) is also satisfied.

Therefore, $(\mathcal{R}, \mathcal{P})$ is indeed a free k-local graded $r \times p$ battlefield around a_0. Moreover, the egg Δ_1 contains all atoms a of G such that $d_{\mathcal{T}}(a_0, a) \leq k - 3(r + p)$ since $\Delta_1 = \Delta_1'$. □

Let us remark that because the construction of the battlefield is performed within a k-zone and Lemma 4.19 gives us control over what happens after we clear the zone, we can build a battlefield around many distant faces independently. Such battlefields are often used in a fashion similar to Lemma 3.12 in the following situation: Let G be an induced subgraph of a larger graph G_0, and suppose that G is drawn in a surface so that for with each component H of $G_0 - V(G)$, there exists a face f_H of G such that all neighbors of H in G are contained in the boundary of f_H. We are able to argue that such a component H (possibly together with some nearby parts of G) contains a substructure S we want to use in some construction. An issue is that we need S to be linked to the rest of the construction in other parts of G_0, and the process used to find S does not give us any control over the connectivity to the rest of G_0. Performing the construction inside a battlefield around f_H often gives us a way to overcome this issue.

Let us give a specific example. Given a cyclic sequence $s_1, \ldots, s_m$ and distinct indices $i_1 < i_2$ and $i_3 < i_4$ in $[m]$, we say that the sets $\{s_{i_1}, s_{i_2}\}$ and $\{s_{i_3}, s_{i_4}\}$ are *interlaced* in the sequence if $i_1 < j < i_2$ holds for exactly one $j \in \{i_3, i_4\}$. Given a graph H and a cyclic sequence S of vertices of H, an *S-cross* in H consists of two vertex-disjoint paths Q_1 and Q_2 with ends in S such that the ends of Q_1 are interlaced with the ends of Q_2.

Lemma 4.22 *Let (G, H) be a separation of a graph G_0, where G has a cellular drawing on a surface Σ with the vertices of $V(G \cap H)$ drawn in the boundary of a face a_0 of G. Suppose that $(\mathcal{R}, \mathcal{P})$ is a 3×4 battlefield around a_0 in G, where $\mathcal{R} = C_1, C_2, C_3$. Let $\Delta_3 \subset \Sigma$ be the disk bounded by C_3 containing a_0 and let $G_3 = (G \cap \Delta_3) \cup H$. Let Z be the sequence of the ends of the paths of $\mathcal{P}$ on C_3 in cyclic order around C_3. If G_3 contains a $V(C_3)$-cross, then $G_3 - (V(C_3) \setminus V(Z))$ contains a Z-cross.*

Let us remark that using Theorem 4.21, the battlefield can be chosen to be free and 43-local. The original $V(C_3)$-cross (which may be separated from the rest of

$G_0 - (V(G_3) \setminus V(C_3))$ by a small cut, e.g., it can attach to a subpath of C_3 with no neighbors outside of Δ_3) thus can be "upgraded" to one with ends in the set Z which is free even after clearing the zone Δ_3, and thus well-linked to other parts of G.

Proof of Lemma 4.22 Consider a $V(C_3)$-cross (Q_1, Q_2) in G_3 with as few vertices in $V(C_3)\setminus Z$ as possible. Let $Z' = Z\cup(V(Q_1\cup Q_2)\cap V(C_3))$; we claim that $Z' = Z$, which implies the conclusion of the lemma. We can assume that the paths Q_1 and Q_2 only intersect Z' in their ends, since if an interior vertex z of say Q_1 belonged to Z', then Q_2 together with one of the subpaths of Q_1 from z to an end of Q_1 would also form a Z'-cross, contradicting the choice of (Q_1, Q_2). Suppose for a contradiction that there exists a vertex $z \in Z' \setminus Z$, say an end of Q_1.

Let $Z = \{z_1, \dots, z_4\}$, where the labels are chosen so that z_1, z_2, z, z_3, and z_4 appear on C_3 in order. For $i \in [2]$, let B_i be the path consisting of

- the segments of the paths of $\mathcal{P}$ from z_i and z_{5-i} to their first intersections u_i and v_i with C_i, respectively, and
- a segment of C_i between u_i and v_i chosen so that B_i separates z from a_0 in Δ_3.

Note that since the sieve $(\mathcal{R}, \mathcal{P})$ is orderly, the paths B_1 and B_2 are vertex-disjoint. The path Q_1 must intersect H, since otherwise it would be drawn in Δ_3 and the vertex-disjoint path Q_2 would not be able to cross it. Consequently, the subgraph $Q_1 \cup Q_2$ (in fact, Q_1) intersects both B_1 and B_2.

Since $Z' \neq Z$ and the paths Q_1 and Q_2 intersect C_3 only in their ends, there exists $i \in [2]$ and an end u of B_i such that $u \notin V(Q_1 \cup Q_2)$. Let B be the subpath of B_i from u to the first intersection v with $Q_1 \cup Q_2$, and let $j \in [2]$ be the index such that $v \in V(Q_j)$. Let Q'_j and Q''_j be the two paths consisting of B and a subpath of Q_j from v to an end of Q_j. Observe that Q_{3-j} together with one of the paths Q'_j and Q''_j forms a Z'-cross in G_3 with fewer vertices in $V(C_3) \setminus Z$ than (Q_1, Q_2), which is a contradiction. □

The next application of battlefields will be useful when considering the linkage problem on general surfaces, and it will directly give us a generalization of Corollary 4.4 from the disk to the cylinder. Consider a graph G_0 with a cellular drawing on the sphere Σ_0 with a respectful tangle $\mathcal{T}$. Let S_1 and S_2 be sets of vertices incident with distinct faces f_1 and f_2 of G_0, and let $\mathcal{S}$ be a root partition of $S_1 \cup S_2$ such that each part $T \in \mathcal{S}$ consists of a vertex of S_1 and a vertex of S_2. Suppose that $\mathcal{S}$ is topologically feasible in $\Sigma_0 \setminus (f_1 \cup f_2)$, i.e., the pairs of $\mathcal{S}$ intersect the boundaries of f_1 and f_2 in the same cyclic order. If the sets S_1 and S_2 are $\mathcal{T}$-free, then by Lemma 2.44, there exists an (S_1, S_2)-linkage $\mathcal{L}$ in $\mathcal{T}$. Clearly, $\mathcal{L}$ is not necessarily an $\mathcal{S}$-linkage, since its paths may not connect the right pairs of vertices of S_1 and S_2. However, if there exists a large battlefield separating f_1 and f_2, it should be intuitively clear that we can reroute the paths inside the battlefield to turn $\mathcal{L}$ into an $\mathcal{S}$-linkage. Let us now make this intuition precise, strengthening the claim a bit at the same time.

Let H be a graph drawn on the cylinder Π, let β_1 and β_2 be the two components of the boundary of Π, let S_0 be a set of vertices of H incident with β_1 and let Z be a set of vertices of H incident with β_2. We say that a root partition $\mathcal{S}$ with $\bigcup\mathcal{S} \subseteq S_0 \cup Z$ is *(S_0, Z)-simple* if each part $T \in \mathcal{S}$ satisfies $T \cap S_0 \neq \emptyset$ and $|T \cap Z| \leq 1$. We say that the set Z is *S_0-linkage-universal* in H if for every (S_0, Z)-simple root partition $\mathcal{S}$ topologically feasible in Π, there exists an $\mathcal{S}$-linkage in H intersecting β_2 only in the vertices of $Z \cap \bigcup\mathcal{S}$.

Lemma 4.23 *For all positive integers $m \leq m'$, there exists an integer k such that the following claim holds. Let G be a graph with a cellular drawing on a surface Σ, let $\mathcal{T}$ be a respectful tangle of order $\theta > 2k$, let a_0 be a face of G, and let S_0 be a $\mathcal{T}$-free set of m vertices incident with a_0. Then there exists a cycle C bounding a k-zone $\Delta \subset \Sigma$ around a_0 and a set $Z \subseteq V(C)$ of size m' such that Z is*

- *S_0-linkage-universal in the subgraph of G drawn in the cylinder $\Delta \setminus a_0$, and*
- *$\mathcal{T}'$-free in the tangle $\mathcal{T}'$ obtained from $\mathcal{T}$ by clearing Δ.*

Proof Let $r = 3m+3$. Without loss of generality, we can assume that $m' > 2m$. By Theorem 4.21, for sufficiently large k, there exists a free k-local $r \times m'$ battlefield formed by a sequence of cycles $\mathcal{R} = C_1, \ldots, C_r$ and a $(V(C_1), V(C_r))$-linkage $\mathcal{P}$ of size m'. For $i \in [r]$, let $\Delta_i \subset \Sigma$ be the closed disk bounded by C_i and containing a_0. Let Z be the set of ends of paths of $\mathcal{P}$ in C_r, and let $\Delta = \Delta_r$ and $C = C_r$. We claim that Z is S_0-linkage-universal in $G \cap (\Delta \setminus a_0) = G \cap \Delta$. Indeed, consider any (S_0, Z)-simple root partition $\mathcal{S}$ topologically feasible in $\Delta \setminus a_0$. Let $Z' = Z \cap \bigcup\mathcal{S}$. We need to show that there exists an $\mathcal{S}$-linkage in $G \cap \Delta$ intersecting C only in Z'.

Let B be the set of the first points of the paths of $\mathcal{P}$ in C_{m+1}. We claim that there exists an (S_0, B)-linkage $\mathcal{P}_1$ of size m in $G_{m+1} = G \cap \Delta_{m+1}$ such that $\mathcal{P}_1$ intersects C_{m+1} only in the ends of its paths: By Menger's theorem, it suffices to verify that for every set $X \subseteq V(G_{m+1})$ of size less than m, there exists a path in G_{m+1} from S_0 to B disjoint from $X \cup (V(C_{m+1}) \setminus B)$. Since both S_0 and Z are $\mathcal{T}$-free, Lemma 2.44 implies that $G - X$ contains a path Q_1 from S_0 to Z, and the initial segment Q_1' of this path till C_m is disjoint from C_{m+1}. There exists $i \in [m]$ such that C_i is disjoint from X. Finally, since $|B| = m' \geq m$, there exists a vertex $v \in B$ such that the initial segment Q_2 of the path of $\mathcal{P}$ ending in v is disjoint from X. Then $Q_1' \cup C_i \cup Q_2$ connects a vertex of S_0 to a vertex of B in $G_{m+1} - (X \cup (V(C_{m+1}) \setminus B))$.

Let B_1 be the set of ends of paths of $\mathcal{P}_1$ in B, and let $\mathcal{S}_1$ be obtained from $\mathcal{S}'$ by replacing each vertex of S_0 by the vertex in B_1 to which it is joined by a path in $\mathcal{P}_1$. Observe that a topological realization of $\mathcal{S}$ in $\Delta \setminus a_0$ can be continuously deformed into a topological realization I_1 of $\mathcal{S}_1$ in the cylinder $\Gamma = \overline{\Delta \setminus \Delta_{m+1}}$. It suffices to show that there exists an $\mathcal{S}_1$-linkage in $G \cap \Gamma$ intersecting C only in Z', since such a linkage can be combined with $\mathcal{P}_1$ to an $\mathcal{S}$-linkage in $G \cap \Delta$.

Such an $\mathcal{S}_1$-linkage in $G \cap \Gamma$ is easy to construct, since its roots are ends of pillars of an orderly $(2m+3) \times m'$ sleeve and $m' > 2m$. In case that $Z' = \emptyset$, we can directly apply Theorem 4.2 in the disk obtained from Γ by patching the hole bounded by C. Otherwise, fix any simple curve γ in I_1 from a vertex $z \in Z'$ to a vertex $s \in S_0$ and otherwise disjoint from the boundary of Γ. Let P_1 be the segment of a pillar of $\mathcal{P}$ from z to C_{2m+2}, and let P_2 be the segment of a pillar of $\mathcal{P}$ from s to C_{2m+2}. Let z' and s' be the ends of P_1 and P_2 in C_{2m+2}, and let Q be an arc of C_{2m+2} between z' and s' intersecting at most $(m'-2)/2$ pillars different from those containing P_1 and P_2. We cut $G \cap \Gamma$ along the path $P_1 \cup Q \cup P_2$ and observe that Theorem 4.2 applies to the resulting linkage problem in the disk obtained by cutting Γ along γ, with the topological realization obtained by cutting I_1 along γ.

Another possibility is to construct an $\mathcal{S}_1$-linkage directly in the cylindrical grid obtained as a minor of the sleeve. We leave the details up to the reader. □

Let us note the following observation on topological feasibility in the cylinder.

Observation 4.24 *Let $\mathcal{S}$ be a root partition of vertices drawn in the boundary of the cylinder Π, and let γ be a simple closed curve drawn in the interior of Π so that γ separates the two components of the boundary of Π. If $\mathcal{S}$ is topologically feasible in Π, then there exists a topological realization I of $\mathcal{S}$ such that for each $T \in \mathcal{S}$, if T intersects both components of the boundary of Π, then the star of I representing T intersects γ exactly in its center, and otherwise it is disjoint from γ.*

Proof Let γ_1 and γ_2 be the components of the boundary of Π, let $\mathcal{S}_0$ consist of the sets of $\mathcal{S}$ intersecting both γ_1 and γ_2, and let us fix a topological realization I' of $\mathcal{S}$.

- If $\mathcal{S}_0 = \emptyset$, then the graph I' in its drawing on Π has a unique face f intersecting both γ_1 and γ_2, and this face separates γ_1 from γ_2. Hence, there exists a simple closed curve $\gamma' \subset f$ separating γ_1 from γ_2.
- If $\mathcal{S}_0 \neq \emptyset$, then let $f_1, \ldots, f_m$ be the faces of I' intersecting both γ_1 and γ_2. Observe that for each $i \in [m]$, the face f_i has exactly two incidences with the centers of the stars of I representing sets from $\mathcal{S}_0$, and let γ_i' be a simple curve in f_i joining these incidences. Let γ' be the concatenation of the curves $\gamma_1, \ldots, \gamma_m$. Observe that γ' is a simple closed curve separating γ_1 from γ_2 and that γ' intersects I' exactly in the centers of the stars representing the sets of $\mathcal{S}$ intersecting both components of the boundary of Π.

There exists a homeomorphism of Π fixing γ_1 and γ_2 and mapping γ' to γ. This homeomorphism maps I' to the desired topological realization I. □

We can now generalize Corollary 4.4 to the cylinder: We only need to add the assumption that the boundaries of the cylinder are far apart in the metric derived from the tangle!

Corollary 4.25 *For every positive integer m, there exists an integer θ such that the following claim holds. Let G be a graph with cellular drawing on the sphere Σ, let $\mathcal{T}$ be a respectful tangle in G of order at least θ, let f_1 and f_2 be a faces of G, and let $\mathcal{S}$ be a root partition of a subset Z of at most m vertices of G incident with $f_1 \cup f_2$. If $\mathcal{S}$ is topologically feasible in $\Sigma \setminus (f_1 \cup f_2)$, the set Z is $\mathcal{T}$-free, and $d_{\mathcal{T}}(f_1, f_2) \geq \theta$, then G contains an $\mathcal{S}$-linkage.*

Proof For $i \in [2]$, let S_i be the set of vertices of Z incident with f_i and apply Lemma 4.23 at f_i with $m' = m$, obtaining a cycle C_i bounding a k-zone $\Delta_i \subset \Sigma$ around f_i and a set $Z_i \subseteq V(C_i)$ of size m which is S_i-linkage-universal in $G \cap (\Delta_i \setminus f_i)$ and free in the tangle obtained by clearing the zone Δ_i. Let G' and $\mathcal{T}'$ be obtained from G and $\mathcal{T}$ by clearing both Δ_1 and Δ_2. Note that since f_1 and f_2 are far apart, the disks Δ_1 and Δ_2 are disjoint and the last part of Lemma 4.19 implies that the sets Z_1 and Z_2 are both $\mathcal{T}'$-free.

Let $\mathcal{S}_0$ be the subset of $\mathcal{S}$ consisting of the sets intersecting boundaries of both f_1 and f_2, and let $m_0 = |\mathcal{S}_0|$. Let γ be the boundary of Δ_1 and let I be the topological realization of $\mathcal{S}$ in $\Sigma \setminus (f_1 \cup f_2)$ obtained using Observation 4.24. Thus, I intersects γ in m_0 points. By applying a homeomorphism to $\Sigma \setminus (f_1 \cup f_2)$ (fixing the boundary of the cylinder and mapping γ to itself), we can assume that these points form a subset Z'_1 of Z_1. For each $T \in \mathcal{S}_0$, let z_T be the vertex of Z'_1 corresponding to the center of the star of I representing T. Let $\mathcal{S}_1$ be the root partition consisting of the parts $T \in \mathcal{S}$ with $T \subseteq S_1$ and of the parts $(T \setminus S_2) \cup \{z_T\}$ for $T \in \mathcal{S}_0$. Note that $\mathcal{S}_1$ is (S_1, Z_1)-simple, and $I \cap \Delta_1$ shows that $\mathcal{S}_1$ is topologically feasible in $\Delta_1 \setminus f_1$. Since Z_1 is S_1-linkage-universal in $G \cap (\Delta_1 \setminus f_1)$, there exists an $\mathcal{S}_1$-linkage $\mathcal{P}_1$ in $G \cap \Delta_1$ intersecting C_1 only in Z'_1.

Let Z'_2 be a subset of Z_2 of size m_0. The sets Z'_1 and Z'_2 are both $\mathcal{T}'$-free, and thus by Lemma 2.44, there exists a (Z'_1, Z'_2)-linkage $\mathcal{P}_0$ in G' of size m_0. Let γ' be the boundary of Δ_2 and let I' be the topological realization of $\mathcal{S}$ in $\Sigma \setminus (f_1 \cup f_2)$ obtained using Observation 4.24 with γ' playing the role of γ. By applying a homeomorphism to $\Sigma \setminus (f_1 \cup f_2)$ (fixing the boundary of the cylinder and mapping γ' to itself), we can assume that for each $T \in \mathcal{S}_0$, the center of the star of I' representing T is the vertex $z'_T \in Z'_2$ joined to z_T by a path of $\mathcal{P}_0$.

Let $\mathcal{S}_2$ be the root partition consisting of the parts $T \in \mathcal{S}$ with $T \subseteq S_2$ and of the parts $(T \setminus S_1) \cup \{z'_T\}$ for $T \in \mathcal{S}_0$. Note that $\mathcal{S}_2$ is (S_2, Z_2)-simple, and $I' \cap \Delta_2$ shows that $\mathcal{S}_2$ is topologically feasible in $\Delta_2 \setminus f_2$. Since Z_2 is S_2-linkage-universal in $G \cap (\Delta_2 \setminus f_2)$, there exists an $\mathcal{S}_2$-linkage $\mathcal{P}_2$ in $G \cap \Delta_2$ intersecting C_2 only in Z'_2. Finally, observe that $\mathcal{P}_1$, $\mathcal{P}_0$, and $\mathcal{P}_2$ combine to an $\mathcal{S}$-linkage in G. □

Let us remark that the assumption that Z is $\mathcal{T}$-free can be checked at each face separately, as easily follows from Lemma 4.5.

Observation 4.26 *Let $\mathcal{T}$ be a respectful tangle of order θ in a graph G with a cellular drawing on a surface Σ, and let Z_1 and Z_2 be non-empty sets of vertices of G such that $d_{\mathcal{T}}(z_1, z_2) \geq 2(|Z_1| + |Z_2|)$ for every $z_1 \in Z_1$ and $z_2 \in Z_2$. If Z_1 and Z_2 are $\mathcal{T}$-free, then $Z_1 \cup Z_2$ is $\mathcal{T}$-free.*

Proof Let $Z = Z_1 \cup Z_2$, and note that $|Z| \leq \frac{1}{2} d_{\mathcal{T}}(z_1, z_2) < \theta$ for every $z_1 \in Z_1$ and $z_2 \in Z_2$. Suppose for a contradiction that Z is not $\mathcal{T}$-free, and thus by Lemma 4.5, there exists a 2-edge-connected confined strictly outerplanar cactus $F \subseteq R_G$ such that $|Z \cap \text{ins}_{\mathcal{T}}(F)| > |F \cap G|$. Since the sets Z_1 and Z_2 (and thus also all their subsets) are $\mathcal{T}$-free, we have $Z \cap \text{ins}_{\mathcal{T}}(F) \not\subseteq Z_1$ and $Z \cap \text{ins}_{\mathcal{T}}(F) \not\subseteq Z_2$. Therefore, there exist vertices $z_1 \in Z_1$ and $z_2 \in Z_2$ with $\{z_1, z_2\} \subset \text{ins}_{\mathcal{T}}(F)$. Let $F' \subseteq F$ be a strictly outerplanar cactus consisting of cycles $K_i \subseteq F$ with $z_i \in \text{ins}_{\mathcal{T}}(K_i)$ for

$i \in [2]$ and of a shortest path between K_1 and K_2 in F. Note that the perimeter of F' traverses each vertex of F' at most twice, and thus it has length at most $4|F' \cap G| \leq 4|F \cap G| < 4|Z|$. Therefore, $d_{\mathcal{T}}(z_1, z_2) < 2|Z|$, which is a contradiction. □

4.4 Linking Across a Surface

Finally, let us turn our attention to the linkage problem in graphs drawn on an arbitrary (fixed) surface. The statement of the main result in this setting is a straightforward generalization of Corollary 4.25.

Theorem 4.27 *For every positive integer m and surface Σ, there exists an integer θ such that the following claim holds. Let G be a graph with cellular drawing on Σ, let $\mathcal{T}$ be a respectful tangle in G of order at least θ, let $\mathcal{F}$ be a set of faces of G, let Z be a set of at most m vertices of G incident with faces in $\mathcal{F}$, and let $\mathcal{S}$ be a root partition of Z. If $\mathcal{S}$ is topologically feasible in $\Sigma \setminus \bigcup \mathcal{F}$, the set Z is $\mathcal{T}$-free, and $d_{\mathcal{T}}(f, f') \geq \theta$ for all distinct $f, f' \in \mathcal{F}$, then G contains an $\mathcal{S}$-linkage.*

Theorem 4.27 is proved in [6, 8]. The proof is quite long and technical; let us give only a basic outline: Without loss of generality, we can assume that each face in $\mathcal{F}$ is incident with a vertex of Z, and thus $|\mathcal{F}| \leq m$.

- First, we use a variant of Lemma 4.23 to gain a lot of freedom in the placement of roots at each of the faces in $\mathcal{F}$, as well as to ensure that all these faces are bounded by cycles. More precisely, we fix $m' \gg m$, and for each $f \in \mathcal{F}$, we find an $O((m')^2)$-zone Δ_f around f bounded by a cycle K_f and a set Z_f of m' vertices of K_f so that the following conditions are satisfied. Let G' and $\mathcal{T}'$ be obtained from G and $\mathcal{T}$ by clearing the zones Δ_f for $f \in \mathcal{F}$:
  - For each $f \in \mathcal{F}$, the set Z_f is $\mathcal{T}'$-free.
  - Let $\mathcal{S}'$ be any root partition obtained from $\mathcal{S}$ by, for each $f \in \mathcal{F}$, replacing the vertices incident with f by any vertices of Z_f that appear in K_f in the same cyclic order. Then any $\mathcal{S}'$-linkage in G' can be turned to an $\mathcal{S}$-linkage in G.

  Let us remark that we cannot quite use Lemma 4.23 directly, since $\mathcal{S}$ can contain parts T_1 and T_2 consisting only of vertices incident with a single face $f \in \mathcal{F}$ such that T_1 and T_2 cannot be realized within a disk surrounding f at the same time (i.e., the vertices of T_1 and T_2 are interlaced in the cyclic order around f, and thus one needs to use the handles or crosscaps of the surface to realize them). Thus, in contrast to the proof of Corollary 4.25, we cannot completely get rid of parts of $\mathcal{S}$ incident with only one face of $\mathcal{F}$.
- Let $K = \bigcup_{f \in \mathcal{F}} K_f$ and let $\mathcal{F}'$ be the set of the faces of G' bounded by the cycles K_f for $f \in \mathcal{F}$. Let T be a subgraph of $K \cup R_{G'} - \rho_{G'}(\mathcal{F}')$ such that the drawing of T is cellular, $K \subseteq T$, and T has exactly one face in addition to those in $\mathcal{F}'$. Such a subgraph exists, since we can keep deleting edges between distinct faces not belonging to $\mathcal{F}'$ from the graph $K \cup R_{G'} - \rho_{G'}(\mathcal{F}')$ until it has only

one such face, and observe that deleting an edge that separates two distinct faces preserves the cellularity of the drawing. Let us furthermore choose the graph T with the described properties so that $|E(T)|$ is minimum possible. In particular, this implies that T does not have vertices of degree one.

Let T' be the graph obtained from T by suppressing the vertices of degree two, and let the *length* of each edge of T' be the length of the corresponding path in T. Since all vertices of T' have degree at least three (or T' is just a loop), generalized Euler's formula shows that $|T'| = O(|\mathcal{F}| + g) = O(m + g)$ and $\|T'\| = O(m + g)$, where g is the Euler genus of Σ. For $f \in \mathcal{F}$, let K'_f be the cycle in T' corresponding to K_f, and let $K' = \bigcup_{f \in \mathcal{F}} K'_f$.

- Choose an integer μ such that $m, g \ll \mu \ll m' \ll \theta$. For each $f \in \mathcal{F}$, let us say that an edge of K'_f is *flexible* if it contains at least μ vertices of Z_f. Since $|Z_f| = m'$ and $|T'| = O(m+g)$, observe that at least one edge of K'_f is flexible. We say that an edge of T' not contained in K' is *long* if it has length more than μ. Note that since G has a respectful tangle of order θ, it has representativity at least θ, and thus every non-contractible cycle in R_G has length at least 2θ. Since T has only one face not in $\mathcal{F}$, every cycle in T distinct from those in K is non-contractible, and since $\|T'\| \ll \mu \ll \theta$, we conclude that every cycle in T' not contained in K' contains a long edge. Similarly, since $d_{\mathcal{T}}(f, f') = \theta$ for all distinct $f, f' \in \mathcal{F}$, observe that every path in T' between different components of K' contains a long edge. Let T'' be the graph obtained from T' be deleting the long edges; we observe that every cycle in T'' belongs to K' and that each component of T'' contains at most one cycle.
- Let I be a topological realization of $\mathcal{S}$ in $\Sigma \setminus \bigcup \mathcal{F}$. Observe that I can be chosen so that the number of its intersections with T' is bounded by a function of m and g (since the number of combinatorially different choices for T' and $\mathcal{S}$ is bounded by a function of m and g).

  Imagine cutting the surface $\Sigma \setminus \bigcup \mathcal{F}'$ along T'', and observe that the resulting surface is homeomorphic to a surface obtained from $\Sigma \setminus \bigcup \mathcal{F}$ by possibly drilling additional holes (corresponding to the components of T'' disjoint from K'). The topological realization I of $\mathcal{S}$ in $\Sigma \setminus \bigcup \mathcal{F}$ can be shifted so that it avoids these holes, and thus it intersects $T' - E(K')$ only in the long edges. Moreover, since the number of intersections of I with T' is bounded by a function of m and g and since $m, g \ll \mu$, we can additionally shift I so that it intersects $T' - E(K')$ only in the vertices of G drawn in the central part of each long edge. Finally, we can shift I so that for each $f \in \mathcal{F}$, it intersects Δ_f only in simple curves joining the roots incident with f to vertices of Z_f contained in a single flexible edge of K'_f. See Fig. 4.1 for the illustration of the current state of the construction.

  Let I' be the intersection of I with $\Sigma \setminus \bigcup \mathcal{F}'$, which we can view as a topological realization of a root partition $\mathcal{S}'$ of a subset of $\bigcup_{f \in \mathcal{F}} Z_f$. As we have seen at the beginning of the argument, it suffices to show that G' contains an $\mathcal{S}'$-linkage, since such a linkage extends to an $\mathcal{S}$-linkage in G.
- Let us cut the surface $\Sigma \setminus \bigcup \mathcal{F}'$ along T. Since the drawing of T is cellular and it has only one face not belonging to $\mathcal{F}'$, the resulting surface is a disk Δ. Let

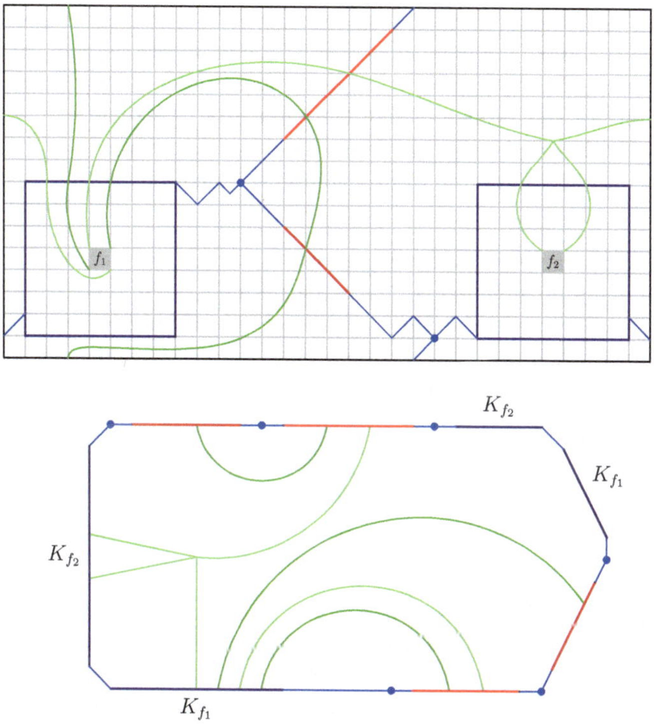

Fig. 4.1 An illustration of the proof of Theorem 4.27. The graph G (in gray) is drawn on the torus obtained by identifying the opposite edges of the rectangle, the two faces of $\mathcal{F}$ are emphasized. The graph T is shown in blue with the central parts of long edges shown in red. The topological realization I is shown in green. The corresponding linkage problem in the disk is depicted at the bottom

us also cut the graph G' and the topological realization I' along T, obtaining a graph G'' drawn in Δ and a topological realization I'' of a root partition $\mathcal{S}''$ of a subset of vertices of G'' drawn in the boundary of Δ. This linkage problem in the disk Δ is illustrated in the bottom part of Fig. 4.1.
Using the fact that T was chosen with minimum number of edges and based on the choice of I, it is possible to argue that the assumptions of Theorem 4.2 are satisfied by G'' and $\mathcal{S}''$. Therefore, there exists an $\mathcal{S}''$-linkage in G'', and gluing back the vertices of T, this gives us the desired $\mathcal{S}'$-linkage in G'.

A basic consequence of Theorem 4.27 is that we can test the presence of an $\mathcal{S}$-linkage in a graph drawn on a fixed surface in polynomial time.

Theorem 4.28 *For every surface Σ and every positive integer m, there exists a polynomial-time algorithm that given a graph G with a cellular drawing on Σ and a root partition $\mathcal{S}$ of m of its vertices, finds an $\mathcal{S}$-linkage in G or correctly decides that it does not exist.*

The idea of the algorithm is as follows: Let $\mathcal{F}$ be a smallest set of faces of G such that each vertex of $\mathcal{S}$ is incident with a face of $\mathcal{F}$. We say that the *complexity* of the instance is the triple (g, k, m), where g is the Euler genus of Σ and $k = |\mathcal{F}|$. By induction, we can assume that an algorithm from Theorem 4.28 exists for all instances of lexicographically smaller complexity. Note that if $g = 0$ and $k \leq 2$, then we can use the algorithm from Theorem 4.3, and thus assume that $g > 0$ or $k \geq 3$.

- The basic operation we are going to use is that of *splitting* a vertex v of the instance along a simple curve γ passing through v: We replace v by two new vertices v_1 and v_2, with the edges to v on one side of γ redirected to v_1 and those on the other side to v_2; let G' be the resulting graph. Note that any $\mathcal{S}$-linkage $\mathcal{L}$ in G corresponds to an $\mathcal{S}'$-linkage in G', where
  - $\mathcal{S}'$ is obtained from $\mathcal{S}$ by splitting a set $S \in \mathcal{S}$ into two (not necessarily non-empty) parts S_1 and S_2 and replacing S by $(S_1 \setminus \{v\}) \cup \{v_1\}$ and $(S_2 \setminus \{v\}) \cup \{v_2\}$, in case that $\mathcal{L}$ passes through v, or
  - $\mathcal{S}'$ is obtained from $\mathcal{S}$ by adding parts $\{v_1\}$ and $\{v_2\}$, in case that $\mathcal{L}$ is disjoint from v.

  Note that the number of root partitions $\mathcal{S}'$ that can be obtained in these ways is bounded by a function of m. Hence, splitting v gives us a bounded number of new instances $(G', \mathcal{S}')$, with the property that G has an $\mathcal{S}$-linkage if and only if there exists an $\mathcal{S}'$-linkage in G' for at least one of these instances (and such an $\mathcal{S}'$-linkage in G' can be easily turned into an $\mathcal{S}$-linkage in G by gluing v_1 and v_2 back into v). Thus, splitting a vertex leads to the natural approach where we try to solve each of the resulting instances recursively. Of course, we need to be careful in the choice of the split vertices to ensure that the total number of recursive calls is bounded.
- Let θ be the constant from Theorem 4.27 for the given Σ and m. Let us first consider the case that R_G contains a non-contractible cycle K of length less than 2θ. We split the vertices of $V(G) \cap V(K)$ one by one along the curve tracing K. As described above, this leads to several new instances; the number of these instances is bounded by a function of m and θ, and θ is in turn bounded by a function of m and g. Consider any of the resulting instances $(G', \mathcal{S}')$. We cut Σ along the cycle K and patch the resulting hole(s), obtaining a surface or a union of two surfaces Σ' of Euler genus smaller than g. The instance $(G', \mathcal{S}')$ considered with its drawing on Σ' has lexicographically smaller complexity than the instance $(G, \mathcal{S})$, since the Euler genus of Σ' (or both surfaces forming Σ') is less than g. Moreover, the number of roots of $\mathcal{S}'$ is at most $k + 2\theta$, and thus the increase in the number of roots is also bounded. Hence, each of the new instances can be solved in polynomial time by the induction hypothesis, giving us the desired polynomial-time algorithm.

  Therefore, we can assume that no such cycle K exists; i.e., if Σ is not the sphere, then R_G has edgewidth at least 2θ, and thus Theorem 2.91 gives us a respectful tangle $\mathcal{T}$ in G of order θ. From now on, we are going to describe the cutting arguments in less detail, focusing only on arguing that the instances that we obtain have lexicographically smaller complexity than (g, k, m).

- If Σ is the sphere, then recall that $k \geq 3$. If there exists a cycle K in R_G of length less than 2θ such that either each of the open disks of $\Sigma - K$ contains at most one vertex of $\rho_G(\mathcal{F})$, or each of them contains more than one vertex of $\rho_G(\mathcal{F})$, then we can split and cut the instance along K, resulting in a bounded number of simpler instances (with roots incident with less than k faces). Similarly, if there exists a confined theta graph F such that each face of F contains at most one vertex of $\rho_G(\mathcal{F})$, then splitting and cutting the instance along F reduces the problem to simpler instances (with roots incident with at most 2 faces).
  Otherwise, for each cycle K in R_G of length less than 2θ, let us define $\text{ins}(K)$ to be the disk bounded by K whose interior contains at most one vertex of $\rho_G(\mathcal{F})$, and observe that ins is a slope of order θ. By Lemma 2.87, we obtain a respectful pretangle $\mathcal{T}$ of order θ in G, which is easily seen to be a tangle by Lemma 2.90.
- If there exist distinct faces $f, f' \in \mathcal{F}$ with $d_{\mathcal{T}}(f, f') < \theta$, then note that by Lemma 4.8, there exists an optimal $(\rho_G(f), \rho_G(f'))$-restraint F which is a $(\rho_G(f), \rho_G(f'))$-tie. Observe that splitting and cutting the instance along F results in instances in the same surface but with roots incident with less than k faces (corresponding to the outer face of F), and in instances in the sphere with the roots incident with at most k faces. In case that $g > 0$, the latter instances are simpler than $(G, \mathcal{S})$. Otherwise the surface Σ is the sphere, and observe that the definition of $\mathcal{T}$ in this case implies that in these instances the roots are incident with $2 < k$ faces, and thus the instances are again simpler than $(G, \mathcal{S})$.
  Therefore, we can assume that $d_{\mathcal{T}}(f, f') = \theta$ for all distinct $f, f' \in \mathcal{F}$.
- If the set Z of roots is not free, then observe that splitting and cutting along the subgraph F from Lemma 4.5 results in instances in the sphere with all roots incident with at most two faces, and in instances (corresponding to the outer face of F) in Σ with roots incident with k faces, but with fewer than m roots. All of these instances are simpler than the original instance $(G, \mathcal{S})$.

In the end, we end up with an instance satisfying the assumptions of Theorem 4.27, and thus (subject to topological feasibility) having an $\mathcal{S}$-linkage. It is possible (though relatively complicated) to extract an algorithm to find the $\mathcal{S}$-linkage from the proof of Theorem 4.27. A simpler approach is to perform the reductions described above with a slightly larger value of θ, and then observe that we can use Lemma 4.19 to clear a zone far apart from $\mathcal{F}$ to decrease the number of vertices while preserving the existence of an $\mathcal{S}$-linkage. We can then repeat the process, iteratively removing such irrelevant parts until one of the reductions described above becomes possible.

Note that this sketch glances over a number of issues that need to be resolved. We need to be able to decide whether the instance is topologically feasible (once Σ and m are fixed, there are only finitely many types of instances that the algorithm can encounter, and thus a list of topologically feasible root partitions can be hard-coded in the algorithm). We need to be able to find a short non-contractible cycle, or a small restraint, or the subgraph certifying that the roots are not $\mathcal{T}$-free (some ideas can be found in [1]). There is also a number of technical details in how we represent the tangles. Overall, the described algorithm is quite complicated and hard to implement with a good time complexity. A simpler (though also not practically useful) algorithm for a more general problem will be described in Sect. 5.3.

4.5 Constructions of Minors

The main application of Theorem 4.27 is in finding (rooted) minors in graphs that are (partially) embedded on a surface with a respectful tangle $\mathcal{T}$ of large order. In this section, we give several useful examples of such constructions that we are going to need throughout the rest of the book.

It is sometimes important to know that $\mathcal{T}$ "points towards" the constructed minor (i.e., that we do not accidentally obtain the minor in an unrelated part of the graph), in the following sense. We say that a tangle $\mathcal{T}$ in a graph G ***controls*** a model μ of a graph H in G if for every separation $(A, B) \in \mathcal{T}$ of order at most $|H|$, we have $\mu(v) \not\subseteq A$ for every $v \in V(H)$. We say that G contains H as a *minor controlled by* $\mathcal{T}$ if there exists a model of H in G controlled by $\mathcal{T}$. As an example, let us give a version of Corollary 3.2 using this concept.

Lemma 4.29 *Let H be a simple graph and let ρ be a bijection between a subset* $\mathrm{dom}(\rho)$ *of $V(H)$ and a set R of vertices of a graph G. Let $\mathcal{T}$ be a tangle in G of order more than $|R|$, and let ν be a model of a clique K with at least* $2|R|+|H|+1$ *vertices in G controlled by $\mathcal{T}$. Let $(A, B) \in \mathcal{T}$ be a separation with $R \subseteq V(A)$ of smallest possible order and subject to that with $|B|$ minimum. Then every vertex $v \in B \setminus A$ is (H, ρ)-irrelevant.*

Proof Consider any separation $(C, D) \in \mathcal{T}$ of order at most $|R|$ with $R \subseteq V(C)$. There exists $v \in V(K)$ such that $V(\nu(v)) \cap V(C \cap D) = \emptyset$, and since $\mathcal{T}$ controls ν, it follows that $\nu(v) \subseteq D - V(C)$. Consequently, (C, D) separates ν from the roots. Conversely, suppose that a separation (C, D) of order at most $|R|$ separates ν from the roots, i.e., that $R \subseteq V(C)$ and there exists $u \in V(K)$ such that $\nu(u) \subseteq D - V(C)$. Since $\mathcal{T}$ controls ν, it follows that $(D, C) \notin \mathcal{T}$, and thus $(C, D) \in \mathcal{T}$.

Therefore, a separation (C, D) of G of order at most $|R|$ separates ν from the roots if and only if $R \subseteq V(C)$ and $(C, D) \in \mathcal{T}$. Consequently, (A, B) is a separation of G that separates ν from the roots, with $|A \cap B|$ smallest possible and subject to that with $|B|$ smallest possible. The claim then follows from Corollary 3.2. □

Let us remark that to apply Lemma 4.29, we only need to deal with the tangle $\mathcal{T}$, without having to construct the model ν (it suffices to be able to prove that one exists). For example, Theorem 4.27 shows the existence of a minor, but as we have seen in the previous section, turning the proof of this theorem into a polynomial-time algorithm to find the minor is rather non-trivial, especially if we want to obtain a good bound on the time complexity. Lemma 4.29 and similar arguments sometimes enable us to sidestep this issue.

The following lemma enables us to easily show that the minors we are about to construct are controlled.

Lemma 4.30 *Let G_0 be a graph and let G be a subgraph of G_0 with a cellular drawing on a surface Σ, and let μ be a model of a graph H in G_0. Let $\mathcal{T}$ be a respectful tangle in G of order more than $2|H|$ conformal with a tangle $\mathcal{T}_0$ in G_0. Suppose that for every $v \in V(H)$, there exist vertices x and y in the same component of $\mu(v) \cap G$ such that $d_{\mathcal{T}}(x, y) > 2|H|$. Then $\mathcal{T}_0$ controls μ.*

Proof Suppose for a contradiction that $\mathcal{T}_0$ does not control μ, and thus there exists a separation $(A_0, B_0) \in \mathcal{T}_0$ of order at most $|H|$ and a vertex $v \in V(H)$ such that $\mu(v) \subseteq A_0$. Let $A = A_0 \cap G$ and $B = B_0 \cap B$; since $\mathcal{T}$ is conformal with $\mathcal{T}_0$, we have $(A, B) \in \mathcal{T}$. By Lemma 2.89, there exists a strictly outerplanar cactus $F \subseteq R_G$ such that $|F \cap G| \leq |A \cap B| \leq |A_0 \cap B_0| \leq |H|$ and $\mu(v) \cap G \subseteq A \subseteq \text{ins}_{\mathcal{T}}(F)$. Consider any vertices x and y in the same component Q of $\mu(v) \cap G$. Observe that F has a component F_Q such that $Q \subseteq \text{ins}_{\mathcal{T}}(F_Q)$. Let F' be a minimal connected subgraph of F_Q such that $x, y \in \text{ins}_{\mathcal{T}}(F')$. Observe that F' is a tie, and thus the perimeter of F' visits each vertex of F' at most twice. Consequently, the perimeter of F' has length at most $4|F' \cap G| \leq 4|H|$, and thus $d_{\mathcal{T}}(x, y) \leq 2|H|$. This contradicts the assumptions. □

Our aim now will be to use Theorem 4.27 to obtain a sufficient condition for the existence of a fixed graph H as a minor in a graph G drawn on a surface. We are going to achieve this by first picking the edges of G representing the edges of H, then connecting them using the theorem. In order to succeed, the ends of the chosen edges must be free and far apart; such a choice is possible according to the following simple observation.

Lemma 4.31 *Let G be a graph with a cellular drawing on a surface Σ and let $\mathcal{T}$ be a respectful tangle of order θ in G. Let m and θ' be positive integers such that $(m-1)(\theta'+7)+11 \leq \theta$. Then there exists a set $\mathcal{F}$ of m faces of G such that*

- *$d_{\mathcal{T}}(f, f') \geq \theta'$ for distinct $f, f' \in \mathcal{F}$, and*
- *for each $f \in \mathcal{F}$, there exist distinct vertices u and v joined by an edge incident with f such that the set $\{u, v\}$ is $\mathcal{T}$-free.*

Proof Let e be any edge of G, and let e' be an edge of G such that $d_{\mathcal{T}}(e, e') = \theta$ which exists by Theorem 4.14. Let P be a shortest path between e and e' in G (the graph G is connected, since its drawing is cellular). For $i \in [\theta]$, let e_i be the first edge of P with $d_{\mathcal{T}}(e, e_i) \geq i$, and let e'_i be the preceding edge of P. Note that for $1 \leq i < j \leq \theta$, the triangle inequality gives $d_{\mathcal{T}}(e, e_j) \leq d_{\mathcal{T}}(e, e'_i) + d_{\mathcal{T}}(e'_i, e_i) + d_{\mathcal{T}}(e_i, e_j)$, and thus

$$d_{\mathcal{T}}(e_i, e_j) \geq d_{\mathcal{T}}(e, e_j) - d_{\mathcal{T}}(e, e'_i) - d_{\mathcal{T}}(e'_i, e_i) \geq j - (i-1) - 4 = j - i - 3.$$

Hence, letting f_i be a face of G incident with e_i, we have

$$d_{\mathcal{T}}(f_i, f_j) \geq d_{\mathcal{T}}(e_i, e_j) - d_{\mathcal{T}}(e_i, f_i) - d_{\mathcal{T}}(e_j, f_j) \geq j - i - 7.$$

Moreover, suppose that $e_i = u_i v_i$ and the set $Z_i = \{u_i, v_i\}$ is not $\mathcal{T}$-free. By Lemma 4.5, we conclude that there exists a cycle K in R_G of length two with a vertex of Z_i contained in the interior of $\text{ins}_{\mathcal{T}}(K)$, and thus with $\rho_G(e_i) \subset \text{ins}_{\mathcal{T}}(K)$. Consider $j \in \{1, \theta\}$, and observe that if $\rho_G(e_j) \not\subseteq \text{ins}_{\mathcal{T}}(K)$, then $V(K) \cap V(G)$ contains a vertex of the part of P between e_i and e_j. Since $|K \cap G| = 1$, we conclude that there exists $j \in \{1, \theta\}$ such that $\rho_G(e_j) \subset \text{ins}_{\mathcal{T}}(K)$, and thus $1 \geq d_{\mathcal{T}}(e_i, e_j) \geq |j - i| - 3$. It follows that if $6 \leq i \leq \theta - 5$, then Z_i is $\mathcal{T}$-free.

Therefore, we can let $\mathcal{F} = \{f_6, f_{13+\theta'}, \ldots, f_{6+(m-1)(\theta'+7)}\}$. □

We are now ready to extend Theorem 4.27 to arbitrary rooted minors.

Theorem 4.32 *For every graph H and surface Σ, there exists an integer θ such that the following claim holds. Let G be a graph with cellular drawing on Σ, let $\mathcal{T}$ be a respectful tangle in G of order at least θ, let $\mathcal{F}$ be a set of faces of G, and let ρ be a bijection from a set* $\text{dom}(\rho) \subseteq V(H)$ *to a set Z of vertices of G incident with the faces in $\mathcal{F}$. If*

- *there exists a drawing of H on $\Sigma \setminus \bigcup \mathcal{F}$ with each vertex $v \in \text{dom}(\rho)$ represented by the point at which $\rho(v)$ is drawn in G,*
- *the set Z is $\mathcal{T}$-free, and*
- *$d_{\mathcal{T}}(f, f') \geq \theta$ for all distinct $f, f' \in \mathcal{F}$,*

then H is a ρ-rooted minor of G controlled by $\mathcal{T}$.

Proof Without loss of generality, we can assume that

(a) the vertices of $\text{dom}(\rho)$ have degree one and are pairwise non-adjacent in H (as we can shift each vertex of $\text{dom}(\rho)$ slightly from its original position in the drawing of H on $\Sigma \setminus \bigcup \mathcal{F}$ and replace it by an adjacent vertex of degree one drawn at the original position), and
(b) the vertices $v \in V(H) \setminus \text{dom}(\rho)$ have degree at least two (as we can add a loop at v drawn in the neighborhood of v on $\Sigma \setminus \bigcup \mathcal{F}$).

Let g be the Euler genus of Σ. We choose the parameters θ' and θ so that $g, |H|, \|H\| \ll \theta' \ll \theta$. We can assume that $|\mathcal{F}| \leq |H|$.

Let H' be the graph obtained from H by subdividing each edge twice and let M be the set of the middle edges of the resulting 3-edge paths. Note that $H' - M$ is a union of stars and each vertex of $\text{dom}(\rho)$ belongs to a different component of $H' - M$. By Lemma 4.31, there exists a set $\mathcal{F}_1'$ of $|M| + |\mathcal{F}|$ faces of G such that $d_{\mathcal{T}}(f, f') \geq 2\theta'$ for distinct $f, f' \in \mathcal{F}_1'$ and for each $f \in \mathcal{F}_1'$, there exist distinct adjacent vertices u_f and v_f incident with f such that $\{u_f, v_f\}$ is $\mathcal{T}$-free. Since the $d_{\mathcal{T}}$-distance between different faces of $\mathcal{F}$ is at least $\theta > 2\theta'$, each face of $\mathcal{F}$ is at $d_{\mathcal{T}}$-distance less than θ' from at most one face of $\mathcal{F}_1'$. Therefore, there exists a subset $\mathcal{F}_1$ of $\mathcal{F}_1'$ of size $|\mathcal{F}_1'| - |\mathcal{F}| = |M|$ such that $d_{\mathcal{T}}(f, f') \geq \theta'$ for all distinct $f, f' \in \mathcal{F} \cup \mathcal{F}_1$.

For each edge $uv \in M$, choose a distinct face $f \in \mathcal{F}_1$ and define $\eta(u) = u_f$ and $\eta(v) = v_f$. For each $z \in Z$, let $\eta(z) = z$. Let $Z_1 = \{u_f, v_f : f \in \mathcal{F}_1\} \cup Z$; Observation 4.26 implies that Z_1 is $\mathcal{T}$-free. Let $\mathcal{S}$ be the root partition of Z_1

containing for each component C of $H' - M$ the part $\eta(V(C))$. Observe that we can continuously deform the drawing of H' so that Z is fixed, the resulting drawing of H' is disjoint from $\bigcup(\mathcal{F} \cup \mathcal{F}_1)$, and every edge $uv \in M$ is drawn in the same way as $\eta(u)\eta(v)$ in G (by "dragging" the middle part of each edge over $\Sigma \setminus \bigcup\mathcal{F}$, "sweeping" the interfering parts of H' in front of the edge). The drawing of the graph $H' - M$ then shows that $\mathcal{S}$ is topologically feasible in $\Sigma \setminus \bigcup(\mathcal{F} \cup \mathcal{F}_1)$.

Therefore, Theorem 4.27 implies that G contains an $\mathcal{S}$-linkage. This $\mathcal{S}$-linkage together with the edges $u_f v_f$ for $f \in \mathcal{F}_1$ forms a ρ-rooted model μ of the graph obtained from $H' - M$ by contracting each component to a single vertex, and this graph is isomorphic to H. Moreover, note that the conditions (a) and (b) from the beginning of the proof imply that each set $S \in \mathcal{S}$ contains vertices $x, y \in S$ such that $d_{\mathcal{T}}(x, y) \geq \theta' - 2 > 2|H|$, and thus by Lemma 4.30, the model μ is controlled by $\mathcal{T}$. □

In particular, we have the following easy consequence, generalizing Corollary 2.34 to non-planar graphs.

Corollary 4.33 *For every graph H which can be drawn on a surface Σ, there exists an integer θ such that if G is a graph with a cellular drawing on Σ and G has a respectful tangle $\mathcal{T}$ of order θ, then H is a minor of G controlled by $\mathcal{T}$.*

It is natural to combine this result with Theorem 2.91.

Corollary 4.34 *For every graph H which can be drawn on a surface Σ other than the sphere, there exists an integer θ such that every H-minor-free graph drawn on Σ has representativity less than θ.*

Next, let us consider the case that the graph G drawn on a surface Σ is a subgraph of some larger graph G_0, and let us show how parts of G_0 outside of G can be used to form minors. In the following definitions, let $\mathcal{T}$ be a respectful tangle in G.

- An ***eye*** (in G_0, over G) is the union of two vertex-disjoint paths P_1 and P_2 in G_0 intersecting G exactly in their endpoints u_1, v_1 and u_2, v_2, such that the set $\{u_1, v_1, u_2, v_2\}$ is $\mathcal{T}$-free and there exists a face f of G along whose boundary the vertices appear in the cyclic order u_1, u_2, v_1, v_2. The face f forms the *foundation* of the eye.
  Eyes can be used to realize graphs that are drawn on Σ with crossings as minors in G_0. Let us remark that Lemma 4.22 applied with a free battlefield can be useful in ensuring that $\{u_1, v_1, u_2, v_2\}$ is $\mathcal{T}$-free.
- A ***jump*** (in G_0, over G) is a path in G_0 intersecting G exactly in its ends u and v. The *foundation* of the jump is formed by u and v. The jump is *θ-long* if $d_{\mathcal{T}}(u, v) \geq \theta$. Jumps can be used to realize edges of the minor that are not drawn on Σ.
- A ***spider*** (in G_0, over G) is a connected subgraph S of G_0 such that $V(S) \cap V(G)$ is an independent set in S and the subgraph $S - V(G)$ (the *body* of the spider) is connected. The elements of $V(S) \cap V(G)$ are the *legs* of the spider and the spider is *θ-wide* if $d_{\mathcal{T}}(u, v) \geq \theta$ for any distinct legs u and v of S. Spiders with many legs can be used to realize vertices of the minor that are not drawn on Σ.

The following lemma confirms the intuitive interpretations described in the definitions.

Lemma 4.35 *For every graph H and a surface Σ, there exist integers t and θ such that the following claim holds. Let G_0 be a graph and let G be a subgraph of G_0 with a cellular drawing on Σ. Let $\mathcal{T}$ be a respectful tangle in G of order at least 2θ conformal with a tangle $\mathcal{T}_0$ in G_0. Suppose that G_0 contains*

- *a set A of a eyes over G, such that $d_{\mathcal{T}}(f_1, f_2) \geq \theta$ for the foundations f_1 and f_2 of any distinct eyes in A,*
- *a set B of b θ-long jumps over G such that $d_{\mathcal{T}}(x, y) \geq \theta$ whenever x is a foundation of a jump in B and y is a foundation of either a different jump in B or an eye in A, and*
- *a set C of c 2θ-wide spiders over G with t legs,*

such that the elements of $A \cup B$ are pairwise vertex-disjoint, and the bodies of the spiders in C are pairwise vertex-disjoint and vertex-disjoint from the elements of $A \cup B$. If there exists a set $C_0 \subseteq V(H)$ of size at most c and a set $B_0 \subseteq E(H)$ of size at most b such that the graph $H_0 = H - C_0 - B_0$ can be drawn in Σ with at most a crossings, then H is a minor of G_0 controlled by $\mathcal{T}_0$.

Proof Without loss of generality, we can assume that C_0 is an independent set in H (otherwise, we can subdivide the edges of H joining the vertices of C_0). Let $t_0 = \|H\|$. Clearly we can assume that $a \leq \|H\|^2$, $b \leq \|H\|$, and $c \leq |H|$, and thus t can be chosen so that $t \geq a + 2b + ct_0$. The value of θ is chosen large enough for the application of Theorem 4.32 below.

We process the spiders $S \in C$ one by one, reducing the number of their legs to t_0 as follows: Since the $d_{\mathcal{T}}$-distance between different legs of S is at least 2θ, any atom of G is at $d_{\mathcal{T}}$-distance less than θ from at most one of them. Hence, the spider S has at least $t - a - 2b - (c - 1)t_0 \geq t_0$ legs that are at $d_{\mathcal{T}}$-distance at least θ from the foundations of the eyes in A and the jumps in B, as well as from the legs of the previously processed spiders. We delete all the legs of S except for t_0 with this property.

We modify H_0 and its drawing as follows:

- For each crossing q in H_0, choose a distinct eye $A_q \in A$. Subdivide the edges close to the crossing and let u_1, u_2, v_1, v_2 be the resulting vertices in order around the crossing. Let γ be a simple curve in Σ from q to the foundation f of A_q that avoids the foundations of other eyes. Continuously deform the drawing of H_0, dragging q along γ so that the vertices u_1, u_2, v_1, v_2 are mapped to the ends of the eye as drawn in G in cyclic order around f; let $\rho(u_1)$, $\rho(u_2)$, $\rho(v_1)$, and $\rho(v_2)$ denote these ends. Moreover, we deform the drawing so that only the two edges of H_0 crossing in q are drawn inside f, and then delete the two edges from H_0.
- For each edge $e = uv \in B_0$ not incident with a vertex of C_0, add vertices u' and v' of degree one adjacent to u and v to H_0 and draw them without crossings next to u and v, also avoiding the foundations of all eyes. Choose a distinct jump

$B_e \in B$ and denote the vertices of the foundation of B_e by $\rho(u')$ and $\rho(v')$. Deform the drawing of H_0 so that u' is mapped to $\rho(u')$ and v' is mapped to $\rho(v')$ as drawn in G, by dragging u' and v' along simple curves avoiding all eyes and vertices with previously fixed locations.

- For each vertex $v \in C_0$ and each neighbor u of v in H, we add a vertex u' of degree one adjacent to u in H_0 and draw it without creating new crossings next to u, also avoiding the foundations of all eyes. Choose a distinct spider $C_v \in C$ and deform the drawing of H_0 so that for each neighbor u of v, the vertex u' is mapped to a distinct leg $\rho(u')$ of C_v as drawn in G. Again, we perform the deformation carefully so that we avoid eyes and vertices with previously fixed locations.

Consider the injective mapping ρ from the set $\mathrm{dom}(\rho) \subseteq V(H_0)$ to $V(G)$ defined by this construction, and observe that by the definition of an eye and by Observation 4.26, the set $Z = \mathrm{img}(\rho)$ is $\mathcal{T}$-free. By Theorem 4.32, we conclude that G contains the modified graph H_0 is a ρ-rooted minor controlled by $\mathcal{T}$. We add to the minor the eye A_q for each crossing q, the jump B_e for each edge $e \in B_0$ not incident with C_0, and the spiders C_v for each $v \in C_0$. Then, we contract the added subgraphs as well as the vertices that we added in the construction of H_0. This gives us a minor of H in G_0 controlled by $\mathcal{T}_0$. □

While finding an eye, a jump, or a spider in isolation is generally not too demanding, the condition that they must be vertex-disjoint is harder to enforce. In the rest of this section, let us develop some tools for dealing with this issue, based on Mader's H-Wege theorem. For a system $\mathcal{S}$ of pairwise disjoint subsets of vertices of a graph G, an *$\mathcal{S}$-path* is a path in G with ends in different parts of $\mathcal{S}$ and otherwise disjoint from $\bigcup \mathcal{S}$. For disjoint sets $U, U_0 \subseteq V(G)$, let $\partial_{U_0,\mathcal{S}}U$ be the set of vertices of U that either belong to $\bigcup \mathcal{S}$ or have a neighbor in $V(G) \setminus (U \cup U_0)$.

Theorem 4.36 (Mader [4]) *Let G be a graph and let $\mathcal{S}$ be a system of pairwise disjoint subsets of vertices of a graph G. The maximum number of pairwise vertex-disjoint $\mathcal{S}$-paths in G is equal to the minimum of*

$$|U_0| + \sum_{i=1}^{m} \lfloor \tfrac{1}{2} |\partial_{U_0,\mathcal{S}} U_i| \rfloor \tag{4.2}$$

over all partitions $U_0, U_1, \ldots, U_m$ of $V(G)$ such that every $\mathcal{S}$-path in $G - U_0$ contains an edge of $G[U_1] \cup \ldots \cup G[U_m]$.

Note that (4.2) is clearly an upper bound on the number of pairwise vertex-disjoint $\mathcal{S}$-paths, as each $\mathcal{S}$-path P either intersects U_0 or contains two vertices of $\partial_{U_0,\mathcal{S}}U_i$ for some $i \in [m]$. The converse is harder to prove; a short proof can be found in [10]. We are only going to need the following Erdős-Pósa style consequence.

Corollary 4.37 *Let G be a graph and let $\mathcal{S}$ be a system of pairwise disjoint subsets of vertices of a graph G. For every positive integer k, either G contains k pairwise* *vertex-disjoint $\mathcal{S}$-paths, or there exists a set $X \subseteq V(G)$ of size less than $2k$ such that every $\mathcal{S}$-path in G intersects X.*

Proof Suppose that G does not contain k pairwise vertex-disjoint $\mathcal{S}$-paths, and thus by Theorem 4.36, there exists a partition $U_0, U_1, \ldots, U_m$ of $V(G)$ such that every $\mathcal{S}$-path contains either a vertex of U_0 or at least two vertices of $\partial_{U_0,\mathcal{S}} U_i$ for some $i \in [m]$, and

$$|U_0| + \sum_{i=1}^{m} \lfloor \tfrac{1}{2} |\partial_{U_0,\mathcal{S}} U_i| \rfloor < k.$$

Let X consist of U_0 and all but one vertex of $\partial_{U_0,\mathcal{S}} U_i$ for each $i \in [m]$. Then every $\mathcal{S}$-path in G intersects X and

$$|X| = |U_0| + \sum_{i=1}^{m} (|\partial_{U_0,\mathcal{S}} U_i| - 1) \leq |U_0| + \sum_{i=1}^{m} 2\lfloor \tfrac{1}{2} |\partial_{U_0,\mathcal{S}} U_i| \rfloor < 2k.$$

□

A spider S is a ***horn*** if S is a subdivision of a star and the legs of S are the leaves. We say that the center of the star is the *tip* of the horn. Horns naturally arise as obstructions to disjointness, as seen e.g. in the proof of the following lemma.

Lemma 4.38 *For every graph H with k edges and every surface Σ, there exists an integer θ such that the following claim holds for every positive integer t. Let G_0 be a graph and let G be a subgraph of G_0 with a cellular drawing on Σ. Let $\mathcal{T}$ be a respectful tangle in G of order at least θ conformal with a tangle $\mathcal{T}_0$ in G_0. Suppose that G_0 contains a set A of more than $k^2 + 14kt$ eyes over G, such that $d_{\mathcal{T}}(f_1, f_2) \geq \theta$ for the foundations f_1 and f_2 of any distinct eyes in A. If H is not a minor of G_0 controlled by $\mathcal{T}_0$, then G_0 contains a $(\theta - 4)$-wide horn over G with t legs.*

Proof Let θ' be the constant from Lemma 4.35 for H and Σ, and let $\theta = \theta' + 4$. For each eye $Y \in A$, let S_Y be the set of the four ends of the paths forming Y, and let $\mathcal{S} = \{S_Y : Y \in A\}$. Let $G_1 = G_0 - E(G) - (V(G) \setminus \bigcup \mathcal{S})$, and observe that each $\mathcal{S}$-path in G_1 is a $(\theta - 4)$-long jump over G.

Suppose first that there exist a set B' of $7k$ pairwise vertex-disjoint $\mathcal{S}$-paths in G_1. Let F be an auxiliary graph with vertex set B' and with two paths of B' adjacent if their ends belong to the same eye. Note that F has maximum degree at most 6, and thus it is 7-colorable and contains an independent set B of size $k = \|H\|$. Since B is an independent set in H, distinct jumps in B have foundations incident with foundations of different eyes of A, and thus the $d_{\mathcal{T}}$-distance between them is at least $\theta - 4 = \theta'$. By Lemma 4.35 applied with $B_0 = E(H)$, we conclude that G_0 contains H as a minor controlled by $\mathcal{T}_0$.

Therefore, we can assume that G_1 does not contain $7k$ pairwise vertex-disjoint $\mathcal{S}$-paths, and by Corollary 4.37, there exists a set $X \subseteq V(G_1)$ of size less than $14k$

intersecting all $\mathcal{S}$-paths in G_1. Suppose first that there exists a subset A' of A of size k^2 such that the eyes in A' are disjoint from X. Note that if $Y, Y' \in A'$ are distinct eyes, then Y is vertex-disjoint from Y', as otherwise their union would contain an $\mathcal{S}$-path in G_1 disjoint from X. Note that H has a drawing (in the plane, and thus also on Σ) in which any two edges of H cross at most once, and thus the number of crossings is at most $\binom{k}{2} < |A'|$; by Lemma 4.35, we conclude that H is a minor of G_0 controlled by $\mathcal{T}_0$.

Hence, we can assume that less than k^2 of the eyes in A are disjoint from X. Let $A_1 \subseteq A$ consist of the eyes $Y \in A$ such that Y intersects X, but the ends of the paths forming the eye Y do not belong to X. Then $|A_1| > |A| - k^2 - |X| > |X|(t-1)$. For each eye $Y \in A_1$, let P_Y be a minimal subpath of Y from an end of Y to a vertex $x_Y \in X$. By the pigeonhole principle, there exists $x \in X$ and a subset A_2 of A_1 of size t such that $x_Y = x$ for each $Y \in A_2$. Note that for distinct $Y, Y' \in A_2$, we have $P_Y \cap P_{Y'} = x$, as if P_Y and $P_{Y'}$ intersected in another vertex, their union would contain an $\mathcal{S}$-path in G_1 disjoint from X. We conclude that $\bigcup_{Y \in A_2} P_Y$ is a $(\theta-4)$-wide horn with tip x and with t legs. □

It may not be immediately obvious that getting a horn instead of intersecting eyes is a progress. However, a key property of horns is that they give rise to minors even if they are not disjoint. For a graph H and a surface Σ, let the ***Σ-apex number*** $a_\Sigma(H)$ of H be the minimum size of a set $C_0 \subseteq V(H)$ such that $H - C_0$ can be drawn on Σ.

Lemma 4.39 *For every graph H and every surface Σ, there exist integers t_0 and θ such that the following claim holds. Let G_0 be a graph and let G be a subgraph of G_0 with a cellular drawing on Σ. Let $\mathcal{T}$ be a respectful tangle in G of order at least 2θ conformal with a tangle $\mathcal{T}_0$ in G_0. Suppose that G_0 contains a set C of $a_\Sigma(H)$ 2θ-wide horns, each with at least t_0 legs. If the tips of the horns in C are pairwise different, then G_0 contains H as a minor controlled by $\mathcal{T}_0$.*

Proof Let t and θ be the constants from Lemma 4.35 for H and Σ. Let $k = \|H\|$, $c = a_\Sigma(H) = |C|$ and $t_0 = c(2k+t)$. Let T be the set of tips of the horns in C. Since the $d_\mathcal{T}$-distance among the legs of any horn $J \in C$ is at least 2θ, any atom of G is at $d_\mathcal{T}$-distance less than θ from at most one of them. Moreover, less than c of the paths from the tip v of J to a leg contain a vertex of $T \setminus \{v\}$. Hence, as in the proof of Lemma 4.35, we can delete all but $2k-1+t$ legs from each horn to obtain a set C' of 2θ-wide horns with the additional property that all legs x and y of distinct horns from C' satisfy $d_\mathcal{T}(x,y) \geq \theta$, and each horn in C' intersects T only in its tip.

Let $\mathcal{S}$ be the system consisting of all single-element sets $\{u\}$ such that u is a leg of a horn on C', and let $G_1 = \left(\bigcup C'\right) \setminus T$. If G_1 contains $k = \|H\|$ pairwise vertex-disjoint $\mathcal{S}$-paths, then observe that these $\mathcal{S}$-paths are θ-long jumps with foundations at $d_\mathcal{T}$-distance at least θ from one another, and thus Lemma 4.35 applied with $B_0 = E(H)$ implies that H is a minor of G_0 controlled by $\mathcal{T}_0$.

Otherwise, Corollary 4.37 implies that a set $X \subseteq V(G_1)$ of size at most $2k-1$ intersects all $\mathcal{S}$-paths. For each horn $J \in C'$, delete from J every path from the tip of

J to a leg of J which is intersected by X, and let C'' be the resulting set of 2θ-wide horns over G. Observe that each horn in C'' has at least t legs, intersects T only in its tip, and is disjoint from X. Distinct horns $J, J' \in C''$ are vertex-disjoint, as otherwise $(J \cup J') - T \subseteq G_1 - X$ would contain an $\mathcal{S}$-path. Therefore, Lemma 4.35 applied with C_0 implies that H is a minor of G_0 controlled by $\mathcal{T}_0$. □

This suggests the following strategy: We somehow find many eyes (or jumps, or spiders) with distant foundations, and use Lemma 4.38 (or an analogous statement) to either get H as a minor, or to obtain a horn. We delete the tip of the horn from the graph, and repeat the process. Eventually, either the number of horns with distinct tips is so large that we can apply Lemma 4.39, or we reach the point where the graph without the deleted tips is "nearly drawn in Σ" in the sense that there are only few distant eyes or jumps. This should be reminiscent of what happens in the Minor Structure Theorem: After deleting a bounded number of apex vertices, we reach a graph that is nearly drawn in a bounded genus surface; and indeed, we will see this idea applied towards the proof of the Minor Structure Theorem in the following chapter.

For now, let us show how we can form horns from spiders and jumps. For spiders, the result is particularly simple, as it suffices to deal with a single spider.

Lemma 4.40 *For every graph H with k edges and every surface Σ, there exists an integer θ such that the following claim holds for every positive integer t. Let G_0 be a graph and let G be a subgraph of G_0 with a cellular drawing on Σ. Let $\mathcal{T}$ be a respectful tangle in G of order at least θ conformal with a tangle $\mathcal{T}_0$ in G_0. Suppose that G_0 contains a θ-wide spider S over G with at least $2k(t+1)$ legs. If G_0 does not contain H as a minor controlled by $\mathcal{T}_0$, then there exists a θ-wide horn $J \subseteq S$ over G with t legs.*

Proof Let θ be the constant from Lemma 4.35 for H and Σ. If $k = 0$, then G_0 contains H as a minor controlled by $\mathcal{T}_0$ by Lemma 4.35 with $B_0 = C_0 = \emptyset$, and thus suppose that $k \geq 1$.

Let $\mathcal{S}$ be the system consisting of all single-element sets $\{u\}$ such that u is a leg of S. If S contains $k = \|H\|$ pairwise vertex-disjoint $\mathcal{S}$-paths, then note that these $\mathcal{S}$-paths are θ-long jumps with foundations at $d_{\mathcal{T}}$-distance at least θ from one another, and thus Lemma 4.35 applied with $B_0 = E(H)$ implies that H is a minor of G_0 controlled by $\mathcal{T}_0$.

Hence, suppose that this is not the case, and thus by Corollary 4.37, there exists a set $X \subseteq V(S)$ of size less than $2k$ intersecting all $\mathcal{S}$-paths. Let L be the set of legs of S that do not belong to X; we have $|L| \geq 2kt \geq 2$. Since $S - V(G)$ is connected and X intersects all $\mathcal{S}$-paths in S, it follows that $S - V(G)$ contains at least one vertex of X. Consequently, for each leg $u \in L$ we can choose a shortest path P_u in S from u to X and intersecting $V(G)$ only in u. Note that the paths P_u and $P_{u'}$ for distinct $u, u' \in L$ may intersect only in their end in X, as otherwise $P_u \cup P_{u'}$ would contain an $\mathcal{S}$-path disjoint from X. By the pigeonhole principle, there exists $x \in X$ and a set $U \subset L$ of size t such that for each $u \in U$, the path P_u ends in x. Then $J = \bigcup_{u \in U} P_u \subseteq S$ is a θ-wide horn over G with t legs. □

For jumps, we could just straightforwardly use (a simpler version of) the argument from the proof of Lemma 4.38. However, there is an additional issue that can be dealt with at the same time. In applications, it will generally be possible to ensure that the starting vertices of the jumps are far apart (by choosing just one jump starting in each carefully selected area), but we have little control over the ending vertices (they may end up close to one another, or to other starting vertices). We show that this is good enough to at least get a horn, though in this case this comes at the cost of having to sacrifice a bounded-diameter region in the subgraph drawn in the surface.

Lemma 4.41 *For every graph H with k edges and every surface Σ, there exists an integer θ_0 such that the following claim holds for all integers $\theta \geq \theta_0$ and $t \geq 1$. Let G_0 be a graph and let G be a subgraph of G_0 with a cellular drawing on Σ. Let $\mathcal{T}$ be a respectful tangle in G of order more than $4\theta + 18$ conformal with a tangle $\mathcal{T}_0$ in G_0. Then at least one of the following claims holds:*

(i) H is a minor of G_0 controlled by $\mathcal{T}_0$; or
(ii) G_0 contains a θ-wide horn with t legs over G; or
(iii) there exists a $(2\theta+9)$-zone Δ around a vertex of G such that G_0 contains a θ-wide horn with t legs over the subgraph of G obtained by clearing Δ (where the wideness is measured in the metric of the tangle obtained from $\mathcal{T}$ by clearing Δ); or
(iv) there exists a set Y of at most $12k^2(t+1)(t+2)$ vertices of G such that every θ-long jump in G_0 over G has both ends at $d_{\mathcal{T}}$-distance less than $9\theta+36$ from Y.

Proof Let θ_0 be the maximum of the constants θ from Lemmas 4.35 and 4.40 for H and Σ. Let Y be an inclusionwise-maximal set of vertices of G such that each vertex $y \in Y$ is an end of a θ-long jump P_y in G_0 over G and $d_{\mathcal{T}}(y_1, y_2) \geq 9\theta+36$ for distinct $y_1, y_2 \in Y$. Note that every θ-long jump in G_0 over G has both ends at $d_{\mathcal{T}}$-distance less than $9\theta + 36$ from Y, and thus if $|Y| \leq 12k^2(t+1)(t+2)$, then (iv) holds. Therefore, suppose that $|Y| > 12k^2(t+1)(t+2)$.

For $y \in Y$, let t_y be the end of P_y different from y. Let $T = \{t_y : y \in Y\}$. Let Y_1 be an inclusionwise-maximal subset of T such that $d_{\mathcal{T}}(t_1, t_2) \geq 2\theta$ for distinct $t_1, t_2 \in Y_1$. Hence, $d_{\mathcal{T}}(t_y, Y_1) < 2\theta$ for every $y \in Y$.

Suppose first that $|Y_1| < 6k(t+2)$. By the pigeonhole principle, there exists $v \in Y_1$ and a subset Y' of Y of size more than $2k(t+1)$ such that $d_{\mathcal{T}}(t_y, v) < 2\theta$ holds for every $y \in Y'$. By Theorem 4.21 (applied with $2\theta + 9$ playing the role of k) there exists a $(2\theta + 9)$-local 2×1 battlefield around v whose egg contains $\{t_y : t \in Y'\}$. Let C_1 and C_2 be the cycles of the battlefield and let Δ be the $(2\theta + 9)$-zone bounded by C_2. Since the $d_{\mathcal{T}}$-distance between distinct vertices of Y is at least $9\theta + 36 > 2(2\theta + 9)$, at most one vertex $y_0 \in Y$ is contained in Δ. Let G' and $\mathcal{T}'$ be obtained from G and $\mathcal{T}$ by clearing Δ. Note that for distinct $y_1, y_2 \in Y' \setminus \{y_0\}$, Lemma 4.19 gives $d_{\mathcal{T}'}(y_1, y_2) \geq d_{\mathcal{T}}(y_1, y_2) - 8\theta - 36 \geq \theta$. Since the drawing of G is cellular, the graph G is connected, and thus the subgraph G_1 of G drawn in the egg the battlefield (including the cycle C_1) is connected. Let

$S = G_1 \cup \bigcup_{y \in Y' \setminus \{y_0\}} P_y$; then S is a θ-wide spider over G' and the set $Y' \setminus \{y_0\}$ of its legs has size at least $2k(t+1)$. By Lemma 4.40, we conclude that (i) or (iii) holds.

Therefore, we can assume that $|Y_1| \geq 6k(t+2)$. Let Y_2 be a subset of Y containing for each $t \in Y_1$ exactly one vertex $y \in Y$ such that $t_y = t$; hence, $|Y_2| = |Y_1|$. Let F be the auxiliary graph with vertex set Y_2 such that distinct $y_1, y_2 \in Y_2$ are adjacent if $d_{\mathcal{T}}(y_1, t_{y_2}) < \theta$ or $d_{\mathcal{T}}(y_2, t_{y_1}) < \theta$. Note that for $y \in Y_2$, at most one vertex of Y_2 is at $d_{\mathcal{T}}$-distance less than θ from t_y (since the $d_{\mathcal{T}}$-distance between distinct elements of Y_2 is at least 2θ), and similarly at most one vertex of Y_1 is at $d_{\mathcal{T}}$-distance less than θ from y. Hence, F has maximum degree at most two, and thus it is 3-colorable and contains an independent set Y_2' of size at least $|Y_2|/3 \geq 2k(t+2)$. Let $R = Y_2' \cup \{t_y : t \in Y_2'\}$ and note that $d_{\mathcal{T}}(u, v) \geq \theta$ for any distinct $u, v \in R$.

Let $\mathcal{S}$ consist of the single-element sets $\{u\}$ for $u \in R$ and let $G_2 = \bigcup_{y \in Y_2'} P_y$. If G_2 contains at least $k = \|H\|$ pairwise vertex-disjoint $\mathcal{S}$-paths, then note that these paths are θ-long jumps over G; by Lemma 4.35, we conclude that H is a minor of G_0 controlled by $\mathcal{T}_0$ and (i) holds.

Otherwise, by Corollary 4.37, there exists a set X of less than $2k$ vertices of G_2 intersecting all $\mathcal{S}$-paths in G_2. For each $y \in Y_2'$ such that neither y nor t_y belongs to X, let x_y be the first vertex of P_y belonging to X, and let P_y' be the subpath of P_y from y to x_y. By the pigeonhole principle, there exists a vertex $x \in X$ and set $Y_3 \subset Y_2'$ of size t such that for each $y \in Y_3$, neither y nor t_y belongs to X and $x_y = x$. Note that for distinct $y_1, y_2 \in Y_3$, the paths P_{y_1}' and P_{y_2}' intersect only in x, as otherwise their union would contain an $\mathcal{S}$-path disjoint from X. Therefore, $\bigcup_{y \in Y_3} P_y$ is a θ-wide horn over G, and (ii) holds. □

Let us recall that the general plan is to accumulate horns until either we have enough horns to use Lemma 4.39 or the graph without the tips of the horns has simple structure. That is, we would like to iterate Lemma 4.41 to eliminate the outcomes (ii) and (iii). During the iteration, we need to be a bit careful so that when we clear the zone in the outcome (iii), we do not destroy the previously found horns. The following lemma shows that clearing the zone only decreases the width of the horns a bit.

Lemma 4.42 *Let G_0 be a graph and let G be a subgraph of G_0 with a cellular drawing on a surface Σ. Let θ be a positive integer, let $\mathcal{T}$ be a respectful tangle in G of order at least $\theta_0 > 2\theta$, and let G' and $\mathcal{T}'$ be obtained from G and $\mathcal{T}$ by cleaning a θ-zone Δ around an atom a_0 of G. If G_0 contains a θ_0-wide horn J over G with t legs, then G_0 also contains a $(\theta_0 - 4\theta)$-wide horn J' over G' with t legs and the same tip.*

Proof Note that at most one leg of J is contained in Δ. If J has such a leg u, then let J' be obtained from J by adding a shortest path in G from u to a vertex u' in the boundary cycle of Δ; otherwise, let $J' = J$. Then J' is a horn over G' with t legs and the same tip as J. Moreover, Lemma 4.19 implies that if v_1 and v_2 are distinct legs of J different from u, then $d_{\mathcal{T}'}(v_1, v_2) \geq d_{\mathcal{T}}(v_1, v_2) - 4\theta \geq \theta_0 - 4\theta$. Finally, if J has a leg u in Δ, f_0 is the face of G' obtained by clearing Δ, and v is any leg of J different from u, then $d_{\mathcal{T}'}(u', v) \geq d_{\mathcal{T}'}(f_0, v) - 1 \geq d_{\mathcal{T}}(a_0, v) - 2\theta - 1 \geq d_{\mathcal{T}}(u, v) - 3\theta - 1 \geq \theta_0 - 4\theta$. Therefore, the horn J' is $(\theta_0 - 4\theta)$-wide. □

Hence, we can iterate Lemma 4.41 and combine it with Lemma 4.39 to ensure that long jumps only occur between several bounded-radius areas.

Lemma 4.43 *For every graph H and every surface Σ, there exist integers θ and p such that the following claim holds. Let G_0 be a graph and let G be a subgraph of G_0 with a cellular drawing on Σ. Let $\mathcal{T}$ be a respectful tangle in G of order at least $(2a_\Sigma(H) + 1)\theta$ conformal with a tangle $\mathcal{T}_0$ in G_0. If G_0 does not contain H as a minor controlled by $\mathcal{T}_0$, then there exist*

- *a subgraph G' of G with a respectful tangle $\mathcal{T}'$ such that G' and $\mathcal{T}'$ are obtained from G and T by less than $a_\Sigma(H)$ repetitions of clearing a θ-zone, and*
- *sets $X \subseteq V(G_0) \setminus V(G')$ and $Y \subset V(G')$ such that $|X| < a_\Sigma(H)$ and $|Y| \leq p$,*

such that every θ-long jump in $G_0 - X$ over G' has both ends at $d_{\mathcal{T}'}$-distance less than θ from Y.

Proof (♀→) Let $c = a_\Sigma(H)$. Choose positive integers t, p, θ_0, …, and θ_c so that $|H|, \|H\| \ll t \ll p \ll \theta_c \ll \theta_{c-1} \ll \ldots \ll \theta_0$, and let $\theta = \theta_0$. Let $G'_0 = G$, $\mathcal{T}'_0 = \mathcal{T}$, and $X_0 = \emptyset$. For $i = 1, 2, \ldots$, we are going to either construct a subgraph G'_i with a tangle $\mathcal{T}'_i$ and a set $X_i \subseteq V(G_0) \setminus V(G'_i)$ such that

- G'_i and $\mathcal{T}'_i$ are obtained from G and $\mathcal{T}$ by at most i repetitions of clearing a θ-zone, and
- each vertex of X_i is a tip of a θ_i-wide horn over G'_i with t legs;

or argue that $G' = G'_{i-1}$, $\mathcal{T}' = \mathcal{T}'_{i-1}$, and $X = X_{i-1}$ satisfy the conclusions of the lemma. Note that by Lemma 4.19, the order of the tangle $\mathcal{T}'_i$ is at least $(2c+1-2i)\theta$. If the construction reaches the step $i = c$, then Lemma 4.39 applied to G'_c and $\mathcal{T}'_c$ with $C = X_c$ implies that G_0 contains H as a minor controlled by $\mathcal{T}$.

Suppose that we have already constructed G'_{i-1}, $\mathcal{T}'_{i-1}$, and X_{i-1}. Let us apply Lemma 4.41 to $G_0 - X_{i-1}$ and G'_{i-1}, with θ_i playing the role of θ. The outcome (i) gives us H as a minor of $G_0 - X_{i-1}$ controlled by $\mathcal{T}_0 - X_{i-1}$, and thus also as a minor of G_0 controlled by $\mathcal{T}_0$. In the case (ii), let x be the tip of a θ_i-wide horn with t legs over G'_{i-1}; we let $G'_i = G'_{i-1}$, $\mathcal{T}'_i = \mathcal{T}'_{i-1}$, and $X_i = X_{i-1} \cup \{x\}$. The invariants are clearly satisfied.

In the case (iii), let Δ be a $(2\theta_i + 9)$-zone in G'_{i-1} such that $G_0 - X_{i-1}$ contains a θ_i-wide horn with t legs over the subgraph G'_i of G'_{i-1} obtained by clearing Δ, and let x be the tip of the horn. Let $\mathcal{T}'_i$ be the tangle obtained from $\mathcal{T}'_{i-1}$ by clearing Δ. Note that $2\theta_i + 9 \leq \theta$, i.e., Δ is also a θ-zone, and thus G'_i and $\mathcal{T}'_i$ are obtained from G and $\mathcal{T}$ by at most i repetitions of clearing a θ-zone. Since $\theta_{i-1} \gg \theta_i$, Lemma 4.42 implies that the elements of X_{i-1} are tips of θ_i-wide horns with t legs over G'_i, and thus we can let $X_i = X_{i-1} \cup \{x\}$, finishing the construction.

Finally, in the case (iv), there exists a set Y of at most p vertices of G'_{i-1} such that every θ_i-long (and thus also θ-long) jump in $G_0 - X_{i-1}$ over G'_{i-1} has both ends at $d_{\mathcal{T}'_{i-1}}$-distance less than $\theta_i \leq \theta$ from Y, and the conclusion of the lemma holds with $G' = G'_{i-1}$, $\mathcal{T}' = \mathcal{T}'_{i-1}$, and $X = X_{i-1}$. □

Note that the vertices of the set Y can be fairly close to each other, and in particular, θ-zones around them can overlap. This could cause complications in the applications of the lemma. To deal with this issue, we use the following standard clustering lemma, saying that any set (given with a metric) can be divided into clusters of relatively nearby elements, where the clusters are much farther apart then their diameter. Given a metric d on a set Y, a set $U \subseteq Y$ is *a-wide* if $d(u, v) > a$ for every distinct $u, v \in U$, and it *b-dominates* Y if $d(U, y) \leq b$ for every $y \in Y$.

Lemma 4.44 *For every function $f : \mathbb{N} \to \mathbb{N}$, there exists a function $\psi_f : \mathbb{N}^2 \to \mathbb{N}$ such that the following claim holds. Let d be a metric on a set Y. For all positive integers k and a, either Y contains an a-wide subset of size k, or there exists an integer ψ such that $a \leq \psi \leq \psi_f(k, a)$ and an $f(\psi)$-wide ψ-dominating set $U \subseteq Y$ of size less than k.*

Proof We can assume that Y is non-empty, as otherwise we can let $U = \emptyset$. Moreover, we can assume that $k \geq 2$, since any single-element subset of Y is a-wide. For given $k \geq 2$ and a, we let $\psi_{k-1} = a$, and for $i = k-2, \ldots, 1$, we let $\psi_i = \psi_{i+1} + f(\psi_{i+1})$; and we define $\psi_f(k, a) = \psi_1$. Let U' be an inclusionwise-maximal a-wide subset of Y. If $|U'| \geq k$, then Y contains an a-wide subset of size k.

Hence, suppose that $|U'| \leq k - 1$. The inclusionwise-maximality of U' implies that U' a-dominates Y, and thus it also $\psi_{|U'|}$-dominates Y. Let U be a smallest subset of Y which $\psi_{|U|}$-dominates Y, and let $i = |U|$ and $\psi = \psi_i$. We need to argue that U is $f(\psi)$-wide. Suppose for a contradiction that there exist distinct $u, v \in U$ such that $d(u, v) \leq f(\psi)$. Let $U_1 = U \setminus \{v\}$. Note that if $y \in Y$ is at distance at most ψ from v, then it is at distance at most $\psi + f(\psi) = \psi_{i-1} = \psi_{|U_1|}$ from u, and thus U_1 $\psi_{|U_1|}$-dominates Y. This contradicts the minimality of $|U|$. Therefore, U is an $f(\psi)$-wide ψ-dominating subset of Y of size less than k. □

Thus, Lemma 4.43 has the following corollary.

Corollary 4.45 *For every graph H and every surface Σ, there exist integers θ and p such that for every function $g : \mathbb{N} \to \mathbb{N}$, there exists an integer κ_0 such that the following claim holds. Let G_0 be a graph and let G be a subgraph of G_0 with a cellular drawing on Σ. Let $\mathcal{T}$ be a respectful tangle in G of order at least $(2a_\Sigma(H) + 1)\theta$ conformal with a tangle $\mathcal{T}_0$ in G_0. If G_0 does not contain H as a minor controlled by $\mathcal{T}_0$, then there exist*

- *a subgraph G' of G with a respectful tangle $\mathcal{T}'$ such that G' and $\mathcal{T}'$ are obtained from G and $\mathcal{T}$ by less than $a_\Sigma(H)$ repetitions of clearing a θ-zone,*
- *a set $X \subseteq V(G_0) \setminus V(G')$ of size less than $a_\Sigma(H)$, and*
- *an integer $\kappa \leq \kappa_0$ and a $g(\kappa)$-wide (in the metric $d_{\mathcal{T}'}$) set $U \subseteq V(G')$ of size at most p,*

such that every θ-long jump in $G_0 - X$ over G' has both ends at $d_{\mathcal{T}'}$-distance less than κ from U.

Proof We apply Lemma 4.44 to the set Y from the outcome of Lemma 4.43 with $a = 1$, $k = p + 1$, and $f(\psi) = g(\psi + \theta)$, letting $\kappa_0 = \psi_f(p + 1, 1) + \theta$. Let ψ and $U \subseteq Y$ be the integer and the set from the outcome of the Lemma, and let $\kappa = \psi + \theta$. Note that U is $f(\psi)$-wide, and thus $g(\kappa)$-wide. Moreover, the set U ψ-dominates Y, and thus every vertex at $d_{\mathcal{T}'}$-distance less than θ from Y is also at $d_{\mathcal{T}'}$-distance less than $\psi + \theta = \kappa$ from U. □

The outcome of this corollary is still perhaps a bit disappointing, in that we are not able to intersect all long jumps by the set X. This cannot be avoided; e.g., suppose that H cannot be drawn on the torus, G_0 can be drawn on the torus with large representativity, and G is a cylindrical grid in G_0 wrapped around the torus. Let us view G as drawn on the sphere, and let $\mathcal{T}$ be the canonical tangle in G, so that the two boundary cycles of G are far apart in the $d_{\mathcal{T}}$-distance. Then G_0 can contain arbitrarily many pairwise vertex-disjoint (long) jumps between the boundary cycles of G, and yet H is not a minor of G_0. This turns out to be essentially the only issue: If we cannot intersect all long jumps by few vertices and H is not a minor of G_0, then G can be transformed into a subgraph of G_0 drawn with high representativity on a surface created from Σ by adding a handle or a crosshandle. The proof of this claim is based on the following lemma.

Lemma 4.46 *For all integers $p \geq 0$ and $\theta, \kappa, \theta_1 \geq 1$ and for any surface Σ, there exist positive integers p' and δ_0 such that the following claim holds. Let G_0 be a graph and let G be a subgraph of G_0 with a cellular drawing on a surface Σ. Let $\mathcal{T}$ be a respectful tangle in G of order at least $(4p+1)\delta_0$ and let U be a $(2\delta_0+\theta)$-wide set of at most p vertices of G such that every θ-long jump in G_0 over G has both ends at $d_{\mathcal{T}}$-distance less than κ from U. Let G' and $\mathcal{T}'$ be obtained from G and $\mathcal{T}$ by clearing δ_0-zones $\{\Delta_y : y \in U\}$ around the elements of U. Then either*

(i) there exists a set X of less than p' vertices of G_0 intersecting all δ_0-long (in $d_{\mathcal{T}'}$-distance) jumps in G_0 over G', or
(ii) there exist distinct elements $y_1, y_2 \in U$, $\mathcal{T}'$-free sets $Z_1, Z_2 \subset V(G')$ of size θ_1, where for $i \in [2]$, Z_i is contained in the boundary of Δ_{y_i}, and G_0 contains θ_1 pairwise vertex-disjoint jumps from Z_1 to Z_2 over G'.

Proof (⇸) Let $p' = \binom{p}{2}\theta_1$. Let us choose r and δ_0 so that $p, \theta, \theta_1, \kappa \ll r \ll \delta_0$. For each $y \in U$, use Theorem 4.21 to find a free δ_0-local graded $r \times \theta_1$ battlefield $(\mathcal{R}_y, \mathcal{P}_y)$ around y whose egg contains all vertices of G at $d_{\mathcal{T}}$-distance at most κ from y. Let $\mathcal{R}_y = C_{y,1}, \ldots, C_{y,r}$, let Δ_y be the δ_0-zone around y bounded by $C_{y,r}$, let f_y be the face of G' obtained by clearing Δ_y, let $\Delta'_y \subset \Delta_y$ be the disk bounded by $C_{y,1}$, and let Z_y be the $\mathcal{T}$-free set of ends of the paths of $\mathcal{P}_y$ in $C_{y,r}$. Let $\Delta' = \bigcup_{y \in U} \Delta'_y$.

For any distinct $y_1, y_2 \in U$, if G_0 contains θ_1 pairwise vertex-disjoint jumps from Z_{y_1} to Z_{y_2} over G', then the second outcome of the lemma holds with $Z_1 = Z_{y_1}$ and $Z_2 = Z_{y_2}$. Otherwise, by Menger's theorem, there exists a set X_{y_1,y_2} of less than θ_1

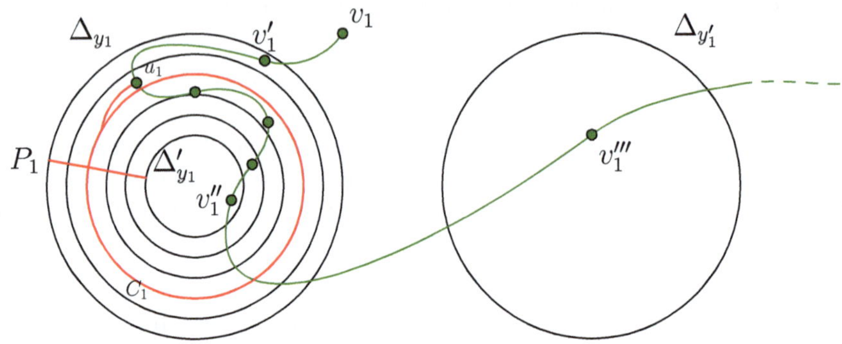

Fig. 4.2 An illustration of the proof of Lemma 4.46. The (initial segment of) the path Q is drawn in green. The subgraphs added to form Q' are drawn in red

vertices of G_0 intersecting all jumps from Z_{y_1} to Z_{y_2} over G'. Let X be the union of the sets X_{y_1,y_2} for all distinct $y_1, y_2 \in U$; clearly, $|X| < p'$.

We claim that X intersects all δ_0-long jumps in G_0 over G', and thus the first outcome holds. See Fig. 4.2 for an illustration of this part of the argument. Suppose for a contradiction that Q is a δ_0-long jump in G_0 over G' disjoint from X, with ends $v_1, v_2 \in V(G')$. By Lemma 4.19, we have $d_{\mathcal{T}}(v_1, v_2) \geq d_{\mathcal{T}'}(v_1, v_2) \geq \delta_0 > \theta$. Since every θ-long jump in G_0 over G has both ends in $G \cap \Delta'$ and Q does not, it follows that Q cannot be a jump over G, and thus $(V(Q) \cap V(G)) \setminus \{v_1, v_2\} \neq \emptyset$. For $i \in [2]$, let v_i' be the intersection of $Q - \{v_1, v_2\}$ with G nearest to v_i in the path Q, and let $y_i \in U$ be such that $v_i' \in \Delta_{y_i}$. The subpath of Q between v_i and v_i' is a jump over G, and since $v_i \notin \Delta'$, we conclude that $d_{\mathcal{T}}(v_i, v_i') \leq \theta$. Consequently, $d_{\mathcal{T}'}(v_i, f_{y_i}) \leq \theta$. Since $d_{\mathcal{T}'}(v_1, f_{y_1}) + d_{\mathcal{T}'}(f_{y_1}, f_{y_2}) + d_{\mathcal{T}'}(f_{y_2}, v_2) \geq d_{\mathcal{T}'}(v_1, v_2) \geq \delta_0$, it follows that $y_1 \neq y_2$.

For $i \in [2]$, let v_i''' be the vertex of $Q - \{v_1, v_2\}$ nearest to v_i belonging to $\bigcup_{y \in U \setminus \{y_i\}} \Delta_y$, and let $y_i' \in U \setminus \{y_i\}$ be the vertex such that $v_i''' \in \Delta_{y_i'}$. Note that such a vertex v_i''' exists, since v_{3-i}' is contained in $\Delta_{y_{3-i}}$. Let v_i'' be the last vertex of Q_i preceding v_i''' which is contained in Δ_{y_i}. Since $y_i' \neq y_i$, U is $(2\delta_0 + \theta)$-wide, and Δ_{y_i} and $\Delta_{y_i'}$ are δ_0-zones, we have

$$\begin{aligned} d_{\mathcal{T}}(v_i'', v_i''') &\geq d_{\mathcal{T}}(y_i, y_i') - d_{\mathcal{T}}(v_i'', y_i) - d_{\mathcal{T}}(v_i''', y_i') \\ &\geq (2\delta_0 + \theta) - 2\delta_0 = \theta, \end{aligned}$$

and thus the subpath of Q between v_i'' and v_i''' is a θ-long jump over G. Since every θ-long jump over G has both ends in Δ', we conclude that $v_i'' \in \Delta_{y_i}'$.

Consider the subpath Q_i of Q between v_i' and v_i''. All the intersections of this path with G belong to Δ_{y_i}. Moreover, since all θ-long jumps over G have both ends in Δ', until this path reaches Δ_{y_i}', the $d_{\mathcal{T}}$-distance between consecutive intersections with G is less than θ. Since $\theta \ll r$, v_i is outside of Δ_{y_i}, and the battlefield $(\mathcal{R}_{y_i}, \mathcal{P}_{y_i})$

is graded, Q_i must have many intersections with G contained in $\Delta_{y_i} \setminus \Delta'_{y_i}$. In particular, since we also have $|X| \ll r$, observe that there exist consecutive cycles C_i and C'_i of $\mathcal{R}_{y_i}$ different from $C_{y_i,r}$ such that Q has an intersection u_i with G contained in the closed annulus $\Lambda_i \subseteq \Delta_{y_i}$ between C_i and C'_i and Λ_i is disjoint from X.

Let Q' be the connected subgraph of G_0 consisting of $Q-\{v_1, v_2\}$ and for $i \in [2]$, a path $P_i \in \mathcal{P}_{y_i}$ disjoint from X_{y_1,y_2} (which exists since $|\mathcal{P}_{y_i}| = \theta_1 > |X_{y_1,y_2}|$), the cycle C_i, and a path of G between u_i and C_i in Λ_i (which exists, since the drawing of G is cellular). This subgraph is disjoint from X_{y_1,y_2} and intersects G' exactly in the ends of P_1 and P_2 which belong to Z_{y_1} and Z_{y_2}. This is a contradiction, since X_{y_1,y_2} intersects all jumps over G' from Z_{y_1} to Z_{y_2}. □

In the first outcome of the lemma, note that we can additionally clear small zones around vertices of $X \cap V(G')$ to ensure that X is disjoint from G'. In the second outcome of the lemma, let $\mathcal{Q}$ be the set of θ_1 pairwise vertex-disjoint jumps from Z_1 to Z_2, order these jumps according to the order of their ends in $\mathrm{bd}(\Delta_{y_1})$, and consider the order of their ends in $\mathrm{bd}(\Delta_{y_2})$. Erdős-Szekeres theorem implies that we can select a subset $\mathcal{Q}' \subseteq \mathcal{Q}$ of size at least $\sqrt{\theta_1}$ such that the ends of paths of $\mathcal{Q}'$ in $\mathrm{bd}(\Delta_{y_2})$ are in order; in case that Σ is orientable, this order can be the same or the opposite to the order of their ends in $\mathrm{bd}(\Delta_{y_1})$. Hence, $G' \cup \bigcup \mathcal{Q}'$ can be naturally drawn on the surface obtained from Σ by drilling holes in the interiors of Δ_{y_1} and Δ_{y_2} and joining the boundaries of the holes by a handle or a crosshandle. Moreover, since Z_1 and Z_2 are $\mathcal{T}'$-free and far apart, it is easy to see that the resulting drawing has large representativity, and more precisely, the tangle in $G' \cup \bigcup \mathcal{Q}'$ induced by $\mathcal{T}'$ is respectful after truncating it to order $\lceil\sqrt{\theta_1}\rceil$. Hence, we obtain the final form of Lemma 4.43.

Corollary 4.47 *For every graph H, every surface Σ, and every positive integer θ', there exist integers θ_0, δ, and p_0 such that the following claim holds. Let G_0 be a graph and let G be a subgraph of G_0 with a cellular drawing on Σ. Let $\mathcal{T}$ be a respectful tangle in G of order $\theta'_0 \geq \theta_0$ conformal with a tangle $\mathcal{T}_0$ in G_0. If G_0 does not contain H as a minor controlled by $\mathcal{T}_0$, then there exists a subgraph G' of G_0 and a tangle $\mathcal{T}'$ in G' conformal with $\mathcal{T}_0$ such that either*

- *$\mathcal{T}'$ has order at least $\theta'_0 - \delta$ and is respectful for a cellular drawing of G' on Σ, and there exists a set $X \subset V(G_0) \setminus V(G')$ of size at most p_0 intersecting all δ-long jumps in G_0 over G', or*
- *$\mathcal{T}'$ has order at least θ' and is respectful for a cellular drawing of G' on a surface obtained from Σ by adding a handle or a crosshandle.*

Proof (⇸) Let θ and p be as in Corollary 4.45 for H and Σ. For every positive integer κ, let p_κ and δ_κ be the integers p' and δ_0 from Lemma 4.46 applied for p, θ, κ, and $\theta_1 = (\theta')^2$. Let us define $g(\kappa) \gg p_\kappa, \delta_\kappa, p, \theta, \theta', \kappa$ and let κ_0 be the integer from Corollary 4.45 for H, Σ, and g. We choose $\theta_0 \gg \delta \gg a_\Sigma(H), \theta, p, \kappa_0$, and $g(\kappa)$ for $\kappa \leq \kappa_0$, and let p_0 be the maximum of $a_\Sigma(H) + p_\kappa$ for $\kappa \leq \kappa_0$.

We first apply Corollary 4.45 to G_0 and G to obtain

- a subgraph G_1 of G with a respectful tangle $\mathcal{T}_1$ such that G_1 and $\mathcal{T}_1$ are obtained from G and $\mathcal{T}$ by less than $a_\Sigma(H)$ repetitions of clearing a θ-zone,
- a set $X_1 \subseteq V(G_0) \setminus V(G_1)$ of size less than $a_\Sigma(H)$, and
- an integer $\kappa \le \kappa_0$ and a $g(\kappa)$-wide (in the metric $d_{\mathcal{T}_1}$) set $U \subseteq V(G_1)$ of size at most p,

such that every θ-long jump in $G_0 - X_1$ over G_1 has both ends at $d_{\mathcal{T}_1}$-distance less than κ from U.

Then, we apply Lemma 4.46 to $G_0 - X_1$ and G_1, and let G_2 and $\mathcal{T}_2$ denote the graph G' and the tangle $\mathcal{T}'$ obtained from G_1 and $\mathcal{T}_1$ by clearing the zones as described in the statement of the lemma.

- In the outcome (i) of Lemma 4.46, let X_2 denote the set X of at most p_κ vertices of $G_0 - X_1$ from the outcome. Let G' and $\mathcal{T}'$ be obtained from G_1 and $\mathcal{T}_1$ by clearing small zones around the vertices of $X_2 \cap V(G_1)$. Since G' and $\mathcal{T}'$ is obtained from G and $\mathcal{T}$ by clearing less than $a_\Sigma(H)$ θ-zones, at most p δ_κ-zones, and at most p_κ $O(1)$-zones, the choice of δ implies that the order of $\mathcal{T}'$ is at least $\theta_0' - \delta$. Observe that the first outcome of Corollary 4.47 holds for G', $\mathcal{T}'$, and the set $X = X_1 \cup X_2$.
- In the outcome (ii) of Lemma 4.46, let $G' = G_2$, $\mathcal{T}' = \mathcal{T}_2$, and proceed as described before the statement of this corollary to conclude that the second outcome of Corollary 4.47 holds.

□

The tools developed in this section so far enable us to constrain long jumps. We also need a way to deal with short (but non-trivial) jumps; the following lemma is useful in turning them into eyes. Given a subgraph M of a graph G, an ***M*-bridge** in G is a subgraph B of G that either

- consists of an edge of G with both ends in M, but not contained in $E(M)$, or
- consists of a component K of $G - V(M)$ together with all vertices of M with a neighbor in K and the edges from them to K.

The vertices in $V(B \cap M)$ are called the *attachments* of the M-bridge.

Lemma 4.48 *Let θ and κ be positive integers, let G_0 be a graph and let G be a subgraph of G_0 with a cellular drawing on a surface Σ, and let $\mathcal{T}$ be a respectful tangle in G of order at least κ. Let v be a vertex of G such that there is no θ-long jump in G_0 over G with an end at $d_{\mathcal{T}}$-distance at most κ from v, and let $(\mathcal{R}, \mathcal{P})$ be a free κ-local graded $(r \times p)$ battlefield around v, where $\mathcal{R} = C_1, \ldots, C_r$ and $p \ge 4$. For $i \in [r]$, let $\Delta_i \subset \Sigma$ be the zone around v bounded by C_i. Let G' and $\mathcal{T}'$ be obtained from G and $\mathcal{T}$ by clearing Δ_r and let f_0 be the resulting face. Let i_0 be an index such that $2 \le i_0 \le r - \theta - 2$. If the C_{i_0}-bridge of G_0 containing v also contains C_{i_0+1}, then G_0 contains an eye over G' with foundation f_0.*

Proof Let B be the set of vertices $u \in V(G)$ such that there exists a path from C_{i_0+1} to u in G disjoint from C_{i_0}. By the cellularity of the drawing of G, all vertices drawn in $\overline{\Sigma \setminus \Delta_{i_0+1}}$ belong to B and no vertex in Δ_{i_0} does, and the vertices of $\bigcup \mathcal{P}$ in $\Sigma \setminus \Delta_{i_0}$ belong to B.

Let $A = V(G) \setminus (B \cup V(C_{i_0}))$ be the set of vertices of G separated from C_{i_0+1} by C_{i_0}. Let F be the union of $G[A]$ and the subgraphs $J - V(C_{i_0})$ for all G-bridges J of G_0 with an attachment in A. By the assumptions of the lemma, there exists a path from $v \in V(F)$ to $V(C_{i_0+1}) \subset B$ in G_0 disjoint from C_{i_0}, and thus $V(F) \cap B \neq \emptyset$. Since no θ-long jump has an end in $A \subset \Delta_{i_0+1} \subset \Delta_r$ and the battlefield is graded, we have $V(F) \cap B \subset \Delta_{i_0+\theta+1} \subseteq \Delta_{r-1}$.

It follows that there exists a path Q in $F \cup (G \cap \Delta_{r-1})$ from C_1 to C_{r-1} vertex-disjoint from C_{i_0}. Moreover, such a path Q can be chosen so that the end y of Q on C_{r-1} is contained in a path $P \in \mathcal{P}$ and Q does not intersect any path of $\mathcal{P} \setminus \{P\}$ in a vertex of B. Choose distinct $P_1, P_2, P_3 \in \mathcal{P} \setminus \{P\}$ so that the cyclic order of their ends on C_r is P_1, P_2, P_3, P. Let Q_1 be the path consisting of the segments of the paths P_1 and P_3 starting in C_r and ending in vertices $z_1 \in V(C_{i_0} \cap P_1)$ and $z_3 \in V(C_{i_0} \cap P_3)$, together with the segment of C_{i_0} between z_1 and z_3 disjoint from P_2. Let Q_2 be the path consisting of P_2, a segment of C_1 between the ends of P_2 and Q, the path Q, and the segment of P from y to C_r. Note that the paths Q_1 and Q_2 are vertex-disjoint. Since the battlefield is free, we conclude that $Q_1 \cup Q_2$ is an eye over G' with foundation f_0. □

References

1. S. Cabello, B. Mohar, Finding shortest non-separating and non-contractible cycles for topologically embedded graphs. Discrete Comput. Geom. **37**, 213–235 (2007)
2. H.B. Griffiths, The fundamental group of two spaces with a common point. Quart. J. Math. **5**, 175–190 (1954)
3. U.H. Karimov, D. Repovš, On the union of simply connected planar sets. Topol. Appl. **122**, 281–286 (2002)
4. W. Mader, Über die Maximalzahl kreuzungsfreier H-Wege. Arch. Math. **31**, 387–402 (1978)
5. N. Robertson, P.D. Seymour, Graph minors. VI. Disjoint paths across a disc. J. Combin. Theory Ser. B **41**, 115–138 (1986)
6. N. Robertson, P.D. Seymour, Graph minors VII. Disjoint paths on a surface. J. Combin. Theory Ser. B **45**, 212–254 (1988)
7. N. Robertson, P.D. Seymour, Graph minors. XI. Circuits on a surface. J. Combin. Theory Ser. B **60**, 72–106 (1994)
8. N. Robertson, P.D. Seymour, Graph minors. XII. Distance on a surface. J. Combin. Theory Ser. B **64**, 240–272 (1995)
9. A. Schrijver, Disjoint circuits of prescribed homotopies in a graph on a compact surface. J. Combin. Theory Ser. B **51**, 127–159 (1991)
10. A. Schrijver, A short proof of Mader's S-paths theorem. J. Combin. Theory Ser. B **82**, 319–321 (2001)

Chapter 5
Towards the Structure Theorem

In the previous chapter, we have developed a number of tools that will be useful both in the proof of the Minor Structure Theorem and in its applications. One missing piece is a way to either find one of the "non-planarities" (an eye or a jump) or to show that the graph can be nearly drawn on a surface. In this chapter, we are going to fill in this gap and combine all the pieces to a sketch of the proof of the Minor Structure Theorem.

5.1 Disjoint Crossed Paths

As our first step, we investigate the conditions that enable us to find an eye. To this end, we need to understand when there exist two vertex-disjoint paths that attach to a cyclically ordered boundary of a graph in the interlaced order. Let us give some definitions needed to state the problem precisely.

Recall that a bordered graph is a graph G with boundary $\partial G \subseteq V(G)$, and that a society is a bordered graph where a cyclic ordering is specified for ∂G. Consider a society G with $\partial G = v_1, v_2, \ldots, v_m$; for distinct $i, j \in [m]$, we define

$$\partial_{[v_i,v_j]}G = \begin{cases} \{v_i, v_{i+1}, \ldots, v_j\} & \text{if } i < j \text{ and} \\ \{v_i, v_{i+1}, \ldots, v_m, v_1, \ldots, v_j\} & \text{if } j < i. \end{cases}$$

We also define $\partial_{(v_i,v_j)}G = \partial_{[v_i,v_j]}G \setminus \{v_i, v_j\}$ and $\partial_{[v_i,v_j)}G = \partial_{[v_i,v_j]}G \setminus \{v_j\}$. A ***cross*** in a society G is the union of two vertex-disjoint paths P_1 and P_2 in G, where for $i \in [2]$ the ends u_i and v_i of P_i belong to ∂G, and one end of P_2 belongs to $\partial_{(u_1,v_1)}G$ and the other end to $\partial_{(v_1,u_1)}G$.

The result that we are aiming for is essentially that if a society G does not contain a cross, then we can draw G in the disk without crossings and with ∂G drawn in

© The Author(s), under exclusive license to Springer Nature Switzerland AG 2025

Z. Dvořák, *Graph Minors*, Springer Monographs in Mathematics,
https://doi.org/10.1007/978-3-031-87469-7_5

order in the boundary of the disk; we call such a society ***rural***. Thus, either the society is useful in getting an eye, or we obtain its well-behaved drawing on a surface. However, as stated, the claim is not quite true: If a non-planar piece G' of G is separated from the boundary of G by a vertex cut of size at most three, then we cannot use the non-planarity to obtain a cross, since at most one path can be routed through the interior of G'. To avoid this issue for the moment, let us first focus on societies that do not have such cuts. A bordered graph G is *k-connected* if there exists no vertex separation (A, B) of G of order less than k such that $\partial G \subseteq A$ and $B \setminus A \neq \emptyset$. It is *internally k-connected* if it is $(k-1)$-connected and $|B \setminus A| \leq 1$ for every vertex separation (A, B) of G of order $k-1$ with $\partial G \subseteq A$, i.e., the only reason that G is not k-connected is that it contains vertices of degree $k-1$.

To prove the result, we are going to need two important tools. The first of them is a variant of a lemma of Tutte [14] on the way pieces of a graph attach to a specified topological minor. Let π be a topological model of a graph H in a graph G, and let $M = \pi(H)$ be the corresponding subgraph of G isomorphic to a subdivision of H. We say that an M-bridge B of G is *π-edge-local* if all its attachments belong to the path $\pi(e)$ for some edge $e \in E(H)$. In many constructions, it is desirable to avoid having π-edge-local $\pi(H)$-bridges, and as Tutte observed, this is always possible in 3-connected graphs.

Lemma 5.1 *Let H be a simple graph and let π be a topological model of H in a simple bordered graph G such that $\partial G = \pi(V(H))$ is the set of the branch vertices of π. If G is 3-connected, then there exists a topological model π' of H in G with $\pi' \restriction V(H) = \pi \restriction V(H)$ such that no $\pi'(H)$-bridge is π'-edge-local. In particular, for every $e \in E(H)$, the path $\pi'(e)$ in G is induced.*

Proof ($\looparrowright$) For a topological model π' of H in G with $\pi' \restriction V(H) = \pi \restriction V(H)$, we say that a vertex $v \in V(G) \setminus V(\pi'(H))$ is *π'-good* if it is contained in a non-π'-edge-local $\pi'(H)$-bridge. Let π' be chosen so that the number of π'-good vertices is maximal. We claim that no $\pi'(H)$-bridge is π'-edge-local.

Suppose for a contradiction that there is an edge $e \in E(H)$ such that the set L of $\pi'(H)$-bridges with attachments only in $\pi'(e)$ is non-empty. We say that a bridge $B \in L$ *covers* a vertex $v \in V(\pi'(e))$ if B has an attachment both strictly before and strictly after v in the path $\pi'(e)$. In particular, the ends of $\pi'(e)$ are not covered by any bridge in L. Note that since G is simple and 3-connected, every bridge in L covers at least one vertex of $\pi'(e)$. Let P be a maximal subpath of $\pi'(e)$ formed by vertices covered by bridges from L, and let u and v be the vertices preceding and following P in $\pi'(e)$. Let $L' \subseteq L$ consist of the bridges that cover a vertex of P.

Observe that each bridge in L' has all attachments in $V(P) \cup \{u, v\}$ and every bridge in $L \setminus L'$ is disjoint from $V(P)$. Let $D = V(P) \cup \{u, v\} \cup \bigcup_{B \in L'} V(B)$ and $C = V(G) \setminus (D \setminus \{u, v\})$. Note that $\partial G = \pi'(V(H)) \subseteq C$ and $C \cap D = \{u, v\}$. Since G is 3-connected, (C, D) is not a vertex separation of G, and thus there exists an edge xy of G with $x \in C \setminus D$ and $y \in D \setminus C$. Since L' consists of $\pi'(H)$-bridges with attachments only in $V(P) \cup \{u, v\}$, we conclude that $y \in V(P)$ and the edge xy belongs to a non-π'-edge-local bridge B_0.

Since $y \in V(P)$, there exists a bridge $B \in L'$ covering y. Let u' and v' be the first and last attachment of B on the path $\pi'(e)$. Let π'' be a topological model of H in G obtained from π' by replacing the segment Q of the path $\pi'(e)$ between u' and v' by a path Q' in B intersecting $\pi'(e)$ exactly in its ends.

We claim that there are more π''-good vertices than π'-good vertices, giving us a contradiction with the choice of π'. Indeed, consider any non-$\pi'(H)$-local bridge A. If no attachment of A is an internal vertex of Q, then A is also a non-π''-edge-local $\pi''(H)$-bridge. If an internal vertex of Q is an attachment of A, then there exists a $\pi''(H)$-bridge A' such that $A \cup Q \subseteq A'$, and A' is not $\pi''(H)$-local (this is clear if $Q \neq \pi'(e)$, and follows from the fact that H is simple if $Q = \pi'(e)$). Therefore, every π'-good vertex is also π''-good. Moreover, because of B_0, the non-π'-good vertex y is π''-good. □

Another useful fact follows from an inspection of Ford-Fulkerson algorithm used to find the maximum number of pairwise vertex-disjoint paths between specified subsets A and B of vertices. The algorithm works by augmenting an arbitrary initial (A, B)-linkage $\mathcal{P}_0$, and the ends of the paths of $\mathcal{P}_0$ remain to be ends of the paths in the augmented solution $\mathcal{P}$ (though in $\mathcal{P}$, the paths may join different vertices than in $\mathcal{P}_0$). This gives us the following corollary.

Lemma 5.2 *Let G be a (directed or undirected) graph and let A and B be subsets of vertices of G. Let $\mathcal{P}_0$ be a strict (A, B)-linkage in G of size k_0. If there exists an (A, B)-linkage in G of size $k \geq k_0$, then there also exists a strict (A, B)-linkage $\mathcal{P}$ of size k such that every end of a path of $\mathcal{P}_0$ is also an end of a path of $\mathcal{P}$.*

As an application of Lemma 5.2, let us see how a cross can be obtained from another kind of non-planarity. Let G be a society. A *tripod* in G consists of a theta-graph formed by three paths P_1, P_2, and P_3 in G intersecting exactly in their ends and a strict $(V(P_1 \cup P_2 \cup P_3), \partial G)$-linkage $\{Q_1, Q_2, Q_3\}$, such that for $i \in [3]$, either the path P_i intersects ∂G in exactly one internal vertex and Q_i is formed by this vertex, or P_i is disjoint from ∂G and Q_i ends in an internal vertex of P_i.

Lemma 5.3 *Let G be a society and let $(P_1, P_2, P_3, Q_1, Q_2, Q_3)$ be a tripod in G. If G is internally 4-connected, then G contains a cross.*

Proof For $i \in [3]$, let s_i be the end of Q_i in P_i. Since G is internally 4-connected and $V(P_1 \cup P_2 \cup P_3)$ contains at least two vertices not belonging to ∂G (the common ends of the paths), Menger's theorem implies that there exists a $(V(P_1 \cup P_2 \cup P_3), \partial G)$-linkage of size four. By Lemma 5.2, there exists such a linkage $\{Q'_1, Q'_2, Q'_3, Q\}$ which is strict and s_i is an end of Q'_i for $i \in [3]$. By symmetry, we can assume that Q has an end s in the path P_1 (where s might be one of the ends of P_1). Let t_1, t_2, t_3, and t be the ends of Q'_1, Q'_2, Q'_3, and Q in ∂G, where by symmetry, we can assume that $t_2 \in \partial_{(t_1,t_3)}G$.

If $t \in \partial_{(t_1,t_2)}G$, i.e., t_1, t, t_2, and t_3 appear in ∂G in order, then G has a cross with one path consisting of Q, the path in $P_1 \cup P_3$ from s to s_3 disjoint from s_1, and Q'_3; and the other path consisting of Q'_1, the path in $P_1 \cup P_2$ from s_1 to s_2 disjoint from s, and Q'_2. The case that $t \in \partial_{(t_3,t_1)}G$ is symmetric. Finally, if $t \in \partial_{(t_2,t_3)}G$, i.e, t_1,

t_2, t, and t_3 appear in ∂G in order, then G has a cross with one path consisting of Q, the subpath of P_1 between s and s_1, and Q_1'; and the other path consisting of Q_2', the path in $P_2 \cup P_3$ disjoint from s, and Q_3'. □

Let us now give the first form of the main result of this section.

Lemma 5.4 *If an internally* 4*-connected society* G *does not contain a cross, then* G *is rural.*

Proof We prove the claim by induction on the number of vertices of G. Without loss of generality, we can assume that G is simple, since suppressing loops and parallel edges affects neither the existence of a cross nor rurality.

If $V(G) = \partial G$ and each edge of G joins consecutive vertices of ∂G, then G is rural. If $|\partial G| \leq 3$, then since G is internally 4-connected, we have $|G \setminus \partial G| \leq 1$ and G is rural. Suppose now that G does not fall into either of these special cases. Thus, if $V(G) = \partial G$, then an edge of G joins non-consecutive vertices of ∂G; let P' be the path formed by this edge. Otherwise, since G is 3-connected, it contains three paths from a vertex $x \in V(G) \setminus \partial G$ to ∂G intersecting only in x; and since $|\partial G| \geq 4$, there are two of these paths such that the ends of the path P' formed by their union are non-consecutive in ∂G.

In conclusion, G contains a path P' joining non-consecutive vertices u and v of ∂G and otherwise disjoint from ∂G. By Lemma 5.1 applied with H consisting of ∂G and the edge uv, it follows that G contains a path P with ends u and v and otherwise vertex-disjoint from ∂G such that letting $M = \partial G \cup P$, no M-bridge in G has attachments only in P. In particular, P is an induced path in G.

If an M-bridge has an attachment both in $\partial_{(u,v)}G$ and in $\partial_{(v,u)}G$, then a path in M together with P forms a cross. Hence, suppose that this is not the case. Let G_1 be the union of P and the M-bridges with an attachment in $\partial_{(u,v)}G$, with ∂G_1 consisting of $\partial_{(u,v)}G$ in the same order as in ∂G and the vertices of P in order starting from v. Let G_2 be the union of P and the M-bridges with an attachment in $\partial_{(v,u)}G$, with ∂G_2 consisting of $\partial_{(v,u)}G$ in the same order as in ∂G and the vertices of P in order starting from u. Note that since G is internally 4-connected, both G_1 and G_2 are internally 4-connected.

Suppose first that both G_1 and G_2 are rural. For $i \in [2]$, consider the drawing of G_i in the disk Δ_i with the vertices of ∂G_i drawn in the boundary of Δ_i in order. Without loss of generality, we can assume that the whole path P is drawn in the boundary of Δ_i. Thus, we can obtain a drawing of G in the disk by gluing the drawings of G_1 and G_2 along P, implying that G is rural.

Hence, suppose that say G_1 is not rural. By the induction hypothesis, the society G_1 contains a cross consisting of paths S_1 and S_2. Without loss of generality, we can assume that S_1 and S_2 intersect ∂G only in their ends, as otherwise they have subpaths with this property forming a cross. Let r be the number of ends of S_1 and S_2 in P, and choose the cross so that r is minimal. If $r = 0$, then S_1 and S_2 form a cross in G. If $1 \leq r \leq 2$, then we can extend S_1 or S_2 or both along P to obtain a cross in G. See Fig. 5.1 for illustrations of the following cases.

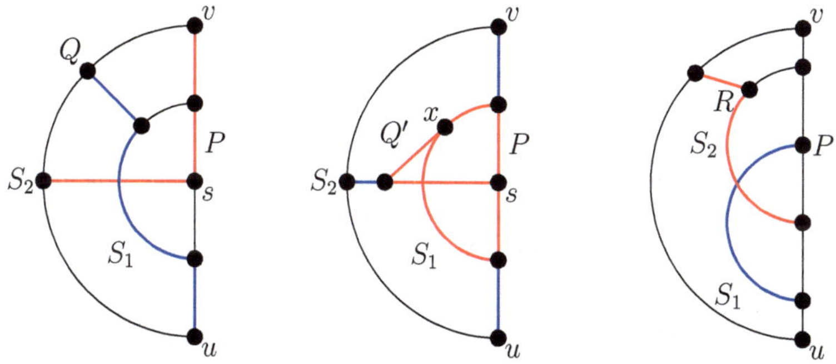

Fig. 5.1 The cases from the proof of Lemma 5.4

Suppose that $r = 3$, and thus say S_1 has both ends in P and S_2 has an end s in P and the other end t in $\partial_{(u,v)}G$. The M-bridge containing S_1 does not attach only to P, and thus there exists a path Q in G_1 from an internal vertex x of S_1 to $\partial_{(u,v)}G$ and disjoint from ∂G_1 except for its end in $\partial_{(u,v)}G$. If Q is disjoint from S_2, then G contains a cross with one path formed by Q, a subpath of S_1, and a subpath of P, and the other path formed by S_2 and a subpath of P. If Q intersects S_2, then let Q' be the initial segment of Q from x to a vertex $y \in V(S_2)$ and disjoint from $S_1 \cup S_2$ except for its ends; note that Q' is also disjoint from $\partial_1 G$, except possibly for the vertex y which can be equal to t. Observe that the three paths in $Q' \cup S_1 \cup S_2 \cup P$ between x and s form a tripod in G (together with the subpaths of S_2 and P connecting them to t, u, and v). By Lemma 5.3, G contains a cross.

Finally, suppose that $r = 4$, and thus both S_1 and S_2 have ends in P. Since the M-bridge(s) containing S_1 and S_2 do not attach only to P, there exists a path R in G_1 from $\partial_{(u,v)}G$ to $S_1 \cup S_2$, with only the first vertex in $\partial_{(u,v)}G$ and only the last vertex in $S_1 \cup S_2$, say in S_2. Then S_1 together with R and a subpath of S_2 forms a cross in G_1 with only three vertices in P, contradicting the minimality of r. □

Let us remark that Lemma 5.4 has the following interesting consequence.

Corollary 5.5 *Every non-planar* 4*-connected graph G with at least four vertices is* 2*-linked.*

Proof Let s_1, t_1, s_2, and t_2 be distinct vertices of G. Let us set $\partial G = s_1, s_2, t_1, t_2$ in cyclic order. Then G is an internally 4-connected society, and since G is not planar, it is not rural. By Lemma 5.4, there exists a cross in G, necessarily consisting of vertex-disjoint paths from s_1 to t_1 and from s_2 to t_2. Therefore, G contains an $\{(s_1, t_1), (s_2, t_2)\}$-linkage. Since this holds for any 4-tuple (s_1, t_1, s_2, t_2) of distinct vertices of G, we conclude that G is 2-linked. □

Since every simple planar graph has average degree less than 6, we obtain the result promised in Sect. 3.2.

Corollary 5.6

$$d_{\text{link}}(2) = 6.$$

Next, we would like to generalize Lemma 5.4 to societies that are not internally 4-connected. Here, the cross-free society again has nearly-planar structure, but we need to allow the parts that are split off by vertex (≤ 3)-cuts to be arbitrary. This motivates the following definitions. A ***segregation*** of $\mathcal{S}$ of a graph G is a system of societies such that

- $G = \bigcup_{A \in \mathcal{S}} A$ and the societies in $\mathcal{S}$ are pairwise edge-disjoint, and
- $V(A_1 \cap A_2) = \partial A_1 \cap \partial A_2$ for any distinct $A_1, A_2 \in \mathcal{S}$.

That is, $\mathcal{S}$ gives a decomposition of G into pieces that are connected only through their boundary vertices. In particular, for any element $A \in \mathcal{S}$, $\big(A, \bigcup(\mathcal{S} \setminus \{A\})\big)$ is a separation of G of order at most $|\partial A|$. If G is a society, we additionally require that

- $\partial G \subseteq \bigcup_{A \in \mathcal{S}} \partial A$.

An element $A \in \mathcal{S}$ is a ***cell*** if $|\partial A| \leq 3$. We say that $\mathcal{S}$ is *simple* if all its elements are cells. The ***projection*** $\pi(A)$ of $A \in \mathcal{S}$ is the graph which

- is empty if $\partial A = \emptyset$,
- consists just of the vertex of ∂A if $|\partial A| = 1$,
- consists of the vertices of ∂A and an edge between them if $|\partial A| = 2$, and
- is the cycle on vertices of ∂A in their prescribed cyclic order if $|\partial A| \geq 3$.

The projection $\pi(\mathcal{S})$ is defined as the union of the projections of elements of $\mathcal{S}$, preserving parallel edges in this union; if G is a society, $\pi(\mathcal{S})$ is also defined to be a society, with $\partial\pi(\mathcal{S}) = \partial G$.

Our aim now will be to generalize Lemma 5.4 by showing that if a society G does not contain a cross, then it has a simple segregation with rural projection. Actually, we are going to show slightly more, finding such a segregation whose projection is a minor of G; this will be needed in applications of this result and further results derived from it. More precisely, consider a segregation $\mathcal{S}$ of a graph or a society G. We say that $A \in \mathcal{S}$ is ***semi-solid*** if $\pi(A)$ is a ∂A-rooted minor of A. We say that the segregation $\mathcal{S}$ is *semi-solid* if all its elements are semi-solid. Note that if $\mathcal{S}$ is semi-solid, then gluing together the rooted models of the projections of the elements of $\mathcal{S}$ gives us a model $\mu_{\mathcal{S}}$ of $\pi(\mathcal{S})$ in G rooted in $V(\pi(\mathcal{S}))$.

On a more technical note, in order to prove the aforementioned generalization of Lemma 5.4, we need to find a segregation avoiding certain degeneracies. Suppose for example that $F = K_{3,m}$ is a subgraph of a society G and F attaches to the rest of G only through the set M of three vertices forming one side of the bipartite graph F. In a simple segregation $\mathcal{S}$ of G, it is quite natural for the whole F to form a single cell with boundary M. It is less natural, but still possible, that $\mathcal{S}$ contains more than one cell, each formed by M and a subset of $V(F) \setminus M$ of size at least two (so that the cell is semi-solid); or even that $\mathcal{S}$ contains cells formed by single edges of F. It turns out to be preferable to avoid these cases by requiring $\mathcal{S}$ to be "compact" in the following sense. We say that a simple segregation $\mathcal{S}$ is *reduced* if for every $A \in \mathcal{S}$ with $|\partial A| = 3$,

- there is no $A' \in \mathcal{S} \setminus \{A\}$ with $\partial A' = \partial A$, and
- there is no vertex $v \in V(\pi(\mathcal{S})) \setminus (\partial A \cup \partial G)$ such that every $A' \in \mathcal{S}$ with $v \in \partial A'$ satisfies $\partial A' \subseteq \{v\} \cup \partial A$.

Lemma 5.7 *Every society has a semi-solid reduced simple segregation with internally 4-connected projection.*

Proof Let G be a society. A separation (A, B) of G with $\partial G \subseteq V(B)$ is *feasible* if it has order at most three, A is connected, and

- $V(A \cap B) = \emptyset$ and A is non-empty, or
- $V(A \cap B) = \{v\}$ and A either consists of v with a single loop, or $|A| > 1$ and there is no loop on v in A, or
- $V(A \cap B) = \{u, v\}$ and A either consists of u, v, and a single edge between them, or $|A| > 2$ and $V(A \cap B)$ is an independent set in A, or
- $|A \cap B| = 3$, $V(A \cap B)$ is an independent set in A, and A contains as a minor the triangle rooted in $V(A \cap B)$.

For a feasible separation (A, B), we view A as a society with boundary $V(A \cap B)$ (cyclically ordered arbitrarily). We say that A is a *feasible subsociety* of G if there exists a feasible separation (A, B) of G such that $\partial A = V(A \cap B)$. A *consolidation* of G is a segregation of G consisting only of feasible subsocieties; note that every vertex and edge of G belongs to a feasible subsociety of G (consisting of just this vertex or edge), and thus G has a consolidation. Clearly, any consolidation is a simple and semi-solid segregation.

Let $\mathcal{S}$ be a consolidation of G with $|\pi(\mathcal{S})| + |\mathcal{S}|$ minimum. Observe that $\mathcal{S}$ is reduced, since if $A \in \mathcal{S}$ with $|\partial A| = 3$ violated the condition from the definition of reducedness, then we could replace A in $\mathcal{S}$ either by $A \cup A'$ in the first case, or by $A \cup \bigcup_{A' \in \mathcal{S}, v \in \partial A'} A'$ in the second case. Hence, it suffices to argue that $\pi(\mathcal{S})$ is internally 4-connected.

Let (C, D) with $\partial G \subseteq D$ be a vertex separation of $\pi(\mathcal{S})$ of smallest order, and suppose that $|C \cap D| \leq 3$ and $C \nsubseteq D$. Since the projection of each cell is a clique, if $A \in \mathcal{S}$ and $\partial A \cap (C \setminus D) \neq \emptyset$, then $\partial A \subseteq C$. Let $\mathcal{S}' = \{A \in \mathcal{S} : \partial A \cap (C \setminus D) \neq \emptyset\}$ and let $A_0 = \bigcup \mathcal{S}'$. Let C' be the subgraph of $\pi(\mathcal{S})$ obtained from the one induced by C by deleting the edges between vertices of $C \cap D$. The minimality of $|C \cap D|$ implies that C' is connected, and in particular each vertex of $C \cap D$ is incident with an edge of C'. Note that C' is a subgraph of $\pi(\mathcal{S}')$, and since each element of $\mathcal{S}'$ is connected, we conclude that A_0 is connected and $C \cap D \subsetneq V(A_0)$. Turn A_0 into a society by letting ∂A_0 be $C \cap D$ with an arbitrary cyclic ordering. Then $\mathcal{S}_0 = (\mathcal{S} \setminus \mathcal{S}') \cup \{A_0\}$ is a segregation of G with $V(\pi(\mathcal{S}_0)) = D$, and thus by the minimality of $|\pi(\mathcal{S})| + |\mathcal{S}|$, it is not a consolidation.

Therefore, A_0 is not a feasible subsociety. Note that since A_0 is a union of feasible subsocieties $A \in \mathcal{S}'$ with $\partial A \nsubseteq \partial A_0$, ∂A_0 is an independent set in A_0. Thus, the only way A_0 could be infeasible is if $|\partial A_0| = 3$ and A_0 does not contain the triangle rooted in ∂A_0 as a minor. In particular, at this point we can conclude that $\pi(\mathcal{S})$ does not have a vertex separation (C, D) with $\partial G \subseteq D$ of order less than 3 such that $C \not\subseteq D$, and thus $\pi(\mathcal{S})$ is 3-connected.

Consequently, C' taken as a bordered graph with $\partial C' = C \cap D$ is 3-connected. If $|C \setminus D| > 1$, then Lemma 1.4 implies that C' contains as a minor the triangle rooted in $C \cap D$. Since $\mathcal{S}'$ is semi-solid, C' is a minor of A_0 rooted in $C \cap D = \partial A_0$, and thus A_0 also contains as a minor the triangle rooted in ∂A_0. It follows that A_0 is feasible, which is a contradiction. Therefore, every vertex separation (C, D) of $\pi(\mathcal{S})$ of order 3 with $\partial G \subseteq D$ satisfies $|C \setminus D| \leq 1$, and thus $\pi(\mathcal{S})$ is internally 4-connected. □

Note that we can view a graph as a society with empty boundary, and thus Lemma 5.7 also says that every graph has a semi-solid reduced simple segregation with essentially 4-connected projection.

We can furthermore slightly improve the connectivity properties of the cells in the segregation, at the expense of the connectivity of the projection. A cell A is ***solid*** if for every $v \in \partial A$, the cell A contains $|\partial A| - 1$ paths from v to $\partial A \setminus \{v\}$ which intersect only in v. Observe that this implies that A is semi-solid. A simple segregation is *solid* if all its cells are solid.

Observation 5.8 *Every semi-solid cell F has a solid simple segregation $\mathcal{S}_F$ with a 3-connected projection. Moreover, the projection can be drawn in the disk with all vertices of ∂F drawn in the boundary of the disk.*

Proof If $|\partial F| \leq 2$, then semi-solidity implies solidity, and we can let $\mathcal{S}_F = \{F\}$. Hence, suppose that $|\partial F| = 3$. Since F is semi-solid, there exists a 2-connected block C of F such that F contains a strict $(\partial F, V(C))$-linkage $\mathcal{P}$ of size three. Let Z be the set of ends of $\mathcal{P}$ in C. For each $v \in \partial F$, let $v' \in Z$ be the end of path of $\mathcal{P}$ starting in v and let C_v be the component of $F - (V(C) \setminus \{v'\})$ containing v and v'. In this case, we let $\mathcal{S}_f$ consist of the cells C_v for $v \in \partial F$ with $\partial C_v = \{v, v'\}$ and of the cell C' consisting of C and all other parts of F not contained in $\bigcup_{v \in \partial F} C_v$, with $\partial C' = \{v' : v \in \partial F\}$. Note that some of the cells C_v may be trivial (consist only of $v = v'$), in which case they of course do not have to be included in $\mathcal{S}_F$. □

Let $\mathcal{S}$ be a segregation of a graph G. An ***arrangement*** of $\mathcal{S}$ on a surface Σ is a drawing of $\pi(\mathcal{S})$ in Σ such that for each $A \in \mathcal{S}$ with $|\partial A| \geq 3$, the cycle $\pi(A)$ bounds a distinct face, which we denote by $\overline{\pi}(A)$. For cells $A \in \mathcal{S}$ such that $|\partial A| \leq 2$, we define $\overline{\pi}(A)$ to be the drawing of $\pi(A)$ induced by $\pi(\mathcal{S})$, excluding the ends of the edge when $|\partial A| = 2$. If G is a society and Σ is the disk, we additionally require that the vertices of ∂G are drawn in the boundary of Σ in the given cyclic order. Note that because we require the existence of the faces $\overline{\pi}(A)$, this is a slightly stronger condition than just requiring that $\pi(\mathcal{S})$ is rural.

We can now easily relax the connectivity assumption of Lemma 5.4. In the proof, we are going to need the following additional definitions. Let $\mathcal{S}$ be a segregation of a graph G. For a path Q in G with both ends in $V(\pi(\mathcal{S}))$ and with all edges in cells of $\mathcal{S}$, let $\pi(Q)$ denote the corresponding path in $\pi(\mathcal{S})$, obtained by contracting the subpaths of Q inside cells to the edges of the projection. We define $\pi(C)$ for a cycle C in G with at least three vertices in $V(\pi(\mathcal{S}))$ and with all edges in cells of $\mathcal{S}$ analogously.

Corollary 5.9 *Let G be a society. The following claims are equivalent:*

(i) G does not contain a cross.
(ii) There exists a semi-solid reduced simple segregation of G with internally 4-connected projection and an arrangement in the disk.
(iii) There exists a solid simple segregation of G with a 3-connected projection and an arrangement in the disk.
(iv) There exists a simple segregation of G with an arrangement in the disk.

Proof Consider any simple segregation $\mathcal{S}$ of G. Note that any cross $Q_1 \cup Q_2$ in G corresponds to a cross $\pi(Q_1) \cup \pi(Q_2)$ in $\pi(\mathcal{S})$. If $\mathcal{S}$ has an arrangement in the disk, then $\pi(\mathcal{S})$ does not have a cross, and thus G does not have a cross, either. Hence, (iv) implies (i).

Suppose now that G does not have a cross. By Lemma 5.7, there exists a semi-solid reduced simple segregation $\mathcal{S}$ of G such that $\pi(\mathcal{S})$ is internally 4-connected. If $\pi(\mathcal{S})$ had a cross, then recall that $\pi(\mathcal{S})$ is a minor of G rooted in ∂G (since $\mathcal{S}$ is semi-solid) and consequently G would have a cross as well. Therefore, $\pi(\mathcal{S})$ does not contain a cross. Since $\pi(\mathcal{S})$ is internally 4-connected, Lemma 5.4 implies that $\pi(\mathcal{S})$ has a drawing in a disk Δ with the vertices of ∂G drawn in the boundary of Δ in the given cyclic order. Consider any cell $A \in \mathcal{S}$ with $|\partial A| = 3$. No vertex of $\pi(\mathcal{S})$ can be drawn inside the subdisk of Δ bounded by the cycle $\pi(A)$, since $\pi(\mathcal{S})$ is internally 4-connected and $\mathcal{S}$ is reduced. Moreover, since $\mathcal{S}$ is reduced, there is no other cell $A' \in \mathcal{S} \setminus \{A\}$ such that $\partial A' = \partial A$. Consequently, we can rearrange the parallel edges in the drawing of $\pi(\mathcal{S})$ so that each such cycle $\pi(A)$ bounds a face, obtaining an arrangement of $\mathcal{S}$ in the disk Δ. Hence, (i) implies (ii).

In case that $\mathcal{S}$ is a semi-solid simple segregation of G with internally 4-connected projection and an arrangement in the disk, we can split each cell of $\mathcal{S}$ as in Observation 5.8, then merge together any cells S_1 and S_2 with $|\partial S_1|, |\partial S_2| \leq 2$ intersecting in a vertex belonging neither to ∂G nor to any other cell, thus obtaining a solid simple segregation $\mathcal{S}'$ with a 3-connected projection. Observe that we can also turn the arrangement of $\mathcal{S}$ in the disk to an arrangement of the segregation $\mathcal{S}'$ in the disk. Hence, (ii) implies (iii). Finally, (iii) implies (iv) trivially. □

Let us remark that the solidity of the segregation is important when relating substructures of $\pi(\mathcal{S})$ to those in the original graph. The existence of disjoint paths guaranteed by the solidity has the following obvious consequence.

Observation 5.10 *Let H be a simple triangle-free graph and let $\mathcal{S}$ be a solid simple segregation of another graph G. Let ρ be an injective function from a set* $\mathrm{dom}(\rho) \subseteq$ *$V(H)$ to $V(\pi(\mathcal{S}))$. If H is a ρ-rooted topological minor of $\pi(\mathcal{S})$, then it is also a ρ-rooted topological minor of G.*

Proof Let ν be a ρ-rooted topological model of H in $\pi(\mathcal{S})$. By the assumption that H is simple and triangle-free, for each cell $S \in \mathcal{S}$ with $|\partial S| = 3$, the topological model ν uses at most two edges vv_1 and vv_2 of $\pi(\mathcal{S})$. The solidity of $\mathcal{S}$ guarantees that S contains paths from v to v_1 and v_2 intersecting only in v. By replacing the edges of ν by these paths, we obtain a ρ-rooted topological model of H in G. □

In view of this observation, it might seem desirable to define the solidity of a cell S with $|\partial S| = 3$ as having a cycle passing through all three vertices of ∂S, so that we can drop the assumption that H is triangle-free; indeed, this would slightly simplify some of the applications. However, it is not always possible to further subdivide cells to achieve this property. More precisely, it is true that if S does not contain a cycle through ∂S, then it has a non-trivial simple segregation $\mathcal{S}_0$, as follows from a characterization of graphs without a cycle passing through three prescribed vertices given in [3, 15]. However, it is not always possible to find such a segregation $\mathcal{S}_0$ with an arrangement in the disk with all vertices of ∂S in the boundary (e.g., if $S = K_{2,3}$ with ∂S formed by the three vertices of degree two), and thus we cannot replace S in $\mathcal{S}$ by $\mathcal{S}_0$ while preserving the property that $\mathcal{S}$ has an arrangement in the disk.

Let us now give an application of Corollary 5.9 to creating eyes, in a setting similar to Lemma 4.48.

Lemma 5.11 *Let G_0 be a graph and let G be a subgraph of G_0 with a cellular drawing on a surface Σ, and let $\mathcal{T}$ be a respectful tangle in G of order at least κ. Let $(\mathcal{R}, \mathcal{P})$ be a free κ-local $(r \times p)$ battlefield around a vertex v of G, where $\mathcal{R} = C_1, \ldots, C_r$ and $p \geq 4$. For $i \in [r]$, let $\Delta_i \subset \Sigma$ be the disk bounded by C_i and containing v. Let G' and $\mathcal{T}'$ be obtained from G and $\mathcal{T}$ by clearing Δ_r and let f_0 be the resulting face. Let i_0 be an index such that $2 \leq i_0 \leq r-1$ and the C_{i_0}-bridge M of G_0 containing v* does not *contain C_{i_0+1}. Let $\mathcal{P}'$ consist of the segments of the paths of $\mathcal{P}$ between C_r and C_{i_0}. Let G_1 be the society $C_{i_0} \cup M$ with boundary ∂G_1 consisting of the ends of the paths of $\mathcal{P}'$ in order along C_{i_0}. Then either G_0 contains an eye over G' with foundation f_0, or G_1 has a solid simple segregation $\mathcal{S}$ with 3-connected projection and an arrangement in the disk Δ such that the cycle $\pi(C_{i_0}) \subset \pi(\mathcal{S})$ is drawn along the boundary of Δ.*

Proof Since $C_{i_0+1} \not\subseteq M$, the paths of $\mathcal{P}'$ intersect $C_{i_0} \cup M$ only in their ends in C_{i_0}. If G_1 contains a cross, then since the battlefield is free, combining this cross with paths of $\mathcal{P}'$ gives us an eye of G_0 over G' with foundation f_0. Otherwise, Corollary 5.9 implies that G_1 has a solid simple segregation $\mathcal{S}$ with 3-connected projection and an arrangement in the disk Δ_0. Let us remark that the boundary of Δ_0 contains the vertices of ∂G_1, but not necessarily all other vertices of $V(C_{i_0}) \cap V(\pi(\mathcal{S}))$, and thus the conclusion of the lemma does not immediately follow. Let $\Lambda \subset \Delta_0$ be the open disk bounded by the cycle $\pi(C_{i_0})$.

For consecutive vertices $x, y \in \partial G_1$, let $Q_{x,y}$ denote the subpath of C_{i_0} between x and y disjoint from $\partial G_1 \setminus \{x, y\}$. For every vertex $u \in V(M) \setminus V(C_{i_0})$, there exists a path from u to C_1 in $M - V(C_{i_0})$, and thus by considering the subpaths of $\mathcal{P}$ contained in G_1, we conclude that the following claim holds.

($\star$) For every vertex $u \in V(M) \setminus V(C_{i_0})$ and any consecutive vertices $x, y \in \partial G_1$, there exists paths in $M - V(Q_{x,y})$ from u to all vertices of $\partial G_1 \setminus \{x, y\}$.

Consider any consecutive vertices $x, y \in \partial G_1$. Suppose for a contradiction that a vertex $u \in V(\pi(\mathcal{S})) \setminus V(C_{i_0})$ is drawn in a connected part Λ' of $\Delta_0 \setminus \pi(Q_{x,y})$ different from the one containing Λ. Then every path Q' from u to ∂G_1 in M satisfies $V(\pi(Q') \cap \pi(Q_{x,y})) \neq \emptyset$, and thus $V(Q' \cap Q_{x,y}) \neq \emptyset$, contradicting ($\star$). Hence, no such vertex exists, and since this is the case for all consecutive vertices $x, y \in \partial G_1$, we conclude that all vertices of $V(\pi(\mathcal{S})) \setminus V(C_{i_0})$ are drawn in Λ. Moreover, there is at least one such vertex v_0, since the part of $\mathcal{P}$ in G_1 gives at least four pairwise vertex-disjoint paths from C_1 to ∂G_1, and thus C_1 cannot be contained in a single cell of $\mathcal{S}$.

Suppose now that for some consecutive vertices $x, y \in \partial G_1$, an edge $e \in E(\pi(\mathcal{S})) \setminus E(\pi(C_{i_0}))$ is drawn in a connected part Λ' of $\Delta_0 \setminus \pi(Q_{x,y})$ different from the one containing Λ. Since Λ' does not contain any vertex of $\pi(\mathcal{S})$, it follows that $e = u_1u_2$ for vertices $u_1, u_2 \in V(\pi(C_{i_0}))$. Let A be the cell of $\mathcal{S}$ such that $e \in E(\pi(A))$. If there exists another vertex $u_3 \in \partial A \setminus \{u_1, u_2\}$, then observe that since $\pi(A)$ bounds a face of $\pi(\mathcal{S})$, the vertex u_3 has to be drawn in the closure of Λ', and thus $u_3 \in V(\pi(C_{i_0}))$ as well; therefore, we have $\partial A \subseteq V(\pi(C_{i_0}))$. This is also clearly the case if $|\partial A| = 2$. Since $\mathcal{S}$ is solid, A contains a path R between u_1 and u_2 intersecting ∂A only in its ends. Since $e \notin E(\pi(C_{i_0}))$, R is not a subpath of C_{i_0}, and thus R has a subpath R' connecting distinct vertices of C_{i_0} and otherwise disjoint from C_{i_0}. Note that C_{i_0} is an induced cycle in G_1, and thus R' has length at least two. Therefore, there exists a vertex $u \in V(A) \setminus \partial A$, namely an internal vertex of R'. Every path from u to v_0 in G_1 must intersect $\partial A \subseteq V(C_{i_0})$, which is a contradiction, since u and v_0 are in the same C_{i_0}-bridge M.

Therefore, all vertices and edges of $\pi(\mathcal{S})$ are drawn in the closure of Λ, and we can let $\Delta = \overline{\Lambda}$. □

Let us remark that the claim would be false in the seemingly more natural formulation, with G_1 having the boundary formed by vertices of C_{i_0} in order. An issue here is that there can exist non-planar pieces of G_1 involving segments of C_{i_0}, and when such pieces are separated from the rest of the graph by cuts of size three, they do not give rise to an eye. However, such pieces then must be contained inside cells of the segregation of G_1, making it impossible for all vertices of C_1 to be drawn in the boundary of Δ.

5.2 Flat Wall Theorem

Using the tools developed so far, we can prove the Flat Wall Theorem (also called the Weak Minor Structure Theorem): Every H-minor-free graph of large treewidth contains a part which is "nearly planar" (in the sense of Corollary 5.9), still has large treewidth and thus contains a large wall as a subgraph, and the interior of this nearly planar part attaches to the rest of the graph only through a bounded number of apex vertices. This result can be seen as a "toy version" of the full Minor Structure Theorem, enabling us to showcase some of the main proof ideas in a simpler setting. However, it also has a number of important algorithmic applications, and in particular it is a key ingredient in the polynomial-time algorithm for testing the presence of a fixed graph as a minor of the input graph.

Let us now give the most abstract formulation of the Flat Wall Theorem. For a graph H, we define the ***apex number*** $a(H)$ as $a_{\text{sphere}}(H)$, i.e., the minimum number of vertices we need to delete from H to make it planar. Let G_0 be a graph with a tangle $\mathcal{T}_0$, let C be a cycle in G_0, and let M be a C-bridge of G_0. A ***flat region of order*** ψ in G_0 is a quadruple $(C, F, \mathcal{S}, \mathcal{T})$, where

- F is a society with the underlying graph $C \cup M$ such that ∂F is a subset of $V(C)$ of size ψ ordered along C,
- $\mathcal{S}$ is a solid simple segregation of F with 3-connected projection,
- $\mathcal{S}$ has an arrangement in a disk Δ with $\pi(C)$ drawn along the boundary of Δ, and
- $\mathcal{T}$ is a tangle of order ψ in $\pi(\mathcal{S})$ $\mu_{\mathcal{S}}$-conformal with $\mathcal{T}_0$ such that ∂F is $\mathcal{T}$-free.

The Flat Wall Theorem basically states that for any H-minor-free graph of large treewidth, we can delete less than $a(H)$ vertices to obtain a subgraph that contains a flat region of large order. In some applications, it will be useful to have more than one flat region, and thus we actually prove the following slightly stronger form of the claim.

Theorem 5.12 *For every graph H and positive integers $\psi \geq 4$ and m, there exists an integer θ such that the following claim holds. If $\mathcal{T}_0$ is a tangle of order θ in a graph G_0 that does not control a minor of H in G_0, then there exists a set $A \subseteq V(G_0)$ of size less than $a(H)$ and m pairwise vertex-disjoint flat regions of order ψ in $G_0 - A$.*

Proof Let us choose parameters $p, r, k, \theta_1, \ldots, \theta_4$, and θ so that

$$|H|, \|H\|, \psi, m \ll p, \theta_4, r \ll k \ll \theta_3 \ll \theta_2 \ll \theta_1 \ll \theta.$$

Since the tangle $\mathcal{T}_0$ has order $\theta \gg \theta_1$, Corollary 2.38 implies that there exists a subgraph W of G_0 isomorphic to a subdivision of a wall and a tangle $\mathcal{T}_1$ of order θ_1 in W conformal with $\mathcal{T}_0$. Fix a drawing of W in the sphere; then $\mathcal{T}_1$ is automatically respectful.

By Lemma 4.43, there exists a subgraph G_2 of W and a respectful tangle $\mathcal{T}_2'$ of order at least $\theta_2 + a(H)$ obtained from W and $\mathcal{T}_1$ by clearing less than $a(H)$ θ_4-zones and sets $A \subseteq V(G_0) \setminus V(G_2)$ of size less than $a(H)$ and $Y \subseteq V(G_2)$ of size at most p such that every θ_4-long jump in $G_0 - A$ over G_2 has both ends at $d_{\mathcal{T}_2'}$-distance less than θ_4 from Y. Note that by Lemma 4.19, $\mathcal{T}_2'$ is conformal with $\mathcal{T}_1$, and thus also with $\mathcal{T}_0$. Let $\mathcal{T}_2$ be the truncation of $\mathcal{T}_2'$ to order θ_2; then the tangle $\mathcal{T}_2$ is conformal with $\mathcal{T}_0 - A$.

By Lemma 4.31, we conclude that G_2 has a set U_0 of at least $p + \|H\|^2 + m$ vertices pairwise at $d_{\mathcal{T}_2}$-distance at least $2\theta_3$. Since each vertex of U_0 is at $d_{\mathcal{T}_2}$-distance less than θ_3 from at most one vertex of Y, we can choose a subset $U \subseteq U_0$ of size $\|H\|^2 + m$ such that $d_{\mathcal{T}_2}(y, u) \geq \theta_3$ for every $y \in Y$ and $u \in U$. It follows that no θ_4-long jump in $G_0 - A$ over G_2 has an end at $d_{\mathcal{T}_2}$-distance at most k from a vertex of U.

For each $u \in U$, Theorem 4.21 gives a free k-local graded $r \times r$ battlefield $(\mathcal{R}_u, \mathcal{P}_u)$ around u. Let $\mathcal{R}_u = C_{u,1}, \ldots, C_{u,r}$ and for $i \in [r]$, let $\Delta_{u,i} \subset \Sigma$ be the closed disk bounded by $C_{u,i}$ containing u, let G_u be the subgraph of G_2 obtained by clearing $\Delta_{u,r}$, and let f_u be the resulting face of G_u. Since the battlefields are k-local, for distinct $u, u' \in U$, we have $\Delta_{u,r} \cap \Delta_{u',r} = \emptyset$.

Suppose first that for at least $\|H\|^2$ vertices $u \in U$, there exists an eye in $G_0 - A$ over G_u with foundation f_u. Note that the eyes for distinct vertices $u, u' \in U$ are pairwise vertex-disjoint, as otherwise $G_0 - A$ would contain a (necessarily θ_4-long) jump with ends in $\Delta_{u,r}$ and $\Delta_{u',r}$, contradicting the choice of U. By Lemma 4.35, it follows that $G_0 - A$ contains H as a minor controlled by $\mathcal{T}_0 - A$, and this also gives a minor of H in G_0 controlled by $\mathcal{T}_0$. This is a contradiction, and thus there exists a set $U' \subseteq U$ of size m such that for $u \in U'$, $G_0 - A$ does not contain an eye over G_u with foundation f_u.

Let $i_0 = 2\psi + 1$ and consider $u \in U'$. By Lemma 4.48, the C_{u,i_0}-bridge M_u of $G_0 - A$ containing u is disjoint from C_{u,i_0+1}. By Lemma 5.11, the graph $F_u = C_{u,i_0} \cup M_u$ with ∂F_u consisting of the last vertices of $\mathcal{P}_u$ in C_{u,i_0} in order along C_{u,i_0} has a solid simple segregation $\mathcal{S}_u$ with 3-connected projection and an arrangement in the disk Δ such that $\pi(C_{u,i_0})$ is drawn along the boundary of Δ.

For distinct $u, u' \in U'$, the graph $G_0 - A$ does not contain a jump from $\Delta_{u,r}$ to $\Delta_{u',r}$, and thus the subgraphs F_u and $F_{u'}$ are pairwise vertex-disjoint. Hence, it suffices to show that for each $u \in U'$, F_u can be turned into a flat region of order ψ. That is, we need to construct a tangle $\mathcal{T}_u$ of order ψ in $\pi(\mathcal{S}_u)\mu_{\mathcal{S}_u}$-conformal with $\mathcal{T}_0 - A$ such that ∂F_u has a $\mathcal{T}_u$-free subset B_u of size ψ; then, letting F'_u be the society obtained from F_u by restricting its boundary to B_u, we conclude that $(C_{u,i_0}, F'_u, \mathcal{S}_u, \mathcal{T}_u)$ is a flat region of order ψ in $G_0 - A$.

Since the battlefield $(\mathcal{R}_u, \mathcal{P}_u)$ is free, Lemma 2.45 implies that there exists a $\mathcal{T}_2$-free set $Z_u \subseteq \partial F_u$ of size at least $r/2 \geq 2\psi$. Lemma 2.48 implies that Z_u is $(\mathcal{T}_0 - A)$-free. Let $s_1, \ldots, s_\psi, t_\psi, t_{\psi-1}, \ldots, t_1$ be the vertices of Z_u in order. Observe that $F_u \cap \left(\bigcup \mathcal{P}_u \cap \bigcup_{i=1}^{\psi} C_{u,i}\right)$ contains an $\{(s_i, t_i) : i \in [\psi]\}$-linkage $\{R_1, \ldots, R_\psi\}$. Let X be the $\psi \times \psi$ grid. We can contract subpaths of the paths $\pi(R_1), \ldots, \pi(R_\psi)$ and subpaths of the cycles $\pi(C_{u,\psi+1}), \ldots, \pi(C_{u,2\psi})$ to obtain X as a minor of $\pi(\mathcal{S}_u)$; let μ be the corresponding model of the grid X in $\pi(\mathcal{S}_u)$. Let $\mathcal{T}'_u$ be the standard tangle of order ψ in the grid X, and let $\mathcal{T}_u$ be the tangle of order ψ in $\pi(\mathcal{S}_u)$ μ-induced by $\mathcal{T}'_u$. Let μ' be the model of X in G_0 obtained by composing μ and $\mu_{\mathcal{S}_u}$. Lemma 2.46 implies that $\mathcal{T}'_u$ is μ'-conformal with $\mathcal{T}_0 - A$, and thus $\mathcal{T}_u$ is $\mu_{\mathcal{S}}$-conformal with $\mathcal{T}_0 - A$. Observe that the set $B_u = \{s_1, \ldots, s_\psi\} \subset \partial F_u$ of the starting vertices of the paths $R_1, \ldots, R_\psi$ is $\mathcal{T}'_u$-free, and thus also $\mathcal{T}_u$-free by Lemma 2.48. □

The formulation and proof of Theorem 5.12 that we chose to give somewhat obfuscates the reason for the name "Flat Wall Theorem". In a more standard formulation, one only operates on the subdivided wall W that we used to start the process, finding a more explicit form of the flat region: By a *subwall* of W, we mean a subdivided wall formed by subpaths of consecutive rows and columns of W, and the *perimeter* of the subwall is the cycle bounding it. One can then find a set $A \subseteq V(G_0)$ of size less than $a(H)$ and a large subwall W' of W with perimeter K such that the K-bridge M of $G_0 - A$ containing $W' - V(K)$ is disjoint from the rest of W, and such that $K \cup M$ has a simple segregation with an arrangement in the disk, with $\pi(K)$ drawn along the boundary of the disk.

Indeed, this is the formulation and argument used by Robertson and Seymour in [10] (the proof of Theorem 5.12 follows the same idea in a more abstract setting). However, there is not much gained by explicitly working with subwalls. If we actually need a wall in a flat region, it is trivial to obtain one using Lemmas 2.44 and 2.65, similarly to the proof of Lemma 2.68. And, in order to prove the Flat Wall Theorem by only considering (sub)walls, one needs to introduce wall versions of concepts such as clearing a zone, the respectful tangle metric, and so on. Admittedly, these wall-specific notions are simpler than their general versions, but nevertheless somewhat superfluous when we have already developed the general theory.

Let us note the following connection between the tangle of a flat region and the ambient tangle $\mathcal{T}_0$ in the whole graph, which may be useful in case $\mathcal{T}_0$ is chosen to have some specific property (e.g., in Erdős-Pósa theory arguments in the style of the second proof of Lemma 2.57, where the tangle is chosen so that the small parts do not contain any of the subgraphs we are trying to hit). For notational convenience, we give the statement in terms of vertex tangles.

Lemma 5.13 *Let $\mathcal{T}_0$ be a vertex tangle of order θ in a graph G_0, let A be a subset of vertices of G_0, let $\psi \le \theta - |A|$ be a positive integer, and let $(C, F, \mathcal{S}, \mathcal{T})$ be a flat region of order ψ in $G_0 - A$ (with $\mathcal{T}$ taken as a vertex tangle). For every $B \in \mathcal{S}$ with $V(B \cap C) \subseteq \partial B$, the corresponding vertex separation $(V(B) \cup A, V(G_0) \setminus (V(B) \setminus \partial B))$ of G_0 belongs to $\mathcal{T}_0$.*

Proof Let $D = V(G_0 - A) \setminus (V(B) \setminus \partial B)$. It suffices to show that the vertex separation $(V(B), D)$ of $G_0 - A$ belongs to $\mathcal{T}_0 - A$. Suppose for a contradiction that $(D, V(B)) \in \mathcal{T}_0 - A$. Since $\mathcal{T}$ is $\mu_{\mathcal{S}}$-conformal with $\mathcal{T}_0 - A$, we have $(\mu_{\mathcal{S}}^{-1}(D), \mu_{\mathcal{S}}^{-1}(V(B))) \in \mathcal{T}$. However, $\mu_{\mathcal{S}}^{-1}(D) = V(\pi(\mathcal{S}))$, contradicting (T1). □

As we have already mentioned, the Flat Wall Theorem was developed for an application in testing the presence of a fixed graph as a (rooted) minor in the input graph, and we give an outline of this algorithm in the next section. Before that, let us give two simpler applications.

The first one is a generalization of Theorem 2.73, concerning the fact that the treewidth of graphs of bounded genus is bounded by a function of their radius. This is not true for H-minor-free graphs for every graph H; e.g., consider a graph G consisting of a large grid and a single vertex v adjacent to all vertices of the grid. Then G has radius one and arbitrarily large treewidth, yet $G - v$ is planar and consequently G is K_6-minor-free. However, there is the following nice characterization of which forbidden minors H enforce this property. We say that a hereditary graph class $\mathcal{G}$ has ***locally bounded treewidth*** if there exists a function f such that every graph $G \in \mathcal{G}$ of radius r has treewidth at most $f(r)$.

Theorem 5.14 *The class of H-minor-free graphs has locally bounded treewidth if and only if $a(H) \leq 1$. Consequently, a minor-closed class $\mathcal{G}$ has locally bounded treewidth if and only if there exists a graph $H \notin \mathcal{G}$ with $a(H) = 1$.*

Proof If $a(H) \geq 2$, then H is not a minor of the "grid plus one vertex" graphs discussed above, and thus the class of H-minor-free graphs does not have locally bounded treewidth.

Suppose now that $a(H) \leq 1$. By Theorems 2.25 and 5.12, there exists a function f such that for every positive integer r, every H-minor-free graph of treewidth more than $f(r)$ contains a flat region of order $4r + 4$ (since $a(H) \leq 1$, the set A of removed vertices must be empty). We claim that every H-minor-free graph G of radius r has treewidth at most $f(r)$, and thus the class of H-minor-free graphs has locally bounded treewidth.

Suppose for a contradiction that $\mathrm{tw}(G) > f(r)$. Hence, there exists a flat region $(C, F, \mathcal{S}, \mathcal{T})$ of order $4r+4$ in G. Let us view $\pi(\mathcal{S})$ as drawn on the sphere and let f_0 be the face of $\pi(\mathcal{S})$ bounded by $\pi(C)$. Then $\mathcal{T}$ is a respectful tangle in $\pi(\mathcal{S})$, and by Theorem 4.14, there exists an edge $e \in E(\pi(\mathcal{S}))$ such that $d_{\mathcal{T}}(f_0, e) = 4r + 4$. Let v be a vertex incident with e, let Q_0 be a shortest path in G from v to C, and let $Q = \pi(Q_0)$. Note that Q is a path from v to $\pi(C)$ in $\pi(\mathcal{S})$ and that Q is at most as long as the path Q_0. But $4r + 4 = d_{\mathcal{T}}(e, f_0) \leq 2\|Q\| + 3$, and thus $\|Q_0\| \geq \|Q\| > 2r$. Therefore, G has diameter more than $2r$, and thus radius more than r, which is a contradiction.

Finally, if a minor-closed graph class $\mathcal{G}$ contains all graphs H with $a(H) = 1$, then it contains all "grid plus one vertex" graphs, and thus it does not have locally bounded treewidth. Otherwise, there exists a graph $H \notin \mathcal{G}$ with $a(H) = 1$, and thus $\mathcal{G}$ is a subclass of the class of H-minor-free graphs; and we conclude that $\mathcal{G}$ has locally bounded treewidth. □

Let us remark that the bound on the treewidth of H-minor-free graphs relative to their radius which follows from the proof of Theorem 5.14 is rather poor (though polynomial), since the proof of Theorem 5.12 relies on the Grid Theorem. The bound can be strengthened to linear one [2]; the proof is an application of the Minor Structure Theorem and we give it in Chap. 8.

We now give the second application of the Flat Wall Theorem, in algorithmic design. For this, the reader needs to observe that the proof of Theorem 5.12 can be turned into a polynomial-time algorithm to find a flat region. This is mostly straightforward; one might have doubts about the usage of Mader's H-Wege theorem, but it is known that this theorem can be algorithmized, see e.g. [1].

The application we are going to show is a fixed-parameter algorithm for ***domination number***, i.e., determining the minimum size $\gamma(G)$ of a dominating set in an input graph G (a set S is *dominating* if every vertex of $V(G) \setminus S$ has a neighbor in S). It is believed not to be possible to determine whether a general n-vertex graph has a dominating set of size at most k in time $f(k) \cdot \mathrm{poly}(n)$, for any function f and any polynomial whose degree does not depend on k. However, such an algorithm exists for H-minor-free graphs, based on the following observation on irrelevance of certain vertices. Given a set $A \subseteq V(G)$, the *A-type* of a vertex $v \in V(G) \setminus A$ is the set of neighbors of v in A, and the *extended A-type* of v is the pair (t, T), where t is the A-type of v and T is the set (not multiset!) of A-types of neighbors of v in $V(G) \setminus A$. Note that if $|A| = a$, then there are only 2^{a+2^a} different extended A-types.

Lemma 5.15 *Let A be a subset of vertices of a graph G and let $v_0, \ldots, v_k$ be vertices in $V(G) \setminus A$ of the same extended A-type that are pairwise at distance at least five in $G - A$. Then $\gamma(G) < k$ if and only if $\gamma(G - v_0) < k$.*

Proof For $i \in [k]$ and a positive integer d, let $N_{i,d}$ be the set of vertices at distance at most d from v_i in $G - A$.

Suppose first that $D \subseteq V(G)$ is a dominating set in G of size $\gamma(G) < k$. Since the sets $N_{1,2}, \ldots, N_{k,2}$ are pairwise disjoint, there exists $i \in [k]$ such that $D \cap N_{i,2} = \emptyset$. Therefore, every neighbor of v_i in $G - A$ has a neighbor in $A \cap D$. Since v_0 and v_i have the same extended A-type, every neighbor of v_0 in $G - A$ also has a neighbor in $A \cap D$. Moreover, every neighbor of v_0 in A is also adjacent to v_i. If $v_0 \in D$, then let $D_0 = (D \setminus \{v_0\}) \cup \{v_i\}$, otherwise let $D_0 = D$. Observe that D_0 is a dominating set in $G - v_0$, and thus $\gamma(G - v_0) \le |D_0| = |D| < k$.

Conversely, suppose that $D' \subseteq V(G - v_0)$ is a dominating set in $G - v_0$ of size $\gamma(G - v_0) < k$. Since the sets $N_{1,1}, \ldots, N_{k,1}$ are pairwise disjoint, there exists $i \in [k]$ such that $D' \cap N_{i,1} = \emptyset$. Therefore, v_i has a neighbor in $A \cap D'$. Since v_0 and v_i have the same A-type, v_0 also has a neighbor in $A \cap D'$. Therefore, D' is a dominating set in G, and $\gamma(G) \le |D'| < k$. □

Corollary 5.16 *For every graph H, there exists a function f and an algorithm that, given an H-minor-free n-vertex graph G and a positive integer k, decides in time $f(k) \cdot \mathrm{poly}(n)$ whether $\gamma(G) < k$.*

Proof For given k, choose $\psi(k)$ and $\theta(k)$ so that $k, |H|, \|H\| \ll \psi(k) \ll \theta(k)$. Let $f(k)$ be chosen so that Theorem 2.8 gives an algorithm with time complexity $O(f(k) \cdot n)$ to determine whether a graph of treewidth at most $\theta(k)$ has domination number less than k.

If $\mathrm{tw}(G) > \theta(k)$, then by Theorem 2.25 and using the algorithm underlying the proof of Theorem 5.12, we find a set $A \subseteq V(G)$ of size less than $a(H)$ and a flat region $(C, F, \mathcal{S}, \mathcal{T})$ of order $\psi(k)$ in $G - A$. Let us view $\pi(\mathcal{S})$ as drawn on the sphere and let f_0 be the face of $\pi(\mathcal{S})$ bounded by $\pi(C)$. Then $\mathcal{T}$ is a respectful tangle in $\pi(\mathcal{S})$.

Lemma 4.31 can be used to obtain a set U of vertices of $\pi(\mathcal{S})$ of size more than $k2^{a(H)+2^{a(H)}}$ such that $d_{\mathcal{T}}(u, v) > 8$ for distinct $u, v \in U$, and $d_{\mathcal{T}}(u, f_0) > 8$ for each $u \in U$. This implies that the graph distance in $G - A$ between distinct vertices of U is at least five. Since the number of extended A-types is at most $2^{|A|+2^{|A|}}$, the pigeonhole principle implies that there exist $k + 1$ distinct vertices $v_0, \ldots, v_k \in U$ of the same extended A-type.

By Lemma 5.15, we have $\gamma(G) < k$ if and only if $\gamma(G - v_0) < k$. Therefore, we can delete v_0 from G and repeat the process described above, until we reach an equivalent subgraph with treewidth at most $\theta(k)$. The desired algorithm then follows from Theorem 2.8. □

Let us remark that the algorithm from Corollary 5.16 is by far not the simplest or most efficient way to design a fixed-parameter algorithm for domination number in proper minor-closed classes. Indeed, there is a substantially simpler and more general meta-algorithmic result that applies to all nowhere-dense graph classes [6], where the notion of *nowhere-dense graph classes* is a far-reaching generalization of proper minor-closed classes. Nevertheless, the proof of Corollary 5.16 gives a simple illustration of a way the Flat Wall Theorem can be used in algorithmic design for more complicated problems, and serves as a warm up for the next section.

5.3 Testing the Presence of a Minor

Perhaps the most important application of the Flat Wall Theorem is in the polynomial-time algorithm to test the presence of a fixed (rooted) minor. For this application, we need a variant of the Flat Wall Theorem in which the pieces of the segregation have bounded treewidth (a reader questioning why we bothered to obtain m flat regions rather than just one in Theorem 5.12 will find the answer here).

Theorem 5.17 *For every graph H and every positive integer $\psi > \max(|H|, 3)$, there exists an integer t such that the following claim holds. For every graph G_0 of treewidth at least t, either*

- *there exists a tangle in G_0 of order ψ controlling a minor of H, or*
- *there exists a set $A \subseteq V(G_0)$ of size less than $a(H)$ and a flat region $(C, F, \mathcal{S}, \mathcal{T})$ of order ψ in $G_0 - A$ such that $\mathrm{tw}(S) < t$ for every $S \in \mathcal{S}$.*

Proof Let $\theta \geq \psi$ be the constant from Theorem 5.12 for $m = a(H) + 3$. Let t be large enough so that Theorem 2.25 implies that every graph of treewidth at least t contains a tangle of order θ.

Let L_0 be the null graph. For $i = 1, 2, \ldots$, we construct a separation (L_i, G_i) of G_0 of order less than m such that $\mathrm{tw}(G_i) \geq t$, as follows: By Theorem 2.25, G_{i-1} contains a tangle $\mathcal{T}_{i-1}$ of order θ. If $\mathcal{T}_{i-1}$ controls a model μ of H in G_{i-1}, then the tangle in G_0 induced by $\mathcal{T}_{i-1}$ also controls μ as a model of H in G_0, and the first outcome of the theorem holds (after truncating the tangle to order ψ).

Otherwise $\mathcal{T}_{i-1}$ does not control a minor of H in G_{i-1}, and Theorem 5.12 implies that there exists a set $A \subseteq V(G_{i-1})$ of size less than $a(H)$ such that $G_{i-1} - A$ contains m pairwise disjoint flat regions of order ψ. At least one of the regions, denoted by $(C, F, \mathcal{S}, \mathcal{T})$, is disjoint from $V(L_{i-1} \cap G_{i-1})$, and thus it also is a flat region in $G_0 - A$. If $\mathrm{tw}(S) < t$ for every $S \in \mathcal{S}$, then the second outcome of the theorem holds.

Hence, suppose that there exists $S \in \mathcal{S}$ such that $\mathrm{tw}(S) \geq t$. We let G_i be the subgraph of G_{i-1} consisting of S, A, and all the edges of G_0 between S and A;

observe that there exists a separation (L_i, G_i) of G_0 such that $V(L_i \cap G_i) = \partial S \cup A$. We now repeat the procedure described above to either obtain one of the conclusions of the theorem or a separation (L_{i+1}, G_{i+1}), and so on. Observe that since G_{i-1} contains another flat region disjoint from $(C, F, \mathcal{S}, \mathcal{T})$, we have $|G_i| < |G_{i-1}|$, and thus this procedure eventually terminates. □

Let us remark that we could have actually set $m = a(H) + 4$, obtain two usable disjoint flat regions, and continue with the smaller one of them, decreasing the number of iterations from $O(|G_0|)$ to $O(\log |G_0|)$ and improving the time complexity of the corresponding algorithm. Let us also remark that in comparison to Theorem 5.12, in this formulation of the Flat Wall Theorem we cannot prescribe a specific tangle $\mathcal{T}_0$ "pointing towards" the flat region. On the flip side, since the elements of the segregation have bounded treewidth, Theorem 2.8 can be used to compute their properties efficiently, which is convenient in algorithmic applications.

Furthermore, we could enforce that $\pi(\mathcal{S})$ has treewidth bounded by a function of θ, by restricting F to the interior of a suitably chosen cycle of $\pi(\mathcal{S})$ in case the treewidth is too large initially (we leave the details for the reader to work out). Since F has a star decomposition with leaves corresponding to the elements of $\mathcal{S}$ and with the torso of the central node isomorphic to $\pi(\mathcal{S})$, it follows we could also ensure that the treewidth of F is bounded. Nevertheless, we do not need this strengthening for the purposes of this chapter.

Our goal now will be to describe a polynomial-time algorithm for the following problem. The problem is parameterized by a fixed graph H and a (possibly empty) subset R of $V(H)$. The input to the problem is a graph G_0 and an injective function $\rho : R \to V(G_0)$, and our goal is to decide whether H is a ρ-rooted minor of G_0 and if so, to find a ρ-rooted model of H in G_0. We call this the R-ROOTED H-MINOR problem. The time complexity of the algorithm we describe is polynomial, where the degree of the polynomial is independent on H (a more detailed analysis shows that the time complexity is cubic [10], though it can be improved to quadratic [8]).

Before we dive into the description of the algorithm, let us note that the problem has two important special cases. One of course is the case that $R = \emptyset$, where we just test whether H is a minor of G_0. The other one concerns the case that H is the perfect matching on the set $R = V(H)$, where $|R| = 2k$, in which case we obtain the k-LINKAGE problem of finding k pairwise vertex-disjoint paths with prescribed ends. Let us remark that by analogy with Menger's theorem, the reader perhaps might consider the latter special case to be substantially simpler than minor testing in general. This intuition is incorrect; for example, in case that k is not a fixed parameter, but a part of the input, then the problem becomes NP-complete [7]. Moreover, in directed graphs, even the 2-LINKAGE problem is NP-complete [4]!

Let us also remark that a polynomial-time algorithm for k-LINKAGE implies a polynomial-time algorithm to decide whether an input graph G_0 contains a fixed graph H as a *topological* minor, though in this case the exponent in the time complexity depends on H: We simply test all possible placements of the branch vertices of the topological minor and of the first edges of each of its paths inside G_0 (there are at most $|G_0|^{|H|} \cdot \|G_0\|^{2\|H\|}$ possible choices), and for each of them we use the k-LINKAGE algorithm to test whether the disjoint paths with the prescribed ends forming the topological minor of H actually exist in G_0. A faster algorithm with cubic time complexity was found in [5].

Let us now start working on the algorithm for the R-ROOTED H-MINOR problem. The basic idea of the algorithm is to find a flat region in the input graph G_0 and show that we can delete a vertex from the middle part of the region without affecting the existence of the rooted minor of H. An issue with this idea is that if we are not careful, we might delete or disconnect a part of G_0 containing a rare substructure that does not appear anywhere else in G_0, thus destroying all rooted models of H in G_0. Let us now give definitions and results designed to deal with this issue.

The ***k-folio*** of a bordered graph G is the set of all (up to isomorphism) pairs (F, ρ) where F is a graph such that $|F| + \|F\| \le k$, ρ is a bijection between a subset of $V(F)$ and ∂G, and F is a ρ-rooted minor of G. The k-folio contains all the information about ways how we can use G in constructing small minors, in the following sense.

Observation 5.18 *Suppose (G_0, G_1) and (G_0, G_2) are separations of graphs G'_1 and G'_2, where $V(G_0 \cap G_1) = V(G_0 \cap G_2)$. For $i \in [2]$, let us turn G_i into a bordered graph by setting $\partial G_i = V(G_0 \cap G_i)$. Let H be a graph and ρ an injective function from $\mathrm{dom}(\rho) \subseteq V(H)$ to $V(G_0)$. If G_1 and G_2 have the same $(2|H|+5\|H\|+4|G_0 \cap G_1|)$ folio, then H is a ρ-rooted minor of G'_1 if and only if it is a ρ-rooted minor of G'_2.*

Proof Let $S = V(G_0 \cap G_1)$. Let us consider a ρ-rooted model μ_1 of H in G'_1, chosen so that the subgraph $M = \bigcup_{v \in V(H)} (\mu_1(H) \cap G_1)$ is minimal; in particular, M is a forest. Let F be the graph defined as follows. The vertex set of F consists of S and of the components of $M - S$. For each vertex $x \in S \cup V(M)$, let $\phi(x) = x$ if $x \in S$ and let $\phi(x) \in V(F)$ be the component of $M - S$ containing x otherwise. The edge set of F consists of the edges $\phi(u)\phi(v)$ for each edge $e = uv$ such that either $e \in E(M)$ and e is incident with a vertex of S, or $e \in \mu_1(E(H)) \cap E(G_1)$. Let $\sigma : S \to S$ be the identity function. Clearly, F is a σ-rooted minor of G_1.

By the minimality of M, each component C of $M - S$ contains an end of an edge of $\mu_1(E(H))$, or is equal to $\mu_1(v)$ for a vertex $v \in V(H)$, or there are at least two edges from C to S in M. Since M is a forest, this implies that $M - S$ has less than $2\|H\| + |H| + |S|$ components, and thus $|F| < 2\|H\| + |H| + 2|S|$. Moreover, since M is a forest, observe that $\|F\| < |F| + \|H\|$. Therefore, F is contained in the $(2|H| + 5\|H\| + 4|G_0 \cap G_1|)$-folio of G_1, and thus also in the $(2|H| + 5\|H\| + 4|G_0 \cap G_1|)$-folio of G_2.

It follows that F is a σ-rooted minor of G_2. A combination of a σ-rooted model of F in G_2 and of the part of μ_1 contained in G_0 gives us a ρ-rooted model of H in G'_2. □

For a flat region $(C, F, \mathcal{S}, \mathcal{T})$ in a graph $G_0 - A$ and any cell $S \in \mathcal{S}$, let $S + A$ denote the bordered graph obtained from the union of S and A by adding all edges of G_0 between S and A, with the boundary $\partial(S + A) = A \cup \partial S$. Suppose that we are testing the presence of a fixed rooted minor H in G_0, where the roots are disjoint from the flat region. By Observation 5.18, there exists an integer k depending only on H such that for each cell $S \in \mathcal{S}$ with $S - \partial S$ disjoint from the roots, only the k-folio of $S + A$ is important. For distinct $S, S' \in \mathcal{S}$, we say that S and S' have *the same A-extended k-folios* if there exists a bijection $p : \partial(S+A) \to \partial(S'+A)$ which is identity on A and preserves the orientation (clockwise or counter-clockwise) of ∂S and $\partial S'$ as drawn in the arrangement of $\mathcal{S}$, such that p extends to an isomorphism of the k-folios of $S + A$ and $S' + A$.

It might seem that we are running into a circularity here: We are going to need to compute the A-extended folios for the cells of the flat region in order to design the algorithm for the R-ROOTED H-MINOR problem, but how to do so without having such an algorithm to start with? Theorem 5.17 allows us to sidestep this issue: We can assume that the graphs $S \in \mathcal{S}$ have bounded treewidth, and thus we can compute their A-extended k-folios using Theorem 2.8.

As usual, let us view $\pi(\mathcal{S})$ as drawn on the sphere, with f_0 being the face bounded by $\pi(C)$ and $\mathcal{T}$ considered as a respectful tangle. Let R_0 be a subset of $V(G_0)$. For a vertex $v \in V(\pi(\mathcal{S}))$, we say that *the θ-neighborhood of v is (k, R_0, κ)-homogeneous* if

- $d_{\mathcal{T}}(v, f_0) > \theta + 3$,
- $d_{\mathcal{T}}(v, \partial S) > \theta$ for every $S \in \mathcal{S}$ such that $V(S) \cap R_0 \neq \emptyset$, and
- for every cell $S \in \mathcal{S}$ with $d_{\mathcal{T}}(v, \partial S) \leq \theta$ and every vertex $u \in V(\pi(\mathcal{S}))$ with $d_{\mathcal{T}}(v, u) \leq \theta - \kappa$, there exists a cell $S' \in \mathcal{S}$ with $d_{\mathcal{T}}(u, \partial S') \leq \kappa$ such that S and S' have the same A-extended k-folios.

That is, every A-extended folio that appears in the θ-neighborhood of v at all appears almost everywhere in the neighborhood. An easy argument shows that in a flat region of sufficiently large order, we can find a homogeneous neighborhood that is relatively large compared to κ.

Lemma 5.19 *Let a and k be positive integers and let $f : \mathbb{N} \to \mathbb{N}$ be a non-decreasing function. There exists a positive integer ψ such that the following claim holds. Let G_0 be a graph, let A and R_0 be sets of vertices of G_0 of size at most a, and let $(C, F, \mathcal{S}, \mathcal{T})$ be a flat region in $G_0 - A$ of order ψ. Then there exists $v \in V(\pi(\mathcal{S}))$ and a positive integer κ such that the $f(\kappa)$-neighborhood of v is (k, R_0, κ)-homogeneous.*

Proof Let m be the number of all possible A-extended k-folios; note that m is a function of a and k. Let $\kappa_0 = 1$ and for $i \in [m]$, let $\kappa_i = f(\kappa_{i-1})$. Choose $\psi \gg \kappa_m, a, k$.

Let us view $\pi(\mathcal{S})$ as drawn on the sphere, with f_0 being the face bounded by $\pi(C)$ and $\mathcal{T}$ considered as a respectful tangle. By Lemma 4.31, we can find a vertex $v_m \in V(\pi(\mathcal{S}))$ such that $d_{\mathcal{T}}(v_m, f_0) > \kappa_m + 3$ and $d_{\mathcal{T}}(v_m, \partial S) > \kappa_m$ for every $S \in \mathcal{S}$ such that $V(S) \cap R_0 \neq \emptyset$. For $i = m, \ldots, 1$, we proceed as follows. If the κ_i-neighborhood of v_i is (k, R_0, κ_{i-1})-homogeneous, then we let $v = v_i$ and $\kappa = \kappa_{i-1}$ and we stop. Otherwise, there exists a cell $S \in \mathcal{S}$ with $d_{\mathcal{T}}(v_i, \partial S) \leq \kappa_i$ and a vertex $v_{i-1} \in V(\pi(\mathcal{S}))$ with $d_{\mathcal{T}}(v_i, v_{i-1}) \leq \kappa_i - \kappa_{i-1}$ such that no cell $S' \in \mathcal{S}$

with $d_{\mathcal{T}}(v_{i-1}, \partial S') \leq \kappa_{i-1}$ has the same A-extended k-folio as S. We continue the process with $i-1$.

Note that the process ensures that fewer distinct A-extended k-folios appear in the κ_{i-1}-neighborhood of v_{i-1} than in the κ_i-neighborhood of v_i, and thus the process terminates at latest at $i = 1$. □

It is relatively easy to show that the central vertex of a homogeneous neighborhood is irrelevant for the existence of the rooted minor of a fixed graph H, as long as there exists a model of H that crosses the boundary of the neighborhood only a bounded number of times and no "essential" part of the model is close to the boundary. The argument is somewhat technical and a reader willing to accept this intuitively clear claim should feel free to skip it.

Let us give definitions needed to state this claim precisely. Consider a ρ-rooted model μ of a graph H in a graph G_0, a set $A \subseteq V(G)$, and a flat region $(C, F, \mathcal{S}, \mathcal{T})$ in $G_0 - A$. We say that μ is *minimal* if for every $x \in V(H)$, $\mu(x)$ is a tree and if $|\mu(x)| > 1$, then every leaf of $\mu(x)$ belongs to $\mathrm{img}(\rho)$ or is incident with an edge of $\mu(E(H))$; that is, we cannot delete any edge or vertex from $\mu(x)$ without breaking the model. Let $H' = \mu(E(H)) \cup \bigcup_{x \in V(H)} \mu(x)$. We say that a set $\mathcal{S}_1 \subseteq \mathcal{S}$ *covers nontrivial parts of* μ if $\bigcup \mathcal{S}_1$ contains all branching points in $H' \cap F$, and thus the part of $H' \cap F$ outside of $\bigcup \mathcal{S}_1$ is a linkage; more precisely,

- every edge of $\mu(E(H)) \cup \bigcup_{x \in V(H)} E(\mu(x))$ between $V(F)$ and A is contained in $\bigcup_{S \in \mathcal{S}_1} (S + A)$, and
- letting $Y = \bigcup_{S \in \mathcal{S}_1} \partial S$ and letting Y' be the set of vertices $y \in Y$ belonging to exactly one cell of $\mathcal{S}_1$, every component of

$$\Big(\mu(E(H)) \cup \bigcup_{x \in V(H)} \mu(x)\Big) \cap \bigcup (\mathcal{S} \setminus \mathcal{S}_1)$$

is a path with both ends in $Y' \cup V(C)$ and otherwise disjoint from Y.

Observe that since the model μ is minimal, any minimal subset of $\mathcal{S}$ covering nontrivial parts of μ has size $O(|H| + \|H\| + |A|)$.

Lemma 5.20 (↬) *For every graph H and all positive integers a, m, and κ, there exists an integer θ_0 such that the following claim holds. Let G_0 be a graph, let A be a set of vertices of G_0 of size at most a, let $\theta \geq \theta_0$ be a positive integer, and let $(C, F, \mathcal{S}, \mathcal{T})$ be a flat region in $G_0 - A$ of order at least 4θ. Let ρ be an injective function from $\mathrm{dom}(\rho) \subseteq V(H)$ to $V(G_0)$. Let $k = 2|H| + 5\|H\| + 4a + 12$. Let $v_0 \in V(\pi(\mathcal{S}))$ be a vertex such that the θ-neighborhood of v_0 is $(k, \mathrm{img}(\rho), \kappa)$-homogeneous. Let K be a cycle in $\pi(\mathcal{S})$ bounding a θ-zone Δ around v_0 such that $d_{\mathcal{T}}(v_0, V(K)) \geq \theta - 2$. Let μ be a minimal ρ-rooted minor of H in G_0 and let $\mathcal{S}_1$ be a minimal subset of $\mathcal{S}$ covering nontrivial parts of μ, such that $d_{\mathcal{T}}(v_0, \partial S) < \theta - 3(4m+1)$ for every cell $S \in \mathcal{S}_1$ with $\partial S \cap \Delta \neq \emptyset$. Let $U = V(K) \cap \bigcup_{x \in V(H)} V(\mu(x))$. If $|U| \leq m$, then H is also a ρ-rooted minor of $G_0 - v_0$.*

Proof Let $\mathcal{S}_0$ consist of all cells $S \in \mathcal{S}$ such that $\overline{\pi}(S)$ is drawn in Δ and let $G_1 = \bigcup_{S \in \mathcal{S}_0} S$. Let $Y = \bigcup_{S \in \mathcal{S}_1 \cap \mathcal{S}_0} \partial S$ and let Y' be the set of vertices of Y belonging to exactly one cell of $\mathcal{S}_1$.

By the assumptions, we have $U \cap Y = \emptyset$ and each vertex of U is incident with exactly two edges used by μ; let U_0 consist of the vertices $u \in U$ such that at most one of these edges is contained in G_1. Let $U_1 = U_0 \cup Y'$. Let

$$Q = \Big(\mu(E(H)) \cup \bigcup_{x \in V(H)} \mu(x)\Big) \cap \bigcup(\mathcal{S}_0 \setminus \mathcal{S}_1);$$

since $\mathcal{S}_1$ covers nontrivial parts of μ, each component of Q is a path with both ends in U_1 and otherwise disjoint from Y. Let $\mathcal{B}_0$ be the set of components of Q with an end in U_0, and note that each vertex of U_0 is contained in at most one path of $\mathcal{B}_0$.

Choose for each $S \in \mathcal{S}_1 \cap \mathcal{S}_0$ a cell S' of $\mathcal{S}_0$ with the same A-extended k-folio, so that the chosen cells are far apart from one another, v_0, and K (this is possible by homogeneity of the θ-neighborhood of v_0); let $\mathcal{S}_2 = \{S' : S \in \mathcal{S}_1\}$, and for each $S \in \mathcal{S}_1 \cap \mathcal{S}_0$ and $z \in \partial S'$, let $\sigma(z)$ be the vertex of ∂S corresponding to z. For each $z \in U_0$, let $\sigma(z) = z$. Let $\mathcal{P}$ be the partition of $U_2 = U_0 \cup \bigcup_{S \in \mathcal{S}_2} \partial S$ such that distinct vertices $z_1, z_2 \in U_2$ are in the same part of $\mathcal{P}$ if and only if either $\sigma(z_1) = \sigma(z_2)$, or $\sigma(z_1)$ and $\sigma(z_2)$ belong to the same component of Q (in this case, $\sigma(z_1)$ and $\sigma(z_2)$ are the ends of one of the paths forming Q). Observe that $\sigma(\mathcal{P})$ is topologically feasible in Δ, and thus $\mathcal{P}$ is topologically feasible in Δ as well.

The plan is to use Theorem 4.27 to find a $\mathcal{P}$-linkage in $\pi(\mathcal{S}_0) - v_0$. Since the segregation $\mathcal{S}$ is solid, this also gives a $\mathcal{P}$-linkage Q' in $\big(\bigcup_{S \in \mathcal{S}_0 \setminus \mathcal{S}_2} S\big) - v_0$. We now replace in the model μ

- the paths of Q and the vertices shared by cells of $\mathcal{S}_1 \cap \mathcal{S}_0$ by the corresponding connected subgraphs of Q', and
- for each $S \in \mathcal{S}_1 \cap \mathcal{S}_0$, the part of μ in $S + A$ by an equivalent part in S' (this is possible by Observation 5.18, since S and S' have the same A-extended k-folio).

In this way, we will obtain a ρ-rooted model of H in $G_0 - v_0$, as desired.

The only technical issue here is that the set U_0 is not necessarily free in the tangle in $\pi(\mathcal{S}_0)$ obtained using Lemma 4.20 (let us remark that the sets ∂S for $S \in \mathcal{S}_2$ are free, since $\pi(S)$ is 3-connected and ∂S is chosen to be far from K). To solve this issue, we use the standard "linking through a cylindrical grid" argument. By Theorem 4.21, there exists a $(\theta - 3)$-local $3m \times m$ battlefield $(\mathcal{R}, \mathcal{Q})$ around v_0 in $\pi(\mathcal{S})$ whose egg contains all vertices $v \in V(G_1)$ such that $d_{\mathcal{T}}(v_0, v) \le \theta - 3(4m + 1)$, and in particular all vertices of $U_1 \setminus U_0$. Let $\mathcal{R} = C_1, \ldots, C_{3m}$. Let $\mathcal{B}_0' = \{\pi(B) : B \in \mathcal{B}_0\}$, let $\mathcal{B}_1$ consist of the paths in $\mathcal{B}_0'$ that do not intersect C_m, and let $\mathcal{B}_2 = \mathcal{B}_0' \setminus \mathcal{B}_1$.

Note that the paths of $\mathcal{B}_1$ cannot have any end in $U_1 \setminus U_0$, and thus they connect distinct vertices of U_0. For each $B \in \mathcal{B}_1$, the drawing of B splits Δ to two parts, one of which contains the egg of the battlefield; let Δ_B be the other part. Observe that no path in $\mathcal{B}_2$ can intersect Δ_B, as it needs to reach $U_1 \setminus U_0$ (which is contained in the egg of the battlefield), but cannot cross B. Let $B_1, B_2, \ldots, B_{m'}$ be an ordering of $\mathcal{B}_1$ such that $j < i$ whenever $\Delta_{B_j} \subset \Delta_{B_i}$. We claim that without loss of generality, we can assume that for $i \in [m']$, B_i does not intersect C_{3m-i}. Indeed, we can ensure that this is true by repeating the following operation: Suppose $i \in [m']$ is the smallest index such that B_i intersects C_{3m-i}, and in particular for each $j \in [i-1]$, B_j is disjoint from C_{3m-i+1}. Then we can replace the parts of B_i contained in the interior of the disk bounded by C_{3m-i+1} by the subpath of C_{3m-i+1} in Δ_{B_i}, making it disjoint from C_{3m-i}.

Consequently, we can assume that all paths in $\mathcal{B}_1$ are disjoint from C_{2m}. Let $\mathcal{Q}'$ consist of the segments of the paths of $\mathcal{Q}$ between C_{2m} and C_m, each intersecting C_m in exactly one vertex, and let Z be the set of ends of the paths of $\mathcal{Q}'$ in C_m. Let U_0' consist of the vertices of U_0 incident with a path of $\mathcal{B}_2$, and let $\mathcal{B}_2'$ contain for each vertex in U_0' the initial segment of this path till its first intersection with C_m. Menger's theorem applied to the graph $\bigcup \mathcal{B}_2' \cup \bigcup \mathcal{Q}' \cup \bigcup_{i=m+1}^{2m} C_i$ shows that this graph contains a (U_0', Z)-linkage $\mathcal{Q}''$ of size $|U_0'|$. Let $\mathcal{T}'$ be the tangle of order $\theta' = \lceil \frac{2}{9}(\theta - 3(4m+1)) \rceil$ in the subgraph of $\pi(\mathcal{S})$ drawn in the disk bounded by C_m obtained using Lemma 4.20. Moreover, because of the cycles $C_1, \ldots, C_m$ and the paths of $\mathcal{Q}$, the lemma also implies that Z is $\mathcal{T}'$-free. We now perform the plan described above in the part of $\mathcal{S}_0$ arranged inside the disk bounded by C_m, with the ends of $\mathcal{Q}''$ in Z playing the role of U_0, and in the end we add the paths of $\mathcal{Q}'' \cup \mathcal{B}_1$ to the resulting model. □

It might seem that this statement easily implies that v_0 is irrelevant for the existence of the minor of H in G_0 even without the assumption that $\mathcal{S}_1$ is far from K. Indeed, using Lemma 5.19, we can find a homogeneous θ_0'-neighborhood with θ_0' much larger than the constant θ_0 needed in Lemma 5.20, then choose the radius θ so that $\theta_0 \le \theta \le \theta_0'/4$ and the cycle bounding the θ-zone found using Corollary 4.16 is far from μ-essential vertices. Since H is a fixed graph, its model intuitively must have bounded complexity, and thus it seems likely that (possibly subject to choosing θ carefully and adjusting μ), it should be possible to ensure that K intersects μ in at most m vertices, where m depends only on H.

This last part is true, but surprisingly hard to prove. The general idea is to use the Unique Linkage Theorem in the spirit of Lemma 3.12 (with Lemma 5.20 replacing Lemma 3.11 in the proof), but the details are quite technical. Thus, we refer the reader to [13] for the proof of the following theorem.

Theorem 5.21 *For every graph H and all positive integers a and κ, there exists an integer θ such that the following claim holds. Let G_0 be a graph, let A be a set of vertices of G_0 of size at most a, and let $(C, F, \mathcal{S}, \mathcal{T})$ be a flat region in $G_0 - A$ of order at least $\theta + 3$. Let ρ be an injective function from $\mathrm{dom}(\rho) \subseteq V(H)$ to $V(G_0)$. Let $k = 2|H| + 5\|H\| + 4a + 12$. If the θ-neighborhood of a vertex $v_0 \in V(\pi(\mathcal{S}))$ is $(k, \mathrm{img}(\rho), \kappa)$-homogeneous, then v_0 is (H, ρ)-irrelevant in G_0.*

It is now easy to describe the algorithm for the R-ROOTED H-MINOR problem (subject to the usual caveat that we actually do not go into technical details, and in particular we use algorithmic versions of several of the results that we only stated in an existential form; the reader can easily check that the proofs of these results give rise to polynomial-time algorithms). Let $a = a(K_{2|R|+|H|+1}) - 1$, and for every positive integer κ, let $f(\kappa)$ be the value of θ in Theorem 5.21 for H, a, and κ. Let ψ be the value from Lemma 5.19 applied with $\max(a, |R|), k = 2|H|+5\|H\|+4a+12$, and f. Let t be the constant from Theorem 5.17 with the clique $K_{2|R|+|H|+1}$ playing the role of H. Consider an input instance consisting of a graph G_0 and an injective function $\rho : R \to V(G_0)$:

- While $\mathrm{tw}(G_0) \geq t$:
  - Apply the algorithm underlying the proof of Theorem 5.17, with the clique $K_{2|R|+|H|+1}$ playing the role of H.
    - If the outcome is a tangle $\mathcal{T}$ in G_0 controlling a minor of the clique $K_{2|R|+|H|+1}$, then use Lemma 4.29 to locate a (H, ρ)-irrelevant vertex v_0.
    - Otherwise, the outcome is a set $A \subseteq V(G_0)$ of size at most a and a flat region $(C, F, \mathcal{S}, \mathcal{T})$ of order ψ in $G_0 - A$ such that each element of $\mathcal{S}$ has bounded treewidth.
      - Compute the A-extended k-folio for each $S \in \mathcal{S}$ using Theorem 2.8.
      - Use Lemma 5.19 to find a vertex $v_0 \in V(\pi(\mathcal{S}))$ and a positive integer κ such that the $f(\kappa)$-neighborhood of v_0 is $(k, \mathrm{img}(\rho), \kappa)$-homogeneous, and thus v_0 is (H, ρ)-irrelevant by Theorem 5.21.
  - Delete v_0 from G_0.
- Find a ρ-rooted model of H in G_0 or decide that it does not exists using Theorem 2.8.

Since all steps in the algorithm can be performed in polynomial time and the number of iterations of the main cycle is clearly bounded by $|G_0|$, the time complexity of the algorithm is polynomial.

5.4 Structure Around a Crossing

Let us quickly summarize the argument underlying the proof of the Weak Minor Structure Theorem (Theorem 5.12): Using the assumption that G_0 has large treewidth, we find a subgraph W of G_0 drawn on the sphere with a respectful tangle of large order. By Lemma 4.43, we can assume that (after clearing bounded-radius zones and removing less than $a(H)$ vertices), all long jumps join vertices of a few bounded-radius regions. Moreover, Lemmas 4.48 and 5.11 together with Lemma 4.35 show that non-eliminable short jumps are contained in a few additional bounded-radius regions. Thus, if we simply look anywhere except for these exceptional regions, we find the desired flat region.

In the full Minor Structure Theorem, we want to describe the whole H-minor-free graph G_0, and thus we cannot ignore the exceptional regions. In Corollary 4.47, we demonstrated the idea used to deal with long jumps: Either we can intersect all of them by a small set of vertices, or we can use them to obtain a high-representativity drawing of a subgraph of G_0 in a surface of larger genus. In the latter case, we can iterate the argument; the number of iterations is bounded by Corollary 4.34, as H can be drawn on any surface of sufficiently large genus. Hence, it remains to deal with the non-planar regions formed by short jumps, and in this section, we develop the necessary tools.

For an illustration of what we might want to achieve, let us consider two examples of highly non-planar societies:

- G_1 is the society with $\partial G_1 = v_1 \dots v_{4m}$, $V(G_1) = \partial G_2$, and the edge set $\{v_i v_{i+2m} : i \in [2m]\}$; that is, G_1 is "crosscap-like".
- G_2 is the society with $\partial G_2 = v_1 \dots v_{4m}$, $V(G_2) = \partial G_2$, and the edge set $\{v_{4i-3}v_{4i-1}, v_{4i-2}v_{4i} : i \in [m]\}$; that is, G_2 consists of m consecutive crosses.

Suppose that G is a subgraph of a larger H-minor-free graph G_0, the graph G is drawn on a surface, and $F \in \{G_1, G_2\}$ is another part of G_0 attaching through its boundary to a face of G. If $F = G_1$, then we can naturally view $G \cup G_1$ as a graph drawn on a surface of higher genus, obtained by adding a crosscap. This brings us closer towards our ultimate goal of showing that the whole graph G_0 has a structure which is close to being drawn on a surface—since H is not a minor of G_0, Theorem 4.32 shows that we can increase the genus of the surface in this way only a bounded number of times.

On the other hand, $F = G_2$ cannot be used in this manner. Indeed, G_2 is a prototypical example of a vortex, an unavoidable ingredient of the Minor Structure Theorem. Hence, in the proof of the Minor Structure Theorem, if we see a non-planar region (society) F, we need to distinguish whether it looks more like G_1 or like G_2. It is useful to be able to do so on a global level, only by looking at the connectivity of ∂F across F. The main distinction between G_1 and G_2 from this perspective is that there are many pairwise vertex-disjoint paths in G_1 from one side of ∂G_1 to the other one (say from $\{v_1, \dots, v_{2m}\}$ to $\{v_{2m+1}, \dots, v_{4m}\}$, but this is not the case in G_2.

A ***transaction*** in a society F is a $(\partial_{[u,v)} F, \partial_{[v,u)} F)$-linkage for distinct $u, v \in \partial F$. A society F is ***d-vortex-like*** if it contains no transaction of size more than d. The following lemma substantiates the intuition that societies without large transactions behave similarly to vortices.

Lemma 5.22 *If a society F is d-vortex-like, then F has a vortical decomposition of adhesion at most d.*

Proof Let $v_1, \dots, v_m$ be the vertices of ∂F in order. For $i \in [m]$, since F is d-vortex-like, it does not contain a $(\{v_1, \dots, v_i\}, \{v_{i+1}, \dots, v_m\})$-linkage of size more than d, and thus it has a vertex separation (A_i, B_i) of order at most d with $v_1, \dots, v_i \in A_i$ and $v_{i+1}, \dots, v_m \in B_i$; we choose such a vertex separation of smallest order with A_i maximal, and subject to that with B_i minimal. Consider indices i and j such that $1 \le i < j \le m$. By (1.1), we have

$$|(A_i \cup A_j) \cap (B_i \cap B_j)| + |(A_i \cap A_j) \cap (B_i \cup B_j)| \leq |A_i \cap B_i| + |A_j \cap B_j|.$$

Note that $(A_i \cap A_j, B_i \cup B_j)$ is a vertex separation with $\{v_1, \ldots, v_i\} \subseteq A_i \cap A_j$ and $\{v_{i+1}, \ldots, v_m\} \subseteq B_i \cup B_j$, and since we chose (A_i, B_i) as such a separation of smallest order, we have $|(A_i \cap A_j) \cap (B_i \cap B_j)| \geq |A_i \cap B_i|$. Therefore, $(A_i \cup A_j, B_i \cap B_j)$ is a separation of F of order at most $|A_j \cap B_j|$. Since $\{v_1, \ldots, v_j\} \subseteq A_i \cup A_j$ and $\{v_{j+1}, \ldots, v_m\} \subseteq B_i \cap B_j$, the maximality of A_j implies $A_i \subseteq A_j$, and the minimality of B_j implies $B_j \subseteq B_i$.

We actually show that F has a path (rather than just a cycle) decomposition (P, β) of F of adhesion at most d, where $P = v_1 v_2 \ldots v_m$ and $v_i \in \beta(v_i)$ for every $i \in [m]$. Let $A_0 = \emptyset$ and $B_0 = V(F)$, and for $i \in [m]$, let $\beta(v_i) = A_i \cap B_{i-1}$; clearly, $v_i \in \beta(v_i)$.

We need to verify that this indeed defines a path decomposition. For any vertex $v \in V(F)$, let i_v be the minimum index such that $v \in A_{i_v}$; such a index exists, since $A_m = V(F)$. Since $v \notin A_{i_v - 1}$, we have $v \in B_{i_v - 1}$, and thus $v \in \beta(v_{i_v})$. Moreover, if u is a neighbor of v with $i_u \leq i_v$, then since (A_{i_v-1}, B_{i_v-1}) is a vertex separation and $v \notin A_{i_v-1}$, we have $u \in B_{i_v-1}$, and since $u \in \beta(v_{i_u}) \subseteq A_{i_u} \subseteq A_{i_v}$, we have $u \in A_{i_v} \cap B_{i_v-1} = \beta(v_{i_v})$. We conclude that the condition (D1) is satisfied by (P, β). Since the sequences $A_i : i \in [m]$ and $B_i : i \in [m]$ are inclusionwise-monotone, the set $\{i \in [m] : v \in \beta(v_i)\}$ is an interval, and thus (D2) holds as well. It follows that (P, β) is a path decomposition of F.

Moreover, for $i \in [m-1]$, we have

$$|\beta(v_i) \cap \beta(v_{i+1})| = |A_i \cap B_{i-1} \cap A_{i+1} \cap B_i| \leq |A_i \cap B_i| \leq d,$$

and thus this path decomposition has adhesion at most d. □

Based on the proof of Lemma 5.22, one should wonder why we defined vortical decompositions as cycle decompositions rather than path decompositions. Indeed, one can do so and in many contexts it is actually more convenient; however, there are more refined formulations of the Minor Structure Theorem where working with cycle decompositions is necessary, see e.g. Sect. 10.2. Moreover, using path decompositions would suggest that vortices have natural starting and ending points, which is not the case. Let us remark that a weak converse to Lemma 5.22 also holds.

Lemma 5.23 *If a society F has a vortical decomposition $(\partial F, \beta)$ of adhesion at most d, then F is $2d$-vortex-like.*

Proof Let u and v be distinct vertices of ∂F, and let u_1 and v_1 be the vertices that precede them in the cyclic ordering of ∂F. By Lemma 2.2, any path from $\partial_{[u,v)} F$ to $\partial_{[v,u)} F$ must intersect $(\beta(u_1) \cap \beta(u)) \cup (\beta(v_1) \cap \beta(v))$, a set of size at most $2d$. It follows that any transaction in F has size at most $2d$. □

Let us go back to the motivation presented above, considering the case that in the proof of the Minor Structure Theorem, we see a society F attaching to a subgraph G of the ambient graph, where G is drawn on a surface. It may be the case that F consists not only of the vortex, but also of some planar neighborhood (up to non-planar parts cut off by vertex cuts of size at most three) around it; in this case, this neighborhood should be included in the subgraph drawn in the surface. Moreover, we cannot distinguish between the cases that the central part of F is something like a crosscap (e.g., the society G_1) or a vortex (e.g., the society G_2) just by looking at the size of transactions in F, since it can be possible to route a large transaction through the planar part F. Hence, we need to look at the structure of the transactions.

Consider any transaction $\mathcal{P}$ in the society F. Observe that in the described situation, the paths of $\mathcal{P}$ passing only through the planar part cannot be crossed by the other paths of $\mathcal{P}$, and in particular, one of them has all other paths of the transaction on one side. More precisely, we say that a path $P \in \mathcal{P}$ with ends $u, v \in \partial F$ is *peripheral* if either $\partial_{(u,v)} F$ or $\partial_{(v,u)} F$ is disjoint from the ends of paths in $\mathcal{P}$. To deal with the issue we just described, we will be interested in transactions without peripheral paths; such transactions are called ***crooked***. Let us remark that a society contains a crooked transaction if and only if it contains a cross.

The following claim is the main result of [9]; its proof is related to those of Sect. 5.1, but substantially more technical. A ***bisegregation*** of a graph or society G is a pair $(\mathcal{S}_0, \mathcal{S}_1)$ such that $\mathcal{S}_0 \cup \mathcal{S}_1$ is a segregation of G and all elements of $\mathcal{S}_1$ are cells. Let us remark that elements of $\mathcal{S}_0$ can also be cells (though in general, they will be non-cell). We say that the bisegregation is *solid* if $\mathcal{S}_1$ is solid. By an *arrangement* of the bisegregation, we mean an arrangement of the segregation $\mathcal{S}_0 \cup \mathcal{S}_1$. For non-negative integers m and δ, we say that the bisegregation is (m, δ)*-simple* if $|\mathcal{S}_0| \le m$ and all elements of $\mathcal{S}_0$ are δ-vortex-like and pairwise vertex-disjoint.

Theorem 5.24 *For every positive integer p, there exists an integer δ such that every society without a crooked transaction of size at least p has a solid* $(1, \delta)$*-simple bisegregation with an arrangement in the disk.*

That is, the only reason why a society does not contain a large crooked transaction is that it consists of a vortex-like part surrounded by planar neighborhood (up to non-planar parts cut off by cuts of size at most three). The dependence of δ on p obtained in [9] is fairly reasonable: $\delta = 3p + 9$.

We can now provide a bit more detail on the usage of Theorem 5.24 in the proof of the Minor Structure Theorem. Suppose we are in the following setting: A subgraph G of the ambient graph G_0 has a cellular drawing on a surface Σ and contains a respectful tangle $\mathcal{T}$ of large order. Moreover, there are no long jumps in G_0 over G (we have already destroyed all long jumps by deleting a bounded number of vertices, as discussed at the beginning of this section). Thus, the attachments of every G-bridge of G_0 are close to one another in the $d_{\mathcal{T}}$-distance. We now add the G-bridges to the drawing of G one by one, and for each G-bridge J we essentially obtain one of the following conclusions:

(i) We can draw J in a disk up to 3-separations using Corollary 5.9, and thus we can extend the part of G_0 drawn on Σ by adding an arrangement of a simple segregation of J to it, or

(ii) J gives rise to a cross, and Theorem 5.24 provides a drawing of J in the disk up to 3-separations and a single vortex, and we can again extend the part of G_0 drawn in Σ, now with an additional vortex, or
(iii) J gives rise to a large crooked transaction.

Let us ignore the case (iii) for now. During this procedure, in addition to the subgraph G, we need to keep track of the subgraph G' obtained from G by adding the previously processed G-bridges, and of an arrangement of a segregation $\mathcal{S}$ of G' in Σ arising from the applications of Corollary 5.9 and Theorem 5.24. It is obviously desirable (though unfortunately not automatic, see Sect. 5.6 for more details) that the arrangement of $\mathcal{S}$ is ***consistent with the drawing of*** G in the following sense:

- all vertices in $V(G) \cap V(\pi(\mathcal{S}))$ are drawn as the same point both in G and $\pi(\mathcal{S})$, and
- for each element $S \in \mathcal{S}$, if $E(S) \cap E(G) \neq \emptyset$, then S is a cell and the subgraph $G \cap S$ is drawn in the closure of $\overline{\pi}(S)$.

With this setting in mind, we define a *partial arrangement of* G_0 *on* Σ to be a tuple $(G', G, \mathcal{S})$, where G is a subgraph of G_0 with a cellular drawing on Σ, G' is the union of G with a set of G-bridges of G_0, and $\mathcal{S}$ is a segregation of G' with an arrangement on Σ consistent with the drawing of G. Note that this implies that any G'-bridge of G_0 is also a G-bridge, i.e., it only intersects G' in the vertices of G.

There are two further issues we need to take into account:

- Above, we suggested considering each G-bridge J separately. That would not quite work, as the arrangement of J would not necessarily be consistent with the already existing arrangement of $\mathcal{S}$. Rather, we need to re-process the part of $\mathcal{S}$ close to J together with the G-bridge J. It is furthermore convenient to process all other G-bridges with attachments close to the attachments of J at the same time.
- In the outcomes (ii) and (iii), we need the ends of the paths forming the cross or the crooked transaction to be $\mathcal{T}$-free, e.g., so that they are useful in constructions of minors discussed in Sect. 4.5.

This leads us to the following form of the main result of this section, to be applied with G_1 consisting of G', J, and the G-bridges with attachments close to J, and with $(\mathcal{R}, \mathcal{P})$ being a free battlefield.

Lemma 5.25 *For every integer $p \geq 2$, there exist integers $r \geq 3$, $c \geq \max(7, 2p)$, and δ such that the following claim holds. Let $(G', G, \mathcal{S})$ be a partial arrangement of a graph G_1 on a surface Σ. Let $(\mathcal{R}, \mathcal{P})$ be an $(r \times c)$-battlefield around an atom a in G, where $\mathcal{R} = C_1, \ldots, C_r$, and let Z be the set of the ends of the paths of $\mathcal{P}$ in C_r. For $i \in \{1, r\}$, let $\Delta_i \subset \Sigma$ be the closed disk bounded by C_i and containing a, and let $\Delta'_i \subset \Sigma$ be the disk bounded by $\pi(C_i)$ and containing a. Suppose that every G'-bridge of G_1 has all attachments in Δ_1, and that every element $S \in \mathcal{S}$ with $\overline{\pi}(S) \cap \Delta_r \neq \emptyset$ is a cell containing at most three vertices of Z. Let Q be the society obtained from the graph consisting of $\bigcup_{S \in \mathcal{S}: \overline{\pi}(S) \subset \Delta'_r} S$ and all G'-bridges of G_1 by* *letting ∂Q be $V(\pi(C_r))$ ordered along the cycle, and let Q' be the society with the same underlying graph, but with $\partial Q'$ being Z in order around C_r. Then one of the following claims holds:*

(i) Q has a solid simple segregation with an arrangement in Δ'_r, or
(ii) Q' contains a cross and Q has a solid $(1, \delta)$-simple bisegregation with an arrangement in Δ'_r, or
(iii) Q' contains a crooked transaction of size at least p.

The proof of Lemma 5.25 is a technical exercise in the "linking through a cylindrical grid" technique. Let us remark that the condition that $|Z \cap V(S)| \leq 3$ for each $S \in \mathcal{S}$ with $\overline{\pi}(S) \cap \Delta_r \neq \emptyset$ is important, since it prevents degeneracies such as the whole battlefield being contained inside S. In particular, it implies that $|V(C_r) \cap V(\pi(\mathcal{S}))| \geq \lceil |Z|/3 \rceil \geq 3$, and thus the cycle $\pi(C_r)$ and the disk Δ'_r are actually defined.

While the definition of a crooked transaction is compact, it is also rather abstract. In order to deal with the outcome (iii) of Lemma 5.25, we are going to need a more detailed understanding of large crooked transactions. Let $\mathcal{P} = \{P_1, \ldots, P_p\}$ be a transaction of size $p \geq 4$ in a society F, let $u_1, \ldots, u_p, v_p, v_{p-1}, \ldots, v_1$ be the ends of the paths of this transaction in order in which they appear in ∂F, and let $\pi : [p] \to [p]$ be the permutation such that for $i \in [p]$, the ends of the path P_i are u_i and $v_{\pi(i)}$. We say that $\mathcal{P}$ is

- a ***crosscap transaction*** if $\pi(i) = p + 1 - i$ for $i \in [p]$,
- a ***leap transaction*** if $\pi(i) = i - 1$ for $i \in [p] \setminus \{1\}$ and $\pi(1) = p$, and
- a ***double-cross transaction*** if $\pi(1) = 2, \pi(2) = 1, \pi(p) = p - 1, \pi(p - 1) = p$, and $\pi(i) = i$ for $i \in [p - 2] \setminus \{1, 2\}$;

see Fig. 5.2 for an illustration. Observe that each of these transactions is crooked. Conversely, any crooked transaction contains a large subtransaction of one of these types.

Lemma 5.26 *Let F be a society, let $p \geq 4$ be an integer, and let $\mathcal{P}$ be a transaction of size at least $3p^2$ in F. If $\mathcal{P}$ is crooked, then a subset of $\mathcal{P}$ is a crosscap, leap, or double-cross transaction of size p.*

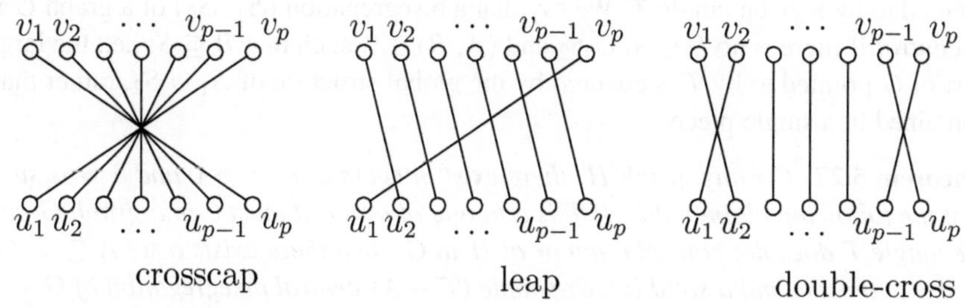

Fig. 5.2 Crooked transaction types

Proof Let $q = |\mathcal{P}|$, let the paths of $\mathcal{P}$ be $P_1, \ldots, P_q$, let $u_1, \ldots, u_q, v_q, v_{q-1}, \ldots, v_1$ be the ends of the paths of $\mathcal{P}$ in order in ∂F, and let $\pi : [q] \to [q]$ be the permutation such that for $i \in [q]$, the ends of the path P_i are u_i and $v_{\pi(i)}$. Let $\{i_1, \ldots, i_k\}$ with $i_1 < i_2 < \ldots < i_k$ be the largest subset of $[q]$ such that the sequence $\pi(i_1), \ldots, \pi(i_k)$ is monotone. Since $q \geq 3p^2$, Erdős-Szekeres theorem implies that either the sequence is decreasing and $k \geq p$, or increasing and $k \geq 3p$. In the former case, the subset $\{P_{i_1}, \ldots, P_{i_p}\}$ of $\mathcal{P}$ forms a crosscap transaction of size p.

Suppose now that the sequence $\pi(i_1), \ldots, \pi(i_k)$ is increasing and $k \geq 3p$. Since the path P_{i_1} is not peripheral in $\mathcal{P}$, there exists a path $Q_1 \in \mathcal{P}$ with an end in $\partial_{(v_{\pi(i_1)}, u_{i_1})} F$. The maximality of the sequence $\pi(i_1), \ldots, \pi(i_k)$ implies that the other end of Q_1 is contained in $\partial_{(u_{i_1}, v_{\pi(i_1)})} F$. Symmetrically, there exists a path Q_k with one end in $\partial_{(u_{i_k}, v_{\pi(i_k)})} F$ and the other end in $\partial_{(v_{\pi(i_k)}, u_{i_k})} F$.

Observe that

- $Q_1, P_{i_1}, \ldots, P_{i_{p-1}}$ is a leap transaction, or
- $Q_k, P_{i_{k-p+2}}, \ldots, P_{i_k}$ is a leap transaction, or
- $Q_1, P_{i_1}, P_{i_{p+1}}, P_{i_{p+2}}, \ldots, P_{i_{2p-4}}, P_{i_k}, Q_k$ is a double-cross transaction.

□

5.5 Local Form of the Minor Structure Theorem

On the global level, the proof of the Minor Structure Theorem follows the argument discussed after Lemma 2.16: We need to construct a tree decomposition of the graph G, with all large torsos nearly drawn on a surface. To do so, we introduce an auxiliary root set R. If R is breakable, we split the graph on the corresponding small cut, process both part recursively, and combine their tree decompositions by introducing the new root node consisting of R and the cut (this is where bounded size bags in Theorem 2.9 come from). If R is unbreakable, we need a result showing that G has a star decomposition such that the torso of the center of the star is nearly-drawn on a surface and consists of the parts of G "well connected" to R.

Given the theory we have developed so far, it is most convenient to state the result in terms of the large order tangle corresponding to R. Thus, we need a local form of the Minor Structure Theorem, describing the structure of the large part of the graph pointed to by a given tangle $\mathcal{T}$. We say that a bisegregation $(\mathcal{S}_1, \mathcal{S}_2)$ of a graph G is ***$\mathcal{T}$-central*** if there is no $S \in \mathcal{S}_1 \cup \mathcal{S}_2$ and $(A, B) \in \mathcal{T}$ such that $B \subseteq S$; i.e., the large part of G pointed to by $\mathcal{T}$ is covered by the global structure of $\mathcal{S}_1 \cup \mathcal{S}_2$, rather than contained in a single piece.

Theorem 5.27 *For any graph H, there exist integers $\alpha, m, \delta \geq 0$ and $\theta > \alpha$ such that the following claim holds. If $\mathcal{T}$ is a tangle of order at least θ in a graph G and the tangle $\mathcal{T}$ does not control a minor of H in G, then there exists a set $A \subseteq V(G)$ of size at most α and a solid (m, δ)-simple $(\mathcal{T} - A)$-central bisegregation of $G - A$ with an arrangement on an H-avoiding surface.*

We could have stated this result much earlier, but at this point the reader has gained more insight into the meaning and origin of all its ingredients:

- The apex vertices A come from cutting long jumps, as in Corollary 4.47.
- The (nearly) simple bisegregation on a surface comes from Corollary 5.9 applied to the parts of G that cannot be used to create eyes.
- The vortex-like parts arise from Lemma 5.25 applied around possible eyes.

It is rather easy to prove the Minor Structure Theorem using Theorem 5.27. At a point in the proof, we will need to delete a subset of vertices of a graph which is nearly drawn on a surface Σ, and argue that it is still nearly drawn on Σ. This is mostly clear, with the following exception: If a deleted vertex is contained in the boundary of a vortex, it is not entirely trivial to keep the structure of the vortex, since each bag of its vortical decomposition is supposed to be associated with a vertex of the boundary. Thus, we need the following lemma.

Lemma 5.28 *Suppose that G_0 is a graph drawn on a surface Σ up to outgrowths $F_1, \ldots, F_m$ with the surface part G, and let R be a set of vertices of G_0. For $i \in [m]$, let $(\partial F_i, \beta_i)$ be a vortical decomposition of F_i of width at most d. Then $G_0 - R$ can be drawn on Σ up to outgrowths $F'_1, \ldots, F'_m$ with surface part $G' \supseteq G - R$, where for $i \in [m]$, F'_i has a vortical decomposition $(\partial F'_i, \beta'_i)$ of width at most $2d + 1$ such that for each $x \in \partial F_i$, there exists $x' \in \partial F'_i$ with $\beta_i(x) \subseteq \beta'_i(x') \cup R$.*

Proof Let $R_0 = R \cap \bigcup_{i=1}^{m} \partial F_i$. First, we delete vertices of $R \setminus R_0$ from the surface part G or from the outgrowths that contain them.

Next, we process vertices $r \in R_0$ one by one. Let F_i be the outgrowth with $r \in \partial F_i$. If there exists a vertex $r' \in \beta_i(r) \setminus \partial F_i$, then we remove r from F_i, replace r by r' in ∂F_i (assigning the bag of r to it in the vortical decomposition), add r' as an isolated vertex to the surface part with r' drawn in the interior of the disk Δ_i in Σ representing F_i, and shrink Δ_i away from r so that r' is contained in the boundary of the resulting disk.

For each $i \in [m]$, let R_i be the set of vertices of R_0 in ∂F_i after the modification described above; i.e., for each $r \in R_i$, we have $\beta_i(r) \subseteq \partial F_i$. Let $F'_i = F_i - R_i$. If $\partial F_i = R_i$, then $V(F'_i) = \bigcup_{r \in R_i} \beta_i(r) \setminus \partial F_i = \emptyset$, i.e., we have completely deleted the outgrowth F_i.

Otherwise, $\partial F'_i$ consists of the vertices $x_1, \ldots, x_t$ of $\partial F_i \setminus R_i$ in order. We define $\beta'_i(x_j) = (\beta_i(x_j) \cup \beta_i(x_{j+1})) \setminus R_i$ for $j \in [t]$, where $x_{t+1} = x_1$. Consider any vertex $r \in R_i$, and let x_j be the nearest vertex of $\partial F'_i$ that precedes it in ∂F_i. We claim that $\beta_i(r) \subseteq \beta'_i(x_j) \cup R$. Indeed, consider any vertex $v \in \beta_i(r) \setminus R$. Since $\beta_i(r) \subseteq \partial F_i$, we have $v = x_{j'}$ for some $j' \in [t]$, and by (V), we have $v \in \beta_i(x_{j'})$. Since we also have $v \in \beta_i(r)$, (D2) implies that $v \in \beta_i(x_j)$ or $v \in \beta_i(x_{j+1})$.

This also implies that $(\partial F'_i, \beta'_i)$ satisfies the condition (D1), and thus it forms a vortical decomposition of F'_i: For any edge $uv \in E(F'_i)$, the condition (D1) for $(\partial F_i, \beta_i)$ implies that there exists a vertex $x \in \partial F_i$ such that $u, v \in \beta_i(x)$. If $x \notin R_i$, then $u, v \in \beta'_i(x)$, since $\beta_i(x) \setminus R_i \subseteq \beta'_i(x)$. If $x \in R_i$, then previous paragraph implies that $u, v \in \beta'_i(x_j)$, where x_j be the nearest vertex of $\partial F'_i$ preceding x in ∂F_i.

Finally, observe that each bag of this vortical decomposition $(\partial F_i', \beta_i')$ has size at most $2(d+1)$, and thus the width of the vortical decomposition is at most $2d+1$. □

We are now ready to derive the Minor Structure Theorem, which we restate for convenience, from its local form.

Theorem 2.9 (Robertson and Seymour [12]) *For every graph H, there exist constants a, m, and d such that every H-minor-free graph G has a tree decomposition in which each torso with more than a vertices can be (a, m, d)-nearly drawn on an H-avoiding surface.*

Derivation from Theorem 5.27 Let α, m, δ, and θ be as in Theorem 5.27. Without loss of generality, we can assume $\theta > \alpha + \max(3, 2\delta + 1)$. Let $a = \max(\alpha + 3\theta - 2, 4\theta - 3)$ and $d = 4\delta + 1$. We are going to prove the following stronger claim by induction on the number of vertices of G:

(⋆) For every H-minor-free graph G and every set $R \subseteq V(G)$ of size at most $3\theta - 2$, the graph obtained from G by adding a clique on R has a tree decomposition (T, β) such that for every node $x \in V(T)$, either

- $|\beta(x)| \leq 4\theta - 3$, or
- the torso of x is (a, m, d)-nearly drawn on an H-avoiding surface.

If $|G| \leq 4\theta - 3$, then we can simply let T be the single-node tree with bag $V(G)$. Hence, suppose that $|G| \geq 4\theta - 2$. We can without loss of generality assume that $|R| = 3\theta - 2$, as otherwise we can add vertices to R. If there exists an R-balanced vertex separation (C, D) of G of order less than θ, then let $R_C = C \cap (R \cup D)$ and $R_D = D \cap (R \cup C)$. Note that

$$R_C \leq \tfrac{2}{3}|R| + \theta - 1 = \tfrac{2}{3}(3\theta - 2) + \theta - 1 < 3\theta - 2,$$

and similarly $|R_D| < 3\theta - 2$. In particular, $C \neq V(G) \neq D$, since neither C nor D contains all vertices of R. Let (T_C, β_C) and (T_D, β_D) be the tree decompositions obtained by applying the induction hypothesis to $G[C]$ with R_C and $G[D]$ with R_D. By Lemma 2.3, there exists nodes $x_C \in V(T_C)$ and $x_D \in V(T_D)$ such that $R_C \subseteq \beta_C(x_C)$ and $R_D \subseteq \beta_D(x_D)$. Let T be the tree obtained from $T_C \cup T_D$ by adding a new node x_0 adjacent to x_C and x_D. Let β match β_C on T_C, β_D on T_D, and let $\beta(x_0) = R \cup (C \cap D)$. Note that (T, β) is a tree decomposition of G together with the clique on R and $|\beta(x_0)| \leq 3\theta - 2 + \theta - 1 = 4\theta - 3$. Moreover, note that the torsos of the nodes in $V(T) \setminus \{x_0\}$ are the same as the torsos of the corresponding nodes of T_C and T_D; e.g., the torso of x_C is the same in T_C and in T, since (T_C, β_C) is a tree decomposition of the graph obtained from $G[C]$ by adding a clique on $R_C = \beta(x_C) \cap \beta(x)$. We conclude that (⋆) holds for G.

Therefore, we can assume that R is θ-unbreakable, and thus

$$\mathcal{T}_R = \{(C, D) : (C, D) \text{ is a vertex separation}, |C \cap D| < \theta, |R \setminus C| > \tfrac{2}{3}|R|\}$$

is a vertex tangle of order θ in G. Let $\mathcal{T}$ be the corresponding tangle in G. By Theorem 5.27, there exists a set $A \subseteq V(G)$ of size at most α and a solid (m, δ)-simple $(\mathcal{T} - A)$-central bisegregation $(\mathcal{S}_0, \mathcal{S}_1)$ of $G - A$ with an arrangement on an H-avoiding surface Σ.

- For each cell $S \in \mathcal{S}_1$, consider the subgraph $S + A$ of G consisting of S, A, and all edges between A and S, and let $R_S = A \cup \partial S \cup (R \cap V(S))$. Note that $S + A$ is cut off from the rest of G by the cut $A \cup \partial S$ of size $\alpha + 3 < \theta$, and since $(\mathcal{S}_0, \mathcal{S}_1)$ is $(\mathcal{T} - A)$-central, we have

$$|R_S| < \alpha + 3 + \tfrac{1}{3}|R| \leq \theta - 1 + \tfrac{1}{3}(3\theta - 2) < 3\theta - 2 = |R|.$$

  In particular, $R \not\subseteq V(S + A)$, and thus we can apply the induction hypothesis to $S + A$ and R_S, obtaining a tree decomposition (T_S, β_S) and a node $x_S \in V(T_S)$ such that $R_S \subseteq \beta(x_S)$.
- Let $\mathcal{S}_0 = \{S_1, \ldots, S_{m'}\}$, where $m' \leq m$, and consider any $i \in [m']$. We can assume that $|\partial S_i| > 3$, as otherwise S_i would be a cell and we could move it to $\mathcal{S}_1$. Let $(\partial S_i, \gamma)$ be a vortical decomposition of S_i of adhesion at most δ obtained using Lemma 5.22. For $v \in \partial S_i$, let v^- and v^+ be the vertices immediately preceding and succeeding v in the cyclic ordering of ∂S_i, and let $S_{i,v}$ be the bordered graph with the underlying graph $S_i[\gamma(v)]$ and the boundary $\partial S_{i,v} = \{v\} \cup (\gamma(v) \cap (\gamma(v^-) \cup \gamma(v^+)))$. Furthermore, let $R_{i,v} = A \cup \partial S_{i,v} \cup (R \cap V(S_{i,v}))$. Note that $S_{i,v} + A$ is cut off from the rest of G by the cut $A \cup \partial S_{i,v}$ of size at most $\alpha + 2d + 1 < \theta$, and since $(\mathcal{S}_0, \mathcal{S}_1)$ is $(\mathcal{T} - A)$-central, we have

$$|R_{i,v}| < \alpha + 2d + 1 + \frac{1}{3}|R| \leq \theta - 1 + \frac{1}{3}(3\theta - 2) < 3\theta - 2 = |R|.$$

  In particular, $R \not\subseteq V(S_{i,v} + A)$, and thus we can apply the induction hypothesis to $S_{i,v} + A$ and $R_{i,v}$, obtaining a tree decomposition $(T_{i,v}, \beta_{i,v})$ and a node $x_{i,v} \in V(T_{i,v})$ such that $R_{i,v} \subseteq \beta(x_{i,v})$. Let F_i be the union of cliques with vertex sets $\partial S_{i,v}$ for $v \in \partial S_{i,v}$, turned into a society by letting $\partial F_i = \partial S_i$. Clearly, F_i has a vortical decomposition $(\partial F_i, \beta_i)$ of width at most 2δ.

Let $G_0 = \pi(\mathcal{S}_1) \cup \bigcup_{i=1}^{m'} F_i$, and note that the graph G_0 is drawn on Σ up to vortices $F_1, \ldots, F_{m'}$ of width at most 2δ and with surface part $\pi(\mathcal{S}_1)$ (for $i \in [m']$, the disk representing F_i is the closure of $\overline{\pi}(S_i)$). Let $G_1 = G_0 - R$; by Lemma 5.28, G_1 is drawn on Σ up to vortices $F_1', \ldots, F_{m'}'$ of width at most $4\delta + 1 = d$ so that the surface part G_1' contains $\pi(\mathcal{S}_1) - R$ and so that for each $i \in [m']$, each bag of the vortical decomposition of F_i is contained in the union of R with a bag of the vortical decomposition of F_i'. Consequently, for each $i \in [m']$ and $v \in \partial S_i$, there exists $v' \in \partial F_i'$ such that $\beta_{i,v}(x_{i,v}) \cap V(G_1) \subseteq \beta_i(v') \setminus R$, which is a clique in G_1. Moreover, for each $S \in \mathcal{S}_1$, $\beta_S(x_S) \cap V(G_1) \subseteq \partial S \setminus R$, and since ∂S forms the clique $\pi(S)$ in $\pi(\mathcal{S}_1)$, $\partial S \setminus R$ forms a clique in G_1.

Let $A' = A \cup R$ and note that $|A'| \leq |A| + |R| \leq \alpha + 3\theta - 2 \leq a$. Let G_2 be obtained from G_1 by adding a clique with vertex set A' and all edges with one end in

A' and the other end in $V(G_1)$. Let (T, β) be the tree decomposition of G (together with the clique on R) obtained from the tree decompositions (T_S, β_S) for $S \in \mathcal{S}_1$ and $(T_{i,v}, \beta_{i,v})$ for $i \in [m']$ and $v \in \partial S_i$ by adding a node x_0 adjacent to x_S for each $S \in \mathcal{S}_1$ and to $x_{i,v}$ for each $i \in [m']$ and $v \in \partial S_i$, and letting $\beta(x_0) = V(G_2)$. Note that G_2 is a supergraph of the torso of x_0, and that G_2 (and thus also the torso of x_0) is (a, m, d)-nearly drawn on Σ. This concludes the proof of ($\star$), and thus also of the Minor Structure Theorem. □

Let us remark that we have actually proved the following slightly stronger form of the Minor Structure Theorem, which simplifies some of its applications. A ***rooted tree decomposition*** is a tree decomposition (T, β) with T being a rooted tree. For a non-root node $x \in V(T)$ with parent y in T, the ***root separator*** of x is $\beta(x) \cap \beta(y)$. For the root of T, the root separator is empty. By a *contractible clique* in a graph drawn on a surface, we mean a clique whose cycles are contractible.

Theorem 5.29 *For every graph H, there exist constants a, m, and d such that the following claim holds. Every H-minor-free graph G has a rooted tree decomposition (T, β) such that for each node $x \in V(T)$ with $|\beta(x)| > a$ and with torso G_x, there exists a set $A_x \subset \beta(x)$ of size at most a containing the root separator of x such that*

- *the graph $G_x - A_x$ is $(0, m, d)$-nearly drawn on an H-avoiding surface,*
- *the surface part of $G_x - A_x$ is a minor of G, and*
- *for each child z of x in T, the set $\beta(x) \cap \beta(z) \setminus A_x$ either induces a contractible clique of size at most three in the surface part of $G_x - A_x$, or forms a subset of a bag of one of the vortices.*

In particular, we can assume that each piece of the decomposition attaches to the parent piece only through the apex vertices; this is clearly convenient for inductive arguments. Let us however remark that even so, the local form of the structure theorem (Theorem 5.27) is stronger than the global form in several aspects:

- It speaks about the structure of the graph G itself, rather than the structure of the torsos G_x of its tree decomposition. The edges added when forming the torso may cause G_x to contain minors or other objects not appearing in G, making Theorem 5.29 harder (or impossible) to use when trying to prove results of form "Every large H-minor-free graph contains ...".
- It describes the structure of the graph with respect to the given tangle, which may be chosen according to a property of interest for an argument (e.g., in Erdős-Pósa property arguments, so that it points towards the objects that we are trying to hit).
- It can be applied even to graphs that may contain H as a minor, as long as the given tangle points away from models of H (does not control them).

Although Robertson and Seymour do not state it, their proof of the Local Minor Structure Theorem actually directly gives the following stronger form. Consider a solid bisegregation $(\mathcal{S}_0, \mathcal{S}_1)$ of a graph F with an arrangement on a surface. Recall that since each cell $S \in \mathcal{S}_1$ is solid, there exists a model μ_S of $\pi(S)$ in S such that $u \in V(\mu_S(u))$ for each $u \in \partial S$; moreover, μ_S can be chosen so that

$\bigcup_{u \in \partial S} V(\mu_S(u))$ contains all vertices of the component of S containing ∂S. Letting $\mu(u) = \bigcup_{S \in \mathcal{S}_1 : u \in \partial S} \mu_S(u)$ for every $u \in \bigcup_{S \in \mathcal{S}_1} \partial S$ gives us a model μ of $\pi(\mathcal{S}_1)$ in F; we say that μ is a ***natural model*** of $\pi(\mathcal{S}_1)$ in F. For each element $S \in \mathcal{S}_0$, the ***vortex face*** of S is the face of $\pi(\mathcal{S}_1)$ containing $\overline{\pi}(S)$. For a tangle $\mathcal{T}$ in F of order at least θ_0, we say that an arrangement of $(\mathcal{S}_0, \mathcal{S}_1)$ on a surface Σ is $(\mathcal{T}, \theta_0)$***-spread*** if

- the drawing of $\pi(\mathcal{S}_1)$ on the surface Σ is cellular and has a respectful tangle $\mathcal{T}_0$ of order θ_0 μ-conformal with $\mathcal{T}$, where μ is a natural model of $\pi(\mathcal{S}_1)$ in F, and
- for distinct $S_1, S_2 \in \mathcal{S}_0$, the $d_{\mathcal{T}_0}$-distance between the vortex faces of S_1 and S_2 is θ_0.

We say that $\mathcal{T}_0$ is the ***surface tangle*** of the spread arrangement. For a graph H and a surface Σ, let $\mathrm{cr}_\Sigma(H)$ be the minimum number of crossings in a drawing of H on Σ.

Theorem 5.30 *For any graph H and any non-decreasing function $\psi : \mathbb{N}^2 \to \mathbb{N}$, there exist integers $\alpha, \delta_0 \geq 0$ and $\theta > \alpha$ such that the following claim holds. Suppose $\mathcal{T}$ is a tangle of order at least θ in a graph G. If $\mathcal{T}$ does not control a minor H of G, then there exists a set $A \subseteq V(G)$ of size at most α such that for some $\delta \leq \delta_0$, $G - A$ has a solid $(\mathrm{cr}_\Sigma(H) - 1, \delta)$-simple bisegregation with a $(\mathcal{T} - A, \psi(|A|, \delta))$-spread arrangement on an H-avoiding surface Σ.*

Robertson and Seymour further strengthened this result in [11], by post-processing the outcome of Theorem 5.27; see Chap. 10 for more details. Somewhat unfortunately, they do not show that the surface tangle $\mathcal{T}_0$ is μ-conformal with $\mathcal{T} - A$, which occasionally complicates papers using their result.

Let us remark that the condition that $\mathcal{T}_0$ is μ-conformal with $\mathcal{T} - A$ implies that $(\mathcal{S}_0, \mathcal{S}_1)$ is $(\mathcal{T} - A)$-central: Indeed, consider any separation $(C, D) \in \mathcal{T} - A$, and suppose for a contradiction that $D \subseteq S$ for some $S \in \mathcal{S}_0 \cup \mathcal{S}_1$. Since $\mathcal{T}_0$ is μ-conformal with $\mathcal{T} - A$, the vertex separation $(\mu^{-1}(V(C)), \mu^{-1}(V(D)))$ belongs to the vertex tangle corresponding to $\mathcal{T}_0$. However, $D \subseteq S$ implies $V(\pi(\mathcal{S}_1)) \subseteq V(C)$, and thus $\mu^{-1}(V(C)) = V(\pi(\mathcal{S}_1))$, which contradicts (T2).

The quantification structure of Theorem 5.30, with α and δ_0 depending on the function ψ, may be a bit confusing at first glance. A more natural first attempt would seem to be "for every H, there exist α and δ_0 such that for all integers ψ, there exists θ such that…". However, this does not work. Consider for example $H = K_8$ and suppose that the constants α and δ_0 independent of ψ existed. Let ψ be large compared to α and δ_0, and for an arbitrarily large θ, let G be a graph of treewidth $\Omega(\theta)$ drawn on the torus with representativity $\psi - 1$. Since G is drawn on the torus, it is H-minor-free, and since it has large treewidth, it contains a tangle of order θ. Because the representativity of G is less than ψ, the drawing of G on the torus cannot have a respectful tangle of order ψ. On the other hand, since the representativity $\psi - 1$ is large compared to α_0 and δ_0, it is not possible to move a part of G to a small number of δ_0-vortex-like parts and delete α_0 apices and end up with a graph that can be drawn on the sphere (or other K_8-avoiding surface with representativity at least ψ). The quantification in Theorem 5.30 with ψ being a function works around this issue: We can let $\alpha = \psi(0, 0)$. If the drawing of G on the torus has representativity at least $\psi(0,0)$, then it satisfies the conclusion with $A = \emptyset$. Otherwise, we can move at most α vertices of G to the apex set A and draw $G - A$ on the sphere. Then the tangle $\mathcal{T} - A$ is automatically respectful, and if we choose $\theta = \alpha + \psi(\alpha, 0)$, it has order at least $\theta - |A| \geq \psi(|A|, 0)$.

5.6 Arranging a Graph on a Surface

Lemma 5.25 gives us a way to segregate and arrange a region of graph that does not give rise to a large crooked transaction. In order to get a segregation with an arrangement on a surface Σ for the whole graph G as in Theorem 5.30, we are going to apply this lemma repeatedly around parts of G that were not arranged yet. An obvious concern is that rearranging a region might introduce inconsistencies with the arrangement of the rest of the graph. A more technical issue is that we need to ensure conformality with the prescribed tangle $\mathcal{T}$ in G, and this is hard to do directly if a large part of the arrangement can change in a poorly controlled way.

To deal with these issues, Robertson and Seymour introduce the notion of a *span*, a subgraph F of G with a fixed drawing on Σ around which the rest of G is arranged. The subgraph F is chosen in a way that its drawing is "locally rigid", in the sense that for any closed disk $\Delta \subset \Sigma$ with F-normal boundary, there is topologically only one way to draw $F \cap \Delta$ so that the intersection with the boundary of Δ is fixed. Thus, any rearrangement within a disk cannot alter the drawing of F, and we can use cycles in F to delimit the region in such a way that the rearranged part is on one side of the cycle and the rest of G is on the other one, ensuring that they do not conflict. Moreover, F is chosen with a tangle conformal with $\mathcal{T}$, and this easily leads to a conformal tangle in the projection of the resulting arrangement.

More precisely, consider a graph F drawn on a surface Σ. A disk $\Delta \subset \Sigma$ with an F-normal boundary is an *F-dial* if either its interior is disjoint from F, or Δ intersects F in a path with both ends in $\mathrm{bd}(\Delta)$. The drawing of F is ***rigid*** if for every F-normal simple closed curve γ in Σ intersecting F at most twice, there exists an F-dial with boundary γ. Note that when the drawing of F has representativity at least three, this is equivalent to saying that F is a subdivision of a 3-connected graph. Let us also remark that rigidity implies that the drawing is cellular. Whitney's theorem [16] states that a 3-connected planar graph has unique drawing in the plane up to homeomorphisms. The following variation on Whitney's result motivates the notion of rigidity.

Lemma 5.31 *Let F be a graph drawn on a surface Σ and let $\Lambda \subset \Sigma$ be an open disk with F-normal boundary. Let F' be another drawing of the same graph on Σ whose restriction to $\Sigma \setminus \Lambda$ matches the drawing of F. If F is rigid, then there exists a homeomorphism f of $\overline{\Lambda}$ to itself fixing* $\mathrm{bd}(\Lambda)$ *such that, letting $f(x) = x$ for $x \in \Sigma \setminus \Lambda$, we have $f(F) = F'$.*

Consider a zone Δ in a rigid drawing of a graph F on a surface Σ, bounded by a cycle K. In the following arguments, we are going to need to consider the subgraph

F' of F obtained by clearing Δ. An issue is that F' is not necessarily rigid, since a pair of vertices of K may form a non-trivial 2-cut in F'. The following definition describes disks bounded by cycles that avoid this issue: We say that Δ is *rigidity-preserving* if for every face f of F disjoint from Δ, the boundary of f intersects K in a connected subgraph.

Lemma 5.32 *Let F be a graph with a rigid drawing on a surface Σ and let $\Delta \subset \Sigma$ be a zone bounded by a cycle K in F. If Δ is rigidity-preserving, then the drawing of the subgraph F' of F obtained by clearing Δ is rigid.*

Proof Consider any F'-normal simple closed curve γ in Σ intersecting F' at most twice. If γ is disjoint from the interior of Δ, then note that the F-dial Λ bounded by γ cannot contain K, and thus Λ is disjoint from the interior of Δ and Λ is also an F'-dial. If γ is contained in Δ, then the subdisk of Δ bounded by γ has interior disjoint from F', and thus it forms an F'-dial.

Hence, suppose that γ intersects both the interior and the complement of Δ. Since $|\gamma \cap F'| \le 2$, we conclude that γ intersects F' exactly in two vertices u and v of K. In particular, the arc of γ between u and v and disjoint from the interior of Δ is drawn inside the same face f of F', and thus also of F. Since Δ is rigidity-preserving, the closure of f contains a subpath P of K with ends u and v. Then γ bounds an F'-dial $\Lambda \subset \Delta \cup f$ such that $F' \cap \Lambda = P$. □

Moreover, every zone can be slightly enlarged to a rigidity-preserving one.

Lemma 5.33 *Let F be a graph with a rigid drawing on a surface Σ and let Δ_1 and Δ_2 be zones in Σ, where Δ_1 is contained in the interior of Δ_2. Then there exists a rigidity-preserving zone Δ_1' contained in the interior of Δ_2 such that $\Delta_1 \subseteq \Delta_1'$. Moreover,*

- *the cycle K_1' bounding Δ_1' intersects the cycle K_1 bounding Δ_1 at least twice, and*
- *every subpath P of K_1' intersecting K_1 exactly in its ends is contained in the boundary of a face of F disjoint from Δ_1'.*

Proof Note that since the drawing of F is rigid, every face of F drawn inside Δ_2 is bounded by a cycle. We obtain Δ_1' by repeating the following operation as long as possible: If there exists a face f disjoint from Δ_1 such that the intersection I of the boundary of f with K_1 is not connected, then let γ be a simple curve such that the ends of γ are vertices u and v of different components of I and γ is otherwise drawn in f. Let Q be the subpath of K_1 between u and v such that the open subdisk Λ of Δ_2 bounded by $Q \cup \gamma$ is disjoint from Δ_1. Let Q' be the subpath of the boundary of f between u and v drawn in the closure of Λ. Replace Q by Q' in K_1; this enlarges the disk bounded by K_1 from Δ_1 to $\Delta_1 \cup \overline{\Lambda} \setminus f$. Note that K_1 and γ separate Q' from the boundary of Δ_2, and thus the modified disk is still contained in the interior of Δ_2.

This procedure clearly finishes after a bounded number of repetitions, resulting in a rigidity-preserving zone Δ_1'; and it is easy to see that K_1' has the properties described at the end of the statement of the lemma. □

By applying this transformation, it is easy to modify the proof of Theorem 4.21 to obtain the following strengthening.

Corollary 5.34 *Let G be a graph with a rigid drawing on a surface Σ, let $\mathcal{T}$ be a respectful tangle of order θ, and let a_0 be an atom of G. Let $r \geq 2$ and $p \geq 1$ be integers and let $k \geq p^2 + 3p + 5r$. If $2k < \theta$, then there exists a free k-local graded $r \times p$ battlefield around a_0 whose egg contains all atoms a of G such that $d_{\mathcal{T}}(a_0, a) \leq k - 3(r + p)$, and such that for each $C \in \mathcal{R}$, the zone around a bounded by C is rigidity-preserving.*

We are now ready to define the key technical notion of a span as mentioned above. Let $\mathcal{T}_0$ be a tangle in a graph F_0 and let Σ be a surface. A pair $(F, \mathcal{T})$ is a ***Σ-span of order*** θ in $(F_0, \mathcal{T}_0)$ if F is a subgraph of F_0 with a rigid drawing on Σ and $\mathcal{T}$ is a respectful tangle in F of order θ conformal with $\mathcal{T}_0$. We say that an arrangement of a bisegregation $(\mathcal{S}_0, \mathcal{S}_1)$ of F_0 is *consistent* with $(F, \mathcal{T})$ if

- the corresponding arrangement of $\mathcal{S}_0 \cup \mathcal{S}_1$ is consistent with the drawing of F,
- for every $S \in \mathcal{S}_0$, the subgraph F intersects S at most in ∂S, and
- for every cell $S \in \mathcal{S}_1$, we have $\big(F \cap S, F \cap \bigcup(\mathcal{S}_0 \cup \mathcal{S}_1 \setminus \{S\})\big) \in \mathcal{T}$.

We can now iterate Lemma 5.25 to give us the initial form of the main result of this section, showing that given a Σ-span with no long jumps, we can either arrange the rest of the graph around the span with a bounded number of vortex-like parts (i.e., obtain the conclusion we desire for the local form of the Minor Structure Theorem), or find a large crooked transaction attaching to a face of the span obtained by clearing a zone.

Lemma 5.35 *Let H be a graph and let Σ be an H-avoiding surface, let ν and ψ be positive integers, and let $\psi_1 : \mathbb{N} \to \mathbb{N}$ be a function. There exists integers θ and δ such that the following claim holds. Let F_0 be a graph and let $\mathcal{T}_0$ be a tangle in F_0 that does not control a minor of H in F_0. Let $(F, \mathcal{T})$ be a Σ-span of order θ in $(F_0, \mathcal{T}_0)$ such that F_0 does not contain any ν-long jump over F. Then either*

(i) there exists a Σ-span $(F_1, \mathcal{T}_1)$ of order at least 2ψ in $(F_0, \mathcal{T}_0)$ and a set $\mathcal{Q}$ of ψ pairwise vertex-disjoint paths in F_0 intersecting F_1 exactly in their ends such that the set Z of ends of paths in $\mathcal{Q}$ is $\mathcal{T}_1$-free, Z is a subset of the vertex set of a cycle K bounding a face of F_1, and ordering Z cyclically along K turns $\mathcal{Q}$ into a crooked transaction; or,

(ii) there exists a Σ-span $(F_1, \mathcal{T}_1)$ of order at least $\psi_1(\delta)$ in $(F_0, \mathcal{T}_0)$ and a solid $(\mathrm{cr}_\Sigma(H) - 1, \delta)$-simple bisegregation $(\mathcal{S}_0, \mathcal{S}_1)$ of F_0 with an arrangement on Σ consistent with $(F_1, \mathcal{T}_1)$ such that for distinct $S_1, S_2 \in \mathcal{S}_0$, the $d_{\mathcal{T}_1}$-distance between the faces of F_1 containing $\overline{\pi}(S_1)$ and $\overline{\pi}(S_2)$ is at least $\psi_1(\delta)$.

Proof Let r, c, and δ be as in Lemma 5.25 for $p = \psi$; without loss of generality, we can assume that $c \geq 20$. Let θ_0 be the constant θ from Lemma 4.35 for H and Σ. Let us choose $\theta \gg \theta_0, \nu, \psi, r, c, \delta, \psi_1(\delta), |H|, \|H\|$.

Let $\mathcal{J}$ be the set of all F-bridges of F_0. Note that since there is no ν-long jump over F, the $d_{\mathcal{T}}$-distance between any two attachments of an F-bridge is less than ν.

Let $\mathcal{J}'$ be an inclusionwise-maximal subset of $\mathcal{J}$ such that $F' = F \cup \bigcup \mathcal{J}'$ has a solid $(0, 0)$-simple bisegregation $(\emptyset, \mathcal{S}')$ with an arrangement on Σ consistent with $(F, \mathcal{T})$.

For any vertex a of F and integer κ, let $\mathcal{J}_{a,\kappa}$ be the set of elements of $\mathcal{J} \setminus \mathcal{J}'$ with all attachments at $d_{\mathcal{T}}$-distance at most κ from a. Suppose that $c, r \ll \kappa \ll \theta$ and that $\mathcal{J}_{a,\kappa} \neq \emptyset$. Let $k = \kappa + 3(r + c)$. By Corollary 5.34, there exists a free k-local $(r \times c)$-battlefield $(\mathcal{R}, \mathcal{P})$ around a whose egg contains all atoms of F at $d_{\mathcal{T}}$-distance at most κ from a and such that letting $\mathcal{R} = C_1, \dots, C_r$, for $i \in [r]$, the zone $\Delta_i \subset \Sigma$ around a bounded by C_i is rigidity-preserving. In particular, Δ_1 contains all attachments of the F-bridges belonging to $\mathcal{J}_{a,\kappa}$. Let $\Delta'_r \subset \Sigma$ be the closed disk bounded by $\pi(C_r)$ containing a, and let $Z_{a,\kappa}$ be the set of ends of the paths of $\mathcal{P}$ in C_r. For every $S \in \mathcal{S}'$ we have $(F \cap S, F \cap \bigcup(\mathcal{S}' \setminus \{S\})) \in \mathcal{T}$, and since $Z_{a,\kappa}$ is $\mathcal{T}$-free, it follows that $|Z_{a,\kappa} \cap V(S)| \leq |\partial S| \leq 3$. Let $F_{a,\kappa} = F' \cup \bigcup \mathcal{J}_{a,\kappa}$. Let us apply Lemma 5.25 with $G_1 = F_{a,\kappa}$ and with its partial arrangement $(F', F, \mathcal{S}')$, and let $\mathcal{Q}_{a,\kappa}$ and $\mathcal{Q}'_{a,\kappa}$ be the societies from the statement of the lemma. Let us consider the possible outcomes:

(i) If $\mathcal{Q}_{a,\kappa}$ has a solid simple segregation $\mathcal{S}_{a,\kappa}$ with an arrangement in Δ'_r, then let $\mathcal{S}'' = \{S \in \mathcal{S}' : \overline{\pi}(S) \not\subseteq \Delta'_r\} \cup \mathcal{S}_{a,\kappa}$. Then $\mathcal{S}''$ is a solid simple segregation of $F' \cup \bigcup \mathcal{J}_{a,\kappa}$ with an arrangement on Σ. Moreover, since the drawing of F is rigid, Lemma 5.31 implies that this arrangement is necessarily consistent with the drawing of F (possibly after a homeomorphic modification inside Δ'_r).
Since each element of $\mathcal{S}'$ contains at most three vertices of $Z_{a,\kappa}$, there exists a set C of at least $|Z_{a,\kappa}|/4 = c/4 \geq 5$ pairwise vertex-disjoint subpaths of C_r with ends in $Z_{a,\kappa}$ and intersecting $\partial \mathcal{Q}_{a,\kappa}$. For each cell $S \in \mathcal{S}_{a,\kappa}$, we have $V(S) \cap \partial \mathcal{Q}_{a,\kappa} \subseteq \partial S$, and thus there are at most $|\partial S| \leq 3$ paths in C with at least one end in S. Therefore, $|Z_{a,\kappa} \cap V(\bigcup(\mathcal{S}'' \setminus \{S\}))| \geq 4$, and since $Z_{a,\kappa}$ is $\mathcal{T}$-free, we have $(F \cap S, F \cap \bigcup(\mathcal{S}'' \setminus \{S\})) \in \mathcal{T}$.
Therefore, the arrangement of the solid $(0, 0)$-simple bisegregation $(\emptyset, \mathcal{S}'')$ is consistent with (F, T), and since $\mathcal{J}_{a,\kappa} \neq \emptyset$, this contradicts the maximality of $\mathcal{J}'$. Hence, this outcome is not possible.

(iii) If $\mathcal{Q}'_{a,\kappa}$ contains a crooked transaction $\mathcal{Q}$ of size at least ψ, then let$Z \subseteq \partial \mathcal{Q}'_{a,\kappa} = Z_{a,\kappa}$ be the set of ends of the paths in $\mathcal{Q}$, and let the Σ-span $(F_1, \mathcal{T}_1)$ be obtained from $(F, \mathcal{T})$ by clearing the zone Δ_r. Since the battlefield is free, we conclude that Z is $\mathcal{T}_1$-free, and thus the outcome (i) holds with $K = C_r$.

Hence, we can assume that the outcome (ii) of Lemma 5.25 holds for any choice of a and κ such that $\mathcal{J}_{a,\kappa} \neq \emptyset$ and $c, r \ll \kappa \ll \theta$. That is, $\mathcal{Q}'_{a,\kappa}$ contains a cross with ends contained in $Z_{a,\kappa}$ and $\mathcal{Q}_{a,\kappa}$ has a solid $(1, \delta)$-simple bisegregation $(\mathcal{S}_{a,\kappa,0}, \mathcal{S}_{a,\kappa,1})$ with an arrangement in Δ'_r.

Let X be the set of attachments of the bridges of $\mathcal{J} \setminus \mathcal{J}'$, viewed as a metric space with the metric $d_{\mathcal{T}}$. Let $\theta'_0 = \theta_0 + 4(\nu + 3(r+c)) \operatorname{cr}_\Sigma(H)$, and suppose first that there exists a θ'_0-wide subset X_0 of X of size $\operatorname{cr}_\Sigma(H)$. For $v \in X_0$, let $\mathcal{J}_v = \mathcal{J}_{v,\nu}$. Since F_0 does not contain any ν-long jumps over F, a bridge of $\mathcal{J} \setminus \mathcal{J}'$ with attachment v is contained in $\mathcal{J}_v$, and thus $\mathcal{J}_v \neq \emptyset$. Moreover, for $v' \in X_0$ distinct from v, any bridge $J \in \mathcal{J}_{v'}$ has all attachments at distance at least $d_{\mathcal{T}}(v, v') - \nu > \nu$ from v;

and thus $\mathcal{J}_v \cap \mathcal{J}_{v'} = \emptyset$. For all $v \in X_0$, let us clear the zone around v bounded by the outer cycle of the battlefield around v in F and $\mathcal{T}$, obtaining a span $(F_1, \mathcal{T}_1)$, and let f_v denote the face of F_1 created by clearing the zone around v. Since $Q'_{v,\nu}$ contains a cross and the battlefields are free, we conclude that F_0 contains pairwise vertex-disjoint eyes with foundations f_v for $v \in X_0$. Moreover, by Lemma 4.19, we have

$$d_{\mathcal{T}_1}(f_v, f_{v'}) \geq d_{\mathcal{T}}(v, v') - 4(\nu + 3(r + c))\,\mathrm{cr}_\Sigma(H) \geq \theta_0$$

for distinct $v, v' \in X_0$. By Lemma 4.35 applied with $\mathrm{cr}_\Sigma(H)$ eyes, it follows that $\mathcal{T}_0$ controls a minor of H in F_0, which is a contradiction.

Therefore, X does not contain a θ'_0-wide subset of size $\mathrm{cr}_\Sigma(H)$. Let $f(\psi') = \psi_1(\delta)+\nu+4(\psi'+3(r+c))\,\mathrm{cr}_\Sigma(H)$. By Lemma 4.44, for some $\psi' \ll \theta$ there exists a set $U \subseteq X$ of size less than $\mathrm{cr}_\Sigma(H)$ which is ψ'-dominating and $f(\psi')$-wide. For each $v \in U$, let $\mathcal{J}_v = \mathcal{J}_{v,\psi'}$. Consider any bridge $J \in \mathcal{J} \setminus \mathcal{J}'$. If J had attachments at $d_{\mathcal{T}}$-distance at most ψ' from distinct $v, v' \in U$, then since F_0 does not contain a ν-long jump over F, we would have $d_{\mathcal{T}}(v, v') \leq 2\psi' + \nu < f(\psi')$, which is a contradiction. Since U is ψ'-dominating the set X containing all attachments of J, we conclude that there exists a unique vertex $v \in U$ such that all attachments of J are at $d_{\mathcal{T}}$-distance at most ψ' from v, and thus $J \in \mathcal{J}_v$. Therefore, $\{J_v : v \in U\}$ is a partition of $\mathcal{J} \setminus \mathcal{J}'$.

For all $v \in U$, let us clear the zone around v bounded by the outer cycle of the battlefield around v in F and $\mathcal{T}$, obtaining a span $(F_1, \mathcal{T}_1)$, and let f_v denote the face of F_1 created by clearing the zone around v. By Lemma 4.19, we have

$$d_{\mathcal{T}_1}(f_v, f_{v'}) \geq f(\psi') - 4(\psi' + 3(r + c))\,\mathrm{cr}_\Sigma(H) \geq \psi_1(\delta)$$

for distinct $v, v' \in U$. Let $\mathcal{S}_1$ be obtained from $\mathcal{S}'$ by replacing for each $v \in U$ the part contained in the closure of f_v by $\mathcal{S}_{v,\psi',1}$, and let $\mathcal{S}_0 = \bigcup_{v \in U} \mathcal{S}_{v,\psi',0}$. Then $(\mathcal{S}_0, \mathcal{S}_1)$ is a solid $(\mathrm{cr}_\Sigma(H) - 1, \delta)$-simple bisegregation F_0 with an arrangement on Σ consistent with $(F_1, \mathcal{T}_1)$, which shows that the outcome (ii) holds. □

While the outcome (ii) corresponds to what we want to achieve in Theorem 5.30, it is not immediately obvious why the outcome (i) is useful for us. By Lemma 5.26, we can forget a part of Q to turn it into a large crosscap, leap, or double-cross transaction. In all the cases, we are going to choose a supergraph F'_1 of F_1 and let $\mathcal{T}'_1$ be the tangle in F'_1 induced by $\mathcal{T}_1$.

- The crosscap case is simplest to deal with: Let $F'_1 = F_1 \cup \bigcup Q$. Using the fact that Z is $\mathcal{T}_1$-free, it is easy to see that $(F'_1, \mathcal{T}'_1)$ is a Σ'-span of large order, where Σ' is the surface obtained from Σ by adding a crosscap. This is a significant improvement, since Corollary 4.33 implies that we can only increase the genus of the surface in this way a bounded number of times.
- In the double-cross case, let F'_1 be obtained from F_1 by adding the paths of Q except for the four crossed ones. Using the fact that Z is $\mathcal{T}_1$-free, it is easy to see that the two faces of F_1 to which the crossed paths of Q attach are far apart

in the $d_{\mathcal{T}'_1}$-distance. Thus, this transformation turns one non-planar region over the span $(F, \mathcal{T})$ into two distant non-planar regions over the span $(F'_1, \mathcal{T}'_1)$. Of course, the distance between non-planar regions over $(F'_1, \mathcal{T}'_1)$ is much smaller than over $(F, \mathcal{T})$, but if we start with a sufficiently large distance, we can iterate the argument from the proof of Lemma 5.35 until we accumulate $\mathrm{cr}_\Sigma(H)$ distant eyes in this way, obtaining a minor of H controlled by $\mathcal{T}_0$ using Lemma 4.35, contradicting the assumptions.

- Finally, in the leap case, if F'_1 is obtained from F_1 by adding the paths of Q except for the one crossing all others, we conclude that there is a long jump over F'_1. Following the idea outlined in Sect. 4.5, the plan is to collect enough jumps so that they can be turned into a horn, and further collect horns until we either get one of the other outcomes or a minor of H (and a contradiction) using Lemma 4.39. We execute this plan in the following section.

In conclusion, we can refine the outcome (i) in Lemma 5.35 as follows. Given Σ-spans $(F, \mathcal{T})$ and $(F', \mathcal{T}')$ in $(F_0, \mathcal{T}_0)$, we say that $(F', \mathcal{T}')$ is a ***λ-rearrangement*** of $(F, \mathcal{T})$ if there exists a Σ-span $(F'', \mathcal{T}'')$ obtained from $(F, \mathcal{T})$ by clearing a λ-zone Δ such that

- F'' is a subgraph of F',
- all vertices and edges of $V(F') \setminus V(F'')$ and $E(F') \setminus E(F'')$ are drawn in Δ, and
- the tangle $\mathcal{T}'$ is induced by $\mathcal{T}''$.

Note that by Lemma 4.19, this implies that $d_{\mathcal{T}'}$-distances between the atoms of F' belonging to F'' are smaller than their $d_{\mathcal{T}}$-distances by at most 4λ (but they can actually be larger, when $E(F') \setminus E(F'') \neq \emptyset$).

Lemma 5.36 *Let H be a graph and let Σ be an H-avoiding surface, let ν and ψ be positive integers, and let $\psi_1 : \mathbb{N} \to \mathbb{N}$ be a function. There exists integers λ, $\theta_2 \geq 2\lambda$, and δ_1 such that the following claim holds. Let F_0 be a graph and let $\mathcal{T}_0$ be a tangle in F_0 that does not control a minor of H in F_0. Let $(F, \mathcal{T})$ be a Σ-span of order at least θ_2 in $(F_0, \mathcal{T}_0)$. Then*

(i) there exists a Σ'-span of order ψ in $(F_0, \mathcal{T}_0)$, where Σ' is the surface obtained from Σ by adding a crosscap, or
(ii) there exists a λ-rearrangement $(F_1, \mathcal{T}_1)$ of $(F, \mathcal{T})$ such that F_0 contains a ν-long jump over F_1, or
(iii) for some $\delta \leq \delta_1$, there exists a Σ-span $(F_1, \mathcal{T}_1)$ of order $\psi_1(\delta)$ in $(F_0, \mathcal{T}_0)$ and a solid $(\mathrm{cr}_\Sigma(H) - 1, \delta)$-simple bisegregation $(\mathcal{S}_0, \mathcal{S}_1)$ of F_0 with an arrangement on Σ consistent with $(F_1, \mathcal{T}_1)$ such that for distinct $S, S' \in \mathcal{S}_0$, the $d_{\mathcal{T}_1}$-distance between the faces of F_1 containing $\overline{\pi}(S)$ and $\overline{\pi}(S')$ is at least $\psi_1(\delta)$.

Let us remark that the outcome (ii) subsumes the assumption of Lemma 5.35 that ν-long jumps are absent. It is also worth mentioning (though probably clear from the proof outline) that the rearrangement radius λ can be much larger than the jump length ν.

A careful reader might suspect that we made a mistake in formulating Lemma 5.36: Following the proof outline given above, it is certainly possible that we encounter a double-cross transaction multiple (but less than $\mathrm{cr}_\Sigma(H)$) times, each time rearranging the span, before we hit a leap transaction. Hence, should the outcome (ii) of Lemma 5.36 actually say that $(F_1, \mathcal{T}_1)$ is obtained from $(F, \mathcal{T})$ by multiple λ-rearrangements? It turns out that this can be avoided and (ii) is correct as stated: Roughly, we can revert the rearrangements except for at most two rearranged regions containing the ends of the jump. If these two regions are far apart, we can observe that after reverting the corresponding rearrangements, there still has to be a long jump between the reverted regions. On the other hand, if the two regions are nearby, then we can view them as a part of a single larger rearranged region.

Compared to Theorem 5.30, the conclusion (iii) of Lemma 5.36 only speaks about the properties of the bisegregation $(\mathcal{S}_0, \mathcal{S}_1)$ with respect to the tangle $\mathcal{T}_1$ in F_1, rather than about a tangle in $\pi(\mathcal{S}_1)$. This is addressed by the following technical lemma, complicated by the fact that the minors F and $\pi(\mathcal{S}_1)$ of F_0 are in general incomparable in the minor relation.

Lemma 5.37 *Let F_0 be a graph with a tangle $\mathcal{T}_0$. Let Σ be a surface and let $(F, \mathcal{T})$ be a Σ-span of order θ in $(F_0, \mathcal{T}_0)$. Let $(\mathcal{S}_0, \mathcal{S}_1)$ be a solid (c, δ)-simple bisegregation of F_0 with an arrangement on Σ consistent with $(F, \mathcal{T})$ such that for distinct $S, S' \in \mathcal{S}_0$, the $d_{\mathcal{T}}$-distance between the faces of F_1 containing $\overline{\pi}(S)$ and $\overline{\pi}(S')$ is θ. Let $\theta' < \theta/18$ be a positive integer. Then there exists a solid $(c, 2\delta+2)$-simple bisegregation $(\mathcal{S}'_0, \mathcal{S}'_1)$ of F_0 with a $(\mathcal{T}_0, \theta')$-spread arrangement on Σ.*

Proof (⇸) First, let us transform $(\mathcal{S}_0, \mathcal{S}_1)$ into a solid $(c, 2\delta + 2)$-simple bisegregation $(\mathcal{S}'_0, \mathcal{S}'_1)$ such that the drawing of $\pi(\mathcal{S}'_1)$ is cellular. Let C be the component of $\pi(\mathcal{S}_1)$ intersecting F. Without loss of generality, we can assume that every component C' of $\pi(\mathcal{S}_1)$ other than C intersects the boundary of an element of $\mathcal{S}_0$: Otherwise, we can combine all elements of $\mathcal{S}_1$ contributing to C' to a single society with empty boundary. Let $\mathcal{S}'_1$ consists of the cells of $\mathcal{S}_1$ that contribute to C. Since the drawing of F is cellular and the arrangement of $\mathcal{S}$ is consistent with the drawing of F, we conclude that the drawing of $C = \pi(\mathcal{S}'_1)$ is cellular.

For each element $S \in \mathcal{S}_0$, let S' be the union of S with the cells of $\mathcal{S}_1$ contributing to the components of $\pi(\mathcal{S}_1)$ distinct from C that intersect S, with $\partial S'$ consisting of the vertices of ∂S contained in C, and let $\mathcal{S}'_0$ be the set of all such societies S'. Clearly, $(\mathcal{S}'_0, \mathcal{S}'_1)$ is a solid bisegregation of F_0 with an arrangement on Σ consistent with $(F, \mathcal{T})$. We need to argue that for each $S \in \mathcal{S}_0$, the society S' is $(2\delta + 2)$-vortex-like. Indeed, consider distinct $u, v \in \partial S'$. By Lemma 5.22, S has a vortical decomposition of adhesion at most δ. Letting A and B be the union of the bags of this decomposition assigned to vertices in $\partial_{[u,v)} S$ and in $\partial_{[v,u)} S$, respectively, observe that (A, B) is a vertex separation of S of order at most 2δ with $\partial_{[u,v)} S \subseteq A$ and $\partial_{[v,u)} S \subseteq B$. Since the drawing of C is cellular, $\overline{\pi}(S)$ is contained in a single face of C, and the vertices u and v belong to C, we conclude that for each component $C' \neq C$ of $\pi(\mathcal{S}_1)$ intersecting S, we have $V(C' \cap S) \subseteq \partial_{[u,v]} S$ or $V(C' \cap S) \subseteq \partial_{[v,u]} S$. It follows that S' has a vertex separation (A', B') of order at most $2\delta + 2$ with $A \cup \{v\} \subseteq A'$, $B \cup \{u\} \subseteq B'$, and $A' \cap B' = (A \cap B) \cup \{u, v\}$.

Consequently, S' does not contain any transaction of size greater than $2\delta + 2$ from $\partial_{[u,v)} S'$ to $\partial_{[v,u)} S'$. Since this holds for all distinct $u, v \in \partial S'$, we conclude that S' is $(2\delta + 2)$-vortex-like.

Next, we need to define a tangle $\mathcal{T}_2$ in $\pi(\mathcal{S}_1') = C$. Let Y be a $\mathcal{T}$-free subset of $V(F)$ of size $6\theta'$. For each $y \in Y \setminus V(C)$, let S_y be the element of $\mathcal{S}_1'$ containing Y and let $P_y = \partial S_y \cap V(F)$. For each $y \in Y \cap V(C)$, let $S_y = y$ and $P_y = \{y\}$. Since $\mathcal{S}_1' \subseteq \mathcal{S}_1$, the definition of consistency gives $(S_y \cap F, \bigcup(\mathcal{S}_0' \cup \mathcal{S}_1' \setminus S_y) \cap F) \in \mathcal{T}$ for each $y \in Y$. Let $L = \bigcup_{y \in Y}(S_y \cap F)$ and $P = \bigcup_{y \in Y} P_y$, and note that $|P_y| \le 3|Y| = 18\theta' < \theta$. By Lemma 2.22, there exists a subgraph R of F such that $V(L \cap R) = P$ and $(L, R) \in \mathcal{T}$. Since Y is $\mathcal{T}$-free, Lemma 2.41 implies $\operatorname{rk}_{\mathcal{T}} P = \operatorname{rk}_{\mathcal{T}}(V(L)) \ge |Y| = 6\theta'$. Hence, by Observation 2.40, there exists a $\mathcal{T}$-free set $Y_0 \subseteq P \subseteq V(C \cap F)$ of size $6\theta'$. Note that since $\mathcal{T}$ is conformal with $\mathcal{T}_0$, the set Y_0 is also $\mathcal{T}_0$-free.

Consider any vertex separation (A, B) of C of order less than $2\theta'$. Let A' be the union of cells $S \in \mathcal{S}_1'$ with $\partial S \subseteq A$ and let B' be the union of those with $\partial S \subseteq B$. Since each $S \in \mathcal{S}_1'$ is a cell, ∂S induces a clique in C, and consequently $\partial S \subseteq A$ or $\partial S \subseteq B$. It follows that (A', B') is a separation of C of order $|A \cap B|$, and thus $(A' \cap F, B' \cap F)$ is a separation of F of order at most $|A \cap B|$. Since Y_0 is $\mathcal{T}$-free and $|Y_0| > 4\theta' > 2|A \cap B|$, exactly one of the subgraphs $A' \cap F$ and $B' \cap F$ contains at most $|A \cap B|$ vertices of Y_0, and thus exactly one of A and B contains at most $|A \cap B|$ vertices of Y_0. Let $\mathcal{T}_2'$ consist of the vertex separations (A, B) of C of order less than $2\theta'$ such that $|A \cap Y_0| \le |A \cap B|$. Then $\mathcal{T}_2'$ is an orientation of vertex $(<2\theta')$-separations of C, and since $|Y_0| = 6\theta'$, the condition (T1) is clearly satisfied by $\mathcal{T}_2'$. Hence, $\mathcal{T}_2'$ is a vertex tangle in C of order $2\theta'$. By definition, we have $\operatorname{rk}_{\mathcal{T}_2'}(Y_0) = 2\theta'$, and thus Y_0 contains a $\mathcal{T}_2'$-free subset Y_2 of size $2\theta'$. Since Y_0 is also $\mathcal{T}_0$-free, Lemma 2.47 implies that the truncation $\mathcal{T}_2$ of $\mathcal{T}_2'$ to order θ' is μ-conformal with $\mathcal{T}_0$, where μ is the natural model of $C = \pi(\mathcal{S}_1')$ in F_0.

Consider any simple closed C-normal curve γ in Σ such that $|\gamma \cap C| < \theta'$. For each $S \in \mathcal{S}_1'$ with $|\partial S| = 3$ such that γ intersects the face $\overline{\pi}(S)$, we can shift γ within the face to make it disjoint from $F \cap \overline{\pi}(S)$. Hence, we can also assume that γ is F-normal, and $|\gamma \cap F| \le |\gamma \cap C| < \theta'$. Since the tangle $\mathcal{T}$ is respectful, Lemma 2.78 implies that γ bounds a disk $\Delta \subset \Sigma$ such that $(F \cap \Delta, F \cap \overline{\Sigma \setminus \Delta}) \in \mathcal{T}$. Since Y_2 is $\mathcal{T}$-free, we have $|Y_2 \cap \Delta| \le |\gamma \cap C|$, and thus also $(C \cap \Delta, C \cap \overline{\Sigma \setminus \Delta}) \in \mathcal{T}_2$. Hence, Lemma 2.78 implies that $\mathcal{T}_2$ is respectful. An analogous argument (modifying an optimal restraint for F into a restraint for C) shows that for distinct $S_1, S_2 \in \mathcal{S}_0'$, if f_1 and f_2 are the faces of C containing $\overline{\pi}(S_1)$ and $\overline{\pi}(S_2)$, then $d_{\mathcal{T}_0}(f_1, f_2) = \theta'$, and thus the arrangement of $(\mathcal{S}_0', \mathcal{S}_1')$ is $(\mathcal{T}_0, \theta')$-spread. □

5.7 Finishing the Proof

We are now ready to put all the pieces together and obtain a proof of the Minor Structure Theorem, by proving its local form given in Theorem 5.30. This final part is complicated by the fact that new jumps may arise through a rearrangement in Lemma 5.36 (without this outcome, the result would follow simply by combining Corollary 4.47 with Lemma 5.36 and noting that by Corollary 4.33, we can only increase the genus of the surface a bounded number of times while avoiding H as a minor). To deal with this issue, we need to keep a more detailed track of long jumps and horns. Let G_0 be a graph with a tangle $\mathcal{T}_0$. For a surface Σ and integers $a, h \geq 0$ and $\sigma, \theta \geq 1$, a $(\Sigma, a, h, \sigma, \theta)$-***animal*** in $(G_0, \mathcal{T}_0)$ is a tuple $(G, \mathcal{T}, A, Y)$, where

- $(G, \mathcal{T})$ is a Σ-span of order θ in $(G_0 - A, \mathcal{T}_0 - A)$,
- $A \subseteq V(G_0) \setminus V(G)$ consists of exactly a vertices, each of which is a tip of a θ-wide horn over G with σ legs, and
- Y is a set of exactly h vertices of G such that each vertex of Y is an end of a θ-long jump in $G_0 - A$ over G and $d_{\mathcal{T}}(y, y') = \theta$ for distinct $y, y' \in Y$.

We refer to the elements of Y as the *hairs* of the animal. Note that the jumps and the horns are not required to be disjoint, and that both ends of the same θ-long jump may belong to Y.

The following claim is essentially a restatement of Lemma 4.46 in terms of animals, with a few minor changes:

- In the second outcome of Lemma 4.46, we apply the comments stated before Corollary 4.47 to turn the jumps into a handle or a crosshandle, obtaining the conclusion (i) of the following lemma.
- We are concluding the existence of Σ-spans, e.g., we require the drawing of G' to be rigid rather than just cellular; to this end, the proof of Lemma 4.46 needs to be adjusted to use Corollary 5.34 instead of Theorem 4.21.
- The outcome (iii) of the following lemma corresponds to the first outcome of Lemma 4.46, but we additionally extend the long jumps from Y through paths in the cleared zones to a set of vertices Y' in the boundaries of the zones. The claim that vertices of A are also tips of horns over G' follow similarly, see also Lemma 4.42.

Let us also remark that the outcome (ii) of the following lemma corresponds to the case that the assumptions of Lemma 4.46 are not satisfied and there exists a θ_1-long jump with an end at $d_{\mathcal{T}}$-distance at least θ_1 from Y.

Lemma 5.38 *For any surface Σ and integers $a, h \geq 0$ and $\theta_1 \geq 1$, there exist integers ν_0, ξ, and $\theta_0 > \nu_0$ such that the following claim holds for every integer σ. Let G_0 be a graph with a tangle $\mathcal{T}_0$. If $\theta \geq \theta_0$ and $(G_0, \mathcal{T}_0)$ contains a $(\Sigma, a, h, \sigma, \theta)$-animal $(G, \mathcal{T}, A, Y)$, then at least one of the following holds:*

(i) $(G_0, \mathcal{T}_0)$ contains a $(\Sigma + \text{handle})$- or $(\Sigma + \text{crosshandle})$-span of order θ_1, or
(ii) $(G_0, \mathcal{T}_0)$ contains a $(\Sigma, a, h + 1, \sigma, \theta_1)$-animal, or
(iii) $(G_0, \mathcal{T}_0)$ contains a $(\Sigma, a, h, \sigma, \theta - \nu_0)$-animal $(G', \mathcal{T}', A, Y')$ and there exists a set $X \subseteq V(G_0) \setminus V(G')$ of size at most ξ such that $A \subseteq X$ and there is no ν_0-long jump over $(G', \mathcal{T}')$ in $G_0 - X$.

Next, we combine this result with Lemma 5.36. In the outcome (ii) of Lemma 5.36, we gain a new long jump J, and thus it seems that we will easily obtain an improvement to the number of hairs of the animal as in (ii) of Lemma 5.38. The situation is actually a bit more complicated, since the λ-rearrangement can also destroy one of the previously existing hairs. However, we are going to observe that *both* ends of J can be added as new hairs (recall that the definition of the animal does not place any constraints on the intersections of the long jumps starting in the hairs, or even require that they are pairwise different), resulting in a net increase in the number of hairs.

Lemma 5.39 *Let H be a graph, let Σ be an H-avoiding surface, let $a, h \geq 0$ and $\theta_1 \geq 1$ be integers, and let $\psi : \mathbb{N}^2 \to \mathbb{N}$ be a non-decreasing function. There exist integers δ_0, ξ, and θ such that the following claim holds for every $\sigma \geq 0$. Let G_0 be a graph with a tangle $\mathcal{T}_0$. If $(G_0, \mathcal{T}_0)$ contains a $(\Sigma, a, h, \sigma + \xi, \theta)$-animal $(G, \mathcal{T}, A, Y)$ and $\mathcal{T}_0$ does not control a minor of H in G_0, then*

- *(i)* *$(G_0, \mathcal{T}_0)$ contains a $(\Sigma+$crosscap$)$-, $(\Sigma+$handle$)$- or $(\Sigma+$crosshandle$)$-span of order θ_1, or*
- *(ii)* *$(G_0, \mathcal{T}_0)$ contains a $(\Sigma, a, h+1, \sigma, \theta_1)$-animal, or*
- *(iii)* *there exists a set $X \subseteq V(G_0)$ of size at most ξ such that for some $\delta \leq \delta_0$, $G_0 - X$ has a solid $(\mathrm{cr}_\Sigma(H)-1, \delta)$-simple bisegregation with a $(\mathcal{T}_0 - X, \psi(|X|, \delta))$-spread arrangement on Σ.*

Proof Let ν_0, ξ, and θ_0 be the constants from Lemma 5.38 for the given Σ, a, h, and θ_1. Let λ, θ_2, and δ_1 be the constants from Lemma 5.36 applied with $\psi = \theta_1$, $\psi_1(\delta) = \max(19\psi(\xi, 2\delta+2), \theta_1)$, and $\nu = \max(\nu_0, \theta_1)$. Let $\delta_0 = 2\delta_1 + 2$ and choose $\theta \gg \theta_1, \nu, \lambda$.

Outcomes (i) and (ii) of Lemma 5.38 give the outcomes (i) and (ii) of the current lemma. Hence, we can assume that the outcome (iii) of Lemma 5.38 holds, and $(G_0, \mathcal{T}_0)$ contains a $(\Sigma, a, h, \sigma+\xi, \theta-\nu_0)$-animal $(G', \mathcal{T}', A, Y')$ and there exists a set $X \subseteq V(G_0) \setminus V(G')$ of size at most ξ such that $A \subseteq X$ and there is no ν_0-long jump over $(G', \mathcal{T}')$ in $G_0 - X$.

Let us now apply Lemma 5.36 with $(F_0, \mathcal{T}_0) = (G_0 - X, \mathcal{T}_0 - X)$ and $(F, \mathcal{T}) = (G', \mathcal{T}')$. The outcome (i) corresponds to the outcome (i) of the current lemma, and the outcome (iii) together with Lemma 5.37 gives the outcome (iii) of the current lemma. Hence, suppose that the outcome (ii) holds, i.e., there exists a λ-rearrangement $(G'_1, \mathcal{T}'_1)$ of $(G', \mathcal{T}')$ and a ν-long jump over G'_1 in $G_0 - X$.

The graph G'_1 is obtained from G' by clearing a λ-zone Δ around an atom x_0 of G' and drawing a new subgraph M of $G_0 - X$ into this zone. Consider a vertex $a \in A$ and a $(\theta - \nu_0)$-wide horn S with tip a and with $\sigma + \xi$ legs over G'. Let S' be obtained from the horn S by removing the paths intersecting $X \setminus \{a\}$ as well as the path to the (at most one) leg at $d_{\mathcal{T}'}$-distance at most $\lambda + \nu_0$ from x_0. We claim that S' is disjoint from M; indeed, otherwise there would exist a jump over G' in $(S'-a) \cup M$ from a leg of S' to a vertex in Δ, and such a jump would be ν_0-long and disjoint from X, a contradiction since there are no such jumps in $G_0 - X$. Hence, a is a tip of a $(\theta - \nu_0 - 4\lambda)$-wide (and thus also θ_1-wide) horn with σ legs over G'_1.

Similarly, let Y_1' be obtained from Y' by removing the (at most one) vertex at $d_{\mathcal{T}'}$-distance at most $\lambda + \nu_0 + \theta_1$ from x_0. Then for each $y \in Y_1'$, the $(\theta - \nu_0)$-long jump from y over G' is disjoint from M, and thus it is a $(\theta - \nu_0 - 4\lambda)$-long (and thus also θ_1-long) jump over G_1'. Let y_1 and y_2 be the ends of a ν-long jump J over G_1' in $G_0 - X$. If y_1 were a vertex of G' at $d_{\mathcal{T}'}$-distance more than $\lambda + \nu_0$ from x_0, then we would similarly see that J is disjoint from M and from the subgraph of G' drawn in Δ. Hence, J would be also a jump over G' in $G_0 - X$, and since no such jump is ν_0-long, we would have $d_{\mathcal{T}'}(y_1, y_2) < \nu_0$. However, since y_1 is far from x_0 in the $d_{\mathcal{T}'}$-distance, it is easy to see that this implies that $d_{\mathcal{T}_1'}(y_1, y_2) < \nu_0 \leq \nu$, which is a contradiction. Therefore, y_1 either belongs to M or is at $d_{\mathcal{T}'}$-distance at most $\lambda + \nu_0$ from x_0. By symmetry, the same claim holds for y_2. Since every vertex $y \in Y_1'$ is at $d_{\mathcal{T}'}$-distance more than $\lambda+\nu_0+\theta_1$ from x_0, observe that $d_{\mathcal{T}_1'}(y, y_i) \geq \theta_1$. Therefore, $(G_1', \mathcal{T}_1', A, Y_1' \cup \{y_1, y_2\})$ is a $(\Sigma, a, h + 1, \sigma, \theta_1)$-animal, and the outcome (ii) holds. □

The rest of the proof of the Local Minor Structure Theorem is straightforward: We repeatedly apply Lemma 5.39, improving the genus or the number of hairs of the animal. When the number of hairs is sufficiently large, we turn them into a new horn using Lemma 4.41. This process eventually stops, since the number of horns is limited by Lemma 4.39 and the genus by Corollary 4.33. As getting the constants right is a bit fiddly, we are now going to show the details of this idea, but the reader should feel free to skip the rest of this section.

First, let us iterate Lemma 5.39 to get as many hairs as needed.

Lemma 5.40 *Let H be a graph, let Σ be an H-avoiding surface, let $a, h, \sigma \geq 0$ and $\theta_1 \geq 1$ be integers, and let $\psi : \mathbb{N}^2 \to \mathbb{N}$ be a non-decreasing function. There exist integers δ_0, σ_0, ξ, and θ such that the following claim holds. Let G_0 be a graph with a tangle $\mathcal{T}_0$. If $(G_0, \mathcal{T}_0)$ contains a $(\Sigma, a, 0, \sigma_0, \theta)$-animal $(G, \mathcal{T}, A, \emptyset)$ and $\mathcal{T}_0$ does not control a minor of H in G_0, then at least one of the following holds:*

- *(i) $(G_0, \mathcal{T}_0)$ contains a $(\Sigma+$crosscap$)$-, $(\Sigma+$handle$)$- or $(\Sigma+$crosshandle$)$-span of order θ_1, or*
- *(ii) $(G_0, \mathcal{T}_0)$ contains a $(\Sigma, a, h, \sigma, \theta_1)$-animal, or*
- *(iii) there exists a set $X \subseteq V(G_0)$ of size at most ξ such that for some $\delta \leq \delta_0$, $G_0 - X$ has a solid $(\mathrm{cr}_\Sigma(H)-1, \delta)$-simple bisegregation with a $(\mathcal{T}_0 - X, \psi(|X|, \delta))$-spread arrangement on Σ.*

Proof (⤙) We prove the claim by induction on h, with the basic case $h = 0$ being trivial, since the outcome (ii) holds with $\sigma_0 = \sigma$ and $\theta = \theta_1$. Suppose now that $h \geq 1$. Let δ_0', ξ', and θ' be the values of δ_0, ξ, and θ from Lemma 5.39 applied with the given H, Σ, a, θ_1, and ψ, and with $h - 1$ playing the role of h. Let δ_0'', σ_0, ξ'', and θ be the values of δ_0, σ_0, ξ, and θ from the current lemma applied inductively with $h - 1$ playing the role of h, the given H, Σ, a, and ψ, and with $\sigma + \xi'$ playing the role of σ and $\max(\theta_1, \theta')$ playing the role of θ_1. Let $\delta_0 = \max(\delta_0', \delta_0'')$ and $\xi = \max(\xi', \xi'')$.

By the induction hypothesis, either the outcomes (i) or (iii) hold, or $(G_0, \mathcal{T}_0)$ contains a $(\Sigma, a, h-1, \sigma+\xi', \theta')$-animal. By Lemma 5.39 applied to this animal, one of the outcomes of the current lemma holds. □

Next, we combine many hairs into a new horn.

Lemma 5.41 *Let H be a graph, let Σ be an H-avoiding surface, let $a, \sigma \geq 0$ and $\theta_1 \geq 1$ be integers, and let $\psi : \mathbb{N}^2 \to \mathbb{N}$ be a non-decreasing function. There exist integers δ_0, σ_0, ξ, and θ such that the following claim holds. Let G_0 be a graph with a tangle $\mathcal{T}_0$. If $(G_0, \mathcal{T}_0)$ contains a $(\Sigma, a, 0, \sigma_0, \theta)$-animal $(G, \mathcal{T}, A, \emptyset)$ and $\mathcal{T}_0$ does not control a minor of H in G_0, then at least one of the following holds:*

(i) $(G_0, \mathcal{T}_0)$ contains a $(\Sigma+$crosscap$)$-, $(\Sigma+$handle$)$- or $(\Sigma+$crosshandle$)$-span of order θ_1, or
(ii) $(G_0, \mathcal{T}_0)$ contains a $(\Sigma, a+1, 0, \sigma, \theta_1)$-animal, or
(iii) there exists a set $X \subseteq V(G_0)$ of size at most ξ such that for some $\delta \leq \delta_0$, $G_0 - X$ has a solid $(\mathrm{cr}_\Sigma(H)-1, \delta)$-simple bisegregation with a $(\mathcal{T}_0 - X, \psi(|X|, \delta))$-spread arrangement on Σ.

Proof (⟜) Let $k = \|H\|$ and $h = 12k^2(\sigma+1)(\sigma+2)+1$. Let θ_0 be the value from Lemma 4.41 applied for H and Σ. Let δ_0, σ_0, ξ, and θ be the values from Lemma 5.40 applied for H, Σ, a, h, σ, and with $\theta_1' = 18\max(\theta_0, \theta_1) + 72$ playing the role of θ_1.

By Lemma 5.40, either the outcomes (i) or (iii) of the current lemma hold, or $(G_0, \mathcal{T}_0)$ contains a $(\Sigma, a, h, \sigma, \theta_1')$-animal $(G', \mathcal{T}', A', Y')$. Let us now apply Lemma 4.41 to $G_0 - A'$, G', and $\mathcal{T}'$ with $\theta' = \max(\theta_0, \theta_1)$ playing the role of θ and with $t = \sigma$. The outcome (i) of Lemma 4.41 does not hold by the assumptions. We claim that (iv) does not hold, either: Otherwise, consider the set $Y \subseteq V(G')$ from this outcome. Every vertex of Y' is an end of a θ'-long jump over G', and thus it is at $d_{\mathcal{T}'}$-distance less than $9\theta'+36$ from Y. Since $|Y'| = h > |Y|$, there exists a vertex $y \in Y$ and distinct vertices $y_1, y_2 \in Y'$ such that $d_{\mathcal{T}'}(y, y_i) < 9\theta'+36$ for $i \in [2]$. However, this implies that $d_{\mathcal{T}'}(y_1, y_2) < 18\theta'+72 = \theta_1'$, which is a contradiction. Note that the outcome (ii) of Lemma 4.41 can be seen as a special case of the outcome (iii) with Δ being empty, and thus we only need to consider the outcome (iii).

By using Corollary 5.34 instead of Theorem 4.21 in the proof of Lemma 4.41, we can also assume that the $(2\theta'+9)$-zone Δ from the outcome (iii) is rigidity-preserving. Let $(G'', \mathcal{T}'')$ be the Σ-span obtained from $(G', \mathcal{T}')$ by clearing the zone Δ and let $a \in V(G_0 - A') \setminus V(G'')$ be the tip of a θ'-wide horn with σ legs over G''. Note that each θ_1'-wide horn over G' with a tip in A' can be transformed into a θ'-wide horn over G'' by extending the (at most one) leg in Δ through a path in $G' \cap \Delta$ to end in the boundary of Δ. Hence, $(G'', \mathcal{T}'', A' \cup \{a\}, \emptyset)$ is a $(\Sigma, a+1, 0, \sigma, \theta_1)$-animal and (ii) holds. □

We can now iterate this lemma to accumulate as many horns as we like.

Lemma 5.42 *Let H be a graph, let Σ be an H-avoiding surface, let $a, \sigma \geq 0$ and $\theta_1 \geq 1$ be integers, and let $\psi : \mathbb{N}^2 \to \mathbb{N}$ be a non-decreasing function. There exist integers δ_0, ξ, and θ such that the following claim holds. Let G_0 be a graph with a tangle $\mathcal{T}_0$. If $(G_0, \mathcal{T}_0)$ contains a Σ-span $(G, \mathcal{T})$ of order θ and $\mathcal{T}_0$ does not control a minor of H in G_0, then at least one of the following holds:*

(i) $(G_0, \mathcal{T}_0)$ contains a (Σ+crosscap)-, (Σ+handle)- or (Σ+crosshandle)-span of order θ_1, or
(ii) $(G_0, \mathcal{T}_0)$ contains a $(\Sigma, a, 0, \sigma, \theta_1)$-animal, or
(iii) there exists a set $X \subseteq V(G_0)$ of size at most ξ such that for some $\delta \leq \delta_0$, $G_0 - X$ has a solid $(\mathrm{cr}_\Sigma(H)-1, \delta)$-simple bisegregation with a $(\mathcal{T}_0 - X, \psi(|X|, \delta))$-spread arrangement on Σ.

Proof (⇝) We prove the claim by induction on a, with the basic case $a = 0$ trivially leading to the outcome (ii) with $\theta = \theta_1$. Suppose now that $a \geq 1$. Let δ_0', σ_0, ξ', and θ' be the values of δ_0, σ_0, ξ, and θ from Lemma 5.41 applied with the given H, Σ, σ, θ_1, and ψ, and with $a - 1$ playing the role of a. Let δ_0'', ξ'', and θ be the values of δ_0, ξ, and θ from the current lemma applied inductively with $a - 1$ playing the role of a, the given H, Σ, and ψ, and with σ_0 playing the role of σ and $\max(\theta_1, \theta')$ playing the role of θ_1. Let $\delta = \max(\delta_0', \delta_0'')$ and $\xi = \max(\xi', \xi'')$.

By the induction hypothesis, either the outcomes (i) or (iii) hold, or $(G_0, \mathcal{T}_0)$ contains a $(\Sigma, a - 1, \emptyset, \sigma_0, \theta')$-animal. By Lemma 5.41 applied to this animal, one of the outcomes of the current lemma holds. □

By applying Lemma 4.39, we can eliminate the horns from the outcome.

Lemma 5.43 *Let H be a graph, let Σ be an H-avoiding surface, let $\theta_1 \geq 1$ be an integer, and let $\psi : \mathbb{N}^2 \to \mathbb{N}$ be a non-decreasing function. There exist integers δ_0, ξ, and θ such that the following claim holds. Let G_0 be a graph with a tangle $\mathcal{T}_0$. If $(G_0, \mathcal{T}_0)$ contains a Σ-span $(G, \mathcal{T})$ of order θ and $\mathcal{T}_0$ does not control a minor of H in G_0, then at least one of the following holds:*

(i) $(G_0, \mathcal{T}_0)$ contains a (Σ+crosscap)-, (Σ+handle)- or (Σ+crosshandle)-span of order θ_1, or
(ii) there exists a set $X \subseteq V(G_0)$ of size at most ξ such that for some $\delta \leq \delta_0$, $G_0 - X$ has a solid $(\mathrm{cr}_\Sigma(H) - 1, \delta)$-simple bisegregation with a $(\mathcal{T}_0 - X, \psi(|X|, \delta))$-spread arrangement on Σ.

Proof (⇝) Let σ and θ_0 be the values of t_0 and θ from Lemma 4.39 for H and Σ, and let $a = a_\Sigma(H)$. Let δ_0, ξ, and θ be the values from Lemma 5.42 applied with a, σ, ψ, and with $\max(\theta_1, 2\theta_0)$ playing the role of θ_1. The outcomes (i) and (iii) of Lemma 5.42 now correspond to the outcomes (i) and (ii) of the current lemma, while the outcome (ii) of Lemma 5.42 would by Lemma 4.39 contradict the assumption that $\mathcal{T}_0$ does not control a minor of H in G_0. □

Similarly, iterating this lemma and combining it with Corollary 4.33 enables us to eliminate the genus increase from the outcome of the lemma.

Lemma 5.44 *For any graph H and any non-decreasing function $\psi : \mathbb{N}^2 \to \mathbb{N}$, there exist integers δ_0, ξ, and θ such that the following claim holds. Let G_0 be a graph with a tangle $\mathcal{T}_0$. If $(G_0, \mathcal{T}_0)$ contains a sphere-span $(G, \mathcal{T})$ of order θ and $\mathcal{T}_0$ does not control a minor of H in G_0, then there exists an H-avoiding surface Σ and a set $X \subseteq V(G_0)$ of size at most ξ such that for some $\delta \le \delta_0$, $G_0 - X$ has a solid $(\mathrm{cr}_\Sigma(H) - 1, \delta)$-simple bisegregation with a $(\mathcal{T}_0 - X, \psi(|X|, \delta))$-spread arrangement on Σ.*

Proof (⇝) If H is planar, then let θ be the value from Corollary 4.33 for the sphere. Then Corollary 4.33 implies that $\mathcal{T}_0$ controls a minor of H in G_0, contradicting the assumptions, and thus the claim of the lemma holds vacuously.

Hence, suppose that H is non-planar, and let $\mathcal{S}$ be the set of all H-avoiding surfaces (which includes at least the sphere). Note that $\mathcal{S}$ is finite, since H can be drawn on any surface of Euler genus at least $2\|H\|$. For each non-H-avoiding surface Π obtained from a surface belonging to $\mathcal{S}$ by adding a crosscap, a handle, or a crosshandle, let θ_Π be the value of θ from Corollary 4.33 applied with Π playing the role of Σ. We next process the surfaces $\Pi \in \mathcal{S}$ in non-increasing order of their Euler genus, and define δ_Π, ξ_Π, and θ_Π to be the values δ_0, ξ, and θ from Lemma 5.43 for H, Π playing the role of Σ, and $\theta_1 = \max(\theta_{\Pi+\text{crosscap}}, \theta_{\Pi+\text{handle}}, \theta_{\Pi+\text{crosshandle}})$. Let $\theta = \theta_{\text{sphere}}$, $\xi = \max\{\xi_\Pi : \Pi \in \mathcal{S}\}$ and $\delta_0 = \max\{\delta_\Pi : \Pi \in \mathcal{S}\}$.

Let $\Sigma \in \mathcal{S}$ be a surface of largest Euler genus such that $(G_0, \mathcal{T}_0)$ contains a Σ-span $(G', \mathcal{T}')$ of order at least θ_Σ (such a surface exists, since the sphere satisfies this condition by the assumptions). We apply Lemma 5.43 to $(G', \mathcal{T}')$. The outcome (ii) corresponds to the outcome of the current lemma. If the outcome (i) holds, then let $\Pi \in \{\Sigma + \text{crosscap}, \Sigma + \text{handle}, \Sigma + \text{crosshandle}\}$ be a surface such that $(G_0, \mathcal{T}_0)$ contains a Π-span of order $\theta_1 \ge \theta_\Pi$. The surface Π is not H-avoiding by the maximality of the Euler genus of Σ, and thus Corollary 4.33 implies that $\mathcal{T}_0$ controls a minor of H in G_0, which is a contradiction. □

The Local Minor Structure Theorem (whose statement we restate for convenience) now easily follows.

Theorem 5.30 *For any graph H, and any non-decreasing function $\psi : \mathbb{N}^2 \to \mathbb{N}$, there exist integers $\alpha, \delta_0 \ge 0$ and $\theta > \alpha$ such that the following claim holds. Suppose $\mathcal{T}$ is a tangle of order at least θ in a graph G. If $\mathcal{T}$ does not control a minor H of G, then there exists a set $A \subseteq V(G)$ of size at most α such that for some $\delta \le \delta_0$, $G - A$ has a solid $(\mathrm{cr}_\Sigma(H) - 1, \delta)$-simple bisegregation with a $(\mathcal{T} - A, \psi(|A|, \delta))$-spread arrangement on an H-avoiding surface Σ.*

Proof Let δ_0, α, and θ_0 be the values of δ_0, ξ, and θ from Lemma 5.44 applied for H and ψ. Let $\theta = f_{2.38}(2\theta_0)$. By Corollary 2.38, there exists a wall subdivision $W \subseteq G$ and a tangle $\mathcal{T}_W$ of order θ_0 in W conformal with $\mathcal{T}$. Note that the unique drawing of W in the sphere is rigid, and thus $(W, \mathcal{T}_W)$ is a sphere-span of order θ_0 in $(G, \mathcal{T})$. The claim then follows from Lemma 5.44. □

References

1. H.Y. Cheung, L.C. Lau, K.M. Leung, Algebraic algorithms for linear matroid parity problems. ACM Trans. Algorithms **10**, 1–26 (2014)
2. E.D. Demaine, M. Hajiaghayi, Equivalence of local treewidth and linear local treewidth and its algorithmic applications, in *Proceedings of the Fifteenth Annual ACM-SIAM Symposium on Discrete Algorithms* (Society for Industrial and Applied Mathematics, Philadelphia, 2004), pp. 840–849
3. H. Fleischner, G.J. Woeginger, Detecting cycles through three fixed vertices in a graph. Inf. Process. Lett. **42**, 29–33 (1992)
4. S. Fortune, J.E. Hopcroft, J. Wyllie, The directed subgraph homeomorphism problem. Theor. Comput. Sci. **10**, 111–121 (1980)
5. M. Grohe, K. Kawarabayashi, D. Marx, P. Wollan, Finding topological subgraphs is fixed-parameter tractable, in *Proceedings of the Forty-Third Annual ACM Symposium on Theory of Computing* (2011), pp. 479–488
6. M. Grohe, S. Kreutzer, S. Siebertz, Deciding first-order properties of nowhere dense graphs. J. ACM **64**, Art. No. 17, 1–32 (2017)
7. R. Karp, Reducibility among combinatorial problems, in *Complexity of Computer Computations*, ed. by R. Miller, J. Thatcher (Plenum Press, New York, 1972), pp. 85–103
8. K. Kawarabayashi, Y. Kobayashi, B. Reed, The disjoint paths problem in quadratic time. J. Combin. Theory Ser. B **102**, 424–435 (2012)
9. N. Robertson, P.D. Seymour, Graph minors. IX. Disjoint crossed paths. J. Combin. Theory Ser. B **49**, 40–77 (1990)
10. N. Robertson, P.D. Seymour, Graph minors. XIII. The disjoint paths problem. J. Combin. Theory Ser. B **63**, 65–110 (1995)
11. N. Robertson, P.D. Seymour, Graph minors: XVII. Taming a vortex. J. Combin. Theory Ser. B **77**, 162–210 (1999)
12. N. Robertson, P.D. Seymour, Graph minors. XVI. Excluding a non-planar graph. J. Combin. Theory Ser. B **89**, 43–76 (2003)
13. N. Robertson, P.D. Seymour, Graph minors. XXII. Irrelevant vertices in linkage problems. J. Combin. Theory Ser. B **102**, 530–563 (2012)
14. W.T. Tutte, *Graph Theory* (Addison-Wesley, Boston, 1984)
15. M.E. Watkins, D.M. Mesner, Cycles and connectivity in graphs. Can. J. Math. **19**, 1319–1328 (1967)
16. H. Whitney, 2-isomorphic graphs. Am. J. Math. **55**, 245–254 (1933)

Chapter 6
Pointers and Sources

Even with a lot of details omitted, the proof of the Minor Structure Theorem is lengthy and somewhat hard to grasp, and I expect that on the first reading, it may be difficult to distinguish important parts from technicalities. To aid the reader, let me point out what I consider to be the parts that need to be understood in order to appreciate the Minor Structure Theorem and to be able to apply it efficiently:

- Almost everything in Chap. 2 is a foundational material that should be known by anyone interested in graph minors theory or even structural graph theory in general. Essentially the only part that can be safely skipped is the proof of the Grid Theorem.
- From Chap. 3, Theorem 3.1 on linking through a clique minor and its consequences, including the linkedness in high-connectivity graphs, are frequently useful. Linked tree decompositions are important in many applications of the Minor Structure Theorem. The reader should be aware of the Unique Linkage Theorem and understand its application shown in Lemma 3.12, though one sees it used much less often.
- From Chap. 4, understanding the metric arising from respectful tangles is a must. Theorem 4.27 and its applications to obtain minors in (nearly) embedded graphs are the most important results of this chapter, and are used in the applications whenever we need to find minors or topological minors in the graphs that satisfy the outcome of the (Local) Minor Structure Theorem. The trick of linking through a cylindrical grid is also frequently useful in this context.
  The reader does not need to fully understand the proof of Theorem 4.27, but definitely should be able to use it to obtain the results analogous to those in Sect. 4.5, of the form "suppose G_0 has a subgraph G drawn on a surface Σ; if H is almost drawn on Σ and the parts of H that are not drawn can be realized by pieces of the ambient graph G_0 that freely attach to distant locations in G, then H is a minor of G_0". Mader's H-Wege theorem and its applications to disentangle the non-embedded parts need to be understood to appreciate the proof of the Minor Structure Theorem, but they are not as often encountered in applications.

© The Author(s), under exclusive license to Springer Nature Switzerland AG 2025

Z. Dvořák, *Graph Minors*, Springer Monographs in Mathematics,
https://doi.org/10.1007/978-3-031-87469-7_6

- From Chap. 5, Lemma 5.4 and Corollary 5.9 are sometimes useful in applications, but they are more important as stepping stones towards the Minor Structure Theorem. The reader should definitely understand the statement of the Flat Wall Theorem (understanding its proof is much less important) and be able to apply it similarly to the examples shown in Sect. 5.2. They should be aware that it is possible to test for a fixed graph whether it is a (rooted) minor of an arbitrary input graph in polynomial time; the details of the algorithm can be safely ignored most of the time.
  Lemma 5.22 and Theorem 5.24 are sometimes important in the applications of the Minor Structure Theorem in cases where we need to alter the structure of the vortices. The local forms of the Minor Structure Theorem (Theorems 5.27 and 5.30) are very important and their statements have to be understood by the reader; they will be needed throughout the next part of the book. One should also understand how the global Minor Structure Theorem arises from them, and how they give the form of the Minor Structure Theorem stated in Theorem 5.29.
  Sections 5.6 and 5.7 that give the main ideas of the proof of the Local Minor Structure Theorem can be safely ignored.

6.1 Sources

Many of the results in Chap. 2 are folklore. The presentation of tangles is largely based on the fundamental paper of Robertson and Seymour [7] (except for my general preference for vertex tangles); I highly encourage everyone interested in the topic of graph minors to read the paper in detail, as it introduces important ideas with many subsequent developments. An excellent overview of the important concepts and proofs of the basic results can also be found in Diestel's book [3]. For more substantial proofs:

- The proof of the Grid Theorem presented in Sect. 2.10 combines ideas of the proofs in [5, 14].
- The proof of the Grid Theorem in planar graphs is from [14].
- The radius-based bound on the treewidth of graphs of bounded genus is essentially taken from the paper Dujmović et al. [4], whose authors acknowledge inspiration from previous proofs of the result.
- The proof of the forbidden tree theorem (Theorem 2.98) is an adaptation of the argument of Diestel [2].

In Chap. 3, the presentation of Theorem 3.1 on linking through a clique minor is based on [10]. The proof of Lemma 3.16 is inspired by the paper of Böhme et al. [1]. Other arguments are generally extracted from the Graph Minors series of papers by Robertson and Seymour.

Chapter 4 is largely based on the Graph Minors series [6, 8, 9]. One part where we deviate is that we skip the intermediate step of proving the "tangle-free" version of Theorem 4.28 as done in [6], thus avoiding many of the technicalities (although since we only provide a vague sketch of the proof of the theorem anyway, this may not be very clear to the reader). The results on (non)disjoint eyes, jumps, and spiders loosely mirror those in [10], though we use simpler arguments leveraging Mader's H-Wege theorem.

The presentation in Chap. 5 generally follows the Graph Minors series [10–13], though we chose a quite different formulation of the Flat Wall Theorem.

References

1. T. Böhme, K. Kawarabayashi, J. Maharry, B. Mohar, Linear connectivity forces large complete bipartite minors. J. Combin. Theory Ser. B **99**, 557–582 (2009)
2. R. Diestel, Graph minors I: a short proof of the path-width theorem. Combin. Probabil. Comput. **4**, 27–30 (1995)
3. R. Diestel, *Graph Theory*. Graduate Texts in Mathematics, vol. 173, 5th edn. (Springer, Heidelberg, 2016)
4. V. Dujmović, P. Morin, D.R. Wood, Layered separators in minor-closed graph classes with applications. J. Combin. Theory Ser. B **127**, 111–147 (2017)
5. A. Leaf, P.D. Seymour, Tree-width and planar minors. J. Combin. Theory Ser. B **111**, 38–53 (2015)
6. N. Robertson, P.D. Seymour, Graph minors VII. Disjoint paths on a surface. J. Combin. Theory Ser. B **45**, 212–254 (1988)
7. N. Robertson, P.D. Seymour, Graph minors. X. Obstructions to tree-decomposition. J. Combin. Theory Ser. B **52**, 153–190 (1991)
8. N. Robertson, P.D. Seymour, Graph minors. XI. Circuits on a surface. J. Combin. Theory Ser. B **60**, 72–106 (1994)
9. N. Robertson, P.D. Seymour, Graph minors. XII. Distance on a surface. J. Combin. Theory Ser. B **64**, 240–272 (1995)
10. N. Robertson, P.D. Seymour, Graph minors. XIII. The disjoint paths problem. J. Combin. Theory Ser. B **63**, 65–110 (1995)
11. N. Robertson, P.D. Seymour, Graph minors. XIV. Extending an embedding. J. Combin. Theory Ser. B **65**, 23–50 (1995)
12. N. Robertson, P.D. Seymour, Graph minors. XV. Giant steps. J. Combin. Theory Ser. B **68**, 112–148 (1996)
13. N. Robertson, P.D. Seymour, Graph minors. XVI. Excluding a non-planar graph. J. Combin. Theory Ser. B **89**, 43–76 (2003)
14. N. Robertson, P.D. Seymour, R. Thomas, Quickly excluding a planar graph. J. Combin. Theory Ser. B **62**, 323–348 (1994)

Part II
Using the Structure Theorem

The Minor Structure Theorem is immensely useful in proving properties of graphs with a fixed forbidden minor and in design of efficient algorithms for such graphs. In this part, we give several examples that should teach the reader enough to be able to apply the Theorem by themselves. The development of most of the applications of the Minor Structure Theorem follows the basic "Five-Step Program":

1. Prove the result for graphs of bounded treewidth. While this is essentially a preparatory step, it gives an insight into the necessary inductive or recursive argument necessary to deal with small separations. Often, this leads to a formulation of a stronger induction hypothesis needed for this purpose; e.g., for coloring problems, postulating that any precoloring of the vertices of the root separator extends to a proper coloring. Moreover, ideas from this step can be useful later when we deal with the vortices.
2. Prove the result for planar graphs and graphs of bounded genus. In this part, we may use techniques based on results for graphs on surfaces we have seen in Sect. 2.12 or Chap. 4, or other general techniques for such graphs, such as discharging arguments.
3. Deal with vortices. This is often the most tricky part, forcing us to handle the interactions between the surface part and the bounded pathwidth part. A common idea is to find a way to separate the distant vertices of the surface from the vortices, exploiting the fact that the vortices together with the subgraph of the surface part near them have bounded treewidth.
4. Deal with apex vertices. This part is usually comparatively straightforward, especially because when adapting the vortex part of the argument, we can often just pretend that the apex vertices belong to all vortices.
5. Finish the proof by an inductive argument dealing with the clique-sums of graphs nearly drawn on surfaces. The ideas from the first step are often sufficient to achieve this.

Of course, there are many variations on this scheme, some of which we are going to demonstrate in this part of the book.

Chapter 7
Low-Treewidth Colorings

Let us start with a relatively simple application of the Minor Structure Theorem, leveraging just its global form stated in Theorem 5.29. A ***layering*** of a graph G is a function $\lambda : V(G) \to \mathbb{N}$ such that for every edge $uv \in E(G)$, we have $|\lambda(u) - \lambda(v)| \leq 1$. That is, λ divides $V(G)$ into layers $\lambda^{-1}(0)$, $\lambda^{-1}(1)$, ..., so that the ends of any edge are either in the same layer or in two consecutive layers. The most natural example is a ***BFS layering***, obtained by selecting a vertex arbitrarily from each component of G and letting $\lambda(v)$ for $v \in V(G)$ be the distance of v from the selected vertices. As our starting point, let us consider a reformulation of Corollary 2.74 in terms of BFS layerings.

Lemma 7.1 *Let G be a graph of Euler genus at most g and let λ be a BFS layering of G. Let ℓ be a positive integer and suppose that a set $L \subset \mathbb{N}$ does not contain more than ℓ consecutive integers. Then* $\mathrm{tw}(G[\lambda^{-1}(L)]) \leq (2g + 3)\ell$.

That is, the union of any ℓ consecutive layers induces a graph whose treewidth is bounded by $f_{\mathcal{G}}(\ell)$ for a function $f_{\mathcal{G}}$ depending only on the considered class $\mathcal{G}$ of graphs (in the case of Lemma 7.1, $\mathcal{G}$ is the class of graphs of Euler genus at most g).

Not all proper minor-closed graph classes have this property: Observe that if a graph G contains a universal vertex u adjacent to all other vertices of G, then any layering of G has at most three non-empty layers, since every vertex must belong either to the layer of u or to one of the two neighboring layers. In particular, if such a graph G has the property that the union of any ℓ consecutive layers induces a graph whose treewidth is bounded by $f_{\mathcal{G}}(\ell)$, then the treewidth of G must be bounded by the constant $f_{\mathcal{G}}(3)$. However, the class of K_6-minor-free graphs contains the graphs obtained from grids by attaching a universal vertex, and these graphs can have arbitrarily large treewidth.

Is there some relaxed version of the property that holds for all proper minor-closed classes? Lemma 7.1 has the following easy corollary, obtained by defining $\gamma(v) = (\lambda(v) \bmod k) + 1$.

© The Author(s), under exclusive license to Springer Nature Switzerland AG 2025

Z. Dvořák, *Graph Minors*, Springer Monographs in Mathematics,
https://doi.org/10.1007/978-3-031-87469-7_7

Corollary 7.2 *For every graph G of Euler genus at most g and every positive integer k, there exists a function $\gamma : V(G) \to [k]$ such that for every $i \in [k]$,*

$$\mathrm{tw}(G - \gamma^{-1}(i)) \leq (2g+3)(k-1).$$

We say that a (not necessarily proper) coloring $\gamma : V(G) \to [k]$ of a graph G by k colors is a *b-low-treewidth* coloring if the deletion of any color class results in a graph of treewidth at most b. In particular, Corollary 7.2 states that for every positive integer k, every graph of Euler genus at most g has a $(2g+3)(k-1)$-low-treewidth coloring by k colors.

We say that a class of graphs $\mathcal{G}$ is ***treewidth-fragile*** if there exists a *bounding function* $f : \mathbb{N} \to \mathbb{N}$ such that for every positive integer k, every graph $G \in \mathcal{G}$ has an $f(k)$-low-treewidth coloring by k colors. For example, Corollary 7.2 shows that the class of graphs of Euler genus at most g is treewidth-fragile with bounding function $f(k) = (2g+3)(k-1)$.

In contrast to bounding the treewidth of consecutive layers in a layering, addition of a bounded number a of universal vertices does not cause any issues for treewidth-fragility: We can color the universal vertices arbitrarily, and this can only increase the treewidth (after deletion of a color class) by at most a.

Observation 7.3 *Suppose $\mathcal{G}$ is a treewidth-fragile class of graphs with bounding function f. Let a be a positive integer and let $\mathcal{G}'$ be the class of graphs G such that $G - A \in \mathcal{G}$ for some set $A \subseteq V(G)$ of size at most a. Then $\mathcal{G}'$ is treewidth-fragile with bounding function $g(k) = f(k) + a$.*

Hence, it is reasonable to ask whether all proper minor-closed classes are treewidth-fragile. Let us show that this is indeed the case.

7.1 Treewidth-Fragility of Minor-Closed Classes

The first step of the Five Step Program, to show treewidth-fragility of graphs of bounded treewidth, is of course trivial. However, we should mention that a somewhat stronger version of Corollary 7.2 holds in this case.

Lemma 7.4 *Every graph G has a coloring γ by $\mathrm{tw}(G)+1$ colors such that for every $k \in [\mathrm{tw}(G)+1]$, the union of any k color classes induces a subgraph of G of treewidth at most $k-1$.*

Proof Let (T, β) be a tree decomposition of G of width $\mathrm{tw}(G)$. Let G' be the supergraph of G with $V(G') = V(G)$ and with two vertices adjacent if and only if they appear in the same bag of (T, β); i.e., we replace every bag of (T, β) by a clique. Note that (T, β) is also a tree decomposition of G', and thus $\mathrm{tw}(G') \leq \mathrm{tw}(G)$. By Lemma 2.10, G' is $\mathrm{tw}(G)$-degenerate, and thus it has a proper coloring γ using at most $\mathrm{tw}(G)+1$ colors.

Consider the union S of some k color classes of γ, and let $\beta_S(x) = \beta(x) \cap S$ for every $x \in V(T)$. Then (T, β_S) is a tree decomposition of $G[S]$. Moreover, by the definition of G' and γ, each bag of this decomposition contains at most one vertex in each of the color classes, and thus it has size at most k. Therefore, the treewidth of $G[S]$ is at most $k - 1$. □

Let us remark that the coloring from the statement of Lemma 7.4 γ is a proper coloring of G, since each color class induces a subgraph of treewidth zero, i.e., an independent set. Moreover, γ is also an ***acyclic coloring*** of G, i.e., it does not contain any bicolored cycle, since the union of any two color classes induces a subgraph of treewidth at most one.

Let us return from our diversion to considering how the Minor Structure Theorem can be used to prove treewidth-fragility of proper minor-closed graph classes. The second step of the Five Step Program, concerning graphs on surfaces, is dealt with in Corollary 7.2. Its proof via the consideration of BFS layerings gives us a guidance for the third step, dealing with vortices. We are going to need the following lemma on integrating the vortices into a tree decomposition of the surface part.

Lemma 7.5 *Suppose that a graph G_0 is drawn on a surface up to vortices $F_1, \ldots, F_m$ of width at most d. Let G be the surface part of G_0 and suppose that for each $i \in [m]$, ∂F_i forms a cycle in G. Then $\operatorname{tw}(G_0) \le (d+1)\operatorname{tw}(G) + d$.*

Proof Let (T, β) be a tree decomposition of G of width $\operatorname{tw}(G)$. For $i \in [m]$, let $(\partial F_i, \beta_i)$ be a vortical decomposition of F_i of width at most d. For $v \in V(G)$, let $\tau(v) = \beta_i(v)$ if $v \in \partial F_i$ for some (unique) $i \in [m]$, and let $\tau(v) = \{v\}$ otherwise. For $x \in V(T)$, let $\beta_0(x) = \bigcup_{v \in \beta(x)} \tau(v)$. We claim that (T, β_0) is a tree decomposition of G_0:

- Consider any edge $e = uv$ of G_0. If $e \in E(G)$, then by (D1) for (T, β), there exists $x \in V(T)$ such that $\{u, v\} \subseteq \beta(x) \subseteq \beta_0(x)$. Otherwise, there exists (unique) $i \in [m]$ such that $e \in E(F_i)$, and thus there exists $z \in \partial F_i$ such that $\{u, v\} \subseteq \beta_i(z) = \tau(z)$. Let x be a node of T such that $z \in \beta(x)$; then $\{u, v\} \subseteq \tau(z) \subseteq \beta_0(x)$. Hence, (T, β_0) satisfies (D1).
- Consider any vertex $v \in V(G_0)$. If $v \in V(G)$, then $S_1 = \beta^{-1}(v)$ induces a connected subtree of T; otherwise, let $S_1 = \emptyset$. If there exists a (unique) $i \in [m]$ such that $v \in V(F_i)$, then $P = \beta_i^{-1}(v)$ induces a connected subgraph of the cycle ∂F_i, and by the assumptions, P also induces a connected subgraph of G; by Lemma 2.2, $S_2 = \beta^{-1}(V(P))$ induces a connected subgraph of T. If no such i exists, then let $S_2 = \emptyset$. Note that if S_1 and S_2 are both non-empty, then $v \in \partial F_i$, $v \in P$, and $S_1 \subseteq S_2$. We conclude that $\beta_0^{-1}(v) = S_1 \cup S_2$ induces a connected subtree of T. Therefore, (T, β_0) satisfies (D2).

Finally, note that for each node $x \in V(T)$, we have

$$|\beta_0(x)| \le |\beta(x)| \cdot (d+1) \le (\operatorname{tw}(G)+1)(d+1),$$

and consequently the width of the tree decomposition (T, β_0) of G_0 is at most $(d+1)\,\mathrm{tw}(G) + d$. □

Let us remark that Lemma 7.5 gives a bound on how large clique minors can appear in nearly drawn graphs.

Corollary 7.6 *Suppose that a graph G_0 is drawn on a surface Σ of Euler genus g up to vortices $F_1, \ldots, F_m$ of width at most d. If K_k is a minor of G_0, then $k \le (d+1)(2g+4m)+d+g+6$. Hence, the class of graphs (a, m, d)-nearly drawn on surfaces of Euler genus at most g has Hadwiger number at most $(d+1)(2g+4m)+d+g+a+6$.*

Proof Let G be the surface part of G_0. Without loss of generality, we can assume that for $i \in [m]$, ∂F_i forms a cycle in G bounding a face f_i (otherwise, we can add such a cycle around the boundary of the disk in Σ representing F_i). Let μ be a model of a clique K in G_0. We can delete from G all vertices contained neither in the model nor in the vortex boundaries. Moreover, suppose that $e \in E(G)$ is an edge of G belonging to $E(\mu(x))$ for some $x \in V(K)$; if e does not have both ends in vortex boundaries, then we can without loss of generality contract e. Therefore, letting Z be the set of vertices of G not contained in the vortex boundaries, we conclude that for every $z \in Z$, there exists $x \in V(K)$ such that $V(\mu(x)) = \{z\}$. In particular, $G[Z]$ is a clique drawn on Σ, and thus $|Z| \le \frac{7+\sqrt{24g+1}}{2} \le g+5$ by Heawood's formula.

Every vertex of $G - Z$ is incident with one of the faces $f_1, \ldots, f_m$, and thus by Corollary 2.77, we have $\mathrm{tw}(G-Z) \le 2g+4m$ (we removed the -1 from the bound so that it is correct even if $m = 0$ and $g = 0$). Note that $G_0 - Z$ is drawn on Σ with vortices $F_1, \ldots, F_m$ and with the surface part $G - Z$, and thus Lemma 7.5 gives $\mathrm{tw}(G_0 - Z) \le (d+1)(2g+4m)+d$. Hence, $\mathrm{tw}(G_0) \le \mathrm{tw}(G_0 - Z) + |Z| \le (d+1)(2g+4m)+d+g+5$. Since K is a minor of G_0, we have $|K| - 1 = \mathrm{tw}(K) \le \mathrm{tw}(G_0)$, giving us the required bound. □

In combination with the layering argument for the surface part, Lemma 7.5 easily implies treewidth-fragility for outgrowths by vortices.

Lemma 7.7 *Let g and d be non-negative integers and let $\mathcal{G}$ be the class of graphs drawn on surfaces of Euler genus at most g up to vortices of width at most d. Then $\mathcal{G}$ is treewidth-fragile with bounding function $f(k) = (d+1)(2g+3)(5k-1)+d$.*

Proof Consider a graph $G_0 \in \mathcal{G}$, drawn on a surface of Euler genus Σ at most g up to vortices $F_1, \ldots, F_m$, and let G be the surface part of G_0. Let $\Delta_1, \ldots, \Delta_m \subset \Sigma$ be pairwise disjoint disks such that for $i \in [m]$, Δ_i intersects G exactly in the vertices of ∂F_i and these vertices are drawn in the boundary of Δ_i in order. Let G' be the graph obtained from G by adding for each $i \in [m]$

- a cycle C_i with vertices ∂F_i in order, such that C_i is drawn along the boundary of the disk Δ_i, and
- a vertex u_i drawn in the interior of Δ_i and adjacent to all vertices of ∂F_i.

Let λ be a BFS layering of G', and for each $i \in [m]$ and $v \in V(F_i) \setminus \partial F_i$, let $\lambda(v) = \lambda(u_i)$. Since u_i is adjacent to all vertices of C_i and λ is a BFS layering of G', we have $|\lambda(x) - \lambda(u_i)| \leq 1$ for every $x \in V(C_i) \cup \{u_i\}$, and thus $|\lambda(v) - \lambda(u_i)| \leq 1$ for every $v \in V(F_i)$. Therefore,

($\star$) For every $i \in F_i$ and $v, v' \in V(F_i)$, we have $|\lambda(v) - \lambda(v')| \leq 2$.

Consider now any positive integer k. For each $v \in V(G_0)$, let us define $\lambda'(v) = \lfloor \lambda(v)/5 \rfloor$ and $\gamma(v) = \lambda'(v) \bmod k$. We are going to show that γ is an $f(k)$-low-treewidth coloring.

Consider any color $c \in \{0, \ldots, k-1\}$ and let $H_0 = G_0 - \gamma^{-1}(c)$. We need to show that $\operatorname{tw}(H_0) \leq f(k)$. To do so, we are actually going to bound a treewidth of a supergraph of H_0. Let $B = \{i \in [m] : V(F_i) \not\subseteq \gamma^{-1}(c)\}$ be the set of indices of the vortices that are not colored entirely by the color c, let $H_0' = H_0 \cup \bigcup_{i \in B}(C_i \cup F_i)$, and let H' be obtained from $H_0 \cap G$ (i.e., the surface part of H_0) by adding the cycles C_i for $i \in B$.

Let L be the set of non-negative integers j such that $\lfloor j/5 \rfloor \bmod k \neq c$. Note that L does not contain more than $5(k-1)$ consecutive integers, and any two maximal segments of consecutive integers in L are separated by a gap of 5 consecutive integers not belonging to L. Let L' be the set of non-negative integers j' such that there exists $j \in L$ satisfying $|j' - j| \leq 2$; then L' does not contain more than $5k - 1$ consecutive integers. Note that a vertex $v \in V(G_0)$ belongs to $V(H_0)$ if and only if $\lambda(v) \in L$. Consider any vertex $v \in V(H_0') \setminus V(H_0)$. Then v belongs to a vortex F_i for some $i \in B$, i.e., such that F_i also contains a vertex v' whose color is not c, and thus $\lambda(v') \in L$. By ($\star$), we have $|\lambda(v) - \lambda(v')| \leq 2$, and thus $\lambda(v) \in L'$. In particular, $V(H') \subseteq \lambda^{-1}(L')$, and Lemma 7.1 implies $\operatorname{tw}(H') \leq (2g+3)(5k-1)$.

Note that H_0' is drawn on Σ up to vortices F_i for $i \in B$, the surface part of H_0' is H', and for each $i \in B$, ∂F_i forms a cycle C_i in H'. By Lemma 7.5, this implies

$$\operatorname{tw}(H_0) \leq \operatorname{tw}(H_0') \leq (d+1)\operatorname{tw}(H') + d \leq (d+1)(2g+3)(5k-1) + d = f(k),$$

as required. □

Altough this is not needed for our application, let us remark that Lemmas 7.5 and 7.7 apply even in the situation where the number of vortices is not bounded. We have already dealt with apices in Observation 7.3, thus finishing the fourth step of the Five Step Program. Note that Observation 7.3 together with Lemma 7.7 show that the basic building blocks of the Minor Structure Theorem are treewidth-fragile.

Corollary 7.8 *For all non-negative integers a, m, d, and g, the class of graphs (a, m, d)-nearly drawn on surfaces of Euler genus at most g is treewidth-fragile with bounding function $f(k) = (d+1)(2g+3)(5k-1) + d + a$.*

For the last step of the Five Step Program, we are going to need to prove a slight strengthening of treewidth-fragility. We say that a class $\mathcal{G}$ of graphs is *strongly treewidth-fragile* with bounding function f if for every graph $G \in \mathcal{G}$, every clique

K in G, and every positive integer k, every function $\gamma_0 : V(K) \to [k]$ extends to an $f(k)$-low-treewidth coloring of G by k colors.

Observation 7.9 *Let $\mathcal{G}$ be a class of graphs of maximum clique number ω. If $\mathcal{G}$ is treewidth-fragile with bounding function f, then $\mathcal{G}$ is also strongly treewidth-fragile with bounding function $f'(k) = f(k) + \omega$.*

Proof Consider any clique K in a graph $G \in \mathcal{G}$ and an $f(k)$-low-treewidth coloring $\gamma : V(G) \to [k]$ of G. Let γ_1 be obtained from γ by recoloring the vertices of K in any way. For any color $c \in [k]$, the graph $G - \gamma_1^{-1}(c)$ is obtained from a subgraph of $G - \gamma^{-1}(c)$ by adding at most $|K| \le \omega$ vertices, and thus $\mathrm{tw}(G - \gamma_1^{-1}(c)) \le f(k) + \omega = f'(k)$. Therefore, γ_1 is an $f'(k)$-low-treewidth coloring of G. □

Note that the assumption on the maximum clique number is not very restrictive, because of the following observation.

Observation 7.10 *If a class $\mathcal{G}$ of graphs is treewidth-fragile with bounding function f, then every graph $G \in \mathcal{G}$ has clique number at most $2f(2) + 2$.*

Proof Consider an $f(2)$-low-treewidth coloring $\gamma : V(G) \to [2]$ of G, and let K be any clique in G. There exists a color $c \in [2]$ such that at most half of the vertices of K have color c, and thus by Lemma 2.3, we have

$$|K|/2 \le \mathrm{tw}(G - \gamma^{-1}(c)) + 1 \le f(2) + 1.$$

Therefore, $\omega(G) \le 2f(2) + 2$. □

Thus, every treewidth-fragile class of graphs is also strongly treewidth-fragile. However, unlike treewidth-fragility, strong treewidth-fragility is preserved by clique-sums with unchanged bounding function.

Lemma 7.11 *If $\mathcal{G}$ is a strongly treewidth-fragile class of graphs with bounding function f, then the class of all graphs $\mathcal{G}'$ obtained from those in $\mathcal{G}$ by clique-sums is also strongly treewidth-fragile with bounding function f.*

Proof We prove that the condition of strong treewidth-fragility with bounding function f holds for all graphs $G \in \mathcal{G}'$ by induction on the number of vertices of G.

If $G \in \mathcal{G}$, then this is true by the assumptions. Otherwise, G is a clique-sum of smaller graphs $G_1, G_2 \in \mathcal{G}'$ on a clique K'. Let K be a clique in the graph G, let k be a positive integer, and let $\gamma_0 : V(K) \to [k]$ be an arbitrary function. Note that each clique in G is also a clique in G_1 or G_2, and by symmetry, we can assume that K is a clique in G_1. By the induction hypothesis, γ_0 extends to an $f(k)$-low-treewidth coloring γ_1 of G_1. Let γ_0' be the restriction of γ_1 to K'. By the induction hypothesis applied to the graph G_2, γ_0' extends to an $f(k)$-low-treewidth coloring γ_2 of G_2. Let γ be the union of γ_1 and γ_2. For every color $c \in [k]$, observe that $G - \gamma^{-1}(c)$ is a clique-sum of the graphs $G_1 - \gamma_1^{-1}(c)$ and $G_2 - \gamma_2^{-1}(c)$, and thus

$$\mathrm{tw}(G - \gamma^{-1}(c)) = \max(\mathrm{tw}(G_1 - \gamma_1^{-1}(c)), \mathrm{tw}(G_2 - \gamma_2^{-1}(c))) \le f(k).$$

Moreover, γ extends γ_0. □

The Minor Structure Theorem (in its simplest form, Theorem 2.9) and the partial results we obtained throughout this section now easily combine to show that proper minor-closed classes are treewidth-fragile.

Theorem 7.12 *For every graph H, the class $\mathcal{G}$ of H-minor-free graphs is (strongly) treewidth-fragile with a bounding function $f(k) = O(k)$.*

Proof Let g be the largest Euler genus of an H-avoiding surface, or $g = 0$ if H is planar. By Theorem 2.9, there exist integers a, d, and m such that every graph in $\mathcal{G}$ is obtained by clique-sums from graphs that can be (a, m, d)-nearly drawn on a surface of Euler genus at most g. By Corollaries 7.6 and 7.8 and Observation 7.9, the class of graphs that can be (a, m, d)-nearly drawn on a surface of Euler genus at most g is strongly treewidth-fragile with bounding function

$$f(k) = (d + 1)(2g + 3)(5k - 1) + (d + 1)(2g + 4m) + 2d + g + 2a + 6.$$

By Lemma 7.11, it follows that $\mathcal{G}$ is strongly treewidth-fragile with the same bounding function. □

7.2 Applications of Treewidth-Fragility

In the rest of this chapter, let us discuss several consequences of Theorem 7.12. Analogously to Corollary 2.76, Theorem 7.12 implies that for a fixed graph H, every H-minor-free graph has treewidth $O(\sqrt{|G|})$. Indeed, the proof of Corollary 2.76 actually does not use the layering results directly, but rather depends only on Lemma 2.75, which asserts a condition substantially weaker than treewidth-fragility. Let us say that a class of graphs $\mathcal{G}$ is ***fractionally treewidth-fragile*** with bounding function f if for every graph $G \in \mathcal{G}$, every assignment w of non-negative weights to vertices of G, and every positive integer k, there exists a set $S \subseteq V(G)$ such that $w(S) \leq \frac{1}{k}w(G)$ and $\mathrm{tw}(G - S) \leq f(k)$. Clearly, if $\gamma : V(G) \to [k]$ is an $f(k)$-low-treewidth coloring, then the color class S of smallest weight has this property, and thus the following claim holds.

Observation 7.13 *If a class of graphs is treewidth-fragile, then it is also fractionally treewidth-fragile with the same bounding function.*

The argument from the proof of Corollary 2.76 now gives the following bound on treewidth of fractionally treewidth-fragile graphs.

Lemma 7.14 *Suppose the class of graphs $\mathcal{G}$ is fractionally treewidth-fragile with bounding function f, where $f(1) = 0$ without loss of generality. Suppose that a graph $G \in \mathcal{G}$ has n vertices, and let k be the largest positive integer such that $kf(k) \leq n$. Then $\mathrm{tw}(G) \leq \frac{2n}{k}$.*

Proof Consider the function w assigning to each vertex the same weight 1. By the definition of fractional treewidth-fragility, there exists $S \subseteq V(G)$ such that $|S| = w(S) \leq n/k$ and $\mathrm{tw}(G - S) \leq f(k)$. Consequently,

$$\mathrm{tw}(G) \leq |S| + \mathrm{tw}(G - S) \leq \frac{n}{k} + f(k) \leq \frac{2n}{k}.$$

□

In particular, when $f(k) = O(k)$ as in Theorem 7.12, we have $k = \Omega(\sqrt{n})$.

Corollary 7.15 *For a fixed graph H, every H-minor-free graph G has treewidth $O(\sqrt{|G|})$, and in particular G has a balanced cut of size $O(\sqrt{|G|})$.*

Since the multiplicative constants in this corollary depend on the constants arising from the Minor Structure Theorem applied for H, they are rather large and non-explicit. Let us remark that, as we are going to see in Chap. 12, a substantially stronger and more explicit result can be obtained directly.

Treewidth-fragility is quite useful in design of approximation algorithms. Let us demonstrate this on a simple example of finding the largest *triangle matching number* $\mu_3(G)$, i.e., the largest number of pairwise vertex-disjoint triangles in the input graph G. Of course, for algorithmic applications, we need *efficient* treewidth-fragility, where given a graph G from the class and a positive integer k, an $f(k)$-low-treewidth coloring of G by k colors can be found in time $\mathrm{poly}(|G|)$. It is easy to see that the proof of Theorem 7.12 gives a polynomial-time algorithm (the decomposition from the Minor Structure Theorem can be found in quadratic time [4]).

Lemma 7.16 *Let $\mathcal{G}$ be an efficiently treewidth-fragile graph class. For every positive integer k, there exists a polynomial-time algorithm that, given a graph $G \in \mathcal{G}$, returns a triangle matching in G of size at least* $(1 - 1/k)\mu_3(G)$.

Proof Let f be the bounding function for the efficient treewidth-fragility of the class $\mathcal{G}$. Thus, we can find an $f(3k)$-low-treewidth coloring $\gamma : V(G) \to [3k]$ in polynomial time. For each $c \in [3k]$, we use Theorem 2.8 to find a triangle matching M_c in $G - \gamma^{-1}(c)$ of optimal size $\mu_3(G - \gamma^{-1}(c))$. We then return the largest of these triangle matchings.

Consider now an optimal triangle matching M in G of size $\mu_3(G)$. Let $c \in [3k]$ be a color chosen uniformly at random. Each vertex $v \in V(G)$ has probability $\frac{1}{3k}$ of having color c, and thus the probability that a triangle in G contains at least one vertex of color c is at most $\frac{1}{k}$. Hence, the expected number of triangles in M containing a vertex of color c is at most $\frac{1}{k}\mu_3(G)$.

Therefore, there exists a color $c_0 \in [3k]$ such that $G - \gamma^{-1}(c_0)$ contains all but at most $\frac{1}{k}\mu_3(G)$ of the triangles of M, and thus $\mu_3(G - \gamma^{-1}(c_0)) \geq (1 - 1/k)\mu_3(G)$. The triangle matching returned by the algorithm is at least as large as the optimal triangle matching M_{c_0} in $G - \gamma^{-1}(c_0)$, and thus it has size at least $(1 - 1/k)\mu_3(G)$, as required. □

Let us remark that (an efficient version) of fractional treewidth-fragility is also sufficient to obtain the conclusion of Lemma 7.16; indeed, as shown in [2], every maximization problem expressible in the first-order logic can be approximated arbitrarily well in this setting. Neither treewidth-fragility nor fractional treewidth-fragility seems enough to deal with minimization problems (such as the minimum dominating set); we are going to discuss a stronger notion overcoming this issue in Sect. 16.2.

Another application of treewidth-fragility concerns testing the existence of a fixed graph F as an induced or non-induced subgraph of an input graph G with n vertices. Without any additional constraints on the graph G, there is an $O\big(n^{\frac{\omega}{3}|H|}\big)$-time algorithm for this problem [7], where ω is the exponent in the time complexity of matrix multiplication. Moreover, since determining the clique number is $W[1]$-hard [1], it is unlikely that an algorithm with time complexity $O(f(H) \cdot n^c)$ for an absolute constant c (independent of H) exists, even if we restrict the graphs H to cliques. However, the situation is different in proper minor-closed classes, or more generally, in efficiently treewidth-fragile graph classes.

Lemma 7.17 *Let $\mathcal{G}$ be an efficiently treewidth-fragile graph class with bounding function f. There exists an algorithm that, given an arbitrary k-vertex graph H and an n-vertex graph $G \in \mathcal{G}$, decides in time $(f(k+1)+1)^{O(k)} \cdot n + \mathrm{poly}(n)$ whether H is an induced or non-induced subgraph of G.*

Proof By the efficient treewidth-fragility, we can in time $\mathrm{poly}(n)$ find an $f(k+1)$-low-treewidth coloring $\gamma : V(G) \to [k+1]$ of G. If H is isomorphic to a subgraph (or induced subgraph) of G, then there exists a color $c \in [k+1]$ that is not used on the vertices of this subgraph, and thus H also appears as a subgraph or an induced subgraph of $G - \gamma^{-1}(c)$. Hence, it suffices to solve the problem separately for the graphs $G - \gamma^{-1}(c)$ of bounded treewidth for $c \in [k+1]$.

Let G' be one of these graphs, and let (T, β) be a rooted tree decomposition of G' of width t. To determine whether a k-vertex graph H is a subgraph (or an induced subgraph) of G', we can use a straightforward dynamic programming algorithm with time complexity $(t+1)^{O(k)} \cdot |T|$: Without loss of generality, we can assume $t \geq 1$, as otherwise G' has no edges and the answer is yes if and only if H has no edges and $k \leq n$. For a node $x \in V(T)$ and an injective function $h : V(H) \to V(G')$, the x-*signature* of h is the function $h_x : V(H) \to (\beta(x) \cup \{\uparrow, \downarrow\})$ such that for $v \in V(H)$, $h_x(v) = h(v)$ if $h(v) \in \beta(x)$, $h(v) = \downarrow$ if $h(v) \notin \beta(x)$ but $h(v) \in \beta(T_x)$, where T_x is the subtree of T rooted in x, and $h_x(v) = \uparrow$ otherwise. Starting from the leaves, for every node $x \in V(T)$ and each of at most $(t+3)^k = (t+1)^{O(k)}$ possible x-signatures σ, we determine whether there exists a function $h : V(H) \to V(G')$ with x-signature σ mapping H to an isomorphic (induced) subgraph of G'. It is easy to see that using the information already computed for the children of x, this can be accomplished in time $(t+1)^{O(k)}$ for each node x, giving us the desired time complexity; see [5] for details. □

Let us however remark that a simpler and more efficient algorithm exists for a much more general family of graph classes with *bounded expansion*. To introduce these classes, let us note that treewidth-fragility in combination with Lemma 7.4 has the following consequence. A *treewidth-s coloring* of a graph G is a coloring such that for every $k \in [s]$, the union of any k color classes induces a subgraph of treewidth at most $k-1$.

Lemma 7.18 *For every treewidth-fragile graph class $\mathcal{G}$ and every positive integer s, there exists an integer m such that every graph from $\mathcal{G}$ has a treewidth-s coloring using at most m colors.*

Proof Let f be the treewidth-fragility bounding function for $\mathcal{G}$. Let $m = (s+1)\big(f(s+1)+1\big)^{s+1}$.

Consider a graph $G \in \mathcal{G}$. By treewidth-fragility, there exists an $f(s+1)$-low-treewidth coloring $\gamma_0 : V(G) \to [s+1]$. For each c, the subgraph $G - \gamma_0^{-1}(c)$ has treewidth at most $f(s+1)$, and thus by Lemma 7.4, it has a treewidth-s coloring γ_c by at most $f(s+1)+1$ colors. We extend γ_c to the whole graph G arbitrarily. For each vertex $v \in V(G)$, let $\gamma(v) = (\gamma_0(v), \gamma_1(v), \ldots, \gamma_{s+1}(v))$. This defines a coloring γ of G using at most m colors.

Consider any $k \le s$ colors $c_1, \ldots, c_k$, where $c_i = (c_{i,0}, \ldots, c_{i,s+1})$ for $i \in [k]$. Let $U = \gamma^{-1}(\{c_1, \ldots, c_k\})$. Since $k \le s$, there exists $a \in [s+1]$ such that $a \ne c_{i,0}$ for every $i \in [k]$, and thus $U \subseteq V(G) \setminus \gamma_0^{-1}(a)$. Let $U' = (V(G) \setminus \gamma_0^{-1}(a)) \cap \gamma_a^{-1}(\{c_{1,a}, \ldots, c_{k,a}\})$. Since U' is the union of at most k color classes of γ_a, it induces a subgraph of treewidth at most $k-1$. Moreover, $U \subseteq U'$, and thus $\operatorname{tw}(G[U']) \le \operatorname{tw}(G[U]) \le k-1$. Since this holds for the union of any $k \le s$ color classes, it follows that γ is a treewidth-s coloring of G. □

Let H be a graph with s vertices, and let γ be a treewidth-s coloring of a graph G. If γ uses only a bounded number of colors, we can use it to determine whether H is a subgraph of G efficiently: For each set S of $s = |H|$ colors, we check whether H appears in the bounded-treewidth subgraph $G[\gamma^{-1}(S)]$. This algorithm is correct, since if H appears as a subgraph of G, then γ assigns at most $|H|$ distinct colors to its vertices, and thus H is also a subgraph of $G[\gamma^{-1}(S)]$ for at least one choice of S. Thus, Lemma 7.18 gives another way to obtain the conclusion of Lemma 7.17.

However, the existence of treewidth-s colorings using a bounded number of colors is not restricted just to proper minor-closed classes (or even just to treewidth-fragile classes). We say that a class $\mathcal{G}$ of graphs has ***bounded expansion*** if for every positive integer s, there exists an integer m such that every graph from $\mathcal{G}$ has a treewidth-s coloring using at most m colors. It turns out that having bounded expansion is a much less restrictive property than being (fractionally) treewidth-fragile. For instance, all proper topological minor-closed classes are known to have bounded expansion. For every Δ, this includes the class of graphs of maximum degree at most Δ, which (for $\Delta \ge 3$) does not have sublinear separators, and thus by Lemma 7.14 is not fractionally treewidth-fragile. And interestingly, there exists a simple linear-time algorithm that finds a treewidth-s coloring (indeed, actually a treedepth-s coloring) using a bounded number of colors for any graph from a class with bounded expansion. This allows one to test presence of a fixed subgraph in linear time in a much more general setting than just for proper minor-closed classes, and completely avoids the usage of the Minor Structure Theorem. Saying more about the intriguing theory of bounded expansion classes is beyond the scope of this book and we refer an interested reader to [6] instead.

7.3 Approximation of the Chromatic Number

As a final application for treewidth-fragility, let us note that it implies a 2-approximation algorithm for chromatic number.

Lemma 7.19 *Let $\mathcal{G}$ be an efficiently treewidth-fragile graph class. There exists a polynomial-time algorithm that for an input graph $G \in \mathcal{G}$ returns a proper coloring of G using at most $2\chi(G)$ colors.*

Proof Let f be the treewidth-fragility bounding function for $\mathcal{G}$. Let $\gamma : V(G) \to [2]$ be an $f(2)$-low-treewidth coloring of G. For $c \in [2]$, we can use Theorem 2.8 to color the subgraph $G - \gamma^{-1}(c)$ optimally using $\chi(G - \gamma^{-1}(c)) \le \chi(G)$ colors in linear time. By using disjoint sets of colors on $G - \gamma^{-1}(1) = G[\gamma^{-1}(2)]$ and $G - \gamma^{-1}(2) = G[\gamma^{-1}(1)]$, we obtain a proper coloring of G using $\chi(G - \gamma^{-1}(1)) + \chi(G - \gamma^{-1}(2)) \le 2\chi(G)$ colors. □

Can this result be improved, at least for proper minor-closed classes? It is unclear whether the approximation ratio 2 is best possible, but it cannot be improved beyond 4/3 (unless P = NP) even for planar graphs: If we had an algorithm with approximation ratio better than 4/3, it would necessarily use at most three colors if and only if the input graph is 3-colorable. But, it is NP-hard to decide whether a planar graph is 3-colorable [3].

One might hope to improve the result to an additive approximation. Because of the Four Color Theorem, it is easy to design an algorithm that for a planar graph G returns a number c such that G is c-colorable and $c \le \chi(G) + 1$: If G is bipartite, then return 2, otherwise return 4. Could this be true for graphs from any proper minor-closed class $\mathcal{G}$? Or at least in a weaker form, returning a number c such that $c \le \chi(G) + \alpha$ for an absolute constant α? Note that this question is only interesting if α is not alloved to depend on the class $\mathcal{G}$, since as follows e.g. from Theorem 1.2, there exists a constant $c_{\mathcal{G}}$ depending on the class $\mathcal{G}$ such that every graph in $\mathcal{G}$ is $c_{\mathcal{G}}$-colorable.

Inspired by the proof of Lemma 7.19, one way to approach this problem would be to show that the vertex set of any graph $G \in \mathcal{G}$ can be split into parts A and B such that $\chi(G[A]) \le \alpha$ and $G[B]$ has bounded treewidth (the bound on treewidth is allowed to depend on the class $\mathcal{G}$). And indeed, it is easy to prove the following promising lemma, completing the first four steps of the Five Step Program for $\alpha = 7$.

Lemma 7.20 *Let a, m, d, and g be non-negative integers. If a graph G_0 is (a, m, d)-nearly drawn on a surface of Euler genus at most g, then there exists* *a partition* $\{A, B\}$ *of* $V(G_0)$ *such that* $G_0[A]$ *is* 6*-degenerate and* $\mathrm{tw}(G_0[B]) \leq (d+1)(2g+4m-1)+d+6g+a$.

Proof Let $X \subseteq V(G_0)$ be a set of size at most a such that $G_0 - X$ is drawn on a surface Σ of Euler genus at most g up to vortices $F_1, \ldots, F_m$ of width at most d, let G be the surface part of $G_0 - X$, and let $G' = G - \bigcup_{i=1}^{m} \partial F_i$. Thus, G' is drawn on the surface Σ.

Using generalized Euler's formula, we see that any subgraph F of G' has less than $3(|F| + g)$ edges, and thus it has average degree less than $6 + 6g/|F|$. Consequently, every subgraph F of G with at least $6g$ vertices has average degree less than 7, and thus it contains a vertex of degree at most six. By repeatedly deleting vertices of degree at most six from G', we conclude that there exists a partition $\{A, B_1\}$ of $V(G')$ such that $G'[A] = G_0[A]$ is 6-degenerate and $|B_1| < 6g$.

Let $B_2 = \bigcup_{i=1}^{m} V(F_i)$. The graph $G_0[B_2]$ consists of the vortices and the subgraph $G[B_0 \cap V(G)]$ of the surface part induced by their boundaries. By Corollary 2.77, $G[B_0 \cap V(G)]$ together with the cycles on ∂F_i for $i \in [m]$ has treewidth at most $2g + 4m - 1$, and by Lemma 7.5, it follows that $\mathrm{tw}(G_0[B_2]) \leq (d+1)(2g+4m-1)+d$. The claim of the lemma follows with $B = B_1 \cup B_2 \cup X$. □

However, although the last step in the Five Step Program is usually routine, in this particular case it fails! Indeed, consider the following construction. Let K be a graph on a linearly ordered set of colors and let G be another graph. By $K \circ G$, we mean the graph with vertices colored by $V(K)$ obtained as follows:

- If K has only one vertex x, then $K \circ G$ is the graph G with all vertices given color x.
- Otherwise, let x be the last vertex in the ordering of K, and let G_x be a copy of G with all vertices of color x. For each vertex $v \in V(G_x)$, we add a copy of $(K - x) \circ G$ and join v with the vertices of this copy whose colors are adjacent to x in K.

See Fig. 7.1 for an example.

Note that $K \circ G$ is obtained by taking several copies of $(K - x) \circ G$, adding an apex vertex to each of them, and then taking their clique-sum with a copy of G. By induction, Observation 2.5 gives the following bound on the largest clique minor in $K \circ G$.

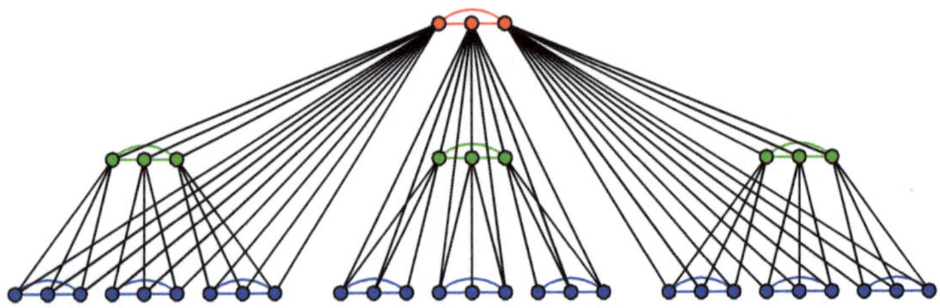

Fig. 7.1 The graph $K_3 \circ C_3$

Observation 7.21 *For any graphs K and G (with the vertex set of K ordered),* $\mathrm{Had}(K \circ G) \le \mathrm{Had}(G) + |K| - 1$.

Let $P_{k,\delta}$ be the graph with vertex set $[k]$ and with vertices $i, j \in [k]$ adjacent if and only if $|i - j| \ge \delta$. The main motivation for the definition of the $\circ$ operation is the following lemma.

Lemma 7.22 *If G is a graph of chromatic number at most δ and m is a positive integer, then*

$$\chi(P_{m\delta,\delta} \circ G) = m \cdot \chi(G).$$

Proof Let $k = m\delta$ and let λ be the coloring of $P_{k,\delta} \circ G$ by the vertex set $[k]$ of $P_{k,\delta}$ from the definition of the $\circ$ operation. For any $c \in [k - \delta + 1]$, the definition of the $\circ$ operation implies that the subgraph of $P_{k,\delta} \circ G$ induced by $\lambda^{-1}(\{c, c + 1, \ldots, c + \delta - 1\})$ is a disjoint union of copies of G, and thus it can be colored using $\chi(G)$ colors. By using disjoint sets of colors on the subgraphs induced by $\lambda^{-1}(\{(i-1)\delta + 1, (i-1)\delta + 2, \ldots, i\delta\})$ for $i \in [m]$, we obtain a coloring of $P_{k,\delta} \circ G$ by $m \cdot \chi(G)$ colors.

On the other hand, let φ be any proper coloring of $P_{k,\delta} \circ G$. Let $b = \chi(G) \le \delta$ and let $S = \{a\delta + j : a \in \{0, \ldots, m-1\}, j \in \{1, \ldots, b\}\}$. For any positive integer i, let $S_i = (S \cap [i-1]) \setminus [i_0]$ for the smallest integer $i_0 \ge 0$ such that $|S_i| < b$; that is, S_i consists of next $b - 1$ elements of S smaller than i if $i \ge b$, and $S_i = \{1, \ldots, i-1\}$ if $i < b$.

For $i = 1, \ldots, k$, we now choose a vertex $v_i \in \lambda^{-1}(k - i + 1)$ and an induced subgraph H_i of $P_{k,\delta} \circ G$ isomorphic to $P_{k-i+1,\delta} \circ G$ as follows. We let $H_1 = P_{k,\delta} \circ G$ and let v_1 be an arbitrary vertex of the "root" copy of G in H_1. For $i \ge 2$, we let H_i be the copy of $P_{k-i+1,\delta} \circ G$ added in the construction of $H_{i-1} = P_{k-i+2,\delta} \circ G$ for the vertex v_{i-1}. Let G_i be the root copy of G in H_i, i.e., the one with vertices of color $k - i + 1$. Choose v_i as a vertex of G_i with $\varphi(v_i)$ different from the colors $\varphi(v_j)$ for $j \in S_i$; note that $|S_i| < b$ and φ uses at least $\chi(G) = b$ colors on G_i, and thus it is always possible to choose v_i.

Consider now any indices $i, j \in S$ such that $j < i$. If $j \in S_i$, then the choice of v_i implies that $\varphi(v_i) \ne \varphi(v_j)$. If $j \notin S_i$, then the choice of the set S implies that $i - j \ge \delta$, and thus v_i is adjacent to v_j in $P_{k,\delta} \circ G$ and again $\varphi(v_i) \ne \varphi(v_j)$. Hence, the coloring φ uses at least $|S| = mb = m \cdot \chi(G)$ different colors. □

In particular, we get the following consequence.

Corollary 7.23 *For any positive integer m and any planar graph G, the graph $G_m = P_{4m,4} \circ G$ satisfies* $\mathrm{Had}(G_m) \le 4m + 3$ *and* $\chi(G_m) = m \cdot \chi(G)$.

Consider now any positive integer α and suppose that we were able to design for every fixed proper minor-closed class $\mathcal{G}$ a polynomial-time algorithm that colors graphs $G \in \mathcal{G}$ using at most $\chi(G) + \alpha$ colors. Let $\mathcal{G}$ be the class of $K_{4\alpha+8}$-minor-free graphs. By Corollary 7.23, for every planar graph G, the graph $G_{\alpha+1}$ belongs to $\mathcal{G}$. Hence, if G is 3-colorable, then the supposed algorithm returns a coloring of

$G_{\alpha+1}$ by at most $\chi(G_{\alpha+1}) + \alpha = (\alpha + 1)\chi(G) + \alpha \leq 4\alpha + 3$ colors. On the other hand, if $\chi(G) = 4$, then $\chi(G_{\alpha+1}) = 4\alpha + 4$, and thus no such coloring exists. Therefore, our supposed algorithm would enable us to decide whether a planar graph G is 3-colorable, which is not possible (unless P = NP).

Let us remark that an argument analogous to the proof of Lemma 7.22 shows that if $\{A, B\}$ is a partition of vertices of $G = K_k \circ (n \times n$ grid), then either $G[A]$ contains a clique of size k or $G[B]$ contains the $n \times n$ grid. Consequently, Lemma 7.20 cannot be extended to all proper minor-closed classes, even if we just desired that $\chi(G_0[A]) \leq \alpha$ for an absolute constant α.

7.4 Degeneracy-Treewidth Partitioning

There turns out to be a quite natural additional assumption that enables us to avoid the obstructions presented in the previous section: We can forbid complete bipartite subgraphs with one side of small size. The argument also showcases a common way of dealing with the apex vertices.

The proof is based on a variant of Lemma 7.20 suitable for inductive purposes. For a graph F and a set $C_0 \subseteq V(F)$, we say that F is *d-degenerate till* C_0 if for every $U \subseteq V(F)$ such that $C_0 \subsetneq U$, there exists a vertex $u \in U \setminus C_0$ with at most d neighbors in U. That is, we can repeatedly delete from F vertices of degree at most d not belonging to C_0 until we reduce F to $F[C_0]$. Let us remark that in the following lemma, the reader should view G_0 as a torso of a node x from the tree decomposition from Theorem 5.29, while F_0 is the subgraph of the original graph induced by the bag of x.

Lemma 7.24 *For integers $a, m, d, g \geq 0$ and $s, r \geq 1$, there exists an integer t such that the following claim holds. Let G_0 be a graph and let X be a set of at most a of its vertices such that $G_0 - X$ is $(0, m, d)$-nearly drawn on a surface of Euler genus at most g. Let F_0 be a spanning subgraph of G_0 such that $K_{s,r} \not\subseteq F_0$. For any partition $\{C_0, D_0\}$ of X, there exists a partition $\{C, D\}$ of $V(G_0)$ such that $C_0 \subseteq C$, $D_0 \subseteq D$, $F_0[C]$ is $(5 + s)$-degenerate till C_0 and $\mathrm{tw}(G_0[D]) \leq t$.*

Proof Let $t = (d + 1)(2g + 4m - 1) + d + 6g + a + \binom{a}{s}(r - 1)$.

Let $F_1, \ldots, F_m$ be the vortices and G the surface part of $G_0 - X$, and let $G' = G - \bigcup_{i=1}^{m} \partial F_i$. Let $M \subseteq V(G')$ consist of the vertices which in F_0 have at least s neighbors in C_0. For each vertex $v \in M$, let $C_v \subseteq C_0$ consist of s arbitrarily chosen neighbors of v in C_0. Since $K_{s,r} \not\subseteq F_0$, for every set $A \subseteq C_0$ of size s, there are at most $r - 1$ vertices $v \in M$ such that $C_v = A$. It follows that $|M| \leq \binom{|C_0|}{s}(r - 1) \leq \binom{a}{s}(r - 1)$.

As in the proof of Lemma 7.20, observe that every subgraph of G' with at least $6g$ vertices contains a vertex of degree at most 6. Let $F' = (G' \cap F_0) - M$. By repeatedly deleting vertices of degree at most 6 from F', we conclude that there exists a partition $\{C_1, D_1\}$ of $V(F')$ such that $F'[C_1] = F_0[C_1]$ is 6-degenerate and

$|D_1| < 6g$. Let $C = C_0 \cup C_1$. Since each vertex of F' has at most $s - 1$ neighbors in F_0 belonging to C_0, we conclude that $F_0[C]$ is $(5 + s)$-degenerate till C_0.

Let $D_2 = \bigcup_{i=1}^{m} V(F_i)$; as in the proof of Lemma 7.20, $\mathrm{tw}(G_0[D_2]) \le (d + 1)(2g + 4m - 1) + d$. The claim of the lemma thus holds for $D = D_0 \cup D_1 \cup D_2 \cup M$. □

The result on graphs from proper minor-closed classes now easily follows by induction.

Theorem 7.25 *For every graph H and positive integers s and r, there exists a positive integer t such that the following claim holds. The vertex set of every H-minor-free graph G that does not contain $K_{s,r}$ as a subgraph has a partition $\{C, D\}$ such that $G[C]$ is $(5 + s)$-degenerate and* $\mathrm{tw}(G[D]) \le t$.

Proof Let a, m, and d be the constants from Theorem 5.29 for H and let g be the maximum Euler genus of an H-avoiding surface (or $g = 0$ if H is planar). Let t be the value from Lemma 7.24.

Let (T, β) be the rooted tree decomposition of G from Theorem 5.29. We construct the sets C and D inductively, starting from the root of T. Consider a node $x \in V(T)$ and suppose that if x is not the root, we have already defined the partition $\{C_y, D_y\}$ of the bag $\beta(y)$ of the parent node y of x. If x is the root, then let $C'_x = \emptyset$, otherwise let $C'_x = C_y \cap \beta(x)$. Let G_x be the torso of x.

- If $|\beta(x)| \le a$, then let $C_x = C'_x$ and $D_x = \beta(x) \setminus C'_x$. Note that $G[C_x]$ is trivially $(5 + s)$-degenerate till C'_x and $\mathrm{tw}(G_x[D_x]) \le a \le t$.
- Otherwise, let A_x be a subset of $\beta(x)$ containing the root separator such that $G_x - A_x$ is $(0, m, d)$-nearly drawn on an H-avoiding surface (of Euler genus at most g). Let $D'_x = A_x \setminus C'_x$. The partition $\{C_x, D_x\}$ of $\beta(x)$ such that $C'_x \subseteq C_x$, $D'_x \subseteq D_x$, $G[C_x]$ is $(5 + s)$-degenerate till C'_x, and $\mathrm{tw}(G_x[D_x]) \le t$ is obtained using Lemma 7.24.

Let $C = \bigcup_{x \in V(T)} C_x$ and $D = \bigcup_{x \in V(T)} D_x$. Note that $G[D]$ is a clique-sum of the graphs $G_x[D_x]$ for $x \in V(T)$, and thus $\mathrm{tw}(G[D]) \le t$ by Observation 2.5.

Moreover, we claim that $G[C]$ is $(5 + s)$-degenerate. Indeed, consider any non-empty set $B \subseteq C$, and let x be a node of T with the root separator S_x such that $B \cap \beta(x) \setminus S_x \neq \emptyset$ and x is as far from the root as possible (such a node x exists, since $\bigcup_{x \in V(T)} (\beta(x) \setminus S_x) = V(G)$). By the maximality of the distance of x from the root, for every vertex $v \in B \cap \beta(x) \setminus S_x$, all neighbors of v in $G[B]$ belong to $\beta(x)$. The set $C \cap \beta(x) = C_x$ was constructed so that $G[C_x]$ is $(5+s)$-degenerate till $C'_x = C \cap S_x$. Therefore, $G[(B \cap C_x) \cup C'_x]$ contains a vertex $v \in B \cap C_x \setminus C'_x = B \cap C_x \setminus S_x$ of degree at most $5 + s$. Since all neighbors of v in $G[B]$ belong to $\beta(x)$, the degree of v in $G[B]$ is at most as large. Hence, we conclude that every subgraph of $G[C]$ has a vertex of degree at most $5 + s$. □

In particular, there exists a polynomial-time algorithm that for any H-minor-free graph G not containing $K_{s,r}$ as a subgraph returns its coloring using at most $\chi(G) + s + 6$ colors. Let us remark that this result applies when G has bounded maximum degree Δ (by choosing $s = 1$ and $r = \Delta + 1$), and when G has girth at least five (choosing $s = r = 2$).

References

1. R. Downey, M. Fellows, Fixed-parameter tractability and completeness. II. On completeness for W[1]. Theor. Comput. Sci. **141**, 109–131 (1995)
2. Z. Dvořák, Approximation metatheorems for classes with bounded expansion, in *18th Scandinavian Symposium and Workshops on Algorithm Theory (SWAT 2022)*. Leibniz International Proceedings in Informatics (LIPIcs), vol. 227 (2022), pp. 22:1–22:17
3. M. Garey, D. Johnson, *Computers and Intractability: A Guide to the Theory of NP-Completeness* (WH Freeman, New York, 1979)
4. M. Grohe, K. Kawarabayashi, B. Reed, A simple algorithm for the graph minor decomposition–logic meets structural graph theory, in *Proceedings of the Twenty-Fourth Annual ACM-SIAM Symposium on Discrete Algorithms* (2013), pp. 414–431
5. D. Marx, M. Pilipczuk, Everything you always wanted to know about the parameterized complexity of Subgraph Isomorphism (but were afraid to ask), in *31st International Symposium on Theoretical Aspects of Computer Science (STACS 2014)*. Leibniz International Proceedings in Informatics (LIPIcs), vol. 25 (2014), pp. 542–553
6. J. Nešetřil, P. Ossona de Mendez, *Sparsity (Graphs, Structures, and Algorithms)*. Algorithms and Combinatorics, vol. 28 (Springer, Berlin, 2012)
7. J. Nešetřil, S. Poljak, Complexity of the subgraph problem. Comment. Math. Univ. Carol. **26**, 415–420 (1985)

Chapter 8
Tighter Grid Theorem

By the Grid Theorem, for every positive integer n there exists an integer t such that every graph of treewidth at least t contains the $n \times n$ grid as a minor. Let $f(n)$ be defined as the minimum integer t with this property. It is not clear what the exact asymptotics of this function is, but as we have discussed in Sect. 2.7, it cannot be better than $\Omega(n^2 \log n)$, as seen by considering random graphs. On the other hand, in Corollary 2.70, we have seen that treewidth $\Omega(n)$ suffices to ensure the existence of the $n \times n$ grid as a minor in a planar graph. Does this result extend to other proper minor-closed classes?

8.1 Linear-Sized Grid Minors

First, let us see that it extends to graphs on surfaces, completing the step two of the Five Step Program. It is possible to give a proof along the lines of the one used to show Corollary 2.70, but a more straightforward one follows from the theory of respectful tangles. For an application in the next step of the argument, we need a slight technical strengthening, showing that a large grid exists even if we avoid using the boundary vertices of several faces.

Lemma 8.1 *Let G be a graph drawn on a surface Σ of Euler genus g and let n be a positive integer. Let $\mathcal{F}$ be a set of at most m faces of G and let R be the set of vertices of G incident with these faces. If $\mathrm{tw}(G) \ge (5g+6)(n+2m+2g)$, then $G - R$ contains the $n \times n$ grid as a minor.*

Proof We prove the claim by induction on g. Clearly we can assume that G is connected (otherwise, we consider just its component of largest treewidth). Let $n' = n + 2m$.

If $g = 0$, then $\mathrm{tw}(G) \ge 6n'$, and by Corollary 2.70, G contains a model μ of the $n' \times n'$ grid H. Since the grid H is a subdivision of a 3-connected graph, it

© The Author(s), under exclusive license to Springer Nature Switzerland AG 2025

Z. Dvořák, *Graph Minors*, Springer Monographs in Mathematics,
https://doi.org/10.1007/978-3-031-87469-7_8

has a unique drawing in the plane. Note that for each face f of G, there exists a facial cycle C_f of H such that every vertex of $\mu(V(H))$ incident with f belongs to $\mu(V(C_f))$. For $f \in \mathcal{F}$, if C_f bounds the outer face of the grid H, then delete C_f from H, obtaining a grid with two fewer rows and columns. If C_f bounds an internal face of H, then delete from H the vertices of C_f and the edges of the two rows and columns intersecting C_f, and suppress the arising vertices of degree two, again obtaining a grid with two fewer rows and columns with a model in G disjoint from the boundary of f. In the end, we are left with a grid with at least n rows and columns, and μ gives rise to its model in $G - R$, as required.

Suppose now that $g > 0$. If there exists a non-contractible simple closed G-normal curve γ intersecting G in a set X of less than $5n'$ vertices, then let Σ' be the surface (or union of two surfaces) obtained by cutting Σ along γ and patching the resulting holes. Consider the graph $G - X$ drawn on Σ', and let $\mathcal{F}'$ consist of the faces of $\mathcal{F}$ whose closure is disjoint from γ (these are also faces of $G - X$) and in case that γ intersects the closure of at least one face of $\mathcal{F}$, also of the (at most two) faces of $G - X$ containing the patches. Note that $|\mathcal{F}'| \le |\mathcal{F}| + 1 \le m + 1$. Moreover, letting R' be the set of vertices of $G - X$ incident with the faces in $\mathcal{F}'$, observe that $R \subseteq R' \cup X$. The Euler genus g' of surface(s) Σ' is smaller than g, and thus

$$\begin{aligned}\operatorname{tw}(G - X) &\ge \operatorname{tw}(G) - 5n' \ge (5g+6)(n'+2g) - 5n'\\ &\ge (5g'+6)(n'+2g) \ge (5g'+6)(n+2(m+2)+2g').\end{aligned}$$

By the induction hypothesis, $G - (X \cup R') \subseteq G - R$ contains the $n \times n$ grid as a minor, and thus the claim of the lemma holds.

Finally, suppose that the drawing of G has representativity at least $5n'$. By Theorem 2.91, G contains a respectful tangle $\mathcal{T}$ of order $5n'$. Let a be an arbitrary atom of G. For $1 \le i \le n'$, let $k_i = n' + 3i$. Corollary 4.16 implies that there exists a cycle C_i bounding a k_i-zone Δ_i around a and such that $k_i - 2 \le d_{\mathcal{T}}(a, v) \le k_i$ for every $v \in V(C_i)$. In particular, the cycles $C_1, \ldots, C_{n'}$ are pairwise vertex-disjoint. By Lemma 4.18, there exists a $(V(C_1), V(C_n))$-linkage in G of size n'. By Lemma 2.64, we conclude that G contains the $n' \times n'$ cylindrical grid as a minor drawn in the annulus between C_1 and $C_{n'}$. As in the planar case, we can remove at most $2m$ rows and columns of this grid minor to make it disjoint from R. □

The third step of the Five Step Program easily follows from Lemma 7.5, even in a somewhat stronger form requiring the presence of the minor in the surface part.

Lemma 8.2 *Suppose that a graph G_0 is drawn on a surface of Euler genus at most g up to vortices $F_1, \ldots, F_m$ of width at most d and let n be a positive integer. If $\operatorname{tw}(G_0) \ge (d+1)(5g+6)(n+2m+2g)+d$, then the surface part G of G_0 contains the $n \times n$ grid as a minor.*

Proof Let G' and G_0' be obtained from G and G_0 by adding the cycles on ∂F_i for $i \in [m]$. By Lemma 7.5, we have $\operatorname{tw}(G_0) \le \operatorname{tw}(G_0') \le (d+1)\operatorname{tw}(G')+d$, and thus $\operatorname{tw}(G') \ge (5g+6)(n+2m+2g)$. By Lemma 8.1, $G' - \bigcup_{i \in [m]} \partial F_i \subseteq G$ contains the $n \times n$ grid as a minor. □

Deleting a bounded number of apex vertices only decreases treewidth by a constant, and thus we get the fourth step of the Five Step Program for free. The claim for all proper minor-closed classes now easily follows from Theorem 5.29 (Theorem 2.9 does not suffice, as it does not give us a way to relate the grid minors in the torsos to those in the whole graph).

Theorem 8.3 *For every graph H, there exist constants α and κ such that for every positive integer n, every H-minor-free graph G of treewidth at least $\alpha n+\kappa$ contains the $n \times n$ grid as a minor.*

Proof Let a, m, and d be the constants from Theorem 5.29 for H, and let g be the maximum Euler genus of an H-avoiding surface (we let $g = 0$ if H is planar). Let $\alpha = (d + 1)(5g + 6)$ and $\kappa = (d + 1)(5g + 6)(2m + 2g) + d + a$.

Let (T, β) be a rooted tree decomposition of G with properties described in Theorem 5.29. Let x be the node of T whose torso G_x has largest treewidth. Note that $\mathrm{tw}(G_x) \ge \mathrm{tw}(G) \ge \alpha n + \kappa$ by Observation 2.5, and in particular $|G_x| > a$. Let A_x be a set of at most a vertices of G_x such that $G_x - A_x$ is $(0, m, d)$-nearly drawn on an H-avoiding surface and the surface part G'_x of $G_x - A_x$ is a minor of G. We have $\mathrm{tw}(G_x - A_x) \ge \alpha n + \kappa - a$, and by Lemma 8.2, G'_x (and thus also G) contains the $n \times n$ grid as a minor. □

Clearly, Theorem 8.3 gives us yet another proof that every H-minor-free graph G has treewidth $O(\sqrt{|G|})$, though again with a very bad dependence of the multiplicative constant in the O-notation on H. More importantly, it has many applications, and in particular plays a key role in the *bidimensionality theory*, see e.g. [1] for an overview. As a simple example, Theorem 8.3 can be used to design parameterized algorithms with subexponential dependency on the parameter. Recall that $\mathrm{fvs}(G)$ is the minimum size of a feedback vertex set in the graph G, i.e., of a set of vertices that intersects all cycles in G.

Lemma 8.4 *Let H be a fixed graph. There exists an algorithm that, given an H-minor-free n-vertex graph G and a positive integer k, decides in time $\exp(O(\sqrt{k}\log k)) \cdot n$ whether $\mathrm{fvs}(G) < k$.*

Proof Leta α and κ be the constants from Theorem 8.3 for H. Let $m = 2\lceil\sqrt{k}\rceil$, and let $t = \alpha m + \kappa$. Using the algorithm of [2], one can in time $\exp(O(t)) \cdot n = \exp(O(\sqrt{k})) \cdot n$ either decide that G has treewidth at least t, or find a rooted tree decomposition (T, β) of G of width at most $2t - 1$. In the former case, Theorem 8.3 implies that G contains an $m \times m$ grid as a minor; observe that G contains at least $\lfloor m/2\rfloor^2 \ge k$ pairwise vertex-disjoint cycles, and thus $\mathrm{fvs}(G) \ge k$.

In the latter case, we use a standard dynamic programming algorithm to determine $\mathrm{fvs}(G)$ exactly in time $\exp(O(t\log t)) \cdot n = \exp(O(\sqrt{k}\log k)) \cdot n$: For every node $x \in T$, let T_x be the subtree of T rooted in x. For every set $S \subseteq \beta(x)$ and every partition $\mathcal{P}$ of $\beta(x) \setminus S$, we determine the minimum size of a feedback vertex set R in $G[\beta(T_x)]$ such that $R \cap \beta(x) = S$ and such that two vertices $u, v \in \beta(x) \setminus S$ are in the same component of $G[\beta(T_x)] - R$ if and only if they are in the same part of $\mathcal{P}$. Observe that the number of choices for S and $\mathcal{P}$ is $\exp(O(|\beta(x)|\log|\beta(x)|))$, and that assuming that we have already computed this information for all children of x, we can compute it for x in time $\exp(O(|\beta(x)| \log |\beta(x)|))$. □

8.2 Linear-Sized Grid Contractions

The key property that enables the application in Lemma 8.4 is that minimum feedback vertex set size in any graph that contains the $n \times n$ grid as a minor is $\Omega(n^2)$. Unfortunately, many other graph parameters lack this property, even if they initially may seem promising. E.g., the domination number of the $n \times n$ grid is $\Omega(n^2)$, but it can appear as a subgraph in a graph G with $\gamma(G) = 1$, if G contains a universal vertex. However, it is easy to see that domination number is monotone under edge contractions; hence, to enable applications such as subexponential parameterized algorithm for domination number, it would suffice to be able to obtain a large grid-like subgraph as a ***contraction***, i.e., by contracting edges and deleting only isolated vertices.

This is not possible in all graph classes with a forbidden minor: The grid plus a universal vertex graphs are K_6-minor-free and every contraction of these graphs still has a universal vertex. To avoid this obstruction, we need to restrict ourselves to the graph classes whose set of forbidden minors contains an ***apex graph***, i.e., a graph planar after removal of a single vertex.

Let us now give the corresponding results more precisely. First, let us note that in the proof of Lemma 8.1, instead of eliminating the rows and columns of the grid minor which intersect the boundaries of the faces in $\mathcal{F}$, we could have instead chosen a large contiguous subgrid of the grid so that the faces of $\mathcal{F}$ are contained in the outer face of the subgrid. The resulting subgrid is smaller (by a $\Theta(m+g)$ factor) than the grid we obtained in Lemma 8.1, but has the pleasing property that it is "drawn planarly" in the surface in the following sense.

Lemma 8.5 *For any non-negative integers g and m, there exists a constant α such that the following claim holds. Let G be a graph drawn on a surface Σ of Euler genus g and let n be a positive integer. Let $\mathcal{F}$ be a set of at most m faces of G. If $\mathrm{tw}(G) \geq \alpha n$, then there exists a disk $\Delta \subseteq \Sigma$ disjoint from the closures of the faces in $\mathcal{F}$ such that the subgraph of G drawn in Δ contains the $n \times n$ grid as a minor.*

We can use this Lemma to prove the analogue of Lemma 8.2, obtaining the grid minor in a part drawn in a disk disjoint from the vortices. The contraction variant of Theorem 8.3 now easily follows. A ***partially triangulated*** $n \times n$ ***grid*** is a plane graph obtained from the $n \times n$ grid by adding diagonals to some of its faces (including the outer one). An *a-apex partially triangulated $n \times n$ grid* is a graph obtained from a partially triangulated $n \times n$ grid by adding at most a vertices with arbitrary neighborhoods.

Theorem 8.6 *For every graph H, there exists a constant α such that for every positive integer n, every H-minor-free graph G of treewidth at least αn has an α-apex partially triangulated $n \times n$ grid as a contraction.*

Proof Without loss of generality, we can assume that G is connected, since the components other than the one of largest treewidth can be contracted to isolated vertices and then deleted. We proceed in the same way as in the proof of Theorem 8.3. Let (T, β) be a rooted tree decomposition of G with properties described in Theorem 5.29, and let x be the node of T whose torso G_x has largest treewidth.

For every component T' of $T - x$, since G_x is the torso of x, $\beta(T') \cap \beta(x)$ is a clique in G_x, and we can contract $G[\beta(T')]$ in G to a subgraph of this clique. Let $G' \subseteq G_x$ be the resulting contraction of G. We make sure that all edges of the surface part G_1 of $G_x - A_x$ are created in these contractions, which is possible since the surface part of $G_x - A_x$ is a minor of G. Hence, $G' - A_x$ is also $(0, m, d)$-nearly drawn on a surface Σ, with the surface part G_1. By Lemma 8.5, we can find the $(2n + 4) \times (2n + 4)$ grid M_0 as a minor of a part of G_1 contained in a disk $\Delta \subseteq \Sigma$ disjoint from the vortices. Let G'' and G_2 be obtained from G' and G_1 by performing the corresponding contractions (but not vertex and edge deletions), so that M_0 is a subgraph of the surface part G_2 of $G'' - A_x$.

Note that the drawing of M_0 in the disk Δ is unique, up to choice of the face of M_0 that contains the boundary of Δ. We can always choose an $(n + 2) \times (n + 2)$ subgrid M of M_0 whose internal faces do not contain $\mathrm{bd}(\Delta)$. Hence, the outer face f_0 of M contains the boundary of Δ.

Consider a face $f \neq f_0$ of M, let C_f be the cycle in M bounding f, and let G_f be the subgraph of G_2 drawn in the interior of f. For each component K of G_f, if K has a neighbor in A_x, we contract it to this neighbor. Otherwise, since G is connected, K has a neighbor in C_f, and we contract it to this neighbor. Note that since $f \subseteq \Delta$, even if we end up contracting different components of G_f to different vertices of C_f, the resulting chords do not cross.

Next, consider the face f_0, and let G_{f_0} be the subgraph of G_2 drawn in the closure of f_0 (i.e., including the cycle C_{f_0} bounding the outer face of M) and all the vortices of $G'' - A_x$. Let K be the component of G_{f_0} containing C_{f_0}. Since G is connected, each component of G_{f_0} other than K has a neighbor in A_x and can be contracted into it. We contract K into a vertex in the outer face of the subgrid $M - C_{f_0}$.

Let F be the graph obtained by these contractions. Observe that $F - A_x$ is a planar graph obtained from the $n \times n$ grid $M - C_{f_0}$ by adding chords to some of its faces, and thus the conclusion of the theorem holds. □

Further improving this result, it is easy to additionally contract the apex vertices into a single one.

Theorem 8.7 *For every graph H, there exists a constant α_0 such that for every positive integer n, every H-minor-free graph G of treewidth at least $\alpha_0 n$ has a 1-apex partially triangulated $n \times n$ grid as a contraction.*

Proof Let α be the constant from Theorem 8.6 for H, and let $\gamma = \lceil\sqrt{\alpha+1}\rceil$ and $\alpha_0 = \alpha\gamma$. By Theorem 8.6, G has a contraction G_1 containing a set A_1 of at most α vertices such that $M = G_1 - A_1$ is a partial triangulation of the $\gamma n \times \gamma n$ grid. We can assume that A_1 is an independent set, as we can contract edges between vertices of A_1 otherwise. Moreover, we can assume that each vertex in A_1 has a neighbor in M, as otherwise it is isolated and we can delete it.

By the choice of γ, M has $\alpha + 1$ pairwise vertex-disjoint (partially triangulated) $n \times n$ subgrids $M_1, \ldots, M_{\alpha+1}$. For each $v \in A_1$, if v has a neighbor in one of these subgrids, then let $i_v \in [\alpha + 1]$ be an index such that v has a neighbor in M_{i_v}; otherwise, let $i_v = 1$. Since $|A_1| \leq \alpha$, there exists $i_0 \in [\alpha + 1]$ such that $i_0 \neq i_v$ for every $v \in A_1$. For every $v \in A_1$, since v has a neighbor in M, the choice of i_v implies that v has a neighbor in $M - V(M_{i_0})$. Therefore, the subgraph of G_1 induced by $(V(M) \setminus V(M_{i_0})) \cup A_1$ is connected, and we can contract it to a single apex vertex over the partially triangulated $n \times n$ grid M_{i_0}. □

Using the results from Sect. 4.5, it is easy to get rid of the apex vertex in case that H is an apex graph.

Corollary 8.8 *For every apex graph H, there exists a constant α_1 such that for every positive integer n, every H-minor-free graph G of treewidth at least $\alpha_1 n$ contains a partially triangulated $n \times n$ grid as a contraction.*

Proof Let θ and t be the constants from Lemma 4.35 applied for H and $\Sigma =$ the sphere. Let α_0 be the constant from Theorem 8.7, choose $\gamma \gg \theta, t$, and let $\alpha_1 = \alpha_0\gamma$.

By Theorem 8.7, G has a contraction G_1 containing a vertex z such that $G_1 - z$ is a partial triangulation of the $\gamma n \times \gamma n$ grid. Let M be the $\gamma n \times \gamma n$ grid subgraph of $G_1 - z$, and let $\mathcal{T}$ be the canonical tangle of order γn in the grid M. We view M as drawn on the sphere; in that case, the tangle $\mathcal{T}$ is automatically respectful. Let S be a maximal set of neighbors of z such that the vertices in S are pairwise at $d_{\mathcal{T}}$-distance at least 2θ. Note that z together with the edges from z to S forms a 2θ-wide spider with $|S|$ legs over M in G_1. Since H is not a minor of G, it also is not a minor of G_1, and since $a(H) \leq 1$, Lemma 4.35 implies that $|S| < t$.

By the maximality of S, every neighbor of z is at $d_{\mathcal{T}}$-distance less than 2θ from S. Since $\gamma \gg |S|, \theta$, observe that there exists an $n \times n$ subgrid M_0 of M at $d_{\mathcal{T}}$-distance at least 2θ from S. Then z has no neighbors in M_0. The subgraph of G_1 induced by $(V(M) \setminus V(M_0)) \cup \{z\}$ is connected (unless z is an isolated vertex, in which case it can be deleted); we contract this subgraph to a single vertex u, then contract an edge between u and a vertex in the boundary of the outer face of M_0. The resulting contraction is a partial triangulation of the $n \times n$ grid M_0. □

As an example application, let us show that domination number has a subexponential parameterized algorithm in apex-minor-free graph classes.

Lemma 8.9 *For every apex graph H, there exists an algorithm that, given an H-minor-free n-vertex graph G and a positive integer k, decides in time* $\exp\left(O\left(\sqrt{k}\right)\right) \cdot$ *n whether G has a dominating set of size less than k.*

Proof Let α_1 be the constant from Corollary 8.8 for H. Let $m = 3\lceil\sqrt{k}\rceil$, and let $t = \alpha_1 m$. Using the algorithm of [2], one can in time $\exp(O(t))\cdot n = \exp(O(\sqrt{k}))\cdot n$ either decide that G has treewidth at least t, or find a tree decomposition (T, β) of G of width at most $2t - 1$.

In the former case, Corollary 8.8 implies that G has a contraction M such that M is a partial triangulation of an $m \times m$ grid. We can divide M into $\lceil\sqrt{k}\rceil^2 \geq k$ disjoint 3×3 subgrids, and observe that a dominating set of M must contain a vertex of each of these subgrids, implying that $\gamma(M) \geq k$. Moreover, observe that contracting an edge or deleting an isolated vertex cannot increase the domination number, and thus $\gamma(G) \geq \gamma(M) \geq k$.

In the latter case, we use a standard dynamic programming algorithm to determine $\gamma(G)$ exactly in time $\exp(O(t)) \cdot n = \exp\left(O\left(\sqrt{k}\right)\right) \cdot n$. □

Recall that in Theorem 5.14, we proved that apex-minor-free graphs have locally bounded treewidth. Corollary 8.8 enables us to strengthen the dependency of the treewidth on the radius to a linear one. Let us present this result in the following form (also including Corollary 2.74).

Theorem 8.10 *Let $\mathcal{G}$ be a minor-closed class of graphs. Then the following claims are equivalent:*

(i) $\mathcal{G}$ has locally bounded treewidth.
(ii) There exists an apex graph H such that $H \notin \mathcal{G}$.
(iii) There exists a constant α such that every graph $G \in \mathcal{G}$ has treewidth at most αr, where r is the radius of G.
(iv) There exists a constant α such that if λ is a BFS layering of a graph $G \in \mathcal{G}$ and r is a positive integer, then the union of any r layers induces a subgraph of treewidth at most αr.

Proof We have argued that (i) ⇒ (ii) holds in Theorem 5.14.

Suppose now that H is an apex graph and every graph in $\mathcal{G}$ is H-minor-free. Let α_1 be the constant from Corollary 8.8 for H and let $\alpha = 4\alpha_1$. Consider a graph $G \in \mathcal{G}$ of radius at most r, where without loss of generality $r \geq 1$. Note that any contraction of G also has radius at most r, and that every partially triangulated $(2r+2) \times (2r+2)$ grid has radius more than r, and thus G does not contain such a grid as a contraction. Therefore, Corollary 8.8 implies $\mathrm{tw}(G) < \alpha_1(2r+2) \leq \alpha r$. It follows that (ii) ⇒ (iii).

The proof of (iii) ⇒ (iv) is the same as the proof of Corollary 2.74, using (iii) instead of Theorem 2.73. The implication (iv) ⇒ (i) is trivial, since any graph of radius r has a BFS layering with $r+1$ layers. □

In particular, Theorem 8.10 implies that the applications of Baker's technique for design of approximation algorithms in planar graphs can usually be generalized to all apex-minor-free graph classes; see Sect. 16.2 for a more detailed discussion.

References

1. E.D. Demaine, M. Hajiaghayi, The bidimensionality theory and its algorithmic applications. Comput. J. **51**, 292–302 (2007)
2. T. Korhonen, A single-exponential time 2-approximation algorithm for treewidth, in *2021 IEEE 62nd Annual Symposium on Foundations of Computer Science (FOCS)* (2022), pp. 184–192

Chapter 9
Topological Minors

What can be said about the structure of graphs avoiding a fixed graph H as a topological minor, rather than a minor? Clearly we cannot get as nice result as the Minor Structure Theorem. For one, not every class of graphs closed under topological minors is characterized by finitely many forbidden topological minors, i.e., the analogue of Theorem 1.6 is false for topological minors. As an example, consider the class $\mathcal{G}$ of graphs where any two vertices of degree at least five are separated by an edge cut of size at most one. Clearly, deleting edges and vertices as well as suppressing vertices of degree two preserves this property, and thus $\mathcal{G}$ is closed under topological minors. For a positive integer k, let H_k be the graph obtained from the path of length k by doubling its edges and attaching three pendant vertices at each of its ends. Then $H_k \notin \mathcal{G}$, but every proper topological minor of H_k belongs to $\mathcal{G}$. And indeed, $\mathcal{G}$ is exactly the class of graphs that avoid all graphs $\{H_k : k \in \mathbb{N}\}$ as topological minors.

Moreover, for every positive integer d, the class $\mathcal{G}_d$ of graphs of maximum degree at most d is topological minor-closed. It is known [1] that there exists a constant $c > 1$ such that the independence number of a subcubic graph cannot be approximated up to a factor of c in polynomial time (unless P $=$ NP), and thus we cannot hope to obtain approximation schemes for proper topological minor-closed classes analogous to the ones from Chap. 7.

However, it turns out that graphs with bounded maximum degree are in a sense the most complicated proper topological minor-closed classes, up to a small generalization. The ***h-index*** of a graph G is the largest integer d such that G has at least d vertices of degree at least d. Clearly, for every positive integer d, the class of graphs of h-index at most d (i.e., with all but at most d vertices having degree at most d) is closed under topological minors. Grohe and Marx [3] proved the following structural theorem for proper topological minor-closed classes.

Theorem 9.1 *For every graph H, there exist integers a, m, d, and g such that every graph G that does not contain H as a topological minor has a tree decomposition*

© The Author(s), under exclusive license to Springer Nature Switzerland AG 2025

Z. Dvořák, *Graph Minors*, Springer Monographs in Mathematics,
https://doi.org/10.1007/978-3-031-87469-7_9

in which each torso of h-index at least a can be (a, m, d)*-nearly drawn on a surface of Euler genus at most* g.

In other words, G is obtained by clique-sums from graphs of h-index less than a and graphs which can be (a, m, d)-nearly drawn on a surface of Euler genus at most g.

Theorem 9.1 and the ideas used to obtain it have several important applications in topological minors theory. Grohe and Marx [3] specifically developed it in order to provide a polynomial-time test for isomorphism of graphs avoiding a fixed topological minor. Another application concerns testing whether a fixed graph H is a topological minor of an input graph G. As we have described in Sect. 5.3, the polynomial-time algorithm for finding linkages leads to an algorithm for testing the presence of a fixed graph H as a (rooted) topological minor with time complexity $|G|^{O(|H|+\|H\|)}$. However, using ideas similar to those from the proof of Theorem 9.1, Grohe et al. [4] developed an algorithm for this problem with cubic time complexity (only the multiplicative constant in the time complexity of the algorithm depends on H).

Somewhat disappointingly, the surfaces in Theorem 9.1 do not need to be H-avoiding. As an example, suppose that H is the double wheel, i.e., H is obtained from a cycle of length $\ell \geq 5$ by adding two vertices adjacent to all vertices of the cycle. Then H is planar, and so no surface is H-avoiding. Note however that in any plane drawing of H, the two vertices of degree ℓ cannot be incident with a common face. Hence, if G is a plane graph with all vertices of degree at least ℓ incident with the outer face, then H is not a topological minor of G. Moreover, such a graph G cannot in general be obtained from graphs of bounded h-index by clique-sums.

Hence, if we want to improve the part of Theorem 9.1 concerning nearly drawn graphs, we need to take into account the properties of H such as the minimum number of faces covering all its vertices of degree greater than three. Such improvements are possible, but their statements in general are rather technical, and thus we refer an interested reader to the papers where they were developed [2, 8]. Here, we show a proof of an interesting special case. We say that a graph H is ***large-degree-outerplanar*** if H has a drawing in the plane with all vertices of degree at least four incident with the outer face.

Theorem 9.2 *For every large-degree-outerplanar graph H, there exists an integer a such that every graph G that does not contain H as a topological minor has a tree decomposition whose torsos have h-index at most a.*

As we have mentioned before, the global form of the structure theorem (even in the stronger version given by Theorem 5.29) is not well suited for arguments such as the proof of Theorem 9.2 which require one to show that specific topological minors or other substructures appear in the graph. The local form, especially in the version given in Theorem 5.30 guaranteeing a respectful tangle of large order, is much more practical, both because it is directly amenable to tools we developed in Sect. 4.5, and because it gives us a better control over the way the rest of the graph attaches to the piece we study.

As a bonus, we are going to obtain a strong local form of Theorem 9.2. Given a vertex tangle $\mathcal{T}$ of order θ in a graph G and a positive integer $\Delta \leq \theta$, a vertex $v \in V(G)$ is $(\mathcal{T}, \Delta)$-*free* if there is no vertex separation $(A, B) \in \mathcal{T}$ of order less than Δ such that $v \in A \setminus B$. Note in particular that if v is $(\mathcal{T}, \Delta)$-free, then $\deg v \geq \Delta$, since otherwise we could let $A = N[v]$ and $B = V(G) \setminus \{v\}$; we have $(A, B) \in \mathcal{T}$ by Observation 2.18, since $(A, V(G)) \in \mathcal{T}$ by (T2).

Theorem 9.3 *For every large-degree-outerplanar graph H, there exist integers a_0 and $\theta \geq a_0 + \Delta(H)$ such that the following claim holds. Let $\mathcal{T}$ be a tangle of order at least θ in a graph G. If H is not a topological minor of G, then there exists a set $X \subseteq V(G)$ of size at most a_0 such that no vertex of $G - X$ is $(\mathcal{T} - X, \Delta(H))$-free.*

Thus, we actually get control of vertices of degree at least $\Delta(H)$, rather than only of those of degree at least a for the larger integer a depending on H as in Theorem 9.2. This is important in applications such as proving well-quasi-ordering on special topological minor-closed classes, where one needs to argue that G is "simpler" than the forbidden topological minor H; see [5–7] for more details.

Let us remark that Theorem 9.2 implies the following degeneracy-like approximate characterization of graphs avoiding a large-degree-outerplanar topological minor. The ∞-*admissibility* of an ordering $v_1, \ldots, v_n$ of vertices of G is the minimum integer d such that for every $i \in [n]$, G contains at most d paths from v_i to $\{v_1, \ldots, v_{i-1}\}$ which intersect only in v_i. The ∞-*admissibility* of a graph G is the minimum of ∞-admissibilities of all orderings of its vertices.

Corollary 9.4 *Let $\mathcal{G}$ be a class of graphs closed under topological minors. The following claims are equivalent:*

(a) $\mathcal{G}$ does not contain all large-degree-outerplanar graphs.
(b) There exists an integer a such that every graph in $\mathcal{G}$ is obtained by clique-sums from graphs of h-index at most a.
(c) There exists an integer d such that every graph in $\mathcal{G}$ has ∞-admissibility at most d.

Proof The implication (a) $\Rightarrow$ (b) clearly follows from Theorem 9.2.

To prove (b) $\Rightarrow$ (c), we show a stronger claim. We say that a graph G has *clique-relative ∞-admissibility* at most d if for every clique K in G, there exists an ordering $v_1, \ldots, v_n$ of vertices of G of ∞-admissibility at most d such that $V(K) = \{v_1, \ldots, v_k\}$. We now prove by induction on the number of vertices that if G is obtained by clique-sums from graphs of h-index at most a, then it has clique-relative ∞-admissibility at most $2a$.

Indeed, if G itself has h-index at most a, then clearly every clique K in G has at most $a + 1$ vertices. Consider an ordering of vertices of G where we put first $V(K)$, then all other vertices of G of degree greater than a, then all other vertices of G. If a vertex $v \in V(G)$ has degree at most a, then there are at most a paths from it intersecting only in v. If v has degree more than a, then there are at most $2a$ vertices preceding it in the ordering. Consequently, it follows that G has clique-relative ∞-admissibility at most $2a$.

Suppose now that G is a clique-sum of graphs G_1 and G_2 on a clique K'. Consider any clique K in G; without loss of generality, we can assume that K is a clique in G_1. By the induction hypothesis, there exists an ordering L_1 of vertices of G_1 of ∞-admissibility at most $2a$ starting with $V(K)$, and an ordering L_2 of vertices of G_2 of ∞-admissibility at most $2a$ starting with $V(K')$. Let L be the ordering of $V(G)$ obtained by concatenating L_1 with $L_2 - V(K')$. Since K' is a clique in G_1, any path P in G with both ends in $V(G_1)$ can be transformed to a path P_1 in G_1 with the same ends and with $V(P_1) \subseteq V(P)$. Moreover, any path in G starting in a vertex $v \in V(G_2) \setminus V(K')$ and ending in a vertex u preceding v in L can be truncated on the first vertex belonging to K' (if any) to obtain a path in G_2 from v to a vertex preceding it in L_2. From these observations, we conclude that L is an ordering of the vertices of G of ∞-admissibility at most $2a$.

Finally, to prove (c) $\Rightarrow$ (a), consider the graph H obtained from a large wall by choosing $(d+1)(d+2)$ edges in the boundary of its outer face that do not share endpoints, subdividing each of them, and adding $d+2$ vertices of degree $d+1$ adjacent to the resulting vertices of degree two in a planar fashion. Then H is large-degree-outerplanar, but it is easy to see that every ordering of $V(H)$ has ∞-admissibility at least $d+1$: Let v be the last of the vertices of degree $d+1$ in the ordering, and observe that H contains paths from v to all other vertices of degree $d+1$ intersecting only in v. Therefore, $H \notin \mathcal{G}$. □

9.1 Untangling the Cuts

To prove Theorems 9.2 and 9.3, we are going to need two Erdős-Pósa type results developed in [6, 9]. First, let us make an observation on free sets in tangles. Let $\mathcal{T}$ be a vertex tangle of order θ in a graph G, let d be a positive integer, and let W be a $\mathcal{T}$-free subset of $V(G)$. We say that a set $W' \supseteq W$ of size less than θ *blocks all non-free d-extensions of* W if for every set $Q \subseteq V(G) \setminus W'$ of size d, if $W \cup Q$ is not $\mathcal{T}$-free, then Q is not $(\mathcal{T} - W')$-free in $G - W'$. Consider a vertex separation (A, B) and a vertex $v \in A \cap B$. We say that a vertex separation (A', B') *gouges off* v *from* (A, B) if $A \subseteq A'$ and $B' \subseteq B \setminus \{v\}$.

Lemma 9.5 (⇝) *For all positive integers k and d, there exists a positive integer k' such that the following claim holds. Let $\mathcal{T}$ be a vertex tangle of order $\theta > k'$ in a graph G. For every $\mathcal{T}$-free subset W of $V(G)$ of size k, there exists a set W' of size at most k' that blocks all non-free d-extensions of W.*

Proof Let $k' = \sum_{i=0}^{d}(k+d)^{i+1}$.

For any vertex separation $(A, B) \in \mathcal{T}$ and any vertex $v \in A \cap B$, if there exists a vertex separation in $\mathcal{T}$ which gouges off v from (A, B) of order less than $k+d$, then let (A_v, B_v) be one of smallest order and subject to that with $|B_v|$ minimal; if no such vertex separation exists, we say that v is *fixed* in (A, B).

Let $\mathcal{A}_0 = \{(W, V(G))\}$. For $i = 1, \ldots, d+1$, let us construct a set $\mathcal{A}_i \subseteq \mathcal{T}$ as follows: For every $(A, B) \in \mathcal{A}_{i-1}$ and every $v \in A \cap B$ not fixed in (A, B), include

(A_v, B_v) in $\mathcal{A}_i$. Since each vertex separation in $\mathcal{A}_{i-1}$ has order less than $k+d$, we have $|\mathcal{A}_i| \le (k+d)|\mathcal{A}_{i-1}|$, and by induction, $|\mathcal{A}_i| \le (k+d)^i$.

Moreover, note that $|A_v \cap B_v| > |A \cap B|$ in this construction when $i \ge 2$. Indeed, (A, B) itself was chosen as (C_w, D_w) for some $(C, D) \in \mathcal{A}_{i-2}$ and $w \in C \cap D$, which implies that there is no vertex separation $(A', B') \in \mathcal{T}$ of order at most $|A \cap B|$ with $A \subseteq A'$ and $B' \subsetneq B$. Since W is $\mathcal{T}$-free, the vertex separations in $\mathcal{A}_1$ have order at least k; by induction, every vertex separation in $\mathcal{A}_i$ has order at least $k+i-1$. In particular, it follows that $\mathcal{A}_{d+1} = \emptyset$.

Let

$$W' = \bigcup_{i=0}^{d} \bigcup_{(A,B)\in\mathcal{A}_i} (A \cap B),$$

and observe that $|W'| \le k'$.

Consider now any set $Q \subseteq V(G) \setminus W'$ of size d such that $W \cup Q$ is not $\mathcal{T}$-free. Hence, there exists a vertex separation $(C, D) \in \mathcal{T}$ of order less than $k+d$ such that $W \cup Q \subseteq C$. Let us choose such a vertex separation of smallest order and subject to that with $|D| - |C|$ minimal. Let $i \in \{0, \dots, d\}$ be maximal such that there exists a separation $(A, B) \in \mathcal{A}_i$ with $A \subseteq C$ and $D \subseteq B$; such i exists, since this holds for $(W, V(G)) \in \mathcal{A}_0$. We claim that

$(\star)$ $\quad A \cap B \subseteq C \cap D.$

Indeed, suppose for a contradiction that there exists a vertex $v \in (A \cap B) \setminus (C \cap D)$. Since $A \subseteq C$, we have $v \notin D$, and thus (C, D) gouges off v from (A, B). This implies that v is not fixed in (A, B), and thus $(A_v, B_v) \in \mathcal{A}_{i+1}$. Note that the vertex separation $(A_1, B_1) = (A_v \cap C, B_v \cup D)$ gouges off v from (A, B); if $|A_1 \cap B_1| < |A_v \cap B_v|$, then $(A_1, B_1) \in \mathcal{T}$ by Observation 2.18 and (A_1, B_1) would contradict the minimality of the order of (A_v, B_v). Therefore, we have $|A_1 \cap B_1| \ge |A_v \cap B_v|$, and by (1.1), the vertex separation $(A_2, B_2) = (A_v \cup C, B_v \cap D)$ has order at most $|C \cap D|$. By (T1), we have $(A_2, B_2) \in \mathcal{T}$. Recall we chose (C, D) as a vertex separation of order less than $k+d$ and with $W \cup Q \subseteq C$ so that $|D| - |C|$ minimal, and since $W \cup Q \subseteq A_2$, it follows that $|B_2| - |A_2| \ge |D| - |C|$. This implies that $A_v \subseteq C$ and $D \subseteq B_v$. However, since $(A_v, B_v) \in \mathcal{A}_{i+1}$, this contradicts the maximality of i. Therefore, $(\star)$ holds.

Since $|A \cap B| \ge k$ and $|C \cap D| < k+d$, we have $|(C \cap D) \setminus W'| \le |(C \cap D) \setminus (A \cap B)| < d$. Therefore, $(C \setminus W', D \setminus W') \in \mathcal{T} - W'$ is a vertex separation in $G - W'$ of order less than d and with $Q \subseteq C \setminus W'$, showing that Q is not $(\mathcal{T} - W')$-free. Therefore, W' indeed blocks all non-free d-extensions of W. □

As a consequence, we obtain an Erdős-Pósa type result on combining free sets.

Corollary 9.6 (↬) *For all positive integers m and d, there exists a positive integer k_1 such that the following claim holds. Let $\mathcal{T}$ be a vertex tangle of order $\theta > k_1$ in a graph G, and let $\mathcal{Q}$ be a system of $\mathcal{T}$-free subsets of $V(G)$ of size d. Then either $\mathcal{Q}$ contains m pairwise vertex-disjoint sets $Q_1, \dots, Q_m$ such that $Q_1 \cup \dots \cup Q_m$ is*

$\mathcal{T}$-free, or there exists a set $W' \subseteq V(G)$ of size at most k_1 such that no set $Q \in \mathcal{Q}$ disjoint from W' is $(\mathcal{T} - W')$-free in $G - W'$.

Proof Let k_1 be the maximum of the constants k' from Lemma 9.5 applied with $k = m'd$ for $m' \in [m-1]$. Let $m' \leq m$ be maximal such that there exist pairwise vertex-disjoint sets $Q_1, \ldots, Q_{m'} \in \mathcal{Q}$ for which $W = Q_1 \cup \ldots \cup Q_{m'}$ is $\mathcal{T}$-free. If $m' = m$, then the first outcome holds. Otherwise, by Lemma 9.5, there exists a set $W' \supseteq W$ of size at most k_1 that blocks all non-free d-extensions of W. Since m' is maximal, for every $Q \in \mathcal{Q}$ disjoint from W', the set $W \cup Q$ is not $\mathcal{T}$-free, and thus Q is not $(\mathcal{T} - W')$-free in $G - W'$. □

Given a graph G and disjoint subsets X and Y of its vertices, a *d-cone* from X to Y is the union of d paths between a vertex $x \in X$ and the set Y, where the paths pairwise interesect only in x and they intersect Y only in their ends. We say x is the *tip* of the cone, and the ends of the paths in Y form the *base* of the cone. From Corollary 9.6, it is easy to obtain an Erdős-Pósa result for cones.

Corollary 9.7 *For all positive integers m and d, there exists a positive integer k_2 such that the following claim holds. Let G be a graph and let $X, Y \subseteq V(G)$ be disjoint. Then either G contains m pairwise vertex-disjoint d-cones from X to Y, or there exists a set $Z \subseteq V(G)$ of size at most k_2 such that $G - Z$ does not contain any d-cone from $X \setminus Z$ to $Y \setminus Z$.*

Proof *(↬)* Let k_1 be the constant from Corollary 9.6 and let $k_2 = \lceil 3k_1/2 \rceil$. Without loss of generality, we can assume that Y is a clique. Moreover, we can assume that $|Y| > k_2$, as otherwise we can let $Z = Y$. Let $\mathcal{T}$ be the set of vertex separations (A, B) of G of order at most k_1 such that $Y \subseteq B$. The argument given at the end of Sect. 2.5 shows that $\mathcal{T}$ is a vertex tangle of order $k_1 + 1$.

A *d-cone top* is a $\mathcal{T}$-free set consisting of a vertex of X and $d-1$ of its neighbors. Let $\mathcal{Q}$ consist of all d-cone tops. Let us consider the two possible outcomes of Corollary 9.6:

- Suppose first that there exist pairwise vertex-disjoint sets $Q_1, \ldots, Q_m \in \mathcal{Q}$ such that $W = Q_1 \cup \ldots \cup Q_m$ is $\mathcal{T}$-free. For any vertex separation (A, B) of G of order at most k_1 with $W \subseteq A$ and $Y \subseteq B$, we have $(A, B) \in \mathcal{T}$ by the definition of $\mathcal{T}$ and $|A \cap B| \geq |W|$ by the $\mathcal{T}$-freeness of W. Hence, Menger's theorem implies that G contains $|W|$ pairwise vertex-disjoint paths from W to Y. For each $i \in [m]$, the star in $G[Q_i]$ with center in X together with the paths from Q_i to Y gives rise to a d-cone, and thus G contains m pairwise vertex-disjoint d-cones from X to Y.
- Next, suppose that there exists a set $Z \subseteq V(G)$ of size at most k_1 such that no set $Q \in \mathcal{Q}$ disjoint from Z is $(\mathcal{T} - Z)$-free in $G - Z$. Suppose that there exists a d-cone C from $X \setminus Z$ to $Y \setminus Z$ in $G - Z$, and let Q consist of the tip of C and $d-1$ of its neighbors in C. Consider any vertex separation $(A, B) \in \mathcal{T} - Z$ with $Q \subseteq A$. Since $(A, B) \in \mathcal{T} - Z$, we have $Y \setminus Z \subseteq B$. Since C contains d pairwise vertex-disjoint paths from Q to $Y \setminus Z$, it follows that $|A \cap B| \geq d$, and thus Q is $(\mathcal{T} - Z)$-free. Moreover, this implies that Q is $\mathcal{T}$-free, and thus $Q \in \mathcal{Q}$. This is a contradiction, and thus no such d-cone exists in $G - Z$.

□

9.2 Structure in Minor-Free Graphs

Our next step will be to prove the Theorem 9.3 under the additional assumption that G avoids another fixed graph H' as a minor controlled by the tangle. This enables us to apply the Local Minor Structure Theorem to G; we are going to need the stronger form stated in Theorem 5.30. First, let us argue that free vertices must be fairly localized.

Lemma 9.8 *For any planar graph H with $\Delta(H) \geq 4$ and any surface Σ, there exist integers μ and ψ_0 such that the following claim holds. Let G be a graph not containing H as a topological minor and let $\mathcal{T}$ be a tangle in G. Let $\psi \geq \psi_0$ be an integer and let $(\mathcal{S}_0, \mathcal{S}_1)$ be a solid bisegregation of G with a $(\mathcal{T}, \psi)$-spread arrangement on Σ and with the surface tangle $\mathcal{T}_0$. There exists a set $U \subseteq V(\pi(\mathcal{S}_1))$ of size less than $|H|$ such that for every $(\mathcal{T}, \Delta(H))$-free vertex $v \in V(G)$, either*

- *there exists $S \in \mathcal{S}_0$ with $v \in V(S) \setminus V(\pi(\mathcal{S}_1))$, or*
- *$v \in V(\pi(\mathcal{S}_1))$ and $d_{\mathcal{T}_0}(U, v) < \mu$.*

Proof Let $D = \Delta(H)$ and $k = D^2+8D+5$. Let H_0 be the matching obtained from H by subdividing all edges twice and deleting the original vertices of H. Let θ be the constant from Theorem 4.32 for H_0 and Σ, and choose $\psi_0 \gg \mu \gg \theta, |H|, \|H\|$.

Suppose first that there exists a set R of $(\mathcal{T}, D)$-free vertices of $\pi(\mathcal{S}_1)$ such that $|R| = |H|$ and $d_{\mathcal{T}_0}(u, v) \geq \mu$ for all distinct $u, v \in R$. For each vertex $v \in R$, we use Theorem 4.21 to find a free k-local $(D+1) \times D$ battlefield $(\mathcal{R}_v, \mathcal{P}_v)$ around v in $\pi(\mathcal{S}_1)$, let Δ_v be the k-zone around v bounded by the outer cycle of the battlefield, and let Z_v the set of ends of the paths of the battlefield in $\mathrm{bd}(\Delta_v)$. Note that the set Z_v is free in the tangle obtained from $\mathcal{T}_0$ by clearing the zone Δ_v, and thus also $\mathcal{T}_0$-free and $\mathcal{T}$-free. Since v is $(\mathcal{T}, D)$-free and Z_v is $\mathcal{T}$-free, there is no separation (A, B) of G of order less than D with $v \in V(A) \setminus V(B)$ and $Z_v \subseteq B$, as neither (A, B) nor (B, A) could belong to $\mathcal{T}$. By Menger's theorem, G contains a system $\mathcal{Q}_v$ of D paths from v to Z_v intersecting only in v. Since $\mathcal{S}_1$ is solid, the cycles and paths of $\mathcal{R}_v$ and $\mathcal{P}_v$ in the minor $\pi(\mathcal{S}_1)$ of G correspond to cycles and paths in G, and the standard "linking through a cylindrical grid" argument shows that we can assume that the paths of $\mathcal{Q}_v$ intersect $\pi(\mathcal{S}_1)$ only in their ends in Z_v and in vertices drawn in the interior of Δ_v.

Let G' be the graph and $\mathcal{T}'$ the respectful tangle obtained from $\pi(\mathcal{S}_1)$ and $\mathcal{T}_0$ by clearing the zones Δ_v for all $v \in R$. Let f_v denote the face obtained by clearing Δ_v. Since $\psi \gg \theta, D$, for distinct $u, v \in R$, we have $d_{\mathcal{T}'}(f_u, f_v) \geq \theta$. By Observation 4.26, the set $\bigcup_{v \in R} Z_v$ is $\mathcal{T}'$-free. Let $\sigma : R \to V(H)$ be an arbitrary bijection. Since H is planar, we can draw it in Σ so that for each $v \in R$, the vertex $\sigma(v)$ is drawn in Δ_v, the edges incident with $\sigma(v)$ intersect Δ_v in their initial segments ending in vertices of Z_v, and all other edges are disjoint from Δ_v. Consider the corresponding drawing of H_0 in $\Sigma \setminus \bigcup_{v \in R} f_v$, and let $\rho : V(H_0) \to \bigcup_{v \in R} Z_v$ be the injective function mapping each vertex of H_0 to the vertex of $\bigcup_{v \in R} Z_v$ represented by the same point. By Theorem 4.32, there is a ρ-rooted model of H_0

in G', and since $\mathcal{S}_1$ is solid, this model corresponds to a linkage Q in G joining for each $xy \in E(H_0)$ the vertex $\rho(x)$ with the vertex $\rho(y)$. Combining Q with $\bigcup_{v \in R} Q_v$ gives us H as a topological minor of G, which is a contradiction.

Therefore, no such set R exists. Hence, letting U be a maximal set of $(\mathcal{T}, D)$-free vertices of $\pi(\mathcal{S}_1)$ such that $d_{\mathcal{T}_0}(x, y) \geq \mu$ for all distinct $x, y \in U$, we have $|U| < |H|$. Consider a $(\mathcal{T}, \Delta(H))$-free vertex $v \in V(G)$. If $v \in V(\pi(\mathcal{S}_1))$, then the maximality of U implies that $d_{\mathcal{T}_0}(U, v) < \mu$. Otherwise, there exists $S \in \mathcal{S}_0 \cup \mathcal{S}_1$ with $v \in V(S) \setminus V(\pi(\mathcal{S}_1))$. Since v is $(\mathcal{T}, D)$-free and $D \geq 4$, it cannot be the case that $S \in \mathcal{S}_1$, as in that case $v \in V(S) \setminus V(\pi(\mathcal{S}_1)) \subseteq V(S) \setminus \partial S$ and $|\partial S| \leq 3 < D$. Therefore, we have $S \in \mathcal{S}_0$, and the first outcome of the lemma holds. □

According to the previous lemma, we need to deal with two kinds of $(\mathcal{T}, \Delta)$-free vertices: Those contained in vortex-like parts, and those in the surface part close to a fixed vertex. This is something which often happens in applications of the Minor Structure Theorem. Conveniently, in the latter case, the bounded-radius piece of the surface part can be encapsulated in a vortex-like piece; and thus we can handle both cases by the same argument. To prove this, we need a variation on Corollary 4.16.

Lemma 9.9 *Let G be a graph with a cellular drawing on a surface Σ and let $\mathcal{T}$ be a respectful tangle in G of order θ. For every vertex or face a of G and every positive integer μ such that $4\mu + 3 < \theta$, there exists a cycle C in G bounding a $(3\mu + 3)$-zone Δ around a such that*

- *every atom b of G with $d_{\mathcal{T}}(a, b) \leq \mu$ is contained in the interior of Δ, and*
- *for every $v_1, v_2 \in V(C)$, there exists a path in $R_G \cap \Delta$ from v_1 to v_2 of length at most $5\mu + 4$.*

Proof (⇸) If there exists a cycle K in R_G of length at most 2μ with $\rho_G(a) \in \mathrm{ins}_{\mathcal{T}}(K)$, then choose such a cycle K with $\mathrm{ins}_{\mathcal{T}}(K)$ maximal; otherwise, let $K = \rho_G(a)$. Let X be the set of vertices of G in $\mathrm{ins}_{\mathcal{T}}(K)$ or at (graph) distance at most 2μ from K in the graph R_G. Clearly, we have $d_{\mathcal{T}_0}(a, x) \leq 3\mu$ for every $x \in X$. By Corollary 4.16, there exists a $(3\mu + 3)$-zone Δ' in G around a bounded by a cycle C' whose interior contains all vertices of G at distance at most 3μ from a; in particular, all vertices of X are drawn in the interior of Δ'.

Let G_0 be the component of $G - X$ containing C', and let f be the face of G_0 containing a. Note that all vertices of X are contained in f. Since G_0 is connected and the face f is contained in the disk Δ', f is a cellular face.

Consider any vertex u of G with $d_{\mathcal{T}}(a, u) \leq \mu$, and let F be a $(\rho_G(a), u)$-restraint of perimeter at most 2μ. If F is contained in the interior of $\mathrm{ins}_{\mathcal{T}}(K)$, then $u \in \mathrm{ins}_{\mathcal{T}}(F) \subseteq \mathrm{ins}_{\mathcal{T}}(K)$ and $u \in X$. Otherwise, note that F contains either $\rho_G(a)$ or a cycle K' of length at most 2μ with $\rho_G(a) \in \mathrm{ins}_{\mathcal{T}}(K')$, and by the maximality of $\mathrm{ins}_{\mathcal{T}}(K)$, in both cases we conclude that F intersects K. Consequently, $V(F) \cap V(G) \subseteq X$, and $F \subseteq f$. Since f is cellular and contained in Δ' and since $\mathrm{ins}_{\mathcal{T}}(F) \subseteq \Delta'$ by Lemma 4.17, we have $u \in \mathrm{ins}_{\mathcal{T}}(F) \subseteq f$. In conclusion, f contains all vertices of G at $d_{\mathcal{T}}$-distance at most μ from a. Observe that any other atom b of G at $d_{\mathcal{T}}$-distance at most μ from a is incident with such a vertex, and thus b is also contained in f.

Note that there exists a cycle C in the boundary of f such that the disk $\Delta \subseteq \Delta'$ bounded by C contains f. The disk Δ is a $(3\mu + 3)$-zone, since it is a subset of Δ'. Consider any vertices $v_1, v_2 \in V(C)$. For $i \in [2]$, let f_i be a face of G incident with v_i and contained in f. Clearly, the face f_i is incident with a vertex $v_i' \in X$, and thus there exists a path P_i in $R_G \cap \Delta$ from v_i to K of length at most $2\mu + 2$. The union of P_1, P_2, and a shortest path in K between the ends of P_1 and P_2 gives a path of length at most $5\mu + 4$ between v_1 and v_2 in $R_G \cap \Delta$. □

We are now ready to turn a bounded-diameter region into a vortex-like society. Given a graph G and its bisegregation $(\mathcal{S}_0, \mathcal{S}_1)$ with an arrangement on a surface Σ, we say that a closed disk $\Delta \subseteq \Sigma$ is $(\mathcal{S}_0, \mathcal{S}_1)$-*normal* if for every $S \in \mathcal{S}_0 \cup \mathcal{S}_1$, either $\overline{\pi}(S) \subseteq \Delta$, or $\overline{\pi}(S) \cap \Delta = \emptyset$. Let us remark that if Δ is a zone in $\pi(\mathcal{S}_1)$, then it is $(\mathcal{S}_0, \mathcal{S}_1)$-normal. Let $G_{(\mathcal{S}_0,\mathcal{S}_1)}[\Delta]$ denote the society whose underlying graph consists of

$$\bigcup_{S \in \mathcal{S}_0 \cup \mathcal{S}_1, \overline{\pi}(S) \subseteq \Delta} S$$

and the vertices of $\pi(\mathcal{S}_0 \cup \mathcal{S}_1)$ drawn in $\mathrm{bd}(\Delta)$, and whose boundary consists of these vertices in order in which they are drawn in $\mathrm{bd}(\Delta)$.

Lemma 9.10 *Let G be a graph, let $\mathcal{T}$ be a tangle in G, and let μ be a positive integer. Let $(\mathcal{S}_0, \mathcal{S}_1)$ be a solid (m, δ)-simple bisegregation of G with a $(\mathcal{T}, \psi)$-spread arrangement on a surface Σ with the surface tangle $\mathcal{T}_0$, where m, δ, and $\psi \geq 6\mu + 7$ are non-negative integers. Let a be a vertex or face of $\pi(\mathcal{S}_1)$. There exists a $(3\mu + 3)$-zone Δ around a in the drawing of $\pi(\mathcal{S}_1)$ such that*

- *every atom of $\pi(\mathcal{S}_1)$ at $d_{\mathcal{T}_0}$-distance at most μ from a is contained in the interior of Δ, and*
- *$G_{(\mathcal{S}_0,\mathcal{S}_1)}[\Delta]$ is $(\frac{5}{2}\mu + \delta + 5)$-vortex-like.*

Proof (⇸) Let Δ be the $(3\mu + 3)$-zone obtained using Lemma 9.9, and let C be the cycle bounding it. If there exists $S_0 \in \mathcal{S}_0$ such that $\overline{\pi}(S_0)$ is contained in Δ, then let f_0 be the face of $\pi(\mathcal{S}_1)$ containing $\overline{\pi}(S_0)$. Note that such S_0 is unique, since the $d_{\mathcal{T}_0}$-distance between any two such faces is at least $\psi \geq 6\mu + 7$, but the $d_{\mathcal{T}_0}$-distance between any atoms in the $(3\mu + 3)$-zone Δ is at most $6\mu + 6$.

Let $F = G_{(\mathcal{S}_0,\mathcal{S}_1)}[\Delta]$. We need to argue that F is $(\frac{5}{2}\mu + \delta + 5)$-vortex-like. Consider distinct $v_1, v_2 \in \partial F = V(C)$, and let P be a path of length at most $5\mu + 4$ between them in $R_{\pi(\mathcal{S}_1)} \cap \Delta$. Let $R = V(F) \cap V(P)$ and note that $|R| \leq \frac{5}{2}\mu + 3$. In case that $\rho_{\pi(\mathcal{S}_1)}(f_0) \in V(P)$, let u_1' and u_2' be the neighbors of $\rho_{\pi(\mathcal{S}_1)}(f_0)$ in P, and observe that for $i \in [2]$, u_i' is incident with the same face of $\pi(\mathcal{S}_0 \cup \mathcal{S}_1)$ as a vertex u_i of ∂S_0. Let $\mathcal{Q}$ be any $(\partial_{[v_1,v_2)} F, \partial_{[v_2,v_1)} F)$-linkage in F. Observe that for each $Q \in \mathcal{Q}$, the path Q either intersects $R \cup \{u_1, u_2\}$, or contains a subpath in S_0 from $\partial_{[u_1,u_2)} S_0$ to $\partial_{[u_2,u_1)} S_0$. Since S_0 is δ-vortex-like, it follows that $|\mathcal{Q}| \leq |R| + 2 + \delta = \frac{5}{2}\mu + \delta + 5$. □

We now aim to show that unless all $(\mathcal{T}, \Delta)$-free vertices in a single vortex can be cut off by a small set of vertices, we can find many disjoint cones out of the vortex, link them out through a battlefield surrounding the vortex to a free set Z in the boundary of the battlefield, and use Theorem 4.32 to obtain the forbidden large-degree-outerplanar graph as a minor. An issue is that we do not have much control over the order of the ends of distinct cones in Z, and to apply Theorem 4.32, we need them to be consecutive. This is achieved in the next two lemmas. First, let us argue that if we take a δ-vortex-like society S_0 (i.e., one without a transaction of size more than δ) and combine it with a planar (up to 3-separations) neighborhood around it, the resulting society still does not have a large transaction consisting only of paths intersecting S_0.

Lemma 9.11 *Let G be a society and let $(\{S_0\}, S_1)$ be a $(1, \delta)$-simple bisegregation of G with an arrangement in the disk. Let $\mathcal{P}$ be a transaction in G. If all paths of $\mathcal{P}$ intersect S_0, then $|\mathcal{P}| \leq \delta + 2$.*

Proof Let u and v be vertices of ∂S such that each path in $\mathcal{P}$ starts in $\partial_{[u,v)}G$ and ends in $\partial_{[v,u)}G$. Let $P_1, \ldots, P_m$ be the initial segments of paths in $\mathcal{P}$ till their first vertex in ∂S_0, ordered according to the order of their ends in $\partial_{[u,v)}G$. Let v_1 and v_m be the ends of P_1 and P_m in ∂S_0. Consider any path $P \in \mathcal{P}$ whose initial segment is neither P_1 nor P_m. Since P ends in $\partial_{[v,u)}G$ and does not cross $P_1 \cup P_m$, observe that $P \cap S_0$ must contain a segment from $\partial_{(v_1,v_m)}S_0$ to $\partial_{(v_m,v_1)}S_0$. Since these segments form a transaction in S_0, there are at most δ of them in total, and thus $|\mathcal{P}| \leq \delta + 2$. □

Next, let us apply this lemma to disentangle cones. Let G be a society. We say that d-cones $C_1, \ldots, C_t$ from $V(G) \setminus \partial G$ to ∂G are *consecutive* if there exist vertices $v_1, \ldots, v_{dt}$ in ∂G in order such that for $i \in [t]$, the base of C_i is $\{v_{d(i-1)+1}, \ldots, v_{di}\}$.

Lemma 9.12 *Let G be a society and let $(\{S_0\}, S_1)$ be a $(1, \delta)$-simple bisegregation of G with an arrangement in the disk Δ, let $d \geq 2$ be an integer, and let $\mathcal{C}$ be a set of pairwise vertex-disjoint d-cones from $V(S_0)$ to ∂G. For any positive integer t, if $|\mathcal{C}| \geq (\delta + 2)t$, then $\mathcal{C}$ contains a subset of t consecutive d-cones.*

Proof Let us choose a point $p \in \mathrm{bd}(\Delta)$ not in ∂G arbitrarily, and for each $C \in \mathcal{C}$, let I_C be a minimal interval of $\mathrm{bd}(\Delta) \setminus \{p\}$ containing the base of C. Let F be the auxiliary graph with vertex set $\mathcal{C}$ in which distinct d-cones $C, C' \in \mathcal{C}$ are adjacent if and only if $I_C \cap I_{C'} \neq \emptyset$.

Suppose first that F has a clique K of size $\delta + 3$. Hence, there exists a point $q \in \mathrm{bd}(\Delta)$ not in ∂G such that for each $C \in K$, I_C contains q. In other words, the base of C has vertices u_C and v_C in different components of $\mathrm{bd}(\Delta) \setminus \{p, q\}$. Note that C contains a path P_C from u_C to v_C passing through the tip of C in S_0. However, $\{P_C : C \in K\}$ is a transaction of size $|K| = \delta + 3$ in G, contradicting Lemma 9.11.

Therefore, F has clique number at most $\delta + 2$. Since F is an interval graph, its chromatic number is equal to its clique number, and thus $\chi(F) \leq \delta + 2$. Therefore, F contains an independent set $\mathcal{C}_0$ of size at least $|\mathcal{C}|/(\delta + 2) \geq t$, and the d-cones in $\mathcal{C}_0$ are consecutive. □

We are now ready to obtain the desired lemma on $(\mathcal{T}, \Delta)$-free vertices in a bounded radius region.

Lemma 9.13 *For all integers $\mu, t, \delta \geq 1$ and $d \geq 2$, there exist integers λ and $\psi_0 > 2\lambda$ such that the following claim holds. Let G be a graph, let $\mathcal{T}$ be a tangle in G, and let $(\mathcal{S}_0, \mathcal{S}_1)$ be a solid (m, δ)-simple bisegregation of G with a $(\mathcal{T}, \psi)$-spread arrangement on a surface Σ with the surface tangle $\mathcal{T}_0$, where m and $\psi \geq \psi_0$ are non-negative integers. Let a be a vertex or face of $\pi(\mathcal{S}_1)$ and let X be a set of $(\mathcal{T}, d)$-free vertices of G such that for each $x \in X$, either $x \in V(\pi(\mathcal{S}_1))$ and $d_{\mathcal{T}_0}(a, x) \leq \mu$, or $x \in V(S)$ for some $S \in \mathcal{S}_0$ such that the vortex face f of S satisfies $d_{\mathcal{T}_0}(a, f) \leq \mu$. Then either*

- *there exists a set $R \subseteq V(G)$ of size at most λ such that no vertex of $X \setminus R$ is $(\mathcal{T} - R, d)$-free in $G - R$, or*
- *there exists a λ-zone Δ around a in the drawing of $\pi(\mathcal{S}_1)$ such that the society $F = G_{(\mathcal{S}_0, \mathcal{S}_1)}[\Delta]$ contains t consecutive pairwise vertex-disjoint d-cones from $V(F) \setminus \partial F$ to ∂F, and the union Z of their bases is $\mathcal{T}'_0$-free, where $\mathcal{T}'_0$ is the tangle obtained from $\mathcal{T}_0$ by clearing the zone Δ.*

Proof Choose $\psi_0 \gg \lambda \gg \beta \gg \kappa \gg \mu, d, t, \delta$.

Since $\psi \gg \lambda$, there exists at most one $S_0 \in \mathcal{S}_0$ such that the vortex face f_0 of S_0 satisfies $d_{\mathcal{T}_0}(a, f_0) \leq \lambda$. If $d_{\mathcal{T}_0}(a, f_0) \leq \mu$ or $d_{\mathcal{T}_0}(a, f_0) > \beta$ or no such S_0 exists, then let $\mu' = \mu$ and $\lambda' = \beta$, otherwise let $\mu' = \beta$ and $\lambda' = \lambda$. Let Δ_0 be the $(3\mu' + 3)$-zone around a obtained using Lemma 9.10, let $F_0 = G_{(\mathcal{S}_0, \mathcal{S}_1)}[\Delta_0]$ and note that $X \subseteq V(F_0) \setminus \partial F_0$ and F_0 is δ'-vortex-like, where $\delta' = \lceil \frac{5}{2}\mu' + \delta + 5 \rceil$.

Let $r = (\delta' + 2)t$. By Theorem 4.21, there exists a free λ'-local $(dr + 1) \times (\kappa + dr)$ battlefield in $\pi(\mathcal{S}_1)$ containing Δ_0 in its egg. Let Δ be the λ'-zone bounded by the outer cycle C of the battlefield and let Z_0 be the $\mathcal{T}'_0$-free set of ends of the paths of the battlefield. Let $F = G_{(\mathcal{S}_0, \mathcal{S}_1)}[\Delta]$; by the choice of μ' and λ', F has a $(1, \delta')$-simple bisegregation $(\{F_0\}, \mathcal{S}'_1)$ with an arrangement in Δ.

Suppose first that F contains a set C of r pairwise vertex-disjoint d-cones from X to $V(C)$. By the standard "linking through a cylindrical grid" argument, we can assume that bases of these d-cones are subsets of Z_0. By Lemma 9.12 applied to F, we can select from C a subset of t consecutive d-cones, implying that the second conclusion of the lemma holds.

Hence, suppose that F does not contain r pairwise vertex-disjoint d-cones from X to $V(C)$, and thus by Corollary 9.7, there exists a set $R \subseteq V(F)$ of size at most $\kappa \leq \lambda$ such that $F - R$ does not contain any d-cone from $X \setminus R$ to $V(C) \setminus R$. We claim that no vertex $x \in X \setminus R$ is $(\mathcal{T} - R, d)$-free in $G - R$, and thus the first conclusion of the lemma holds: Otherwise, consider a $(\mathcal{T} - R, d)$-free vertex $x \in X \setminus R$. Note that the set $Z_0 \setminus R$ of size at least d is $(\mathcal{T} - R)$-free. It follows that $G - R$ has no vertex separation (A, B) of order less than d with $x \in A \setminus B$ and $Z_0 \setminus R \subseteq B$, as neither (A, B) nor (B, A) could belong to $\mathcal{T} - R$. By Menger's theorem, it follows that $G - R$ contains d paths from x to $Z_0 \setminus R$ intersecting only in x. By truncating these paths on the first vertices belonging to C, we conclude that $F - R$ contains a d-cone from $X \setminus R$ to $V(C) \setminus R$, which is a contradiction. □

In combination with Theorem 4.32, we can eliminate the second outcome from Lemma 9.13 for graphs avoiding a large-degree-outerplanar graph as a topological minor.

Corollary 9.14 *For all positive integers μ and δ, every surface Σ, and every large-degree-outerplanar graph H, there exist integers λ and $\psi_0 > 2\lambda$ such that the following claim holds. Let G be a graph not containing H as a topological minor, let $\mathcal{T}$ be a tangle in G, and let $(\mathcal{S}_0, \mathcal{S}_1)$ be a solid (m, δ)-simple bisegregation of G with a $(\mathcal{T}, \psi)$-spread arrangement on Σ with the surface tangle $\mathcal{T}_0$, where m and $\psi \geq \psi_0$ are non-negative integers. Let a be a vertex or face of $\pi(\mathcal{S}_1)$ and let X be a set of $(\mathcal{T}, \Delta(H))$-free vertices of G such that for each $x \in X$, either $x \in V(\pi(\mathcal{S}_1))$ and $d_{\mathcal{T}_0}(a, x) \leq \mu$, or $x \in V(S)$ for some $S \in \mathcal{S}_0$ such that the vortex face f of S satisfies $d_{\mathcal{T}_0}(a, f) \leq \mu$. Then there exists a set $R \subseteq V(G)$ of size at most λ such that no vertex of $X \setminus R$ is $(\mathcal{T} - R, \Delta(H))$-free in $G - R$.*

Proof Without loss of generality, we can assume that H is simple and triangle-free, since subdividing the edges of H preserves the large-degree-outerplanarity. Let V_0 be the set of vertices of H of degree at least four. Let H_0 be the graph obtained from H by, for each vertex $v \in V_0$, subdividing once all edges incident with v and deleting v (the edges between distinct vertices of V_0 are subdivided twice); let T_v be the set of neighbors of v in H_0. Let $T = \bigcup_{v \in V_0} T_v$, and note that all vertices of T have degree one in H_0 and all other vertices of H_0 have degree at most three. Moreover, H_0 has a drawing in the plane such that the vertices of T are incident with the outer face of H_0, and there exists an ordering $v_1, \ldots, v_t$ of V_0 so that in the facial walk of the outer face of H_0 we first encounter all vertices of T_{v_1}, then all vertices of $T_{v_2}, \ldots$, and finally all vertices of T_{v_t}.

Let θ be the constant from Theorem 4.32 for H_0 and Σ. Let λ and ψ_0 be as in Lemma 9.13 for μ, $d = \Delta(H)$, t, and δ; without loss of generality, we assume that $\psi_0 \gg \theta, \lambda$.

The first outcome of Lemma 9.13 matches the desired outcome of this corollary. Hence, suppose that the second outcome holds, and thus there exists a λ-zone Δ around a in the drawing of $\pi(\mathcal{S}_1)$ such that the society $F = G_{(\mathcal{S}_0, \mathcal{S}_1)}[\Delta]$ contains t consecutive pairwise vertex-disjoint $\Delta(H)$-cones $C_1, \ldots, C_t$ from $V(F) \setminus \partial F$ to ∂F, and the union Z of their bases is $\mathcal{T}'_0$-free, where $\mathcal{T}'_0$ is the tangle obtained from $\mathcal{T}_0$ by clearing the zone Δ.

Since H has a plane drawing such that all vertices of V_0 are incident with the outer face (in order $v_1, \ldots, v_t$), we can draw H on Σ so that each vertex of V_0 is drawn in the interior of Δ, for each $i \in [t]$ the initial segment of each edge incident with v_i is drawn in Δ and intersects $\mathrm{bd}(\Delta)$ in a point corresponding to a vertex of the base of C_i, and the drawing is otherwise disjoint from Δ. Consider the corresponding drawing of H_0, and let $\rho : T \to Z$ map each vertex of T to the vertex of Z represented by the same point. Since $\psi \gg \lambda, \theta$, Theorem 4.32 implies that there is a ρ-rooted model of H_0 in the graph obtained from $\pi(\mathcal{S}_1)$ by clearing the zone Δ. Let G' be the union of the cells $S \in \mathcal{S}_1$ such that $\overline{\pi}(S) \not\subseteq \Delta$. Since $\mathcal{S}_1$ is solid, Observation 5.10 implies that H_0 is also a ρ-rooted minor of G', and since H_0 has maximum degree at most three and the vertices of T have degree one, H_0 is actually a ρ-rooted topological minor of G'. In combination with parts of the cones $C_1, \ldots, C_t$, this shows that H is a topological minor of G, which is a contradiction. □

We can now further combine this result with Lemma 9.8, applying Corollary 9.14 separately in the following cases and taking the union of the sets R obtained in each of them:

- For each vertex $a \in U$.
- For each $S \in \mathcal{S}_0$, letting a be the vortex face of S.

This gives the following conclusion.

Corollary 9.15 *For all integers $m \geq 0$ and $\delta \geq 1$, every surface Σ, and every large-degree-outerplanar graph H with $\Delta(H) \geq 4$, there exists an integer λ such that the following claim holds. Let G be a graph not containing H as a topological minor, let $\mathcal{T}$ be a tangle in G, and let $(\mathcal{S}_0, \mathcal{S}_1)$ be a solid (m, δ)-simple bisegregation of G with a $(\mathcal{T}, \psi)$-spread arrangement on Σ, where $\psi \geq \lambda$. Then there exists a set $R \subseteq V(G)$ of size at most λ such that no vertex of $X \setminus R$ is $(\mathcal{T} - R, \Delta(H))$-free in $G - R$.*

Finally, by combining this corollary with the strong form of the Local Minor Structure Theorem, we show that Theorem 9.3 holds if an additional minor is forbidden.

Lemma 9.16 *For every graph H' and every large-degree-outerplanar graph H, there exist integers a_0 and $\theta_0 \geq a_0 + \Delta(H)$ such that the following claim holds. Let $\mathcal{T}$ be a tangle of order at least θ_0 in a graph G, and suppose that $\mathcal{T}$ does not control a minor of H'. If H is not a topological minor of G, then there exists a set $X \subseteq V(G)$ of size at most a_0 such that no vertex of $G - X$ is $(\mathcal{T} - X, \Delta(H))$-free.*

Proof We can assume that H' is non-planar, as otherwise Theorem 2.37 and Corollary 4.33 imply that if G has a tangle $\mathcal{T}$ of large order, then $\mathcal{T}$ controls a minor of H' in G. Similarly, if G has a tangle of large order, then it contains H as a minor, and since it does not contain H as a topological minor, we can assume $\Delta(H) \geq 4$.

Let $\lambda : \mathbb{N} \to \mathbb{N}$ be the function such that for a positive integer δ, $\lambda(\delta)$ is the maximum of the constants λ from Corollary 9.15 for H'-avoiding surfaces Σ, $m = \mathrm{cr}_\Sigma(H') - 1$, and the given H and δ. Let $\psi(a, \delta) = \lambda(\delta)$ for every non-negative integer a. Let α, δ_0, and θ be the constants from Theorem 5.30 for H' and ψ. Let $a_0 = \alpha + \max_{\delta \leq \delta_0} \lambda(\delta)$ and $\theta_0 = \max(\theta, a_0 + \Delta(H))$.

By Theorem 5.30, there exists a set $A \subseteq V(G)$ of size at most α such that for some $\delta \leq \delta_0$, $G - A$ has a solid $(\mathrm{cr}_\Sigma(H') - 1, \delta)$-simple bisegregation with a $(\mathcal{T} - A, \lambda(\delta))$-spread arrangement on an H'-avoiding surface Σ. By Corollary 9.15, there exists a set $R \subseteq V(G - A)$ of size at most $\lambda(\delta) \leq a_0 - \alpha$ such that no vertex of $G - (R \cup A)$ is $(\mathcal{T} - (R \cup A), \Delta(H))$-free. Hence, the claim of the lemma holds with $X = R \cup A$. □

9.3 Dealing with a Clique Minor

The main result in its local form (Theorem 9.3) now easily follows using the linking through a clique minor technique from Sect. 3.1. First, let us restate Theorem 3.1 in terms of tangles controlling a clique minor.

Lemma 9.17 *For any graph H, there exists a positive integer $k \geq |H|$ such that the following claim holds. Let G be a graph with a vertex tangle $\mathcal{T}$ of order at least k that controls a model ν of K_k in G. Let ρ be a bijection between a subset* $\mathrm{dom}(\rho)$ *of $V(H)$ and a set R of vertices of G. If R is $\mathcal{T}$-free, then H is a ρ-rooted minor of G.*

Proof We can assume that H is a simple graph (otherwise, subdivide its edges). Let $k = 2|H|$. Suppose that a vertex separation (C, D) of order less than k separates ν from the roots, i.e., that $R \subseteq V(C)$ and there exists $u \in V(K_k)$ such that $\nu(u) \subseteq D - V(C)$. Since $\mathcal{T}$ controls ν, it follows that $(D, C) \notin \mathcal{T}$, and thus $(C, D) \in \mathcal{T}$. Since R is $\mathcal{T}$-free, it follows that $|C \cap D| \geq |R|$; i.e., every vertex separation that separates ν from the roots has order at least $|R|$. By Theorem 3.1, H is a ρ-rooted minor of G. □

We also need the following claim about $(\mathcal{T}, d)$-free vertices.

Lemma 9.18 *Let $d \geq 2$ be an integer and let $\mathcal{T}$ be a vertex tangle of order at least d in a graph G. If a vertex $v \in V(G)$ is $(\mathcal{T}, d)$-free, then there exists a set $U \subseteq N(v)$ of size $d - 1$ such that $U \cup \{v\}$ is $\mathcal{T}$-free.*

Proof (↬) First, let us show that $\mathrm{rk}_{\mathcal{T}}(N[v]) \geq d$: Otherwise, there would exist a vertex separation $(A, B) \in \mathcal{T}$ of order less than d with $N[v] \subseteq A$. Without loss of generality $v \notin B$, as otherwise we could remove v from B and the resulting separation would still belong to $\mathcal{T}$ by Observation 2.18. But then (A, B) would show that v is not $(\mathcal{T}, d)$-free, contradicting the assumptions.

By Observation 2.40, there exists a $\mathcal{T}$-free set $C \subseteq N[v]$ of size d. If $v \in C$, then we can let $U = C \setminus \{v\}$; hence, suppose that $v \notin C$. For each $u \in U$, let $C_u = (C \setminus \{u\}) \cup \{v\}$. If $\mathrm{rk}_{\mathcal{T}}(C_u) = d$, then we can let $U = C_u$. Hence, we can assume that $\mathrm{rk}_{\mathcal{T}}(C_u) = d - 1$, and thus there exists a vertex separation $(A_u, B_u) \in \mathcal{T}$ of order $d - 1$ with $C_u \subseteq A_u$. Since $\mathrm{rk}_{\mathcal{T}}(C) = d$, C cannot be a subset of A_u, and thus $u \in B_u \setminus A_u$. Hence, the neighbor v of u belongs to $A_u \cap B_u$.

Let us choose a vertex $u_1 \in C$ arbitrarily. By Lemma 2.41, we have $d - 1 = \mathrm{rk}_{\mathcal{T}}(C_{u_1} \setminus \{v\}) \leq \mathrm{rk}_{\mathcal{T}}(A_{u_1}) = \mathrm{rk}_{\mathcal{T}}(A_{u_1} \cap B_{u_1})$. By Lemma 2.44, G contains a set $\mathcal{P}$ of $d - 1$ pairwise vertex-disjoint paths from $C_{u_1} \setminus \{v\}$ to $A_{u_1} \cap B_{u_1}$. Since $|A_{u_1} \cap B_{u_1}| = d - 1$ and $v \in A_{u_1} \cap B_{u_1}$, a vertex u_2 of C_{u_1} is joined by a path

$P \in \mathcal{P}$ to v. Note that $C \setminus \{u_1, u_2\} = C_{u_1} \cap C_{u_2} \setminus \{v\} \subseteq A_{u_1} \cap A_{u_2}$ and $\mathcal{P} \setminus \{P\}$ is a system of $d - 2$ pairwise vertex-disjoint paths from $C \setminus \{u_1, u_2\}$ to B_{u_1} not containing the vertex $v \in A_{u_1} \cap A_{u_2} \cap (B_{u_1} \cup B_{u_2})$. Consequently, the vertex separation $(A_{u_1} \cap A_{u_2}, B_{u_1} \cup B_{u_2})$ of G has order at least $d - 1$. By (1.1), the separation $(A_{u_1} \cup A_{u_2}, B_{u_1} \cap B_{u_2})$ has order at most $d - 1$ and belongs to $\mathcal{T}$ by (T1). However, we have $C \subset C_{u_1} \cup C_{u_2} \subseteq A_{u_1} \cup A_{u_2}$, which is a contradiction since C is $\mathcal{T}$-free. □

We are now ready to finish the proof of the local form of the main result, which we restate for convenience.

Theorem 9.3 *For every large-degree-outerplanar graph H, there exist integers a_0 and $\theta \geq a_0 + \Delta(H)$ such that the following claim holds. Let $\mathcal{T}$ be a tangle of order at least θ in a graph G. If H is not a topological minor of G, then there exists a set $X \subseteq V(G)$ of size at most a_0 such that no vertex of $G - X$ is $(\mathcal{T} - X, \Delta(H))$-free.*

Proof We can assume that $\Delta(H) \geq 4$: The existence of a tangle of large order implies that H is a minor of G by Corollary 2.34, and if $\Delta(H) \leq 3$, this would also imply that H is a topological minor of G. Without loss of generality, we can assume that H does not contain any isolated vertices, as otherwise we can eliminate them by adding pendant edges incident with them.

Let H_0 be the matching obtained from H by subdividing each edge twice and deleting the original vertices of H. Let k be the constant from Lemma 9.17 for H_0, and let $H' = K_k$. Let a'_0 and θ'_0 be the constants a_0 and θ_0 from Lemma 9.16 for H' and H. Let $d = \Delta(H)$. Let k_1 be the constant from Corollary 9.6 for $m = |H|$ and d. Let $\theta = \max(\theta'_0, k_1 + 1)$ and $a_0 = \max(a'_0, k_1)$.

We can assume that the tangle $\mathcal{T}$ controls a minor of H', as otherwise the claim follows from Lemma 9.16. Let $\mathcal{Q}$ be the set of all $\mathcal{T}$-free d-element subsets of $V(G)$ consisting of a vertex and $d - 1$ of its neighbors. Suppose first that there exists a system $\{Q_v \in \mathcal{Q} : v \in V(H)\}$ of pairwise vertex-disjoint sets such that the set $R = \bigcup_{v \in V(H)} Q_v$ is $\mathcal{T}$-free. For each vertex $v \in V(H)$, let $q_v \in Q_v$ be a vertex adjacent in G to all other vertices of Q_v. Let $\rho : V(H_0) \to R$ map for each $v \in V(H)$ the neighbors of v in H_0 injectively to vertices of Q_v so that one of them is mapped to q_v (each vertex of H has at least one neighbor in H_0, since H does not have isolated vertices). By Lemma 9.17, H_0 is a ρ-rooted minor of G, and since H_0 is a matching, we can assume the model of this minor consists exactly of paths in G between vertices of $\mathrm{img}(\rho)$. These paths combine with stars on $G[Q_v]$ for $v \in V(H)$ to a topological minor of H in G, which is a contradiction.

Therefore, the union of any $m = |H|$ pairwise vertex-disjoint sets from $\mathcal{Q}$ is not $\mathcal{T}$-free. By Corollary 9.6, there exists a set $X \subseteq V(G)$ of size at most $k_1 \leq a_0$ such that no set $Q \in \mathcal{Q}$ disjoint from X is $(\mathcal{T} - X)$-free in $G - X$. Thus, no set $Q' \subseteq V(G - X)$ consisting of a vertex and $d - 1$ of its neighbors is $(\mathcal{T} - X)$-free, as otherwise Q' would also be $\mathcal{T}$-free and we would have $Q' \in \mathcal{Q}$. By Lemma 9.18, we conclude that no vertex of $G - X$ is $(\mathcal{T} - X, \Delta(H))$-free. □

9.4 Local to Global

Finally, let us show how Theorem 9.3 implies the global form of the result, Theorem 9.2. Theorem 9.3 implies that almost all large degree vertices can be cut off by small cuts, and it might seem that that is all we need to use the standard way to construct tree decompositions described after the proof of Lemma 2.16. However, an issue is that the cuts separating different vertices may cross and thus not give rise to a star decomposition of G. Hence, some postprocessing to avoid this issue is needed.

Throughout most of this section, we are going to work in the following setting, which we refer to as ($\star$):

- G_0 is a graph and $\mathcal{T}_0$ is a vertex tangle of order θ_0 in G_0.
- $\Delta \le \theta_0$ is a positive integer such that no vertex of G_0 is $(\mathcal{T}_0, \Delta)$-free.
- For every vertex $v \in V(G_0)$ of degree at least Δ, $(A_v, B_v) \in \mathcal{T}_0$ is a vertex separation with $v \in A_v \setminus B_v$ of smallest order, and subject to that with $|A_v|$ minimal; and $W_v = A_v \setminus B_v$.
- $\mathcal{W}$ consists of the sets W_v that are inclusionwise-maximal, i.e., such that there is no vertex v' of degree at least Δ with $W_{v'} \supsetneq W_v$.
- For each $W \in \mathcal{W}$, v_W is a vertex of degree at least Δ such that $W = W_{v_W}$ (if there are several vertices with this property, v_W is chosen among them arbitrarily), and $M = \{v_W : W \in \mathcal{W}\}$.

Note that the maximality of the sets in $\mathcal{W}$ and the fact that for each set we chose only one representative in M has the following consequence.

Observation 9.19 *With the notation as in ($\star$), for distinct $v_1, v_2 \in M$, we have $W_{v_1} \not\subseteq W_{v_2}$.*

Actually, even stronger claim holds.

Lemma 9.20 *Let the notation be as in ($\star$), and suppose that $\theta_0 > 2(\Delta - 1)$. If v_1 and v_2 are distinct vertices of M, then $v_1 \notin W_{v_2}$.*

Proof Let $(A, B) = (A_{v_1} \cap A_{v_2}, B_{v_1} \cup B_{v_2})$ and $(A', B') = (A_{v_1} \cup A_{v_2}, B_{v_1} \cap B_{v_2})$. Since $\theta_0 > 2(\Delta - 1)$, Observation 2.18 and Lemma 2.22 imply that $(A, B), (A', B') \in \mathcal{T}_0$. By (1.1), we have either $|A' \cap B'| < |A_{v_2} \cap B_{v_2}|$, or $|A \cap B| \le |A_{v_1} \cap B_{v_1}|$. Since $v_2 \in A_{v_2} \setminus B_{v_2} \subseteq A' \setminus B'$, the former would contradict the minimality of the order of (A_{v_2}, B_{v_2}). Hence, the latter holds.

If $v_1 \in A \setminus B$, then the choice of (A_{v_1}, B_{v_1}) implies that $|A \cap B| = |A_{v_1} \cap B_{v_1}|$ and $|A_{v_1}| \le |A|$. The latter implies $A_{v_1} \subseteq A_{v_2}$, and thus

$$|A \cap B| = |A_{v_1} \cap (B_{v_1} \cup B_{v_2})| = |A_{v_1} \cap B_{v_1}| + |A_{v_1} \cap B_{v_2} \setminus B_{v_1}|.$$

Since $|A \cap B| = |A_{v_1} \cap B_{v_1}|$, it follows that $A_{v_1} \cap B_{v_2} \subseteq B_{v_1}$. Hence,

$$A_{v_2} \cap B_{v_2} \subseteq (A_{v_1} \cap B_{v_2}) \cup (A_{v_2} \setminus A_{v_1}) \subseteq B_{v_1}.$$

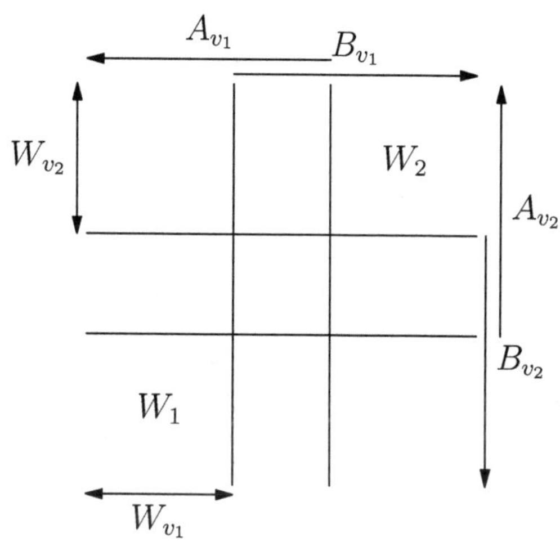

Fig. 9.1 The vertex separations from the proof of Lemma 9.21

We conclude that

$$W_{v_1} = A_{v_1} \setminus B_{v_1} \subseteq A_{v_2} \setminus (A_{v_2} \cap B_{v_2}) = A_{v_2} \setminus B_{v_2} = W_{v_2},$$

which contradicts Observation 9.19.

Therefore, $v_1 \notin A \setminus B$. Since $v_1 \in A_{v_1}$ and $v_1 \notin B_{v_1}$, this is equivalent to $v_1 \notin A_{v_2} \setminus B_{v_2} = W_{v_2}$. □

If the sets in $\mathcal{W}$ were disjoint, we could simply use them as the rays of the desired star decomposition. This is not quite true, but their intersections are restricted.

Lemma 9.21 *Let the notation be as in (⋆), and suppose that $\theta_0 > 2(\Delta - 1)$. If v_1 and v_2 are distinct vertices of M and $W_{v_1} \cap W_{v_2} \neq \emptyset$, then $v_1 \in A_{v_2} \cap B_{v_2}$ or $v_2 \in A_{v_1} \cap B_{v_1}$.*

Proof Suppose for a contradiction that neither is the case. By Lemma 9.20, $v_1 \notin A_{v_2} \setminus B_{v_2}$, and since $v_1 \notin A_{v_2} \cap B_{v_2}$, it follows that $v_1 \notin A_{v_2}$. Symmetrically, we have $v_2 \notin A_{v_1}$.

For $i \in [2]$, let $W_i = W_{v_i} \setminus A_{v_{3-i}}$, let $B_i = V(G_0) \setminus W_i$, and let A_i consist of W_i and the neighbors of W_i in B_i, so that (A_i, B_i) is a vertex separation of G_0, see Fig. 9.1 for an illustration. Observe that

$$|A_1 \cap B_1| + |A_2 \cap B_2| \leq |A_{v_1} \cap B_{v_1}| + |A_{v_2} \cap B_{v_2}|.$$

Indeed, for $i \in [2]$, all neighbors of W_i in B_i belong to $(A_{v_1} \cap B_{v_1}) \cup (A_{v_2} \cap B_{v_2})$, and the vertices with neighbors both in W_1 and W_2 belong to $A_{v_1} \cap B_{v_1} \cap A_{v_2} \cap B_{v_2}$.

By symmetry, we can assume that $|A_1 \cap B_1| \leq |A_{v_1} \cap B_{v_1}|$. Since $W_1 \subseteq W_{v_1}$, we have $A_1 \subseteq A_{v_1}$, and Observation 2.18 implies that $(A_1, B_1) \in \mathcal{T}_0$. Since $v_1 \in W_{v_1}$ and $v_1 \notin A_{v_2}$, we have $v_1 \in W_1 = A_1 \setminus B_1$. The choice of (A_{v_1}, B_{v_1}) thus implies that $|A_1 \cap B_1| = |A_{v_1} \cap B_{v_1}|$ and $|A_1| \geq |A_{v_1}|$. This is a contradiction, since $A_1 \subseteq A_{v_1}$ and $\emptyset \neq W_{v_1} \cap W_{v_2} \subseteq A_{v_1} \setminus B_{v_2} \subseteq A_{v_1} \setminus A_1$. □

This implies that no vertex can be in too many elements of $\mathcal{W}$.

Corollary 9.22 *Let the notation be as in (⋆), and suppose that $\theta_0 > 2(\Delta - 1)$. Every vertex of G_0 belongs to less than 2Δ of the sets of $\mathcal{W}$.*

Proof Consider a vertex $u \in V(G_0)$ and let $S \subseteq M$ consist of the vertices $v \in M$ such that $u \in W_v$. Let $\mathbf{K}$ be the complete graph with vertex set S and with the edge v_1v_2 directed towards v_2 if $v_2 \in A_{v_1} \cap B_{v_1}$ and towards v_1 if $v_1 \in A_{v_2} \cap B_{v_2}$; by Lemma 9.21, each edge of $\mathbf{K}$ is directed in at least one direction. Moreover, the outdegree of each vertex $v \in S$ is at most $|A_v \cap B_v| \leq \Delta - 1$. It follows that $\mathbf{K}$ has at most $(\Delta - 1)|S|$ edges. Since $\mathbf{K}$ is a complete graph, this implies that $\binom{|S|}{2} \leq (\Delta - 1)|S|$, and thus $|S| \leq 2\Delta - 1$. Hence, u is contained in at most $2\Delta - 1$ of the sets of $\mathcal{W}$. □

Moreover, each element of $\mathcal{W}$ is disjoint from almost all others.

Corollary 9.23 *Let the notation be as in (⋆), and suppose that $\theta_0 > 2(\Delta - 1)$. For every vertex $u \in M$, there exist fewer than $2\Delta^2$ vertices $v \in M$ such that $W_u \cap A_v \neq \emptyset$.*

Proof Consider any vertex $v \in M$ such that $W_u \cap A_v \neq \emptyset$. We claim that the set W_v intersects $A_u \cap B_u$. Note that Observation 2.18 implies that $G_0[W_v]$ is connected, as otherwise we could move from A_v to B_v the vertex set of a component of $G_0[W_v]$ not containing v, thus decreasing the size of A_v. Moreover, each vertex of $A_v \cap B_v$ has a neighbor in W_v, as otherwise we could remove the vertex from A_v. If W_v were disjoint from $A_u \cap B_u$, then since $v \in B_u$ by Lemma 9.20, it would follow that $W_v \subseteq B_u \setminus A_u$ and $A_v \subseteq B_u$, contrary to the assumption that $W_u \cap A_v \neq \emptyset$.

Therefore, for every $v \in M$ such that $W_u \cap A_v \neq \emptyset$, the set W_v intersects $A_u \cap B_u$. By Corollary 9.22, there are less than $2\Delta|A_u \cap B_u| < 2\Delta^2$ choices for v. □

We are now ready to prove the result on star decompositions. Let $\mathcal{T}_0$ be a vertex tangle in G_0. A star decomposition (S, β) of G_0 is $\mathcal{T}_0$-*central* if for every ray x of S, we have $(\beta(x), \beta(S - x)) \in \mathcal{T}_0$.

Lemma 9.24 *Let G_0 be a graph, let Δ be a positive integer, and let $\mathcal{T}_0$ be a vertex tangle of order $\theta_0 \geq 2\Delta^3$ in G_0. If no vertex of G_0 is $(\mathcal{T}_0, \Delta)$-free, then G_0 has a $\mathcal{T}_0$-central star decomposition (S, β) such that the torso of the center w of S has maximum degree less than $2\Delta^4$.*

Proof Let the notation be as in (⋆). Let

$$C = \Big(V(G_0) \setminus \bigcup \mathcal{W}\Big) \cup \bigcup_{v \in M} (A_v \cap B_v)$$

be the set of vertices of G_0 obtained by excluding those strictly split off by the vertex separations corresponding to $\mathcal{W}$, but keeping those of them that contribute to the boundaries of the separations. Let $K_1, \ldots, K_m$ be the components of $G_0 - C$. Let S be the star with m rays $r_1, \ldots, r_m$ and center w. For $i \in [m]$, let $\beta(r_i)$ consist of $V(K_i)$ and the neighbors of $V(K_i)$ in C, and let $\beta(w) = C$. Clearly, (S, β) is a star decomposition of G_0.

Consider any vertex $u \in M$. If a vertex z belongs to $A_u \cap C$, then either $z \in A_u \cap B_u$, or $z \in W_u$ and there exists $v \in M$ such that $z \in A_v \cap B_v$. By Corollary 9.23, there are fewer that $2\Delta^2$ choices for v, each of them contributing at most $|A_v \cap B_v| \leq \Delta - 1$ vertices z to $A_u \cap C$. We conclude that $|A_u \cap C| < |A_u \cap B_u| + 2\Delta^2(\Delta - 1) < 2\Delta^3$ for every $u \in M$.

Consider any $i \in [m]$. There exists $u \in M$ such that $V(K_i) \cap W_u \neq \emptyset$, and since K_i is connected and $A_u \cap B_u \subseteq C$ is disjoint from K_i, it follows that $V(K_i) \subseteq W_u$ and $\beta(r_i) \subseteq A_u$. Therefore, $|\beta(r_i) \cap C| \leq |A_u \cap C| < 2\Delta^3$. Since $\beta(r_i) \subseteq A_u$, Observation 2.18 implies that $(\beta(r_i), \beta(S - r_i)) \in \mathcal{T}_0$, and thus the star decomposition (S, β) is $\mathcal{T}_0$-central.

Consider any vertex $z \in C$, and let us bound the degree of z in the torso G' of w. Suppose first that $z \in W_u$ for some $u \in M$, and in particular all neighbors of z in G_0 belong to A_u. If z has a neighbor in K_i for some $i \in [m]$, then since K_i is disjoint from $A_u \cap B_u \subseteq C$, this neighbor belongs to W_u. As we have argued in the previous paragraph, this implies that $V(K_i) \subseteq W_u$, and $\beta(r_i) \cap C \subseteq A_u \cap C$. Therefore, all neighbors of z in G' belong to $A_u \cap C$, and thus $\deg_{G'} z < 2\Delta^3$.

Suppose now that $z \notin \bigcup \mathcal{W}$. Every vertex of degree at least Δ belongs to $\bigcup \mathcal{W}$, and thus $\deg_{G_0} z < \Delta$. Hence, there exist less than Δ indices $i \in [m]$ such that z has a neighbor in K_i, and each of them contributes at most $|\beta(r_i) \cap C| < 2\Delta^3$ to the degree of z in G'. Therefore, $\deg_{G'} z < 2\Delta^4$.

In conclusion, the maximum degree of G' is less than $2\Delta^4$. □

In combination with Theorem 9.3, this gives us the desired star decompositions of topological minor-free graphs.

Corollary 9.25 *For every large-degree-outerplanar graph H, there exist integers a_1 and θ_1 such that the following claim holds. Let $\mathcal{T}$ be a tangle of order at least θ_1 in a graph G. If H is not a topological minor of G, then G has a $\mathcal{T}$-central star decomposition (S, β) such that the torso of the center w of S has h-index at most a_1.*

Proof Let a_0 and θ be the constants from Theorem 9.3. Let $\theta_1 = \theta + 2\Delta^3(H)$ and $a_1 = a_0 + 2\Delta^4(H)$.

Let X be a set of at most a_0 vertices of G such that no vertex of $G - X$ is $(\mathcal{T} - X, \Delta(H))$-free. By Lemma 9.24, there exists a $(\mathcal{T} - X)$-central star decomposition (S, β') of $G - X$ such that the torso of the center w of S has maximum degree less than $2\Delta^4(H)$. For every node $x \in V(S)$, let $\beta(x) = \beta'(x) \cup X$; then (S, β) is a $\mathcal{T}$-central star decomposition of G. Moreover, only the vertices in X can have degree at least $a_0 + 2\Delta^4(H)$ in the torso G' of w in the star decomposition (S, β), and thus G' has h-index at most a_1. □

The global structure theorem now follows by the standard construction of tree decompositions.

Theorem 9.2 *For every large-degree-outerplanar graph H, there exists an integer a such that every graph G that does not contain H as a topological minor has a tree decomposition whose torsos have h-index at most a.*

Proof Let a_1 and θ_1 be the constants from Corollary 9.25, and let $a = a_1 + 4\theta_1$. We are going to prove a strengthening of the theorem by induction on the number of vertices:

(†) For every graph G that does not contain H as a topological minor and for every set $R \subseteq V(G)$ of size at most $3\theta_1 - 2$, the graph obtained from G by adding a clique on R has a tree decomposition (T, β) whose torsos have h-index at most a.

If $|G| \leq 4\theta_1 - 3$, then we can simply let T be the single-node tree with bag $V(G)$. Hence, suppose that $|G| \geq 4\theta_1 - 2$. We can without loss of generality assume that $|R| = 3\theta_1 - 2$, as otherwise we can add vertices to R. If there exists an R-balanced vertex separation (C, D) of G of order less than θ_1, then let $R_C = C \cap (R \cup D)$ and $R_D = D \cap (R \cup C)$. Note that

$$R_C \leq \tfrac{2}{3}|R| + \theta_1 - 1 = \tfrac{2}{3}(3\theta_1 - 2) + \theta_1 - 1 < 3\theta_1 - 2,$$

and similarly $|R_D| < 3\theta_1 - 2$. In particular, $C \neq V(G) \neq D$, since neither C nor D contains all vertices of R. Let (T_C, β_C) and (T_D, β_D) be the tree decompositions obtained by applying the induction hypothesis to $G[C]$ with R_C and $G[D]$ with R_D. By Lemma 2.3, there exist nodes $x_C \in V(T_C)$ and $x_D \in V(T_D)$ such that $R_C \subseteq \beta_C(x_C)$ and $R_D \subseteq \beta_D(x_D)$. Let T be the tree obtained from $T_C \cup T_D$ by adding a new node x_0 adjacent to x_C and x_D. Let β match β_C on T_C, β_D on T_D, and let $\beta(x_0) = R \cup (C \cap D)$. Note that (T, β) is a tree decomposition of G and $|\beta(x_0)| \leq 3\theta_1 - 2 + \theta_1 - 1 = 4\theta_1 - 3 < a$. Moreover, note that the torsos of the nodes in $V(T) \setminus \{x_0\}$ are the same as the torsos of the corresponding nodes of T_C and T_D; e.g., the torso of x_C is the same in T_C and in T, since (T_C, β_C) is a tree decomposition of the graph obtained from $G[C]$ by adding a clique on $R_C = \beta(x_C) \cap \beta(x)$. Hence, (†) holds for G.

Therefore, we can assume that R is θ_1-unbreakable, and thus

$$\mathcal{T}_R = \{(C, D) : (C, D) \text{ is a vertex separation}, |C \cap D| < \theta_1, |R \setminus C| > \tfrac{2}{3}|R|\}$$

is a vertex tangle of order θ_1 in G. By Corollary 9.25, G has a $\mathcal{T}_R$-central star decomposition (S, β_0) such that the torso of the center w of S has h-index at most a_1. For each ray r of S, let $R_r = \beta_0(r) \cap (\beta_0(w) \cup R)$. Since (S, β_0) is $\mathcal{T}_R$-central, we have $|\beta_0(r) \cap \beta_0(w)| < \theta_1$ and $|R \setminus \beta_0(r)| > \frac{2}{3}|R|$, and thus $|R_r| \leq |\beta_0(r) \cap \beta_0(w)| + |\beta_0(r) \cap R| < \theta_1 + \frac{1}{3}|R| < 3\theta_1 - 2 = |R|$. By the induction hypothesis, $G[\beta_0(r)]$ together with the clique on R_r has a tree decomposition (T_r, β_r) whose torsos have h-index at most a. By Lemma 2.3, there exists a node $x_r \in V(T_r)$ such that $R_r \subseteq \beta(x_r)$.

Let (T, β) be the tree decomposition of G together with a clique on R obtained from the disjoint union of the tree decompositions (T_r, β_r) for the rays r of S by adding a node x_0 adjacent to x_r for each ray r and letting $\beta(x_0) = \beta_0(w) \cup R$. Note that torsos of all nodes other than x_0 are the same as in the tree decomposition for the corresponding ray r, since (T_r, β_r) is a tree decomposition for the graph together with the clique on $R_r = \beta(x_r) \cap \beta(x_0)$. Moreover, the torso G' of x_0 is obtained from the torso of w (which has h-index at most a_1) by adding the vertices of $R \setminus \beta_0(w)$ and edges incident with the vertices of R; consequently, G' has h-index at most $a_1 + |R| < a$. □

Let us remark that Theorem 9.1 is proved in exactly the same way; in fact, the proof is a bit simpler, since the subcase considered in Sect. 9.2 (there is an additional forbidden minor H' depending on the forbidden topological minor H) directly corresponds to the desired outcome by the Local Minor Structure Theorem.

References

1. P. Berman, M. Karpinski, On some tighter inapproximability results, in *International Colloquium on Automata, Languages, and Programming* (Springer, Berlin, 1999), pp. 200–209
2. Z. Dvořák, A stronger structure theorem for excluded topological minors. ArXiv, 1209.0129 (2012)
3. M. Grohe, D. Marx, Structure theorem and isomorphism test for graphs with excluded topological subgraphs. SIAM J. Comput. **44**, 114–159 (2015)
4. M. Grohe, K. Kawarabayashi, D. Marx, P. Wollan, Finding topological subgraphs is fixed-parameter tractable, in *Proceedings of the Forty-Third Annual ACM Symposium on Theory of Computing* (2011), pp. 479–488
5. C. Liu, Graph structures and well-quasi-ordering. PhD thesis, Georgia Institute of Technology (2014)
6. C. Liu, Recent progress on well-quasi-ordering graphs, in *Well-Quasi Orders in Computation, Logic, Language and Reasoning: A Unifying Concept of Proof Theory, Automata Theory, Formal Languages and Descriptive Set Theory*, ed. by P.M. Schuster, M. Seisenberger, A. Weiermann (Springer International Publishing, Cham, 2020), pp. 161–188
7. C. Liu, Packing topological minors half-integrally. J. Lond. Math. Soc. **106**, 2193–2267 (2022)
8. C. Liu, R. Thomas, Excluding subdivisions of bounded degree graphs. J. Combin. Theory B **134**, 1–35 (2019)
9. N. Robertson, P.D. Seymour, Graph minors XXIII. Nash-Williams' immersion conjecture. J. Combin. Theory B **100**, 181–205 (2010)

Chapter 10
Minors in Large Connected Graphs

There are aspects of Theorem 5.30 that complicate its applications in more delicate settings. One issue is that the number of apices is only bounded by some constant, rather than the more desirable bound $a_\Sigma(H) - 1$ which we got e.g. in the Flat Wall Theorem; this is rather easy to deal with, as we can use the results from Sect. 4.5 to argue that the attachment points of all but at most $a_\Sigma(H) - 1$ apices must be contained in a bounded number of bounded radius regions. If needed, we can furthermore use Lemma 9.10 to turn these regions to vortices, leaving us with only $a_\Sigma(H) - 1$ apex vertices attaching to the surface part (losing the fine control over the number of vortices as a trade-off).

A more substantial issue concerns the connectivity of vortices. In case that we find desirable substructures in different parts of a vortex, how can we connect them together? For this, it would be convenient for the vortical decomposition to be linked in the sense described in Sect. 3.4. While this property is relatively easy to establish in path decompositions, see Lemma 3.14, in vortices this is complicated by the fact that we need to keep control over their connection to the surface part. On the other hand, we can take advantage of the possibility to include pieces of the surface part in the vortex. The technicalities of this argument were worked out in [9].

An important result whose proof heavily relies on these improvements concerns minors in large sufficiently connected graphs.

Theorem 10.1 (Norin and Thomas [7]) *For every integer $t \geq 5$, there exists n_0 such that every t-connected K_t-minor-free graph with at least n_0 vertices is obtained from a planar graph by adding at most $t - 5$ apex vertices.*

Let us remark that for $t = 5$, this is straightforward to prove from Wagner's characterization of K_5-minor-free graphs [11], but already for $t = 6$ the proof is rather involved [5]. For $t = 6$, Jørgensen [3] conjectured that the claim holds for all 6-connected K_6-minor-free graphs, without the assumption of having at least n_0 vertices. However, for general t, this assumption cannot be dropped: Constructions based on random graphs give examples of $\Theta(t\sqrt{\log t})$-connected K_t-minor-free

© The Author(s), under exclusive license to Springer Nature Switzerland AG 2025

Z. Dvořák, *Graph Minors*, Springer Monographs in Mathematics,
https://doi.org/10.1007/978-3-031-87469-7_10

graphs with $\Theta(t\sqrt{\log t})$ vertices, while adding at most $t-5$ apex vertices to a planar graph results in a graph of minimum degree at most t,

Theorem 10.1 is an exact characterization of large t-connected K_t-minor-free graphs: Because planar graphs are K_5-minor-free, graphs obtained from them by adding at most $t-5$ apex vertices are K_t-minor-free. It is also interesting with respect to Hadwiger's conjecture: Since every planar graph is 4-colorable, graphs obtained from them by adding at most $t-5$ apex vertices are $(t-1)$-colorable. Hence, Theorem 10.1 implies that Hadwiger's conjecture for any given t holds for sufficiently large t-connected graphs. Finally, since every planar graph has a vertex of degree at most five, Theorem 10.1 has the following consequence, which we previously discussed in Sect. 3.4.

Corollary 10.2 *For every integer $t \geq 5$, there exists n_0 such that every t-connected graph with at least n_0 vertices and minimum degree at least $t+1$ contains K_t as a minor.*

In this chapter, we first describe the aforementioned improvements to the Local Minor Structure Theorem, then demonstrate some of the simplest ideas going to the proof of Theorem 10.1 by showing the following much easier variation, whose proof is inspired by [1].

Theorem 10.3 *For any integers $t, n \geq 1$, there exists an integer n_2 such that the following claim holds. If a $2t$-connected graph G of minimum degree at least $30t+1$ has at least n_2 vertices, then it contains $K_{t,n}$ as a minor.*

Let us recall that in Sect. 3.4, we proved this result with the additional assumption that the graph has bounded treewidth. At the end of the chapter, we are going to discuss an application of these results in graph coloring.

10.1 Localizing Apices

Let A be a set of vertices of a graph F_0, and let $F = F_0 - A$. Let $\mathcal{T}$ be a tangle in the graph F, and let $(\mathcal{S}_0, \mathcal{S}_1)$ be a solid bisegregation of F with a $(\mathcal{T}, \psi_0)$-spread arrangement on a surface, with the natural model μ of $\pi(\mathcal{S}_1)$ in F and the surface tangle $\mathcal{T}_0$. Recall that we chose the model μ so that for each $S \in \mathcal{S}_1$, all vertices of the component of S containing ∂S belong to $\bigcup_{u\in\partial S} V(\mu(u))$. Consider a vertex $v \in A$. The ***anchors*** of v are the vertices $u \in V(\pi(\mathcal{S}_1))$ such that v has a neighbor in $\mu(u)$. For positive integers t and ψ, we say that v is (t, ψ)-*global* if there exists a set U of its anchors of size t such that the $d_{\mathcal{T}_0}$-distance between distinct vertices of U, as well as between u and the vortex faces of the elements of $\mathcal{S}_0$, is at least ψ. We say that v is ***vortex-local*** if all neighbors of v in $V(F)$ are contained in $\bigcup_{S\in\mathcal{S}_0}(V(S) \setminus \partial S)$.

We can now state the main result of this section, a version of Theorem 5.30 restricting the number of global apices and forcing all other apices to be vertex-local.

This is in principle a simple combination of Lemma 4.35 to bound the number of global apices and Lemma 9.10 to establish vortices at the anchors of non-global apices, though the process of choosing the centers of newly created vortices is somewhat complicated by the fact that we want them to be pairwise far apart, and also that apices that were originally global may cease to be if we introduce new vortices.

Theorem 10.4 *For any graph H, positive integer t, and a non-decreasing function $\psi : \mathbb{N}^2 \to \mathbb{N}$, there exist integers $\alpha, \delta_0 \geq 0$ and $\theta > \alpha$ such that the following claim holds. Suppose $\mathcal{T}$ is a tangle of order at least θ in a graph G. If $\mathcal{T}$ does not control a minor H of G, then there exists a set $A \subseteq V(G)$ of size at most α such that for some $\delta \leq \delta_0$ and $m \leq \alpha$, $G - A$ has a solid (m, δ)-simple bisegregation with a $(\mathcal{T} - A, \psi(m, \delta + |A|))$-spread arrangement on an H-avoiding surface Σ, and A consists only of vertex-local apices except for at most $a_\Sigma(H) - 1$ apices that are $(t, \psi(m, \delta + |A|))$-global.*

Proof Without loss of generality, we can assume that H is non-planar, as otherwise every tangle of sufficiently large order controls a minor of H. Let t_0 and θ_0 be the maximum of the constants t and θ in Lemma 4.35 over all H-avoiding surfaces Σ. Without loss of generality, we can assume that $t \geq t_0$. Let $f(m, d, \mu) = \max(\theta_0, \psi(m, \lceil \frac{5}{2}\mu \rceil + d + 7)) + (12\mu + 15)m$. Let $\psi_0 : \mathbb{N}^2 \to \mathbb{N}$ be chosen so that $\psi_0(a, d) \gg \theta_0, a, d, t, \|H\|$, and let α', δ'_0, and θ be the constants from Theorem 5.30 for H and ψ_0 playing the role of ψ. Let $\alpha = t\alpha' + \mathrm{cr}_{\mathrm{sphere}}(H)$ and choose $\delta_0 \gg \theta_0, \alpha', \delta'_0, t, \|H\|$.

By Theorem 5.30, there exists a set $A \subseteq V(G)$ of size at most $\alpha' \leq \alpha$ such that for some $\delta' \leq \delta'_0$, $G - A$ has a solid $(\mathrm{cr}_\Sigma(H), \delta')$-simple bisegregation $(\mathcal{S}_0, \mathcal{S}_1)$ with a $(\mathcal{T} - A, \psi_0(|A|, \delta'))$-spread arrangement on an H-avoiding surface Σ with the surface tangle $\mathcal{T}_0$. Without loss of generality, we can assume that there exists a society $S_0 \in \mathcal{S}_0$, as otherwise we can move an element $S_0 \in \mathcal{S}_1$ to $\mathcal{S}_0$ (S_0 is a cell, and thus 1-vortex-like). Moreover, we can assume that $\partial S \neq \emptyset$ and S is connected for every $S \in \mathcal{S}_1$, as otherwise we can move the components K of S with $V(K) \cap \partial S = \emptyset$ to $\mathcal{S}_0$. Hence,

(†) if a vertex $v \in A$ has a neighbor in $S \in \mathcal{S}_1$, then ∂S contains an anchor of v.

Let U_0 be the set of vortex faces for the elements of $\mathcal{S}_0$, let $A_0 = A$ and $\mu_0 = 0$. We now repeat the following procedure for $i = 1, 2, \ldots$:

(i) If there exist distinct $u_1, u_2 \in U_{i-1}$ such that $d_{\mathcal{T}_0}(u_1, u_2) < f(|U_{i-1}|, \delta' + |A|, \mu_{i-1})$, with labels of u_1 and u_2 chosen so that $u_2 \notin U_0$ if possible, then let $U_i = U_{i-1} \setminus \{u_2\}$, $A_i = A_{i-1}$, and $\mu_i = \mu_{i-1} + f(|U_{i-1}|, \delta' + |A|, \mu_{i-1})$.
(ii) Otherwise, if there exists a vertex $v \in A_{i-1}$ and a (possibly empty) set Q_v of less than t anchors of v such that every anchor of v is at $d_{\mathcal{T}_0}$-distance less than $f(|U_{i-1}|, \delta' + |A|, \mu_{i-1})$ from $Q_v \cup U_{i-1}$, then let $U_i = U_{i-1} \cup Q_v$, $A_i = A_{i-1} \setminus \{v\}$, and $\mu_i = \max(\mu_{i-1}, f(|U_{i-1}|, \delta' + |A|, \mu_{i-1}))$.
(iii) Otherwise, let $r = i - 1$ and finish the process.

Note that $r \leq t|A| + |U_0| \leq t|A| + \mathrm{cr}_{\mathrm{sphere}}(H)$, since in each step we remove a vertex from A_{i-1} or U_{i-1} and at most $t-1$ vertices are added for each vertex of A. Similarly, $|U_i| \leq t|A| + \mathrm{cr}_{\mathrm{sphere}}(H)$ for each i, and in particular $|U_r| \leq \alpha$. Note that μ_i is bounded by a function of μ_{i-1}, $|U_{i-1}| \leq t|A| + \mathrm{cr}_{\mathrm{sphere}}(H)$, δ', and θ_0, and since $\psi_0(|A|, \delta'), \delta_0 \gg \theta_0, |A|, \delta', t, \mathrm{cr}_{\mathrm{sphere}}(H)$, by induction (with at most $r \leq t|A| + \mathrm{cr}_{\mathrm{sphere}}(H)$ inductive steps), we conclude that $\psi_0(|A|, \delta'), \delta_0 \gg \mu_i, f(|U_i|, \delta' + |A|, \mu_i)$ for each i. In particular, since the $d_{\mathcal{T}_0}$-distance between distinct faces of U_0 is least $\psi_0(|A|, \delta')$, in (i) we never have $u_1, u_2 \in U_0$, and thus $U_0 \subseteq U_r$.

Moreover, for each apex vertex $v \in A \setminus A_r$, we claim that

(⋆) each anchor u of v is at $d_{\mathcal{T}_0}$-distance less than μ_r from U_r.

Indeed, let i_v be the index for which v was deleted in (ii); clearly, we have $d_{\mathcal{T}_0}(u, U_{i_v}) < \mu_{i_v}$. We claim that $d_{\mathcal{T}_0}(u, U_i) < \mu_i$ for each $i \in \{i_v + 1, \ldots, r\}$. Indeed, if in the step i we perform the operation (ii), then $U_i \supseteq U_{i-1}$ and $\mu_i \geq \mu_{i-1}$, and thus the inequality is preserved. If we perform the operation (i), the inequality is also preserved unless u_2 is the only vertex of U_{i-1} at $d_{\mathcal{T}_0}$-distance less than μ_{i-1} from u; however, in that case $d_{\mathcal{T}_0}(u, U_i) \leq d_{\mathcal{T}_0}(u, u_1) \leq d_{\mathcal{T}_0}(u, u_2) + d_{\mathcal{T}_0}(u_1, u_2) < \mu_{i-1} + f(|U_{i-1}|, \delta' + |A|, \mu_{i-1}) = \mu_i$.

Let $\delta = \lceil \frac{5}{2}\mu_r \rceil + \delta' + 7 \leq \delta_0$. For each $u \in U_r$, let Δ_u be the $(3\mu_r + 3)$-zone around u in the drawing of $\pi(\mathcal{S}_1)$ obtained in Lemma 9.10. Let $\mathcal{S}_1'$ consist of the elements $S \in \mathcal{S}_1$ such that $\overline{\pi}(S) \not\subseteq \bigcup_{u \in U_r} \Delta_u$, for each $u \in U_r$ let $S_u = (G - A)_{(\mathcal{S}_0, \mathcal{S}_1)}[\Delta_u]$, and let $\mathcal{S}_0' = \{S_u : u \in U_r\}$. Then $(\mathcal{S}_0', \mathcal{S}_1')$ is a solid bisegregation of $G - A$. This bisegregation is $(m, \delta - 2)$-simple for $m = |U_r| \leq \alpha$ by Lemma 9.10. Let $G_1 = \pi(\mathcal{S}_0' \cup \mathcal{S}_1')$. The graph G_1 is obtained from $\pi(\mathcal{S}_1)$ by clearing the zones Δ_u for $u \in U_r$; let $\mathcal{T}_1$ be the respectful tangle in G_1 obtained from $\mathcal{T}_0$ by clearing these zones. Since the step (i) does not apply at U_r, we have $d_{\mathcal{T}_0}(u_1, u_2) \geq f(m, \delta' + |A|, \mu_r)$ for all distinct $u_1, u_2 \in U_r$, and thus Lemma 4.19 implies that $d_{\mathcal{T}_1}(f_1, f_2) \geq \psi(m, \delta + |A|) + 3m$ for the faces f_1 and f_2 corresponding to the interiors of Δ_{u_1} and Δ_{u_2}.

A technical issue that we now need to overcome is that G_1 is not necessarily equal to $\pi(\mathcal{S}_1')$: Consider any edge e of the cycle C_u bounding the zone Δ_u for $u \in U_r$. If $e \in E(\pi(S))$ for a cell $S \in \mathcal{S}_1$ such that $\overline{\pi}(S) \subseteq \Delta_u$, then e does not appear in $\pi(\mathcal{S}_1')$. Therefore, $\pi(\mathcal{S}_1')$ is obtained from G_1 by deleting some of the edges of cycles C_u for $u \in U_r$ and the resulting isolated vertices. Moreover, $\pi(\mathcal{S}_1')$ can be disconnected, in case that a face f of G_1 different from the interior f_u of Δ_u is incident with several deleted edges of C_u. To deal with these issues, we proceed as follows: For each $u \in U_r$, let $\Delta_u' \supset \Delta_u$ be a 4-zone around f_u in G_1 obtained using Corollary 4.16. Observe that $\pi(\mathcal{S}_1')$ has exactly one component G_1' not contained in any of the zones Δ_u' for $u \in U_r$. Let $\mathcal{S}_u$ consist of cells $S \in \mathcal{S}_1'$ such that $\overline{\pi}(S) \subseteq \Delta_u'$ and $\pi(S)$ is disjoint from G_1', and let $\mathcal{S}_1'' = \mathcal{S}_1' \setminus \bigcup_{u \in U_r} \mathcal{S}_u$, so that $\pi(\mathcal{S}_1'') = G_1'$. Let S_u' be the society with the graph $S_u \cup \bigcup_{S \in \mathcal{S}_u} S$ and with the boundary $\partial S_u' = \partial S_u \cap V(G_1')$. Let $\mathcal{S}_0'' = \{S_u' : u \in U_r\}$. Since S_u is $(\delta - 2)$-

vortex-like, Lemma 9.11 implies that S'_u is δ-vortex-like. Moreover, it is easy to see that G'_1 has a respectful tangle $\mathcal{T}'_1$ conformal with $\mathcal{T}_1$ such that $d_{\mathcal{T}'_1}(a'_1, a'_2) \geq d_{\mathcal{T}_1}(a_1, a_2) - 3m$ for any atoms a'_1 and a'_2 of G'_1 and the corresponding atoms a_1 and a_2 of G_1. Therefore, $(\mathcal{S}''_0, \mathcal{S}''_1)$ is a solid (m, δ)-simple bisegregation of $G - A$ with a $(\mathcal{T} - A, \psi(m, \delta + |A|))$-spread arrangement in Σ.

For each apex vertex $v \in A \setminus A_r$, the claims (†) and ($\star$) together with Lemma 9.10 imply that if v has a neighbor x in a cell $S \in \mathcal{S}_1$, then there exists $u \in U_r$ such that the corresponding anchor in ∂S is contained in the interior of Δ_u, and thus $x \in V(S_u) \setminus \partial S_u \subseteq V(S'_u) \setminus \partial S'_u$. Therefore, v is a vortex-local apex vertex over $(\mathcal{S}''_0, \mathcal{S}''_1)$.

Consider now an apex vertex $v \in A_r$, and let Q_v be a maximal set of anchors of v such that $d_{\mathcal{T}_0}(z, U_r \cup Q_v \setminus \{z\}) \geq f(m, \delta', \mu_r)$ holds for every $z \in Q_v$. Clearly, every anchor of v is at $d_{\mathcal{T}_0}$-distance less than $f(m, \delta' + |A|, \mu_r)$ from $Q_v \cup U_r$, as otherwise we could add it to Q_v. Since (ii) does not apply to v, we have $|Q_v| \geq t$. Lemma 4.19 implies that the anchors in Q_v are at $d_{\mathcal{T}_1}$-distance at least $\max(\theta_0, \psi(m, \delta + |A|)) + 3m$ from each other and from the faces $\overline{\pi}(S_u)$ for $u \in U_r$, and thus also at $d_{\mathcal{T}'_1}$-distance at least $\max(\theta_0, \psi(m, \delta + |A|))$ from each other and from the vortex faces of $\mathcal{S}''_0$. Therefore, the apex vertices in A_r are $(t, \psi(m, \delta + |A|))$ global over $(\mathcal{S}''_0, \mathcal{S}''_1)$. Moreover, since H is not a minor of G controlled by $\mathcal{T}$, Lemma 4.35 implies that $|A_r| \leq a_\Sigma(H) - 1$. □

Let us remark that the transformation from the proof of Theorem 10.4 is often quite useful in proofs using the structure theorem: If we show that a certain type of irregularities can only occur in a bounded number of regions of bounded radius, we can capture these regions inside vortices and assume that no irregularities appear in the surface part.

Furthermore, by applying the argument used to prove the Minor Structure Theorem (Theorem 2.9) from its local version (Theorem 5.27), we can obtain a version of the Minor Structure Theorem with a constraint on apex vertices. We say that a graph G_1 can be *(a, α, m, d)-nearly drawn on Σ* if there exists a set $A \subseteq V(G_1)$ of size at most a such that $G_1 - A$ can be drawn on Σ up to at most m vortices of width at most d, and at most α of the vertices of A have neighbors in the surface part of $G_1 - A$.

Theorem 10.5 (Dvořák and Thomas [2]) *For every graph H, there exist constants a, m, and d such that every H-minor-free graph G has a tree decomposition in which each torso with more than a vertices can be $(a, a(H)-1, m, d)$-nearly drawn on an H-avoiding surface.*

We leave the derivation of this result from Theorem 10.4 as an exercise to the reader; it suffices to follow the argument given in Sect. 5.5 with the following modification: When the unbreakable set R is added to the apex set, new vortices need to be introduced to contain the neighbors of vertices of R in the surface part.

10.2 Taming Vortices

Next, let us turn our attention to improving the connectivity properties of vortices. For this more refined version of vortices, it is more convenient to associate bags of their cycle decomposition with edges rather than nodes. For a technical reason that we explain later, it will also be necessary to specify the boundaries of the bags, which are naturally associated with the nodes.

For a society F, let us view ∂F as a cycle (which does not have to be a subgraph of F). An *open path* is either a sequence $e_0, v_1, e_1, \ldots, v_m, e_m$ of pairwise distinct vertices and edges of ∂F, with v_i incident with e_{i-1} and e_i for $i \in [m]$, or the cyclic sequence of all vertices and edges of ∂F. Note in particular that there are $|\partial F| + 1$ open paths containing all edges of ∂F, depending on the choice of at most one vertex not belonging to the open path. A ***strong vortical decomposition*** of a society F is a function β assigning

- pairwise edge-disjoint subgraphs of F to edges of ∂F and
- sets of vertices of F to vertices of ∂F,

such that the following conditions are satisfied. For every $v \in V(F)$, let $\beta^{-1}(v) = \{x \in \partial F : v \in \beta(x)\} \cup \{x \in E(\partial F) : v \in V(\beta(x))\}$. Then:

(DE1) $F = \bigcup_{e \in E(\partial F)} \beta(e)$

(DE2) for every vertex $v \in V(F)$, the set $\beta^{-1}(v)$ forms an open path in ∂F, and

(VE) for every vertex $v \in \partial F$ incident with $e_1, e_2 \in E(\partial F)$, the set $\beta^{-1}(v)$ forms the open path $e_1 v e_2$.

Let us remark that we chose the edge-bags of the strong vertical decomposition to be subgraphs rather than subsets of vertices (unlike our definition of a cycle decomposition). Up to this difference, the conditions (DE1) and (DE2) imply that letting B be the linegraph of the cycle ∂F, $(B, \beta \restriction E(\partial F))$ is a cycle decomposition of F. Moreover, if we let B' be the cycle of length $2|\partial F|$ on vertices and edges of ∂F in order, then (B', β) is also a cycle decomposition of F.

The *adhesion* of a strong vortical decomposition is the maximum of $|\beta(z)|$ over $z \in \partial F$. Consider a vertex $z \in \partial F$ incident with edges $e_1, e_2 \in E(\partial F)$. Note that (DE2) implies that $\beta(z) \subseteq V(\beta(e_1) \cap \beta(e_2))$: For each $v \in \beta(z)$, since $\beta^{-1}(v)$ is an open path, we have $v \in V(\beta(e_1))$ and $v \in V(\beta(e_2))$. The condition (DE2) is very close to just saying that $\beta(z) = V(\beta(e_1) \cap \beta(e_2))$, but there is an exception: If a vertex $v \in V(F)$ belongs to $\beta(e)$ for every $e \in E(\partial F)$, then there can exist a single vertex $z \in \partial F$ such that $v \notin \beta(z)$. The main reason for this exception is that we want to be able to suppress vertices of ∂F: Consider an open subpath $e_0 z' e_1 z e_2$ of ∂F, and suppose that there exists a vertex $u \in V(F)$ contained in the bags of all edges of ∂F except for $\beta(e_1)$. We want to be able to replace the open subpath $e_1 z e_2$ by an edge e with bag $\beta(e) = \beta(e_1) \cup \beta(e_2)$. After this replacement, we have $V(\beta(e) \cap \beta(e_0)) \supseteq V(\beta(e_1) \cap \beta(e_0)) \cup \{u\}$, rather than equal to $V(\beta(e_1) \cap \beta(e_0))$ as one would expect. Thus, requiring $\beta(z')$ to be specified explicitly and allowing $\beta^{-1}(u)$ to be an open path containing all edges but not z' after the contraction avoids having to increase the adhesion of the strong vortical decomposition during this operation. For a specific usage, see the definition of β' in Lemma 10.7, which would run into the same problem of potentially increasing adhesion in a hard-to-control manner.

10.2.1 From Vortical Decompositions to Strong Vortical Decompositions

The property (VE) is analogous to the property (V) of vortical decompositions, but additionally requires that each vertex $v \in \partial F$ belongs only to the bags of incident edges e_1 and e_2. Because of the property (VE), there exist societies that have vortical decompositions of small adhesion but not strong vortical decompositions of small adhesion. We say that β is a ***semi-strong vortical decomposition*** of F if it satisfies (DE1), (DE2), and (V).

Observation 10.6 *If a society F has a vortical decomposition $(\partial F, \beta_0)$ of adhesion at most δ, then it also has a semi-strong vortical decomposition β of adhesion at most $\delta + 2$.*

Proof Let $v_1, e_1, v_2, e_2, \ldots, v_m, e_m, v_{m+1} = v_1$ be the vertices and edges of ∂F in order. Let $e_{m+1} = e_1$ and $v_{m+2} = v_2$. For each edge e of G, let $i_e \in [m]$ an arbitrary index such that both ends of e are contained in $\beta(v_{i_e})$, which exists by (D1). For $i \in [m]$, we define $\beta(e_i)$ to be the subgraph of G with vertex set $\beta_0(v_i) \cup \{v_{i+1}\}$ and edge set $\{e \in E(G) : i_e = i\}$, and we define $\beta(v_{i+1}) = \beta(e_i) \cap \beta(e_{i+1})$. By the construction, β satisfies the condition (DE1). By (V) for $(\partial F, \beta_0)$, we have $v_{i+1} \in \beta_0(v_{i+1}) \subseteq V(\beta(e_{i+1}))$, and since $v_{i+1} \in \beta(e_i)$ by definition, we have $v_{i+1} \in \beta(v_{i+1})$; hence, β satisfies (V). The condition (D2) for $(\partial F, \beta_0)$ implies (DE2). Moreover, $|\beta(v_{i+1})| \leq |(\beta_0(v_i) \cap \beta_0(v_{i+1})) \cup \{v_{i+1}, v_{i+2}\}| \leq \delta + 2$. □

We address the lack of property (VE) by incorporating a neighborhood from the surface part in the vortex. To simplify the presentation, we are going to perform all transformations in a bisegregation with only one vortex arranged in a disk, but of course the same arguments can be applied within zones around separate vortices from Theorem 5.30 or 10.4. Given a society G' drawn in the disk Δ with vertices of $\partial G'$ drawn in the boundary, the *boundary distance* for a face f of G' is the minimum number of intersections of a simple curve from f to $\text{bd}(\Delta)$ with G'.

Lemma 10.7 *Let G be a society and let $(\{F\}, S_1)$ be a solid bisegregation of G with an arrangement in the disk Δ such that F has a semi-strong vortical decomposition of adhesion δ and the vortex face f of F has boundary distance $d \geq 2$ in $\pi(S_1)$. Then there exists $S_1' \subseteq S_1$ and a society F' such that $(\{F'\}, S_1')$ is a solid bisegregation of G with an arrangement in Δ, F' has a strong vortical decomposition of adhesion $\delta + 1$, and the vortex face of F' has boundary distance at least $d - 2$.*

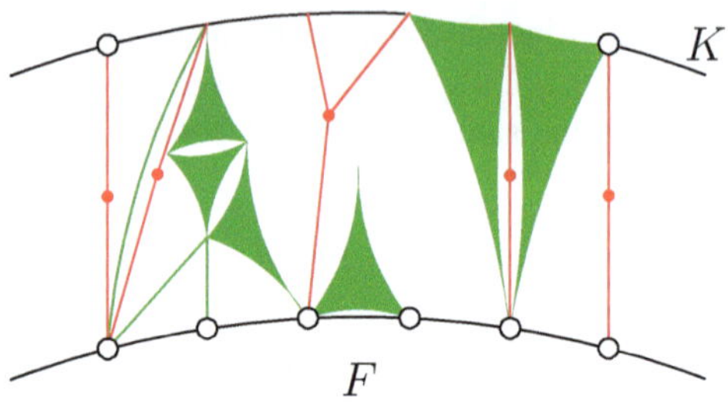

Fig. 10.1 The definitions from the proof of Lemma 10.7. The green faces and edges correspond to the cells in $\mathcal{S}_1''$. The elements of M and the paths P_v for $v \in \partial F'$ are drawn in red

Proof Let $G' = \pi(\mathcal{S}_1) - \partial F$ and let f_0 be the face of G' containing f. Since f has boundary distance $d \geq 2$ in $\pi(\mathcal{S}_1)$, f_0 has boundary distance at least $d - 1 > 0$ in G', and thus there exists a cycle K in the boundary of f_0 separating f_0 from $\text{bd}(\Delta)$. Let $\Lambda \subseteq \Delta$ be the disk bounded by K, let $\mathcal{S}_1''$ consist of the cells $S \in \mathcal{S}_1$ with $\overline{\pi}(S) \subseteq \Lambda$, let $\mathcal{S}_1' = \mathcal{S}_1 \setminus \mathcal{S}_1''$, and let $F' = G_{(\{F\},\mathcal{S}_1)}[\Lambda]$. Clearly $(\{F'\}, \mathcal{S}_1')$ is a solid bisegregation of G with an arrangement in Δ and the vortex face of F' has boundary distance at least $d - 2$. Moreover, note that we can identify K with the cycle $\partial F'$.

See Fig. 10.1 for an illustration of the following definitions. Let $H = \pi(\{F\} \cup \mathcal{S}_1)$ and let M be the set of vertices of R_H corresponding to the faces h of H such that h is incident with a vertex of ∂F and a vertex of K, and there is no $S \in \mathcal{S}_1''$ such that $h = \overline{\pi}(S)$. For each $m \in M$, choose a neighbor $u_m \in \partial F$ of m in R_H arbitrarily. Observe that every vertex $v \in V(K)$ is incident with a face corresponding to a vertex $m \in M$; let P_v be the path vmu_m in R_H. For $e = v_1v_2 \in E(\partial F')$, let f_e be the face of $K \cup \pi(F) \cup P_{v_1} \cup P_{v_2}$ drawn inside Λ and incident with e. Let the labels be chosen so that v_2 is the successor of v_1 in the cyclic ordering of $\partial F'$. Let Δ_e consist of f_e, the edge e, the vertices and edges of P_{v_1} that do not belong to P_{v_2}, and the path P_e in ∂F from the end of P_{v_1} to the end u_2 of P_{v_2}, excluding the vertex u_2. This construction ensures that $\Lambda \setminus \overline{\pi}(F)$ is the disjoint union of the sets Δ_e for $e \in E(\partial F')$.

Let β be a semi-strong vortical decomposition of F of adhesion δ. For every $e = v_1v_2 \in E(\partial F')$, let

$$\beta_1(e) = \{v_1, v_2\} \cup \bigcup_{S \in \mathcal{S}_1'' : \overline{\pi}(S) \subseteq \Delta_e} S \cup \bigcup_{x \in E(P_e) \cup V(P_e)} \beta(x).$$

For every $z \in \partial F'$, let u_z be the end of the path P_z in ∂F and let

$$\beta_1(z) = \{z\} \cup \beta(u_z).$$

We claim that β_1 is a strong vortical decomposition of F'. Clearly, (VE) is ensured by the definition of β_1. Observe that β_1 satisfies (DE1), since each edge of F' is contained either in F or in some cell of $\mathcal{S}_1''$, F satisfies (DE1), each $S \in \mathcal{S}_1'$ satisfies $\overline{\pi}(S) \subseteq \Delta_e$ for exactly one edge $e_S \in E(\partial F')$, and each $e' \in E(\partial F)$ satisfies

$e' \in E(P_e)$ for exactly one edge $e \in E(\partial F')$. The condition (DE2) is clear for vertices $v \in \partial F'$, which appear in $\beta(v)$ and the bags of incident edges, and for vertices in $v \in V(S) \setminus (\partial F \cup \partial F')$ for each $S \in \mathcal{S}_1''$, which appear only in the bag of e_S. Consider now a vertex $v \in V(F)$. Note that $v \in V(\beta_1(e))$ exactly if $\beta^{-1}(v)$ intersects P_e, and $v \in \beta_1(z)$ exactly if $\beta^{-1}(v)$ contains u_z. From this, it is easy to see that $\beta^{-1}(v)$ is an open path in $\partial F'$, and thus (DE2) holds.

Finally, the definition of β' on vertices of $\partial F'$ implies that the adhesion of β' is $\delta + 1$. □

Conversely, observe that if β is a semi-strong vortical decomposition of a society F, then for any distinct $u, v \in \partial F$, if P_1 and P_2 are the two subpaths of ∂F between u and v, then $\beta(u) \cup \beta(v)$ is a cut in F separating $\beta(E(P_1))$ from $\beta(E(P_2))$. From this, we get the following conclusion.

Observation 10.8 *If a society F has a semi-strong vortical decomposition of adhesion δ, then F is 2δ-vortex-like.*

10.2.2 Guardrails Around a Vortex

We are now going to incorporate further pieces of the surface part to the vortex. A natural concern is that doing so repeatedly would end up with the vortex "breaking out", reaching the boundary of the disk in which we operate. This is addressed by the following easy observation.

Lemma 10.9 *Let G be a society and let $(\{F\}, \mathcal{S}_1)$ be a solid $(1, \delta)$-simple bisegregation of G with an arrangement in the disk Δ. Let f be the vortex face of F in $\pi(\mathcal{S}_1)$. Suppose that there exists a $k \times (2\delta + 6)$-battlefield $(\mathcal{R}, \mathcal{P})$ in $\pi(\mathcal{S}_1)$ around f for some $k \geq \delta+3$, let $\mathcal{R} = C_1, \ldots, C_k$, and let Δ' be the disk bounded by $C_{\delta+3}$. Let $\Lambda \subseteq \Delta$ be an $(\{F\}, \mathcal{S}_1)$-normal disk containing f. If $F' = G_{(\{F\},\mathcal{S}_1)}[\Lambda]$ is δ-vortex-like and disjoint from C_k, then Λ is contained in the interior of Δ'.*

Proof Suppose for a contradiction that Λ is not contained in the interior of Δ', and thus there exists a simple curve γ in Λ starting in f and ending at a point p in the boundary of Δ'. Let $P_1, \ldots, P_{2\delta+6}$ be the paths of $\mathcal{P}$ ordered according to their intersections with $C_{\delta+3}$ so that p appears in the closed interval of $\mathrm{bd}(\Delta')$ between the intersections with $P_{\delta+3}$ and $P_{\delta+4}$. For $i \in [\delta+3]$, we can choose a path $Q_i \subseteq P_i \cup C_i \cup P_{2\delta+7-i}$ with ends in C_k so that the paths $Q_1, \ldots, Q_{\delta+3}$ are pairwise vertex-disjoint and separate p from f inside Δ'. This implies that they intersect γ, and thus also F'. Let u and v be the first vertices of Q_1 and $Q_{\delta+3}$ belonging to $\partial F'$. Observe that for $i \in [\delta+2] \setminus \{1\}$, $Q_i \cap F'$ contains a segment starting in $\partial_{(u,v)} F'$ and ending in $\partial_{(v,u)} F'$. This is a contradiction, since we assumed that F' is δ-vortex-like, and thus does not contain a transaction of size $\delta + 1$. □

Together with Observation 10.8, Lemma 10.9 implies that we are free to incorporate pieces of the surface part to the vortex without having to worry about breaking out of the battlefield (and potentially decreasing the distance between distinct vortices too much), as long as we are careful not to increase the adhesion and not to include a large part of the battlefield at once. To simplify the statements of the following lemmas, we are not going to include the presence of the surrounding battlefield in their assumptions.

A somewhat related concern is that the vortex F that we consider could be cut off from the rest of the graph by a small cut (say of size at most three) in the surface part. As we incorporate pieces of the surface part into F, we could eventually reach this cut, resulting in a cell $F' \supseteq F$. While this is not a problem in principle (nothing forbids us from still considering these cells to be vortices), some of the arguments that we are about to show would need to deal with these degenerate cases separately, complicating the presentation. However, this is easy to avoid. Let G be a society and let $(\{F\}, \mathcal{S}_1)$ be a solid bisegregation of G with an arrangement in the disk Δ. For a set $Z \subseteq \partial G$, we say that F is *Z-glued* if there exists a $(V(F), Z)$-linkage in G of size $|Z|$. Let us fix a constant $\gamma \geq 4$. At the beginning of the process (before applying Lemma 10.7), we can proceed as follows at each vortex F_0:

- As in Theorem 4.21, find a free graded $(r \times \gamma)$-battlefield in the surface part around F_0 for sufficiently large r, let Δ be the zone bounded by the outer cycle of the battlefield and let Z be the set of γ ends of the paths of the battlefield in $\mathrm{bd}(\Delta)$.
- Use Lemma 9.10 to incorporate the egg of this battlefield (but not its outer cycle) in the vortex.

The paths of the battlefield imply that the resulting vortex is Z-glued to the free set Z. Enlarging the vortex using Lemma 10.7 (or similar arguments we are about to describe) clearly preserves Z-gluedness. Similarly to the existence of the surrounding battlefield, we are going to refer to the Z-gluedness property only when useful.

10.2.3 Enlarging a Vortex

Our plan is to add pieces of the surface part to the initial vortex F to make the vortex linked. This is very nearly achieved by simply adding as much to F as possible while preserving the fact that it has a strong vortical decomposition of adhesion δ, but there is a technical issue that forces us to actually consider a slightly different measure of the width of the decomposition. The ***warp*** of a (semi-)strong vortical decomposition β of F is the minimum integer p such that

(i) $|\beta(v)| \leq p$ for every $v \in \partial F$, and
(ii) for each $e = uv \in E(\partial F)$, if $\beta(e)$ contains a $(\beta(u), \beta(v))$-linkage of size p, then every such linkage contains a path from u to v.

The warp of the decomposition is very close to its adhesion a, specifically either a or $a + 1$; the latter is the case if (ii) would be violated for some edge e with $p = a$.

The definition of the warp is chosen to deal with the following issue: We are going to pad the decomposition so that $|\beta(v)| = p$ for every $v \in \partial F$. In this situation, ideally we would like that $\beta(e)$ contains a $(\beta(u), \beta(v))$-linkage of size p for every $e = uv \in E(\partial F)$. However, this is not always possible and there will turn out to be edges where we only find a $(\beta(u) \setminus \{u\}, \beta(v) \setminus \{v\})$-linkage of size $p - 1$. The condition (ii) ensures that on the "good" edges with a linkage of size p, we can restrict the linkage to a $(\beta(u) \setminus \{u\}, \beta(v) \setminus \{v\})$-linkage of size $p - 1$ in order to ensure global consistency around the vortex.

Let G be a society and let $(\{F\}, \mathcal{S}_1)$ be a solid bisegregation of G with an arrangement in the disk Δ such that F has a strong vortical decomposition of warp p and the vortex face f of F has boundary distance at least two in $\pi(\mathcal{S}_1)$. We say that F is *warp-p-maximal* if there is no $(\{F\}, \mathcal{S}_1)$-normal disk $\Lambda \subseteq \Delta$ containing f such that the society $F' = G_{(\{F\}, \mathcal{S}_1)}[\Lambda]$ satisfies one of the following conditions:

(a) F' is not equal to F and has a strong vortical decomposition of warp at most p, or
(b) there exists a set $A \subseteq V(F') \setminus \partial F'$ of size at most two such that $F' - A$ has a strong vortical decomposition of warp less than p.

We can easily restrict our attention to warp-p-maximal vortices: We incorporate as much of $\mathcal{S}_1$ into the vortex F as possible while not increasing its warp, ensuring that there is no F' satisfying (a). If some F' satisfies (b), we delete the vertices of A and repeat the process with the society $F' - A$ of smaller warp. After at most p iterations, we arrive at a warp-maximal society. In what follows, we show that warp-maximal societies have the linkedness properties that we desire. At the end of the process, we add the deleted vertices of A back to all bags of the strong vortical decomposition. Let us remark that since the vertices of A are not contained in $\partial F'$, deleting them does not affect the Z-gluedness of the vortex to a set Z of vertices outside of F'.

The possibility to delete vertices in (b) is needed to avoid increasing the warp in certain "degenerate" cases, which give rise to the situation described in the following observation.

Observation 10.10 *Let β be a strong vortical decomposition of a society F, and suppose that $|\beta(v)| \leq p$ for every $v \in \partial F$. Let e_0 be an edge of ∂F, and suppose that the condition (ii) from the definition of warp p holds for all edges in $E(\partial F) \setminus \{e_0\}$, but not for e_0. If $\bigcap_{v \in \partial F}(\beta(v) \setminus \{v\}) \neq \emptyset$, then there exists a set $A \subseteq V(F) \setminus \partial F$ of size at most two such that $F - A$ has a strong vortical decomposition of warp less than p.*

Proof Let x be a vertex contained in $\beta(v) \setminus \{v\}$ for every $v \in \partial F$. The existence of x implies that $|\beta(v)| \geq 2$, and thus $p \geq 2$. Let β_1 be obtained from β by deleting x from every bag. If $p = 2$, then $|\beta_1(v)| = 1$ for every $v \in \partial F$, and β_1 trivially satisfies the condition (ii) for warp 1; hence, we can let $A = \{x\}$.

Suppose now that $p \geq 3$. Let $v_0 \in \partial F$ be a vertex incident with e_0. Since e_0 fails (ii) for warp p, we have $|\beta(v_0)| = p \geq 3$, and thus there exists $x' \in \beta(v_0) \setminus \{v_0, x\}$.

Let $A = \{x, x'\}$ and let β_2 be the strong vortical decomposition of $F - A$ obtained from β by deleting A from every bag. We claim that β_2 has warp at most $p - 1$. Indeed, the condition (i) clearly holds, since we deleted x from all bags. For the condition (i), consider any edge $e = uv$ of ∂F with $|\beta_2(u)| = |\beta_2(v)| = p - 1$. Note that $e_1 \neq e_0$, since $|\beta_2(v_0)| = |\beta(v_0)| - 2 \leq p - 2$. Any $(\beta_2(u), \beta_2(v))$-linkage in $\beta_2(e)$ extends to a $(\beta(u), \beta(v))$-linkage in $\beta(e)$ by adding the path consisting only of x. Therefore, since (ii) holds for β at e, it also holds for β_2 at e. □

Let us now show that the maximality of F implies that all components of $\pi(\mathcal{S}_1)$ intersect $\mathrm{bd}(\Delta)$, and thus the technical issue that we ran into in Theorem 10.4 (see the paragraph defining $(\mathcal{S}_0'', \mathcal{S}_1'')$) does not arise. To prove this, it is convenient to assume that $G' = \pi(\{F\} \cup \mathcal{S}_1)$ is 2-connected as a society, i.e., that it has no vertex separation (A, B) of order at most one such that $\partial G' \subseteq A$ and $B \setminus A \neq \emptyset$. This is without loss of generality: Otherwise, there exists an $(\{F\}, \mathcal{S}_1)$-normal disk $\Lambda \subseteq \Delta$ with $\mathrm{bd}(\Lambda) \cap G'$ consisting of at most one vertex v and with G' having a non-empty intersection with the interior of Λ. In this case, we could replace all $S \in \{F\} \cup \mathcal{S}_1$ with $\overline{\pi}(S) \subseteq \Lambda$ by the society $G_{(F,\mathcal{S}_1)}[\Lambda]$ with the boundary v, taken as 0-vortex-like if it contains F and as a cell otherwise.

We say that a disk $\Lambda \subseteq \Delta$ is 2-*separating* if it is $(\{F\}, \mathcal{S}_1)$-normal, $\mathrm{bd}(\Lambda)$ intersects $V(\pi(\{F\} \cup \mathcal{S}_1))$ in exactly two vertices belonging to ∂F, one arc of $\mathrm{bd}(\Lambda)$ between these two vertices is drawn inside the face $\overline{\pi}(F)$ and the other one outside, and either $R = G_{(\emptyset,\mathcal{S}_1)}[\Lambda]$ is non-empty or a vertex of ∂F is contained in the interior of Λ.

Observation 10.11 *Let G be a society and let $(\{F\}, \mathcal{S}_1)$ be a solid bisegregation of G with an arrangement in the disk Δ, such that $\pi(F)$ is contained in the interior of Δ and the society $\pi(\{F\} \cup \mathcal{S}_1)$ is 2-connected. If there is no 2-separating disk, then the society $\pi(\mathcal{S}_1)$ is 1-connected.*

Let us remark that the absence of 2-separating disks is actually a stronger condition, since it implies that the boundary of every face of $\pi(\{F\} \cup \mathcal{S}_1)$ other than $\overline{\pi}(F)$ shares either at most one vertex or an edge and its ends with the cycle $\pi(F)$.

Lemma 10.12 *Let G be a society and let $(\{F\}, \mathcal{S}_1)$ be a solid bisegregation of G with an arrangement in the disk Δ, such that $\pi(F)$ is contained in the interior of Δ and F has a strong vortical decomposition of warp p. If F is warp-p-maximal, then there is no 2-separating disk.*

Proof Let $f = \overline{\pi}(F)$ and let β be a strong vortical decomposition of F of warp p. Suppose for a contradiction that Λ is a 2-separating disk, and let u and v be the vertices forming the intersection of $\mathrm{bd}(\Lambda)$ with ∂F. Let $R = G_{(\emptyset,\mathcal{S}_1)}[\Lambda]$ and let Q be the subpath of ∂F with ends u and v contained in Λ.

Let $F' = G_{(\{F\},\mathcal{S}_1)}[\overline{f} \cup \Lambda]$, with $\partial F'$ consisting of the part of ∂F not contained in the interior of Λ. Note that $R \subseteq F'$, and thus if R is non-empty, then $F' \neq F$. If R is empty, then the definition of a 2-separating disk implies that the interior of Λ contains a vertex of ∂F, and thus $\partial F' \neq \partial F$, and consequently we have $F' \neq F$ again.

Let e_0 be the edge uv of $\partial F'$. For $x \in V(\partial F') \cup (E(\partial F') \setminus \{e_0\})$, let $\beta'(x) = \beta(x)$. Let $\beta'(e_0) = R \cup \beta(E(Q))$. Clearly, β' is a strong vortical decomposition of F'. If β' has warp p, we obtain a contradiction with the warp-p-maximality of F. Hence, suppose that F' does not have warp p; hence, $p \geq 2$ and e_0 violates (ii).

Therefore, there exists a $(\beta(u), \beta(v))$-linkage $\mathcal{P}$ in $\beta'(e_0)$ of size p such that the path $P \in \mathcal{P}$ starting in u does not end in v. Note that the parts of paths of $\mathcal{P}$ contained in R can be viewed as drawn in $\Lambda_0 = (\Lambda \setminus \overline{f}) \cup V(Q)$ without crossings. Let γ_0 be the arc of $\mathrm{bd}(\Lambda)$ contained in Λ_0. We say that a vertex or edge x of Q is *exposed* if there exists a simple curve in Λ_0 starting in γ_0 and ending in x and disjoint from the paths of $\mathcal{P}$ except for its end in x. Let u' be the last exposed vertex of Q belonging to P. Note that $u' \neq v$, since P does not end in v. Let v' be successor of u' in Q. The edge $e = u'v'$ of Q is exposed: Since u' is exposed and belongs to P, the only path of $\mathcal{P}$ that can separate e from γ_0 is P. Hence, if e were not exposed, then P would have a segment drawn in Λ_0 starting in u' and ending in a farther vertex u'' in Q; but then this vertex u'' would be exposed and contradicting the choice of u'.

Let Q_u and Q_v be the components of $Q - e$ containing u and v, respectively. Consider any path $P' \in \mathcal{P}$, and let $x_{P'}$ be its last vertex in $\beta(E(Q_u))$. Let P_0' be the segment of P' between $x_{P'}$ and the nearest following vertex $y_{P'}$ belonging to $\beta(E(Q_v))$. Note that P_0' cannot be contained in R, since the edge e is exposed. It follows that all edges of P_0' belong to $\beta(e)$. If P_0' has at least one edge, or if P_0' only consists of the vertex $x_{P'}$ and $x_{P'} \in \beta(u') \cap \beta(v')$, then P_0' is a path between $\beta(u')$ and $\beta(v')$ in $\beta(e)$. If this is the case for every $P' \in \mathcal{P}$, then $\mathcal{P}' = \{P_0' : P' \in \mathcal{P}\}$ is a $(\beta(u'), \beta(v')$-linkage in $\beta(e)$ of size p, and since β has warp p, the path $P_0 \subseteq P$ of $\mathcal{P}'$ joins u' with v'. However, this implies that v' is an exposed vertex belonging to P, contradicting the choice of u'.

We conclude that there exists a path $P' \in \mathcal{P}$ such that P_0' only consists of the vertex $x = x_{P'}$ and $x \notin \beta(u') \cap \beta(v')$. Since $x \in \beta(E(Q_u)) \cap \beta(E(Q_v))$, we conclude that $\beta^{-1}(x)$ is an open path not containing e, but containing both u and v. Therefore, $x \in \beta'(z) \setminus \{z\}$ for every $z \in \partial F'$. By Observation 10.10 for F' and e_0, there exists a set $A \subseteq V(F') \setminus \partial F'$ of size at most two such that $F' - A$ has a strong vortical decomposition of warp less than p. This contradicts the warp-p-maximality of F. □

Next, let us make the adhesion of the strong vortical decomposition uniform. The proof is analogous to Lemma 10.7, but instead of adding a full cycle of vertices surrounding the vortex, we only add arcs over boundary vertices of small adhesion. We are actually going to prove a stronger claim for semi-strong vortical decompositions, which we are going to need soon.

Lemma 10.13 *Let G be a society and let $(\{F\}, \mathcal{S}_1)$ be a solid bisegregation of G with an arrangement in the disk Δ, such that the vortex face of F has boundary distance at least two in $\pi(\mathcal{S}_1)$, the society $\pi(\{F\} \cup \mathcal{S}_1)$ is 2-connected, and F has a strong vortical decomposition of warp p. Let β be a* semi-strong *vortical decomposition of F of warp at most p and let Q be a subpath of ∂F of length at least two with ends u_1 and u_2 such that (VE) holds for all vertices in $\partial F \setminus V(Q - \{u_1, u_2\})$. If $|\beta(u)| < p$ for every $u \in V(Q - \{u_1, u_2\})$, then F is not warp-p-maximal.*

Proof ($\Leftarrow$) Suppose for a contradiction that $|\beta(u)| < p$ for every $u \in V(Q - \{u_1, u_2\})$, but F is warp-p-maximal. Let G_0 be the graph obtained from $\pi(\{F\} \cup \mathcal{S}_1)$ by deleting the edges incident with the internal vertices of Q and not belonging to $E(Q)$, and let f_0 be the face of G_0 incident with Q different from the face $f = \overline{\pi}(F)$. Since F is warp-p-maximal, the arrangement of the bisegregation $(\{F\}, \mathcal{S}_1)$ has no 2-separating disk by Lemma 10.12. Observe that this implies that the boundaries of f_0 and f intersect exactly in the path Q. Moreover, since the boundary distance of f is at least two, there exists a cycle K_0 in the boundary of f_0 such that the disk Λ bounded by K_0 contains $\overline{f_0}$. Let P be the path $K_0 - V(Q - \{u_1, u_2\})$ with ends u_1 and u_2. Let $\Delta = \overline{f} \cup \Lambda$; then Δ is an $(\{F\}, \mathcal{S}_1)$-normal disk. Let $F' = G_{(\{F\}, \mathcal{S}_1)}(\Delta)$. Note that $\partial F'$ is obtained from ∂F by replacing Q by P.

Let $H = \pi(\{F\} \cup \mathcal{S}_1)$ and let M be the set of vertices of R_H corresponding to the faces $h \neq f$ of H such that h is incident with a vertex of $V(Q) \setminus \{u_1, u_2\}$ and a vertex of $V(P) \setminus \{u_1, u_2\}$, and there is no $S \in \mathcal{S}_1$ such that $h = \overline{\pi}(S)$. For each $m \in M$, let u_m be the neighbor of m in $Q - u_2$ closest to u_2. Observe that every vertex $v \in V(P) \setminus \{u_1, u_2\}$ is incident with a face corresponding to a vertex $m \in M$; choose such a vertex m so that u_m is as close to u_2 as possible, and let P_v be the path vmu_m in R_H. Let $P_{u_1} = u_1$ and $P_{u_2} = u_2$. For $e = v_1v_2 \in E(P)$, let f_e be the face of $P \cup \pi(F) \cup P_{v_1} \cup P_{v_2}$ drawn inside Λ and incident with e. Let the labels v_1 and v_2 be chosen so that v_2 is the successor of v_1 in the cyclic ordering of $\partial F'$. Let Δ_e consist of f_e, the edge e, the vertices and edges of P_{v_1} that do not belong to P_{v_2}, and the path P_e in ∂F from the end of P_{v_1} to the end z of P_{v_2}, excluding the vertex z. This construction ensures that $\Lambda \setminus \{u_2\}$ is the disjoint union of the sets Δ_e for $e \in E(\partial F')$.

For edges and vertices x of the path $\partial F - (V(Q) \setminus \{u_1, u_2\})$, let $\beta_1(x) = \beta(x)$. For every edge $e = v_1v_2 \in E(P)$, let

$$\beta_1(e) = \{v_1, v_2\} \cup \bigcup_{S \in \mathcal{S}_1 : \overline{\pi}(S) \subseteq \Delta_e} S \cup \bigcup_{x \in E(P_e) \cup V(P_e)} \beta(x).$$

For every $v \in V(P) \setminus \{u_1, u_2\}$, let z be the end of the path P_v in $V(Q) \setminus \{u_1, u_2\}$ and let

$$\beta_1(v) = \{v\} \cup \beta(z).$$

Observe that β_1 is a strong vortical decomposition of F'.

Since we assumed that F is warp-p-maximal and $F' \neq F$, β_1 cannot have warp at most p. Since $|\beta(u)| < p$ for every $u \in V(Q - \{u_1, u_2\})$, we have $|\beta_1(v)| \leq p$ for every $v \in \partial F'$. The condition (ii) of the definition of warp p clearly holds for β_1 at every edge $e \in E(\partial F') \setminus E(P)$, since it holds for β. Therefore, there exists $e = v_1v_2 \in E(P)$ such that $|\beta_1(v_1)| = |\beta_1(v_2)| = p$ and a $(\beta_1(v_1), \beta_1(v_2))$-linkage $\mathcal{P}$ in $\beta_1(e)$ of size p such that the path $P' \in \mathcal{P}$ starting in v_1 does not end in v_2. Let u_1' and u_2' be the ends of P_{v_1} and P_{v_2} in Q.

Suppose first that $v_1 \neq u_1$. Let f' be the face of H incident with e and drawn in Λ. Since P is contained in the boundary of f_0, f' is also incident with a vertex of Q. Recall that u_1' was chosen to be as close to u_2 as possible; we conclude that u_1' is contained in the boundary of f'. Because of the face f', the set $\{u_1', v_2\}$ forms a cut in $\beta_1(e)$ separating v_1 from $\beta_1(v_2)$. Since the path P' cannot pass through u_1', we conclude that it ends in v_2, which is a contradiction.

Therefore, we have $v_1 = u_1$, and in particular the edge $e = u_1v_2$ is the only edge of $\partial F'$ for which (ii) is violated. Since F is warp-p-maximal, Observation 10.10 implies that $\bigcap_{x \in \partial F'}(\beta_1(x) \setminus \{x\}) = \emptyset$.

Consider any vertex $z \in \beta(u_1) \cap \beta(u_2')$. Note that $z \notin \partial F'$, and thus there must exist a vertex $x \in \partial F' \setminus \{u_1, u_2'\}$ such that $z \notin \beta_1(x)$. Note that $\beta^{-1}(z)$ is an open path by (DE2) for β, and this open path does not contain x or the end of the path P_x. Therefore, letting Q' be the segment of Q between u_1 and u_2', we conclude that $Q' \subset \beta^{-1}(z)$. In particular, it follows that $z \in \beta(y)$ for every $y \in V(Q')$.

Since this holds for all $z \in \beta(u_1) \cap \beta(u_2')$, we conclude that for every $y \in V(Q')$, the set $\beta(y)$ separates $\beta(u_1)$ from $\beta(u_2')$ in $\beta(E(Q'))$. If the path P' starts by a segment P_0' drawn in Λ, then let $u \in V(Q') \setminus \{u_1\}$ be the end of this segment; otherwise, let $P_0' = u_1$ and let u be the neighbor of u_1 in Q'. In either case, P' intersects $\beta(u)$, and P_0' together with $\beta(u)$ separates $\beta(u_1)$ from $\beta_1(v_2)$ in $\beta_1(e)$. Consequently, all paths $\mathcal{P} \setminus \{P'\}$ intersect the set $\beta(u) \setminus V(P')$ of size less than $|\beta(u)| \leq p - 1$. This is a contradiction, since $|\mathcal{P} \setminus \{P'\}| = p - 1$. □

In particular, Lemma 10.13 applied to a warp-p-maximal vortex gives the following.

Corollary 10.14 *Let G be a society and let $(\{F\}, \mathcal{S}_1)$ be a solid bisegregation of G with an arrangement in the disk Δ, such that the vortex face f of F has boundary distance at least two in $\pi(\mathcal{S}_1)$ and the society $\pi(\{F\} \cup \mathcal{S}_1)$ is 2-connected. Let β be a strong vortical decomposition of F of warp p. If F is warp-p-maximal, then $|\beta(u)| = p$ for every $u \in \partial F$.*

10.2.4 Establishing a Linkage

Let us now apply Lemma 10.13 to show that a warp-p-maximal vortex must be quite well connected. We say that a strong vortical decomposition β of a society F of warp p is ***internally linked*** if for every $e = uv \in E(\partial F)$, there exists a $(\beta(u) \setminus \{u\}, \beta(v) \setminus \{v\})$-linkage in $\beta(e) - \{u, v\}$ of size $p - 1$.

Lemma 10.15 *Let G be a society and let $(\{F\}, \mathcal{S}_1)$ be a solid bisegregation of G with an arrangement in the disk Δ, such that the vortex face f of F has boundary distance at least two in $\pi(\mathcal{S}_1)$ and the society $\pi(\{F\} \cup \mathcal{S}_1)$ is 2-connected. Let $Z \subseteq \partial G$ have size at least four and suppose that F is Z-glued. If F is warp-p-maximal, then every strong vortical decomposition β of F of warp at most p is internally linked.*

Proof Since F is Z-glued, we have $|\partial F| \geq 4$. Consider any open subpath $e_u u e v e_v$ of ∂F. By Corollary 10.14, we have $|\beta(u)| = |\beta(v)| = p$. We need to show that there exists a $(\beta(u) \setminus \{u\}, \beta(v) \setminus \{v\})$-linkage in $\beta(e) - \{u, v\}$ of size $p - 1$.

Suppose first that there exists a separation (A, B) of $\beta(e)$ of order less than p such that $\beta(u) \subseteq V(A)$, $\beta(v) \subseteq V(B)$, and $u, v \in V(A \cap B)$. Let $\beta_1(e_u) = \beta(e_u) \cup A$, $\beta_1(e_v) = \beta(e_v) \cup B$, $\beta_1(e) = A \cap B$, $\beta_1(u) = \beta_1(v) = V(A \cap B)$, and $\beta_1(x) = \beta(x)$ for $x \in (E(\partial F) \setminus \{e, e_u, e_v\}) \cup (\partial F \setminus \{u, v\})$. Then β_1 is a semi-strong vortical decomposition of F. Note that the warp of β_1 is at most p; the modified bags satisfy the condition (ii) trivially, since we have $|\beta_1(u)|, |\beta_1(v)| < p$. However, this gives a contradiction with Lemma 10.13 for the subpath Q of ∂F with internal vertices u and v. Hence, no such separation exists.

Suppose now that there exists a separation (A, B) of $\beta(e)$ of order less than p such that $\beta(u) \subseteq V(A)$, $\beta(v) \subseteq V(B)$, and $u \in V(A \cap B)$. By the previous paragraph, we have $v \notin V(A)$. Let $\beta_2(e_u) = \beta(e_u) \cup A$, $\beta_2(e) = B$, $\beta_2(u) = V(A \cap B)$, and $\beta_2(x) = \beta(x)$ for $x \in (E(\partial F) \setminus \{e, e_u\}) \cup (\partial F \setminus \{u\})$. Then β_2 is a strong vortical decomposition of F of warp p, and since $|\beta_2(u)| < p$, this gives a contradiction with Lemma 10.13 for the subpath Q of ∂F with internal vertex u. Hence, no such separation exists.

By symmetry between u and v, we conclude that if (C, D) is a vertex separation of $\beta(e)$ of order less than p such that $\beta(u) \subseteq C$ and $\beta(v) \subseteq D$, then $u \in C \setminus D$ and $v \in D \setminus C$. It follows that every vertex separation (C_u, D_u) of $\beta(e) - u$ with $\beta(u) \setminus \{u\} \subseteq C_u$ and $\beta(v) \subseteq D_u$ has order at least $p - 1$, as otherwise $(C_u \cup \{u\}, D_u \cup \{u\})$ would be a vertex separation of $\beta(e)$ of order less than p with u in both parts. Symmetrically, every vertex separation (C_v, D_v) of $\beta(e) - v$ with $\beta(u) \subseteq C_v$ and $\beta(v) \subseteq D_v \setminus \{v\}$ has order at least $p - 1$. By Menger's theorem, we conclude that there exists a $(\beta(u) \setminus \{u\}, \beta(v))$-linkage $\mathcal{P}_u$ in $\beta(e) - u$ of size $p - 1$ and a $(\beta(u), \beta(v) \setminus \{v\})$-linkage $\mathcal{P}_v$ in $\beta(e) - v$ of size $p - 1$.

Suppose that there exists a vertex separation (C, D) of $\beta(e)$ with $\beta(u) \subseteq C$ and $\beta(v) \subseteq D$ of size at most $p - 1$. Clearly both $\mathcal{P}_u$ and $\mathcal{P}_v$ pass through $C \cap D$, and thus $|C \cap D| = p - 1$. In this case, $\beta(e) - \{u, v\}$ contains a $(\beta(u) \setminus \{u\}, \beta(v) \setminus \{v\})$-linkage of size $p - 1$ consisting of the initial segments of the paths of $\mathcal{P}_u$ till $C \cap D$ and the final segments of the paths of $\mathcal{P}_v$ from $C \cap D$, as desired.

Finally, let us consider the case that every vertex separation (C, D) of $\beta(e)$ with $\beta(u) \subseteq C$ and $\beta(v) \subseteq D$ has size at least p. By Menger's theorem, there exists a $(\beta(u), \beta(v))$-linkage $\mathcal{P}$ in $\beta(e)$ of size p. By the condition (ii) of warp, a path P of this linkage joins u with v. Hence, $\mathcal{P} \setminus \{P\}$ is a $(\beta(u) \setminus \{u\}, \beta(v) \setminus \{v\})$-linkage of size $p - 1$ in $\beta(e) - \{u, v\}$. □

Similarly to the linked path decompositions discussed in Sect. 3.4, we can consider the union $\mathcal{L}$ of the linkages in $\beta(e)$ for all $e \in E(\partial F)$. Note that $\mathcal{L}$ is a disjoint union of cycles (or possibly singleton vertices) in $F - \partial F$, and that $\{P \cap \beta(e) : P \in \mathcal{L}\}$ is a $(\beta(u) \setminus \{u\}, \beta(v) \setminus \{v\})$-linkage of size $p - 1$ in $\beta(e) - \{u, v\}$ for every edge $e = uv \in \partial F$. Note however that $\mathcal{L}$ can have fewer than $p - 1$ components: It may happen that if we start in a vertex x of $\beta(u) \setminus \{u\}$ for some $u \in \partial F$ and trace the linkage paths around the vortex until we reach $\beta(u)$ again, we end up in a vertex different from x.

It would be convenient to get a full $(\beta(u), \beta(v))$-linkage of size p for every $e = uv \in E(\partial F)$. This may not be always possible, but we can complete such a linkage using the paths in the surface part. Let G be a graph or a society and let $(\mathcal{S}_0, \mathcal{S}_1)$ be a solid bisegregation of G with an arrangement on a surface Δ (a disk when G is a society). Suppose β is a strong vortical decomposition of $F \in \mathcal{S}_0$ of warp p. We say that β is *linked* in this bisegregation if it is internally linked and for every $e = uv \in E(\partial F)$, either there exists a $(\beta(u), \beta(v))$-linkage in $\beta(e)$ of size p (necessarily connecting u with v), or u and v are adjacent in $\pi(\mathcal{S}_1)$. Since the bisegregation is solid, note that this implies that there exists a cycle C in G containing vertices of ∂F in order such that for every $e = uv \in E(\partial F)$, there exists a $(\beta(u) \setminus \{u\}, \beta(v) \setminus \{v\})$-linkage in $\beta(e) - V(C)$ of size $p - 1$. Thus, the cycle C can be added to the overall linkage $\mathcal{L}$ we discussed in the previous paragraph.

Lemma 10.16 *Let G be a society and let $(\{F\}, \mathcal{S}_1)$ be a solid bisegregation of G with an arrangement in the disk Δ, such that the vortex face f of F has boundary distance at least two in $\pi(\mathcal{S}_1)$ and the society $\pi(\{F\} \cup \mathcal{S}_1)$ is 2-connected. Let $Z \subseteq \partial G$ have size at least four and suppose that F is Z-glued. If F is warp-p-maximal, then every strong vortical decomposition β of F of warp at most p is linked in the bisegregation.*

Proof By Lemma 10.15, β is internally linked. Consider any edge $e = uv \in E(\partial F)$ such that $\beta(e)$ does not contain a $(\beta(u), \beta(v))$-linkage of size p. By Menger's theorem, there exists a separation (A, B) of $\beta(e)$ of order at most $p-1$ with $\beta(u) \subseteq V(A)$ and $\beta(v) \subseteq V(B)$. Since $\beta(e) - \{u, v\}$ contains a $(\beta(u) \setminus \{u\}, \beta(v) \setminus \{v\})$-linkage of size $p - 1$, we have $u, v \notin V(A \cap B)$.

Suppose for a contradiction that u and v are not adjacent in $\pi(\mathcal{S}_1)$. Let f_0 be the face of $\pi(\{F\} \cup \mathcal{S}_1)$ incident with e and different from $\overline{\pi}(F)$. Since there is no 2-separating disk by Lemma 10.12, the boundary of f_0 shares only the edge e and its ends with the cycle $\pi(F)$. Hence, there exists a vertex w incident with f_0 but not contained in ∂F. Let F' be obtained from F by adding w and placing it between u and v in $\partial F'$. Note that $F' = G_{(\{F\},\mathcal{S}_1)}(\Lambda)$ for an $(\{F\}, \mathcal{S}_1)$-normal disk Λ obtained from the closure of $\overline{\pi}(F)$ by adding w and a part of f_0. Let $\beta_1(uw) = A + w$, $\beta_1(wv) = B + w$, $\beta_1(w) = \{w\} \cup V(A \cap B)$, and $\beta_1(x) = \beta(x)$ for $x \in (E(\partial F)\setminus\{e\}) \cup \partial F$. Then β_1 is a strong vortical decomposition of F'. Moreover, β_1 clearly has warp p, since $|\beta_1(w)| \leq p$ and w is an isolated vertex in $\beta_1(uw)$ and $\beta_1(wv)$. Since $F' \neq F$, this contradicts the assumption that F is warp-p-maximal.

We conclude that $e \in E(\pi(\mathcal{S}_1))$ for every such edge $e \in E(\partial F)$, and thus β is linked in the bisegregation $(\{F\}, \mathcal{S}_1)$. □

10.3 Improved Structure Theorem

Let us now put together what we have achieved in this chapter so far. Let $\gamma \geq 4$ be a fixed integer.

- We use Theorem 10.4 to get a solid (m, δ)-simple bisegregation with a $(\mathcal{T} - A, \psi'(m, \delta + |A|))$-spread arrangement $(\mathcal{S}_0, \mathcal{S}_1)$, with A consisting only of vertex-local apices and at most $a_\Sigma(H) - 1$ global apices. Let $\mathcal{T}_0$ be the surface tangle.
- Then, we process each vertex $F \in \mathcal{S}_0$, enlarging it within an $O(\delta + \gamma)$-zone Δ_F bounding a battlefield around it:
  - We use Lemma 9.10 to make F Z-glued for a $\mathcal{T}_0$-free set Z of 2γ vertices of $\pi(\mathcal{S}_1)$ drawn in the boundary of Δ_F.
  - We use Observation 10.6 and Lemma 10.7 to ensure that F has a strong vortical decomposition of warp $O(\delta + \gamma)$.
  - We then enlarge F and remove a set A_F of $O(\delta+\gamma)$ vertices from $V(F) \setminus \partial F$ to make F warp-p-maximal for some $p = O(\delta + \gamma)$.
  - By Lemma 10.16, this implies that the strong vortical decomposition β of F is linked in the resulting bisegregation.
  - We can now add A_F back to all bags of β; the resulting strong vortical decomposition is still linked, since the vertices of A_F form trivial linkage paths by themselves.
  - Moreover, F is still Z-glued, and Lemma 4.19 implies that (if we were careful to choose the boundary of Δ_F to be far enough in the $d_{\mathcal{T}_0}$-distance), the set Z is $\mathcal{T}'_0$-free for the tangle $\mathcal{T}'_0$ obtained from $\mathcal{T}_0$ by clearing all the bits of $\pi(\mathcal{S}_1)$ added to F. By Lemma 2.45, we conclude that $\mathrm{rk}_{\mathcal{T}'_0}(\partial F) \geq \gamma$.

Since the warp of each resulting vortex is $p = O(\delta + \gamma)$, by starting the process with function $\psi'(m, d) = \psi(m, O(d+\gamma)) + O(m(d+\gamma))$, the construction ensures that the vortex faces in the result are at $d_{\mathcal{T}'}$-distance at least $\pi(m, p + |A|)$ from one another.

In conclusion, we get the following claim. Let us say that a bisegregation $(\mathcal{S}_0, \mathcal{S}_1)$ with an arrangement on a surface is (m, p)-*strongly-simple* if $|\mathcal{S}_0| \leq m$ and each $F \in \mathcal{S}_0$ has a strong vortical decomposition of warp at most p which is linked in the bisegregation. It is γ-*free* if $\mathrm{rk}_{\mathcal{T}_0}(\partial F) \geq \gamma$ for every $F \in \mathcal{S}_0$, where $\mathcal{T}_0$ is the surface tangle of the arrangement.

Theorem 10.17 *For any graph H, any positive integers t and γ, and any non-decreasing function $\psi : \mathbb{N}^2 \to \mathbb{N}$, there exist integers α, $p_0 \geq 0$ and $\theta > \alpha$ such that the following claim holds. Suppose $\mathcal{T}$ is a tangle of order at least θ in a graph G. If $\mathcal{T}$ does not control a minor H of G, then there exists a set $A \subseteq V(G)$ of size at most α such that for some $p \leq p_0$ and $m \leq \alpha$, $G - A$ has a solid (m, p)-strongly-simple γ-free bisegregation with a $(\mathcal{T} - A, \psi(m, p + |A|))$-spread arrangement on an H-avoiding surface Σ, and A consists only of vertex-local apices except for at most $a_\Sigma(H) - 1$ apices that are $(t, \psi(m, \delta + |A|))$-global.*

10.4 Finding Large Bipartite Minors

Let us now show the application of Theorem 10.17 in the proof of Theorem 10.3. Vortices are dealt with using Lemma 3.16 in which we considered graphs with a long path decomposition. Let β be a strong vortical decomposition of a society F. The fact that β is fundamentally a cycle decomposition rather than a path decomposition would cause difficulties; to overcome this, we are going to only consider the subgraph $F' = \beta(P)$ of F for a long subpath P of ∂F. Let u and v be the ends of P. There is a small complication: The restriction β_1 of β to P is not necessarily a path decomposition of F', since for a vertex $x \in \beta(u) \cap \beta(v)$, (DE2) allows the open path $\beta^{-1}(x)$ to wrap around F in the arc opposite to P, and thus $\beta_1^{-1}(x)$ can consist of two paths, one containing u and one containing v. This can be avoided by choosing the path P carefully.

Lemma 10.18 *Let β be a strong vortical decomposition of a society F with $|\partial F| \geq 2$, let $a \geq 1$ be the adhesion of β, and let w be an assignment of non-negative weights to vertices of ∂F. There exists a path $P \subset \partial F$ such that $\beta \restriction E(P)$ is a path decomposition of $\beta(P)$ and $w(V(P)) \geq \frac{1}{\max(2,2a-1)} w(\partial F)$.*

Proof Let v_0 be an arbitrary vertex of ∂F. For each $x \in \beta(v_0)$, if $\beta^{-1}(x) \neq \partial F$, then let u_x and v_x be the first and the last vertex on the open path $\beta^{-1}(x)$; otherwise, let $u_x = v_x = v_0$. Let $S_0 = \bigcup_{x \in \beta(v_0)} \{u_x, v_x\}$. If $|S_0| \geq 2$, then let $S = S_0$, otherwise let S be obtained from S_0 by adding an arbitrary vertex of ∂F different from v_0. Since $\beta(v_0)$ contains at most $a - 1$ vertices different from v_0 and since $u_{v_0} = v_{v_0} = v_0$ by (VE), we have $v_0 \in S$ and $|S| \leq \max(2, 2a - 1)$. Therefore, there exists a subpath P of ∂F with one end in S and otherwise disjoint from S such that $w(V(P)) \geq \frac{1}{\max(2,2a-1)} w(\partial F)$.

Let u and v be the ends of P. Let $F' = \beta(P)$ and let β_1 be the restriction of β to the edges of P. We claim that β_1 is a path decomposition of F'. Indeed, consider any vertex $x \in V(F')$. We have $\beta_1^{-1}(x) = \beta^{-1}(x) \cap E(P)$. By (DE2), $\beta^{-1}(x)$ is an open path. If $\beta_1^{-1}(x)$ were disconnected, this would imply that $\beta^{-1}(x)$ contains u, v, and the arc of ∂F between u and v different from P. Since internal vertices of P do not belong to S, we have $S \subseteq V(\beta^{-1}(x))$. Since $v_0 \in S$, this implies $x \in \beta(v_0)$. But then P would contain both u_x and v_x, which is a contradiction, since P was chosen so that it contains only one vertex of S. □

Note that the weights can be used to ensure that "interesting" parts of the vortex are covered by P. We are also going to need the following easy observation.

Observation 10.19 *Let G be a graph with a cellular drawing on a surface Σ and let n be a positive integer. If G contains a respectful tangle $\mathcal{T}$ of order $\theta \geq 2n + 3$, then $|G| \geq n$.*

Proof By Theorem 4.14, there exist edges e_1 and e_2 with $d_{\mathcal{T}}(e_1, e_2) = \theta$. Let P be a path in G starting with e_1 and ending with e_2. The last vertex v of P satisfies $d_{\mathcal{T}}(e_2, v) \leq 2$, and thus $d_{\mathcal{T}}(e_1, v) \geq \theta - 2$. For $3 \leq i \leq \theta - 2$, let v_i be the first vertex of P such that $d_{\mathcal{T}}(e_1, v_i) \geq i$, and let v_i' be the preceding vertex of P. For $3 \leq i \leq j \leq \theta - 2$, we have

$$d_{\mathcal{T}}(v_i, v_j) \geq d_{\mathcal{T}}(e_1, v_j) - d_{\mathcal{T}}(e_1, v_i) \geq d_{\mathcal{T}}(e_1, v_j) - d_{\mathcal{T}}(e_1, v_i') - 2 \geq j - i - 1.$$

Hence, the vertices $v_3, v_5, \ldots, v_{2n+1}$ are pairwise distinct. □

Of course, if the surface Σ is fixed, then $|G| = \Omega(\theta^2)$, as we have seen in Lemma 8.1; however, the bound from Observation 10.19 makes for simpler calculations. We are now ready to prove the main result of this chapter.

Theorem 10.3 *For any integers $t, n \geq 1$, there exists an integer n_2 such that the following claim holds. If a $2t$-connected graph G of minimum degree at least $30t+1$ has at least n_2 vertices, then it contains $K_{t,n}$ as a minor.*

Proof Let $n_0 : \mathbb{N} \to \mathbb{N}$ be a non-decreasing function such that for every $k \geq t$, $n_0(k)$ is at least the constant n_0 from Lemma 3.16 applied with t, n, and k. Let $N(d) = 2d \cdot (2n_0(d)+3)$. Let θ be the constant from Theorem 10.17 for $H = K_{t,n}$, $t = \gamma = 1$, and $\psi(m, d) = 14(mN(d) + 6tn) + 3$. Let n_2 be the constant from Theorem 3.18 for t, n, and $k = \lceil \frac{3}{2}\theta \rceil$.

If G has treewidth at most k, then the claim follows from Theorem 3.18. Hence, suppose that $\mathrm{tw}(G) > k$, and by Theorem 2.25, G has a tangle $\mathcal{T}$ of order θ. Note that $a_\Sigma(K_{t,n}) \leq t - 2$ for every surface Σ, since we can delete $t - 2$ vertices from $K_{t,n}$ to turn it into the planar graph $K_{2,n}$. If $H = K_{t,n}$ is not a minor of G, then Theorem 10.17 implies that there exists a set $A \subseteq V(G)$ and integers p and m such that $G - A$ has a solid (m, p)-strongly-simple 1-free bisegregation with a $(\mathcal{T} - A, \psi(m, p + |A|))$-spread arrangement on a $K_{t,n}$-avoiding surface Σ, and A consists only of vertex-local apices except for a subset $A_0 \subseteq A$ of at most $t - 3$ apices that are $(1, \psi(m, p + |A|))$-global. Let $g \leq tn$ be the Euler genus of Σ.

For each cell $S \in \mathcal{S}_1$, note that $\partial S \cup A_0$ separates $S - \partial S$ from the rest of the graph G. Since $|\partial S \cup A_0| \leq 3 + t - 3 < 2t$ and G is $2t$-connected, we conclude that $V(S) = \partial S$ for every $S \in \mathcal{S}_1$. Since the bisegregation is solid, we conclude that $S = \pi(S)$ for every $S \in \mathcal{S}_1$, and thus $G' = \pi(\mathcal{S}_1)$ is a subgraph of $G - A$. Let W be the set of vertices of G' of degree at most 6. Each vertex $v \in V(G') \setminus \bigcup_{F \in \mathcal{S}_0} \partial F$ has in G neighbors only in $V(G')$ and A_0, and thus $\deg_{G'} v \geq \deg_G v - |A_0| \geq 29t + 4 > 6$. Therefore, $W \subseteq \bigcup_{F \in \mathcal{S}_0} \partial F$. Generalized Euler's formula gives $\|G'\| \leq 3|G'| + 3g \leq 3|G'| + 3tn$. On the other hand, we clearly have $\|G'\| \geq \frac{7}{2}(|G'| - |W|)$, and thus $|W| \geq \frac{1}{7}(|G'| - 6tn)$. Since G' has a respectful tangle of order $\psi(m, p + |A|)$, Observation 10.19 implies that $|G'| \geq 7(mN(p + |A|) + 6tn)$, and thus $|W| \geq mN(p + |A|)$.

Therefore, there exists $F \in \mathcal{S}_0$ such that $|\partial F \cap W| \geq N(p + |A|)$. Let β be a strong vortical decomposition of F of warp $p' \leq p$ which is linked in the bisegregation $(\mathcal{S}_0, \mathcal{S}_1)$. Let F_1 be obtained from F by, for each $e = uv \in \partial F$ such that $uv \in E(G')$, adding the edge uv to F_1, and let β_1 be the strong vortical decomposition of F_1 obtained by adding these edges to the corresponding bags of β. Since F is linked in the bisegregation $(\mathcal{S}_0, \mathcal{S}_1)$, observe that for every

$e = uv \in \partial F_1$, the bag $\beta_1(e)$ contains a $(\beta_1(u), \beta_1(v))$-linkage of size p' such that a path of this linkage joins u with v. Let F_2 be obtained from F_1 by adding $G[A]$ and all edges of G between A and F_1, and let β_2 be a strong vortical decomposition of F_2 of adhesion $p'' = p' + |A|$ obtained by adding A to all bags of β_1 and distributing the edges incident with A among the edge bags containing their ends arbitrarily. The vertices of A can be added to the linkage as single-vertex paths.

By Lemma 10.18, there exists a path $P \subseteq \partial F_2$ such that $|W \cap V(P)| \geq \frac{1}{2p''}|W \cap \partial F| \geq 2n_0(p'') + 3$ and the restriction β_2' of β_2 to the path Q formed by the edges of P is a path decomposition of $F_2' = \beta_2(P)$. Clearly, the path decomposition (Q, β_2') is p''-linked. Since $|W \cap V(P)| \geq 2n_0(p'') + 3$, there exists a set $U \subset W \cap V(P)$ of size $n_0(p'')$ such that vertices of U are non-adjacent in P and have distance at least two from the ends of P. For each $u \in U$, let $e_{u,1}$ and $e_{u,2}$ be the edges of P incident with u. Note that $e_{u,1}$ and $e_{u,2}$ are vertices of Q, different from the vertices $e_{u',1}$ and $e_{u',2}$ for any other vertex $u' \in U$. Let (Q_3, β_3) be the path decomposition obtained from (Q, β_2') by, for each vertex $u \in U$, contracting the edge of Q between $e_{u,1}$ and $e_{u,2}$. Slightly abusing the notation, let us denote the resulting vertex of Q_3 with the bag $\beta_2'(e_{u,1}) \cup \beta_2'(e_{u,2})$ by u.

Consider any $u \in U$:

- Note that $\beta_3^\circ(u) \cap U = \{u\}$, and in particular $\beta_3^\circ(u) \neq \emptyset$.
- For each vertex $v \in \beta_3^\circ(u) \setminus \{u\}$, all neighbors of v in G are contained in $\beta_3(u)$, and thus $\deg_{F_2'} v = \deg_G v \geq 30t + 1$. Moreover, $\deg_{F_2'} u \geq \deg_G u - \deg_{G'} u \geq \deg_G u - 6 \geq 30t - 5 \geq 2t + 1$.
- Since G is $2t$-connected, there is no vertex separation (R, S) of $F_2'[\beta(u)]$ of order less than $2t$ with $\{u\} \cup (\beta(u) \setminus \beta^\circ(u)) \subseteq R$ and $S \not\subseteq R$.

Therefore, (Q_3, β_3) with $X = U$ satisfies the assumptions of Lemma 3.16, and we conclude that $K_{t,n}$ is a minor of $F_2' \subset G$. □

Let us now quickly discuss the main ideas of the proof of Theorem 10.1. Consider a t-connected K_t-minor-free graph G. If G has large treewidth, Theorem 10.17 gives us its bisegregation $(\mathcal{S}_0, \mathcal{S}_1)$ and arrangement on a surface Σ with a set A_0 of at most $a_\Sigma(K_t) - 1 \leq t - 5$ global apex vertices. Since G is t-connected and at most $t - 5$ apex vertices have neighbors in $\bigcup \mathcal{S}_1$, we have $V(S) = \partial S$ for every cell $S \in \mathcal{S}_1$, and thus $\pi(\mathcal{S}_1)$ is a subgraph of G.

- If $|A_0| = t - 5$, then since K_t is not a minor of G and $K_t - A_0 = K_5$ can be drawn on any surface of positive genus, Lemma 4.35 implies that Σ is the sphere. Moreover, since K_5 can be drawn on the sphere with only one crossing, Lemma 4.35 implies that $G - A_0$ contains no long jump over $\pi(\mathcal{S}_1)$, and thus no vortex-local apex vertex can have attachments in two different vortices. Hence, we can incorporate non-global apex vertices into vortices. Similarly, there is no eye over $\pi(\mathcal{S}_1)$, and we can use Corollary 5.9 to show that there actually are no vortices. It follows that G can be obtained from a planar graph by adding $t - 5$ apex vertices, as desired.

- If $|A_0| = t - 6$, then Σ again is the sphere, since $K_t - A_0 = K_6$ can be drawn on any surface of positive genus. The argument here is overall similar to the previous case, but substantially more difficult, since the drawings of K_6 in the sphere have multiple crossings. It is important that K_6 has a drawing where all the crossings are incident with a single face, which can be used to restrict the structure of each vortex. Eventually, one shows that all vortices and non-global apex vertices can be eliminated after removing at most one vertex, again obtaining the desired conclusion.
- Finally, suppose that $|A_0| \le t - 7$. Since G is t-connected, $G - A_0$ has minimum degree at least 7. Since the surface part has average degree close to six or smaller, similarly to the proof of Theorem 10.3 it follows that the vortices of $G - A_0$ have relatively long boundaries. The most technical part of the proof then is an analysis of vortices similar in spirit to Lemma 3.16, though much more complicated, since the setting does not make it possible to find a well-linked subgraph in each piece.

The case of bounded treewidth is handled together with the long vortex case, using the transformation of a tree decomposition to a long path decomposition shown in the proof of Theorem 3.18.

10.5 Coloring K_t-Minor-Free Graphs

Linear Hadwiger's conjecture is a weakening of Hadwiger's conjecture that states that there exists a constant c such that for every positive integer t, every K_t-minor-free graph has chromatic number at most ct. Even this weaker version remains open. However, Theorem 10.3 can be used to show that for sufficiently large c, the conjecture could for any fixed t be in principle proved just by investigation of finitely many graphs.

We say that a graph G is ***minor-χ-critical*** if every proper minor of G has chromatic number less than $\chi(G)$. Clearly, if counterexamples to (linear) Hadwiger's conjecture for any given t exists, then the smallest ones are minor-χ-critical. The linkedness results from Sect. 3.2 can be used to show that minor-χ-critical graphs have large connectivity.

Lemma 10.20 *Let k be a positive integer and let G be a graph of chromatic number greater than $15k$. If G is minor-χ-critical, then G is k-connected.*

Proof Let $d = \chi(G) - 1 \ge 15k$. Since G is minor-χ-critical, G has minimum degree at least d: Indeed, if G contained a vertex v of degree less than d, then a d-coloring of $G - v$ (which exists by the minor-χ-criticality of G) would greedily extend to a d-coloring of G.

Let (A, B) be a vertex separation of G such that $A \not\subseteq B$ and $B \not\subseteq A$ of the smallest possible order. Suppose for a contradiction that $m = |A \cap B| < k$. We claim that

($\star$) for $X \in \{A, B\}$ and any ordering $v_1, \ldots, v_m$ of the vertices of $A \cap B$, the path $v_1 \ldots v_m$ is a $\{v_1, \ldots, v_m\}$-rooted minor of $G[X]$.

Indeed, suppose that say $X = A$. The graph $G[A] - B$ has minimum degree at least $d - k \geq 14k \geq d_{\text{link}}(k) + 4k$ by Theorem 3.4. By Corollary 3.6, $G[A] - B$ has a k-linked subgraph H of minimum degree more than $7k$. Since (A, B) is a separation of smallest order, Menger's theorem implies that $G[A]$ contains an $(A \cap B, V(H))$-linkage $\mathcal{P}$ of size m. For $i \in [m]$, let $P_i \in \mathcal{P}$ be the path starting in v_i and let u_i be the end of P_i in H; without loss of generality, we can assume that P_i intersects H only in u_i. Since H has minimum degree more than $2m$, we can for $i \in [m-1]$ select a neighbor z_i of u_i in H so that the selected vertices are pairwise distinct and also distinct from $u_1, \ldots, u_m$. Since H is k-linked, it contains an $\{(z_i, u_{i+1}) : i \in [m-1]\} \cup \{(u_1, u_1)\}$-linkage $\mathcal{Q}$. Contracting the paths of $\mathcal{P} \cup \mathcal{Q}$ gives us the path $v_1 \ldots v_m$ as the desired minor of $G[A]$.

Let ψ be a coloring of $G[A \cap B]$ using the smallest possible number $c \leq d$ of colors. For $X \in \{A, B\}$, let G_X be the graph obtained from $G[X]$ by, for each $i \in [c]$, contracting the vertices of $\psi^{-1}(i)$ to a single vertex x_i. Since $G[A \cap B]$ cannot be colored using fewer than c colors, $W = \{x_1, \ldots, x_c\}$ is a clique in G_X.

We claim that G_X is a minor of G: Indeed, suppose that say $X = B$. Let $v_1, \ldots, v_m$ be an ordering of the vertices $A \cap B$ such that $\psi(v_i) \leq \psi(v_j)$ for all $1 \leq i < j \leq m$. By ($\star$), we can contract $G[A]$ to (a supergraph of) the path $v_1 \ldots v_m$. By further contracting disjoint monochromatic segments of this path, we obtain G_X as a minor of G.

Since G is minor-χ-critical, G_A and G_B are d-colorable. By permuting the colors if necessary, we can assume that the colorings φ_A of G_A and φ_B of G_B assign the same colors to the vertices of the clique W. Thus, we can combine φ_A and φ_B to a d-coloring of G: Color the vertices of $A \setminus B$ according to φ_A, the vertices of $B \setminus A$ according to φ_B, and for $i \in [c]$, give all vertices of the independent set $\psi^{-1}(i)$ the color $\varphi_A(x_i) = \varphi_B(x_i)$. This is a contradiction, since $\chi(G) > d$. □

In combination with Theorem 10.3, this has the following consequence.

Corollary 10.21 *There exists a function $n_2 : \mathbb{N} \to \mathbb{N}$ such that the following claim holds for every $c \geq 31$ and every positive integer t. If every K_t-minor-free graph with at most $n_2(t)$ vertices is ct-colorable, then every K_t-minor-free graph is ct-colorable.*

Proof Let $n_2(t)$ be the value of n_2 from Theorem 10.3 for $n = t$. Suppose that for positive integers t and $c \geq 31$, there exists a K_t-minor-free graph that is not ct-colorable, and let G be the smallest such graph. Clearly, G is minor-χ-critical, and G is $2t$-connected by Lemma 10.20 applied with $k = 2t$. Moreover, as in the proof of Lemma 10.20, we see that G has minimum degree at least $ct \geq 30t + 1$. Since K_t is a minor of $K_{t,t}$ (obtained by contracting a perfect matching in $K_{t,t}$), the graph G is $K_{t,t}$-minor free. By Theorem 10.3, it follows that G has at most $n_2(t)$ vertices. □

Let us remark that by using Theorem 10.1 instead of Theorem 10.3, we could decrease the lower bound on c to 15. Kawarabayashi and Reed [4] showed that the result is actually true also for Hadwiger's conjecture. That is, for every positive integer t there exists n_0 such that if there exists a K_t-minor-free non-$(t-1)$-colorable graph, then there exists one with at most n_0 vertices.

Note that the proof of Corollary 10.21 has the following (conditional) algorithmic consequence.

Corollary 10.22 *For every $c \geq 31$ and every positive integer t, there exists a polynomial-time algorithm that, given a K_t-minor-free graph G, either finds a proper ct-coloring of G, or finds a minor of G that is not ct-colorable.*

Proof If G contains a vertex of degree at most $ct - 1$, then we delete it, solve the smaller instance $G - v$ recursively, and either return the same non-ct-colorable minor, or extend the returned ct-coloring of $G - v$ to G greedily.

If G is not $2t$-connected, then we consider a vertex separation (A, B) of G such that $A \not\subseteq B$ and $B \not\subseteq A$ of the smallest possible order, construct minors G_A and G_B of G as in the proof of Lemma 10.20, and apply the algorithm to them recursively. If one of the recursive calls returns a non-ct-colorable minor of G_A or G_B, then we return the same minor. Otherwise we obtain ct-colorings of G_A and G_B, and we follow the proof of Lemma 10.20 to combine them to a ct-coloring of G.

Finally, if G is $2t$-connected and has minimum degree at least ct, then Theorem 10.3 implies that G has at most $n_2(t)$ vertices, and we can find its ct-coloring (or decide that G is not ct-colorable) by brute force. □

Kawarabayashi and Reed [4] proved the analogous conditional result for $(t-1)$-colorability. It seems difficult to get an unconditional algorithm, i.e., one which would in the negative case show that the graph G (rather than a minor of G) is not colorable by the given number of colors: Let us say that a graph H is ***nearly-k-apex*** if we can remove k vertices from H so that the resulting graph can be drawn in the plane with only one crossing. Consider any integer $t \geq 5$.

- The clique K_t is nearly-$(t-5)$-apex.
- For $c \geq 2$, let H'_c be a graph obtained from a 4-connected planar graph with at least $c+8$ vertices by adding an edge joining vertices not incident with the same face and crossing only one other edge. Let $H_{t,c}$ be the graph consisting of H'_c together with a clique of $t-5$ vertices completely joined to H'_c. Then $H_{t,c}$ is nearly-$(t-5)$-apex.

Dvořák and Thomas [2] proved that for every $t \geq 5$ and $c \geq 2$, it is NP-complete to decide whether an $H_{t,c}$-minor-free graph is $(t+c-4)$-colorable. From the perspective of the Minor Structure Theorem (even in its strong form, Theorem 10.17), any nearly-$(t-5)$-apex graph behaves rather similarly to K_t. Hence, an unconditional algorithm to decide $(t-1)$-colorability of K_t-minor-free graphs would have to distinguish between $H_{t,3}$ and K_t somewhere in the fine details.

Let us remark that a conditional algorithmic result can also be obtained fairly easily without any substantial vortex analysis. Let us note the following easy observation.

Observation 10.23 *Let G be a connected graph drawn on a surface of Euler genus g, and let X be a non-empty set of vertices of G such that* $\deg v \geq 7$ *for every vertex* $v \in V(G) \setminus X$*. Then every vertex of G is at distance at most* $10g + 13 + \log_{1.1} n$ *from X.*

Proof Consider any vertex $v \in V(G)$. Let S_0 be the set of vertices of G at distance at most $10g$ from v. If $X \cap S_0 \neq \emptyset$, then the claim clearly holds. Otherwise, since G is connected and $X \neq \emptyset$, we have $|S_0| \geq 10g$. For $i \in \mathbb{N}$, let S_i be the set of vertices of G at distance at most i from S_0, and let d be the smallest index such that $S_d \cap X \neq \emptyset$.

Consider any $i \in [d]$. Since $|S_{i-1}| \geq |S_0| \geq 10g$, the generalized Euler's formula implies that $\|G[S_{i-1}]\| \leq 3|S_{i-1}| + 3g \leq 3.3|S_{i-1}|$. The number of edges between S_{i-1} and $S_i \setminus S_{i-1}$ is

$$\Big(\sum_{u \in S_{i-1}} \deg u\Big) - 2\|G[S_{i-1}]\| \geq 7|S_{i-1}| - 6.6|S_{i-1}| = 0.4|S_{i-1}| \geq \frac{0.4}{3.3}\|G[S_{i-1}]\|.$$

Therefore, we have

$$\|G[S_i]\| \geq \|G[S_{i-1}]\| + \frac{0.4}{3.3}\|G[S_{i-1}]\| \geq 1.1\|G[S_{i-1}]\|.$$

Since $\|G[S_0]\| \geq 1$, we have

$$1.1^d \leq \|G[S_d]\| \leq \|G\| \leq 3n + 3g \leq 3.3n,$$

and thus $d \leq \log_{1.1}(3.3n) \leq 13 + \log_{1.1} n$. Therefore, v is at distance at most $10g + 13 + \log_{1.1} n$ from X. □

The algorithm then follows from the following lemma, partially inspired by [4].

Lemma 10.24 *Let* $t \geq 5$ *be an integer and let H be a nearly-*$(t-5)$*-apex graph. There exists a constant* γ *such that if G is an n-vertex H-minor-free graph, then*

(a) G has apex number at most $t - 5$*; or,*
(b) G contains a vertex v such that $\deg v \leq t - 2$*, or* $\deg v = t - 1$ *and the neighborhood of v is not a clique, or* $\deg v = t$ *and the neighborhood of v contains three pairwise non-adjacent vertices; or,*
(c) G has a vertex separation (A, B) *of order at most* $t-2$ *such that* $G[B]$ *contains a clique on* $A \cap B$ *as a rooted minor; or,*
(d) $\mathrm{tw}(G) \leq \gamma \log n$*.*

Proof We give just a quick outline of the proof and we leave the details up to the reader. Choose $\gamma \gg \theta \gg \psi \gg l \gg |H|$. Consider any tangle $\mathcal{T}$ in G of order at least θ and apply Theorem 10.4; let $A \subseteq V(G)$ be a set of size at most α such that $G - A$ has a solid (m, δ)-simple bisegregation $(\mathcal{S}_0, \mathcal{S}_1)$ with a $(\mathcal{T} - A, \psi)$-spread arrangement on an H-avoiding surface Σ, and A consists only of vertex-local apices except for a set $A_0 \subseteq A$ of at most $a_\Sigma(H) - 1$ apices that are (l, ψ)-global. Let $\mathcal{T}_0$ be the surface tangle of the arrangement.

Note that since H is nearly-$(t-5)$-apex, we have $a_\Sigma(H) \leq t-4$, and thus $|A_0| \leq t-5$. Suppose first that there exists an $(\mathcal{S}_0, \mathcal{S}_1)$-normal disk $\Delta \subseteq \Sigma$ whose boundary intersects $\pi(\mathcal{S}_1)$ in a set Y of most three vertices, such that $A'_\Delta = V((G-A)_{(\mathcal{S}_0,\mathcal{S}_1)}[\Delta]) \neq Y$ and a vertex of $A'_\Delta \cap V(\pi(\mathcal{S}_1))$ is at $d_{\mathcal{T}_0}$-distance more than three from the vortex faces of $\pi(\mathcal{S}_1)$. Choose Δ as a maximal such disk; then it is easy to see that Y is $\mathcal{T}_0$-free. Let $A_\Delta = A'_\Delta \cup A_0$ and $B_\Delta = (V(G) \setminus A_\Delta) \cup (Y \cup A_0)$. Then (A_Δ, B_Δ) is a vertex separation of G of order at most $t-2$, and using Lemma 4.35, it is easy to check that $G[B]$ contains a clique on $Y \cup A_0 = A_\Delta \cap B_\Delta$ as a rooted minor. We conclude that (c) holds in this case.

Hence, suppose there is no such disk. Note that this implies that for each cell $S \in \mathcal{S}_1$, if ∂S is at $d_{\mathcal{T}_0}$-distance more than three from the vortex faces of $\pi(\mathcal{S}_1)$, then $V(S) = \partial S$. Suppose now that $|A_0| = t-5$. Since H is nearly-$(t-5)$-apex, we have $a_\Pi(H) \leq t-5$ for any surface Π other than the sphere, and thus in this case Σ must be the sphere. Moreover, if a vertex-local apex gave rise to a long jump or if a vortex could be used to create an eye, Lemma 4.35 would imply that H is a minor of G. Hence, using Corollary 5.9, it is easy to eliminate all vortices (and vortex-local apices), and thus without loss of generality, we can assume that $\mathcal{S}_0 = \emptyset$ and $A = A_0$. Moreover, since there are no vortex faces, we have $V(S) = \partial S$ for every $S \in \mathcal{S}_1$, and thus $\pi(\mathcal{S}_1) = G - A$. We conclude that (a) holds.

Therefore, suppose that $|A_0| \leq t-6$. Suppose now that $\pi(\mathcal{S}_1)$ has a vertex v of degree at most 6 at $d_{\mathcal{T}_0}$-distance at least three from vortex faces. Note that $V(S) = \partial S$ for every cell $S \in \mathcal{S}_1$ with $v \in \partial S$, and thus the neighbors of v in $G - A_0$ are exactly its neighbors in $\pi(\mathcal{S}_1)$. Moreover, note that v is not contained in a separating (or a non-contractible) triangle in $\pi(\mathcal{S}_1)$, and thus the neighbors of v that are non-consecutive in the cyclic ordering around v according to the drawing of $\pi(\mathcal{S}_1)$ are non-adjacent. If $\deg_{\pi(\mathcal{S}_1)} v \leq 4$, then $\deg_G v \leq |A_0| + 4 \leq t-2$. If $\deg_{\pi(\mathcal{S}_1)} v = 5$, then $\deg_G v \leq t-1$ and v has two non-adjacent neighbors. If $\deg_{\pi(\mathcal{S}_1)} v = 6$, then $\deg_G v \leq t$ and v has three pairwise non-adjacent neighbors. In either case, (b) holds.

Finally, suppose that every vertex at $d_{\mathcal{T}_0}$-distance at least three from vortex faces has degree at least 7 in $\pi(\mathcal{S}_1)$. Let X be the set of vertices at $d_{\mathcal{T}_0}$-distance at most three from the vortex faces (if there are no vortex faces, then let X consist of any single vertex). Let $g \leq \|H\|$ be the Euler genus of Σ; by Lemma 10.23, every vertex of $\pi(\mathcal{S}_1)$ is at the graph distance at most $10g + 13 + \log_{1.1} n$ from X. We can now use Lemma 9.10 to incorporate all vertices at $d_{\mathcal{T}_0}$-distance at most three from vortex faces (and thus also non-trivial elements of $\mathcal{S}_1$) into vortices. Then, for each $F \in \mathcal{S}_0$, we use Lemma 5.22 to find a vortical decomposition β_F of F of bounded adhesion, and for each $v \in \partial F$ with neighbors v_1 and v_2 in ∂F, we split off $\beta_F(v)$ and replace it by a clique on the set $\beta(v) \cap (\beta(v_1) \cup \beta(v_2)) \cup \{v\}$. Finally, we add the vertices of A back to the graph. By Theorem 2.73 and Lemma 7.5, we conclude that the resulting graph has treewidth $O(\log n)$. In conclusion, we get the following claim.

($\star$) Suppose that G does not satisfy (a), (b), and (c). For every tangle $\mathcal{T}$ in G of order at least θ, there exists a $\mathcal{T}$-central star decomposition (S, β) of G such that the torso of the center of S has treewidth $O(\log n)$.

A standard argument (the proof method discussed after Lemma 2.16) then shows that $\text{tw}(G) = O(\log n)$, and thus (d) holds. □

From this, it is easy to get the following conditional algorithmic result.

Corollary 10.25 *Let $t \geq 5$ and $c \geq t-1$ be integers and let H be a nearly-$(t-5)$-apex graph. There exists a polynomial-time algorithm that, given an H-minor-free graph G, either returns a c-coloring of G, or a minor of G that is not c-colorable.*

Proof First, go over all $(t-5)$-vertex sets $A \subseteq V(G)$ and check whether $G - A$ is planar. If so, then $G - A$ is 4-colorable by the Four Color Theorem, and a 4-coloring of G can be found in quadratic time using the algorithm described in [10]. Giving each vertex of A a new color results in a $(t-1)$-coloring of G.

Next, suppose that G contains a vertex v such that $\deg v \leq t$ and the neighborhood of v contains an independent set S of size at least $\deg v - t + 3$ (such a vertex can clearly be found in linear time). Let G' be the minor of G obtained by contracting the edges between v and S (and deleting v if $S = \emptyset$), and let s be the vertex resulting from the contraction. We recurse on G'; if the result is a non-c-colorable minor F of G', then we return F, which is also a minor of G. Otherwise, we obtain a c-coloring φ' of G'. Color all vertices of S by the color $\varphi'(s)$, then choose a color for v; this is possible, since at most $\deg v - (|S| - 1) \leq t - 2 < c$ different colors are used on the neighborhood of v. Hence, we obtain a c-coloring of G.

Next, we go over all sets $Z \subseteq V(G)$ of size at most $t-2$ and check whether they form a cut. If so, we also check whether there exists a component K of $G - Z$ such that $G_1 = G - K$ contains the clique on Z as a Z-rooted minor, using the algorithm of [8]. If so, we recurse on the subgraph G_1. If the result is a non-c-colorable minor F of G_1, then we return F. Otherwise, we obtain a c-coloring φ_1 of G_1. Contract G_1 to the clique on Z, then contract all edges of Z joining the vertices assigned the same color by φ_1, and let G_2 denote the resulting minor of G and Z_2 the resulting clique in G_2. Note that φ_1 assigns distinct colors to vertices of Z_2. We recurse on the subgraph G_2, either obtaining a non-c-colorable minor of G_2 (and thus also of G), or a c-coloring φ_2 of G_2. By renaming the colors if necessary, we can assume that φ_2 matches φ_1 on the clique Z_2. Hence, using φ_1 on G_1 and φ_2 on K gives us a c-coloring of G.

Hence, we can assume that the graph G does not satisfy (a), (b), and (c) of Lemma 10.24, and thus $\text{tw}(G) \leq \gamma \log n$, where $n = |G|$. We can now use the algorithm of [6] to find a tree decomposition of G of width at most $2\gamma \log n$, then a standard dynamic programming algorithm to find a c-coloring of G or decide that it is not c-colorable, in time $\exp(O(\text{tw}(G))) = \exp(O(\gamma \log n)) = n^{O(\gamma)}$. □

Of course, a cost for the relative simplicity of this algorithm is its truly horrifying time complexity $n^{O(\gamma)}$, where γ is a constant derived from the Minor Structure Theorem. For context, the algorithm of Kawarabayashi and Reed [4] has quadratic time complexity, though of course the multiplicative constant in the complexity is also derived from the Minor Structure Theorem, making it impractically large.

References

1. T. Böhme, K. Kawarabayashi, J. Maharry, B. Mohar, Linear connectivity forces large complete bipartite minors. J. Combin. Theory B **99**, 557–582 (2009)
2. Z. Dvořák, R. Thomas, List-coloring apex-minor-free graphs. ArXiv, 1401.1399 (2014)
3. L. Jørgensen, Contractions to K_8. J. Graph Theory **18**, 431–448 (1994)
4. K. Kawarabayashi, B. Reed, Hadwiger's conjecture is decidable, in *Proceedings of the Forty-First Annual ACM Symposium on Theory of Computing, STOC '09* (2009), pp. 445–454
5. K. Kawarabayashi, S. Norine, R. Thomas, P. Wollan, K_6 minors in large 6-connected graphs. J. Combin. Theory B **129**, 158–203 (2018)
6. T. Korhonen, A single-exponential time 2-approximation algorithm for treewidth, in *2021 IEEE 62nd Annual Symposium on Foundations of Computer Science (FOCS)* (2022), pp. 184–192
7. S. Norin, R. Thomas, K_t-minors in large t-connected graphs (in preparation, 2009)
8. N. Robertson, P.D. Seymour, Graph Minors. XIII. The disjoint paths problem. J. Combin. Theory B **63**, 65–110 (1995)
9. N. Robertson, P.D. Seymour, Graph Minors: XVII. Taming a vortex. J. Combin. Theory B **77**, 162–210 (1999)
10. N. Robertson, D.P. Sanders, P.D. Seymour, R. Thomas, The four colour theorem. J. Combin. Theory B **70**, 2–44 (1997)
11. K. Wagner, Über eine Eigenschaft der ebenen Komplexe. Math. Ann. **114**, 570–590 (1937)

Chapter 11
Sources

The proof of treewidth-fragility of proper minor-closed classes from Chap. 7 is essentially the one given by DeVos et al. [3]; the construction showing the additive non-approximability of chromatic number in proper minor-closed classes is taken from [4].

The proof of linear grid minor theorem in Chap. 8 is a rather easy exercise thanks to Theorem 5.29 which guarantees that the surface part is actually a minor of the whole graph; though a (not much more complicated) proof without using this strengthening is also possible [2].

Chapter 9 on topological minors draws from [6] for the derivation of the global form of the structure theorem and from [8] for the proof of the local form.

For Chap. 10, the fact that attachments of non-global apices can be captured in vortices can be found in [5], though it has been widely known in the community before. The treatment of vortices is inspired by [9]; we choose to work with strong vortical decompositions from the beginning, whereas the setting of [9] more closely matches semi-strong vortical decomposition setting (with some additional features, such as "empty" nodes that do not correspond to any vertices), and concludes that the vortical decomposition is strong by proving its linkedness. The proof of Theorem 10.3 is based on the ideas from [1]; they prove a somewhat more complicated result postulating the existence of many disjoint copies of $K_{t,n}$ or of a subdivision of $K_{t,n}$, which requires them to assume larger connectivity. The argument from Sect. 10.5 is inspired by [7].

Let me remark that two of the results that we discuss in Chap. 10 have somewhat dubious publication status.

- Theorem 10.1 of Norin and Thomas was announced in 2009, but the paper has not appeared yet. I have been shown a substantial partial manuscript, but while I understand the overall ideas of the argument, I have not verified the details and I cannot vouch for their correctness.
- The result of Kawarabayashi and Reed on decidability and conditional algorithm for $(t-1)$-coloring of K_t-minor-free graphs appeared as an extended abstract

© The Author(s), under exclusive license to Springer Nature Switzerland AG 2025

Z. Dvořák, *Graph Minors*, Springer Monographs in Mathematics,
https://doi.org/10.1007/978-3-031-87469-7_11

in the proceedings of STOC'09, but full version is also not available. On the first reading, it seems like the details of the proof should be reasonably easy to work out (the argument described in the paper only refers to the Flat Wall Theorem and Thomassen's famous result on 5-list-colorability of planar graphs [10]). However, upon closer inspection, it becomes obvious (and was confirmed by Kawarabayashi in private communication) that the proof requires deeper consideration of vortices and consequently it probably also is rather more technical than the extended abstract makes it appear.

Since neither of the groups of authors seems to have any doubts about the correctness of their arguments, I have chosen to present the results as proven; but, it needs to be acknowledged that they have not passed the same level of scrutiny as other results presented in this book.

References

1. T. Bähme, K. Kawarabayashi, J. Maharry, B. Mohar, Linear connectivity forces large complete bipartite minors. J. Combin. Theory, Ser. B **99**, 557–582 (2009)
2. E.D. Demaine, M. Hajiaghayi, Linearity of grid minors in treewidth with applications through bidimensionality. Combinatorica **2**, 19–36 (2008)
3. M. DeVos, G. Ding, B. Oporowski, D. Sanders, B. Reed, P.D. Seymour, D. Vertigan, Excluding any graph as a minor allows a low tree-width 2-coloring. J. Combin. Theory Ser. B **91**, 25–41 (2004)
4. Z. Dvořák, K. Kawarabayashi, Additive non-approximability of chromatic number in proper minor-closed classes, in *45th international colloquium on automata, languages, and programming, ICALP 2018*. LIPIcs, vol. 107 (Schloss Dagstuhl - Leibniz-Zentrum fuer Informatik, Wadern, 2018), pp. 47:1–47:12
5. Z. Dvořák, R. Thomas, List-coloring apex-minor-free graphs (2014). ArXiv, 1401.1399
6. M. Grohe, D. Marx, Structure theorem and isomorphism test for graphs with excluded topological subgraphs. SIAM J. Comput. **44**, 114–159 (2015)
7. K. Kawarabayashi, On the connectivity of minimum and minimal counterexamples to Hadwiger's conjecture. J. Combin. Theory Ser. B **97**, 144–150 (2007)
8. C. Liu, R. Thomas, Excluding subdivisions of bounded degree graphs. J. Combin. Theory Ser. B **134**, 1–35 (2019)
9. N. Robertson, P.D. Seymour, Graph Minors: XVII. Taming a vortex. J. Combin. Theory Ser. B **77**, 162–210 (1999)
10. C. Thomassen, Every planar graph is 5-choosable. J. Combin. Theory Ser. B **62**, 180–181 (1994)

Part III
Avoiding the Structure Theorem

As we have seen, the Minor Structure Theorem is an extremely powerful tool with many applications. However, it should also be clear that it is not the best tool for every circumstance. Firstly, it is not very useful if we are interested in quantitative bounds. E.g., consider Theorem 2.9 for $H = K_t$ and let $a(t)$ be the function giving the dependence of the parameter a on t. The proof we presented involves multiple iterations of steps that each lead to an exponential increase, showing that $a(t)$ is at best a tower function of height depending on t. Even worse, some of the steps of the proof, e.g., Theorem 4.27, do not directly come with explicit bounds, and thus deriving any bound on $a(t)$ is a non-trivial problem. This issue has been largely mitigated in a recent proof of the structure theorem by Kawarabayashi, Thomas and Wollan [2], who proved a slightly weaker version of the Minor Structure Theorem with $a(t) = \exp(O(t^{26} \log t))$. Still, even with this bound, the minor structure theorem is not applicable e.g. in relation to Hadwiger's conjecture: The current best result [1] states that a K_t-minor-free graph has chromatic number $O(t \log \log t)$. It is clear that any attempt to improve this bound using the Minor Structure Theorem is doomed from the beginning, since the Theorem does not say anything at all for graphs with less than $a(t)$ vertices.

Secondly, there are many cases where we can actually obtain a result using the Minor Structure Theorem, but the dependence on the excluded minor is so bad that the result does not have any uses in practice. E.g., we have already seen several proofs that every n-vertex K_t-minor-free graph has a balanced cut of size at most $c(t) \cdot \sqrt{n}$. This is a rather interesting result from algorithmic perspective; for example, as we are going to see in Chap. 12, it implies that many NP-hard problems on K_t-minor-free graphs can be solved in subexponential time. However, since the time complexity involves the constant $c(t)$ arising from the Minor Structure Theorem, it is clear that the algorithm is too slow for practical use. Moreover, to find the balanced cut, one would need to algorithmize the whole proof of the Minor Structure Theorem, which is not too hard theoretically, but certainly not feasible in practice. Fortunately, much simpler proofs of the existence of sublinear separators leading to much better bounds exist, as we are going to see shortly.

Finally, while the Minor Structure Theorem is powerful, it is also rather cumbersome; even conceptually trivial applications may take multiple pages to execute in order to deal with all the technicalities. Moreover, even stating the Theorem requires a non-trivial amount of definitions, if that is needed in the context. Hence, direct arguments avoiding the Structure Theorem are often much simpler and more elegant (though harder to devise).

References

1. M. Delcourt, L. Postle, Reducing linear Hadwiger's conjecture to coloring small graphs (2021). ArXiv, 2108.01633
2. K. Kawarabayashi, R. Thomas, P. Wollan, Quickly excluding a non-planar graph (2020). ArXiv, 2010.12397

Chapter 12
Sublinear Separators

We have already seen several proofs of the following fact: For every H, every H-minor-free n-vertex graph has a balanced cut of size $O(\sqrt{n})$. As the proofs were based on the Minor Structure Theorem, they do not give good bounds on the multiplicative constant of the O-notation. Kawarabayashi and Reed [3] gave a substantially more involved proof, also based on the Structure Theorem, of the following quantitatively stronger claim.

Theorem 12.1 *For every positive integer t, there exists an integer α such that every K_t-minor-free n-vertex graph has a balanced cut of order at most*

$$\alpha + (4t + 7/2)\sqrt{n}.$$

That is, they managed to move the dependence on the Minor Structure Theorem constants to the additive term, though of course from the practical perspective, this is not much of a difference. However, Alon et al. [1] proved a result that is asymptotically slightly weaker, but avoids the Minor Structure Theorem and consequently is much more practical.

Theorem 12.2 *Let t be a positive integer, let G be a graph, and let w be an assignment of non-negative weights to vertices of G. If G is K_t-minor-free, then it has a w-balanced cut of size less than $\sqrt{t^3 n}$.*

Let us remark that Alon et al. [1] actually proved that K_t-minor-free graphs have treewidth less than $\sqrt{t^3 n}$, which is a slightly stronger claim by Lemma 2.15 (though only by a constant factor, as seen from Lemma 2.16). They also conjectured that the bound from Theorem 12.2 can be improved to $O(t\sqrt{n})$. This would be best possible for n-vertex 3-regular expanders, which do not have any separator of size $o(n)$, but also do not have enough edges to contain K_t as a minor for $t = 2\lceil\sqrt{n}\rceil$. Moreover, it would match the bound $O(\sqrt{gn})$ on the size of balanced cuts in n-vertex graphs of genus g (see Corollary 2.76 or [2]), since K_t has genus more than g for $t = \Theta(g^{1/2})$, and thus these graphs are $K_{\Theta(g^{1/2})}$-minor-free. Theorem 12.1 proves this bound for

© The Author(s), under exclusive license to Springer Nature Switzerland AG 2025

Z. Dvořák, *Graph Minors*, Springer Monographs in Mathematics,
https://doi.org/10.1007/978-3-031-87469-7_12

any fixed t, but not for t depending on n (such as in the expander example), since α from the statement of the theorem depends on t.

12.1 Existence of Small Separators

Let us now show a proof of Theorem 12.2. The key ingredient is the following lemma on separating multiple sets of vertices.

Lemma 12.3 *Let r be a real number, let G be a graph with n vertices, and let A_1, …, A_k be subsets of vertices of G. If no connected subgraph of G with at most r vertices intersects all of $A_1, \ldots, A_k$, then there exists a set $Z \subseteq V(G)$ of size at most $(k-1)n/r$ such that no component of $G - Z$ intersects all of $A_1, \ldots, A_k$.*

Proof Let F be the graph obtained from $k-1$ copies $G_1, \ldots, G_{k-1}$ of G by, for $i \in [k-2]$, adding the matching between the two copies of vertices of A_{i+1} in G_i and G_{i+1}. Let A'_1 be the copy of A_1 in G_1 and A'_k the copy of A_k in G_{k-1}.

If there exists a path P in F between A'_1 and A'_k with at most r vertices, then note that P corresponds to a walk in G starting in A_1, visiting $A_2, \ldots, A_{k-1}$, and ending in A_k with at most $|P| \le r$ vertices. Consequently, G contains a connected subgraph with at most r vertices intersecting all of $A_1, \ldots, A_k$.

Hence, suppose that this is not the case, and thus the distance between A'_1 and A'_k in F is at least $\lfloor r \rfloor$. For $0 \le i \le \lfloor r \rfloor$, let Z'_i be the set of vertices of F at distance exactly i from A'_1, and note that Z'_i separates A'_1 from A'_k. Let Z_i be the corresponding set of at most $|Z'_i|$ vertices of G. Observe that no component of $G - Z_i$ intersects all of $A_1, \ldots, A_k$: Otherwise, let W be a walk in $G - Z_i$ containing in order a vertex of A_1, a vertex of $A_2, \ldots,$ and a vertex of A_k (between these vertices, W may pass through $A_1 \cup \ldots \cup A_k$ any number of times). Then W would correspond to a walk in $F - Z'_i$ from A'_1 to A'_k.

Since the sets Z'_i are pairwise disjoint, there exists $i \in \{0, \ldots, \lfloor r \rfloor\}$ such that

$$|Z_i| \le |Z'_i| \le \frac{|F|}{\lfloor r \rfloor + 1} \le \frac{(k-1)n}{r}.$$

Hence, we can let $Z = Z_i$. □

Let G be an n-vertex K_t-minor-free graph and let w be an assignment of non-negative weights to vertices of G. Let $r = \sqrt{tn}$. We now build a w-balanced cut by gradually refining a partition of $V(G)$ to parts A, C, and B satisfying the following invariants:

(a) There are no edges between A and B.
(b) $w(A) \le w(G)/2$.
(c) For some $k \le t$, there exists a model μ of K_k in $G[A \cup C]$ such that $C \subseteq \bigcup_{v \in V(K_k)} V(\mu(v))$ and $|C \cap V(\mu(v))| \le r$ for every $v \in V(K_k)$.

Since G is K_t-minor-free, we must actually have $k \leq t-1$ in (c). Moreover, without loss of generality, we can assume that $C \cap V(\mu(v)) \neq \emptyset$ for every $v \in V(K_k)$, as otherwise we can remove v from the model and decrease k without affecting the validity of the invariants.

For every $v \in V(K_k)$, let N_v be the set of neighbors of $\mu(v)$ in B. The refinement procedure proceeds as follows:

(i) If $G[B]$ contains a connected subgraph D with at most r vertices intersecting all sets N_v for $v \in V(K_k)$, then we replace B by $B \setminus V(D)$, C by $C \cup V(D)$, add D to μ to obtain a model of K_{k+1}, and increase k. Repeat the procedure.

(ii) Otherwise, by Lemma 12.3, there exists a set $Z \subseteq B$ of size at most $|B|(k-1)/r \leq nt/r = r$ such that no component of $G[B] - Z$ intersects all sets N_v for $v \in V(K_k)$.

(ii-a) If $w(K) \leq w(G)/2$ for every component K of $G[B] - Z$, then by (a) and (b), $C \cup Z$ is a w-balanced cut. Moreover, by (c) we have $|C \cup Z| \leq (t-1)r + r = tr = \sqrt{t^3 n}$, as desired; the procedure stops.

(ii-b) Otherwise, $G[B] - Z$ has a (unique) component B_0 with $w(B_0) > w(G)/2$. Let $v \in V(K_k)$ be a vertex such that $N_v \cap V(B_0) = \emptyset$.

(ii-b1) If $N_v = \emptyset$, then we replace A by $A \cup V(\mu(v))$, C by $C \setminus V(\mu(v))$, remove v from the model μ, and decrease k. Note that the weight of A does not increase beyond $w(G)/2$, since $w(G - B_0) < w(G)/2$. Repeat the procedure.

(ii-b2) Otherwise, let $Z_0 \subseteq Z$ be minimal such that Z_0 separates N_v from B_0, and let B_1 be the component of $G[B] - Z_0$ containing B_0. By the minimality of Z_0, for each $u \in Z_0$ there exists a path P_u in $G[B \setminus V(B_1)]$ from u to N_v. Let M consist of $\mu(v)$, vertices of N_v and edges of G from N_v to $V(\mu(v))$, and $\bigcup_{u \in Z_0} P_u$. We replace A by $A \cup (B \setminus (Z_0 \cup V(B_1))) \cup V(\mu(v))$, C by $(C \setminus V(\mu(v))) \cup Z_0$, B by $V(B_1)$, and replace $\mu(v)$ by M in μ.

Note that weight of A does not increase beyond $w(G)/2$, since $w(G - B_1) \leq w(G - B_0) < w(G)/2$. The set Z_0 we added to C separates B_1 from all vertices added to A, and thus (a) is preserved. Moreover, M is connected and its intersection with the new set C is Z_0, and thus has size at most $|Z| \leq r$, showing that (c) holds. Therefore, we can iterate the procedure.

The sublinear separator theorem now follows by a straightforward analysis of this procedure.

Proof of Theorem 12.2 Without loss of generality, we can assume that G is connected. Let v_0 be an arbitrary vertex of G and let $A = \emptyset$, $C = \{v_0\}$, and $B = V(G) \setminus \{v_0\}$; the invariants (a), (b), and (c) are clearly satisfied with $k = 1$. We run the procedure described above. Note that since G is connected, we have $C \neq \emptyset$ throughout the run, and in particular in (i) the subgraph D must be non-empty. Hence, in each iteration we decrease size of B in (i) or (ii-b2), keep B unchanged but decrease k in (ii-b1), or stop in (ii-a). Since in each iteration we also increase k by at most one, we conclude that we can iterate the procedure at most $2|G|$ times. When the procedure ends in (ii-a), we obtain the desired w-balanced cut of size at most $\sqrt{t^3 n}$. □

Let us remark that the described procedure obviously gives a polynomial-time algorithm to obtain the w-balanced cut.

12.2 Applications

A key application of small separators is in design of subexponential-time algorithms and approximation schemes. Although we are mostly motivated by considering proper minor-closed classes, let us present the results in a more general setting of graphs with sublinear separators. For a function $s : \mathbb{N} \to \mathbb{N}$, we say that a graph G has ***s-separators*** if every induced subgraph G' of G has a balanced separator of size at most $s(|G'|)$. A class $\mathcal{G}$ of graphs has *efficient s-separators* if there exists a polynomial-time algorithm that for an induced subgraph G' of a graph $G \in \mathcal{G}$ returns a balanced separator of size at most $s(|G'|)$. Thus, Theorem 12.2 says that the class of K_t-minor-free graphs has efficient s-separators for the function $s(n) = \sqrt{t^3 n}$.

Sublinear separators can be used to solve problems in subexponential time via the natural divide-and-conquer algorithm. Although the idea can be applied more generally, let us describe it in the setting of *constraint satisfaction problems* (CSPs): The instance of the problem is a quadruple $(\mathbf{G}, D, w_V, w_E)$, where $\mathbf{G}$ is a directed graph, D is a finite set, and $w_V : V(\mathbf{G}) \times D \to \mathbb{R} \cup \{\infty\}$ and $w_E : E(\mathbf{G}) \times D^2 \to \mathbb{R} \cup \{\infty\}$ are cost functions. The task is to find an assignment $f : V(G) \to D$ of values from D to vertices of G that minimizes the sum of the costs of vertices and edges, i.e.,

$$\sum_{v \in V(\mathbf{G})} w_V(v, f(v)) + \sum_{e=uv \in E(\mathbf{G})} w_E(e, f(u), f(v)).$$

Examples of problems expressible in this setting are:

- k-colorability of a graph G: Let $\mathbf{G}$ be any orientation of G, let $D = [k]$, for every $v \in V(G)$ and $c \in D$ let $w_V(v, c) = 0$, and for every $e = uv \in E(\mathbf{G})$, let $w_E(e, c_1, c_2) = \infty$ if $c_1 = c_2$ and $w_E(e, c_1, c_2) = 0$ otherwise. Then the CSP $(\mathbf{G}, D, w_V, w_E)$ has solution 0 if G is k-colorable and ∞ otherwise.
- Largest independent set in a graph G: Let $\mathbf{G}$ be any orientation of G, let $D = [2]$, for $v \in V(G)$ let $w_V(v, 1) = 0$ and $w_V(v, 2) = -1$, and for every edge $e = uv \in E(\mathbf{G})$ let $w_E(e, 2, 2) = \infty$ and $w_E(e, i, j) = 0$ if $i = 1$ or $j = 1$. Consider the CSP instance $(\mathbf{G}, D, w_V, w_E)$. Note that for $f : V(G) \to [2]$, the

weight of f is finite exactly if $f^{-1}(2)$ is an independent set in G, and in that case the weight of f is $|f^{-1}(2)|$. Therefore, the solution to the instance is $-\alpha(G)$.

Lemma 12.4 *Let $\mathcal{G}$ be a class of graphs with efficient $O(n^{1-\varepsilon})$-separators for some positive $\varepsilon < 1$ and let D be a set of size $k \geq 2$. There exists an algorithm with time complexity $\exp(O(n^{1-\varepsilon}\log k))$ that solves any CPS instance $(\mathbf{G}, D, w_V, w_E)$, where $\mathbf{G}$ is an orientation of an n-vertex graph $G \in \mathcal{G}$.*

Proof If $V(G) = \emptyset$, then the claim is trivial. Otherwise, we find a balanced vertex separation (A_1, A_2) of G of size $O(n^{1-\varepsilon})$. Let $Z = A_1 \cap A_2$. We go over all functions $f : Z \to D$ and for each of them, we proceed as follows: For $i \in [2]$, let w_E^i be the restriction of w_E to edges of $\mathbf{G}[A_i \setminus Z]$, and for $v \in A_i \setminus Z$ and $c \in D$, let

$$\begin{aligned} w_V^{f,i}(v,c) = & w_V(v,c) + \sum_{e=uv\in E(\mathbf{G}):u\in Z} w_E(e, f(u), c) \\ & + \sum_{e=vu\in E(\mathbf{G}):u\in Z} w_E(e, c, f(u)). \end{aligned}$$

Thus, $w_V^{f,i}(v,c)$ accounts for the cost incurred by v and edges between v and Z when the value c is assigned to v and the values on Z are chosen according to f. Let $c_{f,i}$ be the solution to the instance $(\mathbf{G}[A_i \setminus Z], D, w_V^{f,i}, w_E^i)$, and let

$$c_f = c_{f,1} + c_{f,2} + \sum_{v\in Z} w_V(v, f(v)) + \sum_{e=uv\in E(\mathbf{G}[Z])} w_E(e, f(u), f(v)).$$

Note that c_f is the minimum cost of the assignments $f' : V(\mathbf{G}) \to D$ for the instance $(\mathbf{G}, D, w_V, w_E)$ such that f' extends f.

Therefore, the solution to this instance is the minimum of c_f over all functions $f : Z \to D$. We obtain this solution by recursively solving each of the $\exp(O(n^{1-\varepsilon}\log k))$ pairs of instances described in the previous paragraph, each with at most $\frac{2}{3}n$ vertices. An easy calculation shows that this results in $\exp(O(n^{1-\varepsilon}\log k))$ time complexity. □

Thus, k-colorability of n-vertex K_t-minor-free graphs can be decided in time $\exp(O(t^{3/2}n^{1/2}\log k))$ and largest independent set in such graphs can be found in time $\exp(O(t^{3/2}n^{1/2}))$. Another important property of graphs with sublinear separators is that they can be broken into constant-size components by removal of a small fraction of vertices; this lends itself to design of approximation algorithms, as well as further applications (see e.g. Sect. 14.3).

Lemma 12.5 *Let $\mathcal{G}$ be a class of graphs with efficient $O(n^{1-\varepsilon})$-separators for some positive $\varepsilon \leq 1$. There exists a polynomial-time algorithm that for any graph $G \in \mathcal{G}$ and any positive integer k returns a set $S \subseteq V(G)$ of size at most $\frac{1}{k}|G|$ such that every component of $G - S$ has size $O(k^{1/\varepsilon})$.*

Actually, for many applications the following stronger form is needed. A *cover* for a graph G is a system $\mathcal{C}$ of induced subgraphs of G such that $G = \bigcup \mathcal{C}$. For $C \in \mathcal{C}$, let ∂C consist of the vertices of C that belong to another element of $\mathcal{C}$, and let $\partial \mathcal{C} = \bigcup_{C \in \mathcal{C}} \partial C$. Thus, each component of $G - \partial \mathcal{C}$ is a subgraph of an element of $\mathcal{C}$. It follows that Lemma 12.5 is implied by the following stronger result by taking $S = \partial \mathcal{C}$.

Lemma 12.6 *Let $\mathcal{G}$ be a class of graphs with efficient $O(n^{1-\varepsilon})$-separators for some positive $\varepsilon \le 1$. There exists a polynomial-time algorithm that for any graph $G \in \mathcal{G}$ and any positive integer k returns a cover $\mathcal{C}$ of G such that each graph in $\mathcal{C}$ has $O(k^{1/\varepsilon})$ vertices and*

$$\sum_{C \in \mathcal{C}} |\partial C| \le \tfrac{1}{k}|G|.$$

Proof Let $\gamma \ge 1$ be a constant such that $\mathcal{G}$ has efficient $\gamma n^{1-\varepsilon}$-separators, and let

$$c = \frac{3\gamma}{(3/4)^{1-\varepsilon} + (1/4)^{1-\varepsilon} - 1}.$$

Let $M(k) = \max\bigl((12\gamma)^{1/\varepsilon}, 3(ck)^{1/\varepsilon}\bigr) = O(k^{1/\varepsilon})$. We are actually going to prove a stronger claim: If $|G| \ge (ck)^{1/\varepsilon}$, then there exists a cover $\mathcal{C}$ such that each graph in $\mathcal{C}$ has at most $M(k)$ vertices and

$$\sum_{C \in \mathcal{C}} |\partial C| \le \tfrac{1}{k}|G| - c|G|^{1-\varepsilon}.$$

The algorithm is as follows. Let $n = |G|$:

- If $n \le M(k)$, then return $\mathcal{C} = \{G\}$.
- Otherwise, let (A_1, A_2) be a balanced vertex separation of G of order at most $\gamma n^{1-\varepsilon}$, for $i \in [2]$, let $\mathcal{C}_i$ be the cover of $G[A_i]$ returned by a recursive call, and return $\mathcal{C} = \mathcal{C}_1 \cup \mathcal{C}_2$.

In the former case, note that $\sum_{C \in \mathcal{C}} |\partial C| = 0 \le \frac{1}{k}|G| - c|G|^{1-\varepsilon}$, since $|G| \ge (ck)^{1/\varepsilon}$. In the latter case, we have

$$|A_i| = n - |A_{3-i} \setminus A_i| \ge \tfrac{1}{3}n > \tfrac{1}{3}M(k) \ge (ck)^{1/\varepsilon},$$

and thus the assumption on the minimum size of the considered graph is satisfied in the recursive call. Moreover,

$$|A_i| = |A_i \setminus A_{3-i}| + |A_1 \cap A_2| \le \tfrac{2}{3}n + \gamma n^{1-\varepsilon} < \tfrac{3}{4}n,$$

since $n > M(k)$ implies $\frac{\gamma}{n^{\varepsilon}} < \frac{1}{12}$. Consequently, the recursion is finite, and clearly the returned set $\mathcal{C}$ contains only graphs with at most $M(k)$ vertices. We have $|A_1|, |A_2| \le \frac{3}{4}n$ and $|A_1| + |A_2| \ge n$; observe that this implies that

$$|A_1|^{1-\varepsilon} + |A_2|^{1-\varepsilon} \geq ((3/4)^{1-\varepsilon} + (1/4)^{1-\varepsilon})n^{1-\varepsilon}.$$

Consider any subgraph $C \in \mathcal{C}_i$. Note that if $v \in \partial C$ in $\mathcal{C}$ but not in $\mathcal{C}_i$, then C is the only element of $\mathcal{C}_i$ containing v and $v \in A_1 \cap A_2$. Therefore,

$$\begin{aligned}\sum_{C \in \mathcal{C}} |\partial_{\mathcal{C}} C| &\leq 2|A_1 \cap A_2| + \sum_{i=1}^{2} \sum_{C \in \mathcal{C}_i} |\partial_{\mathcal{C}_i} C| \\ &\leq 2\gamma n^{1-\varepsilon} + \frac{|A_1| + |A_2|}{k} - c(|A_1|^{1-\varepsilon} + |A_2|^{1-\varepsilon}) \\ &\leq \frac{n}{k} + 3\gamma n^{1-\varepsilon} - c((3/4)^{1-\varepsilon} + (1/4)^{1-\varepsilon})n^{1-\varepsilon} \\ &= \frac{n}{k} - cn^{1-\varepsilon},\end{aligned}$$

where the last equality holds by the choice of c. □

Lemma 12.5 lends itself naturally to obtaining efficient approximation algorithms: Just ignore what happens on S, and solve the problem on each component of $G - S$ by brute force. For example, we immediately get the following result.

Corollary 12.7 *Let $\mathcal{G}$ be a class of graphs with efficient $O(n^{1-\varepsilon})$-separators for some positive $\varepsilon \leq 1$. There exists an algorithm that for any n-vertex graph $G \in \mathcal{G}$ and any positive integer k returns in time $O(\text{poly}(n)) + \exp(O(k^{1/\varepsilon}))n$ an independent set of G of size at least $\alpha(G) - n/k$.*

Proof By Lemma 12.5, we can find a set $S \subseteq V(G)$ such that $|S| \leq n/k$ and each component of $G - S$ has size $O(k^{1/\varepsilon})$. By a brute-force algorithm, we can in time $\exp(O(k^{1/\varepsilon}))n$ find a largest independent set in each component of $G - S$ and let A be their union. Then A is an independent set in G and

$$|A| = \alpha(G - S) \geq \alpha(G) - |S| \geq \alpha(G) - \tfrac{n}{k}.$$

□

Let us remark that Lemma 12.6 implies that graphs with sublinear separators have bounded average degree.

Corollary 12.8 *Let $\mathcal{G}$ be a class of graphs with $O(n^{1-\varepsilon})$-separators for some positive $\varepsilon \leq 1$. Then there exists a constant d such that every graph $G \in \mathcal{G}$ has average degree at most d.*

Proof By Lemma 12.6 with $k = 1$, there exists an integer d_0 such that every graph $G \in \mathcal{G}$ has a cover $\mathcal{C}$ by subgraphs of size at most d_0 satisfying

$$\sum_{C \in \mathcal{C}} |\partial C| \leq |G|.$$

Note that

$$\sum_{C\in\mathcal{C}} |C| = \sum_{C\in\mathcal{C}} |V(C)\setminus\partial C| + \sum_{C\in\mathcal{C}} |\partial C| \le 2|G|,$$

since the sets $V(C)\setminus\partial C$ for distinct $C\in\mathcal{C}$ are pairwise disjoint. Hence,

$$\|G\| \le \sum_{C\in\mathcal{C}} \|C\| \le \sum_{C\in\mathcal{C}} \binom{|C|}{2} \le \sum_{C\in\mathcal{C}} \frac{d_0|C|}{2} \le d_0|G|.$$

Consequently, each graph $G\in\mathcal{G}$ has average degree $2\|G\|/|G| \le 2d_0$, and thus we can let $d = 2d_0$. □

Corollary 12.8 implies that every n-vertex graph $G \in \mathcal{G}$ has independence number at least $\frac{n}{d+1}$, and thus Corollary 12.7 gives an independent set A such that

$$\frac{|A|}{\alpha(G)} \ge 1 - \frac{n/k}{n/(d+1)} = 1 - \frac{d+1}{k},$$

i.e., by choosing k large enough, we can make the relative error arbitrarily small.

Lemma 12.5 cannot be used to obtain approximation algorithms for problems where the optimal solution has sublinear size, as there it is possible that e.g. the whole optimal solution is contained in the deleted set S. Moreover, the approaches that use deletion of sublinear separators in general run into difficulties when dealing with problems with constraints on non-adjacent vertices (for example, finding a maximum-size distance-r independent set, i.e., a set of vertices pairwise at distance greater than r), since then the solutions in different components of $G - S$ are not independent. We are going to see techniques without such limitations in Chap. 16.

Let us remark that for $\gamma = t^{3/2}$ and $\varepsilon = 1/2$ corresponding to Theorem 12.2, the proof of Lemma 12.6 gives $M(k) = O(t^3k^2)$, and thus Corollary 12.8 implies that every K_t-minor-free graph has average degree $O(t^3)$. This is of course far from the optimal bound given in Theorem 1.2.

References

1. N. Alon, P.D. Seymour, R. Thomas, A separator theorem for graphs with an excluded minor and its applications, in *Proceedings of the twenty-second annual ACM symposium on Theory of computing* (ACM, 1990), pp. 293–299
2. J. Gilbert, J. Hutchinson, R. Tarjan, A separator theorem for graphs of bounded genus. J. Algorithms **5**, 391–407 (1984)
3. K. Kawarabayashi, B. Reed, A separator theorem in minor-closed classes, in *Proceedings of the 51st annual IEEE symposium on. foundations of computer science* (2010)

Chapter 13
Chordal Partitions

A ***chordal partition*** of a graph G is a sequence $\mathcal{P} = P_1, \ldots, P_m$ of pairwise disjoint subsets of $V(G)$ such that $V(G) = \bigcup_{i=1}^{m} P_i$,

(i) each part of $\mathcal{P}$ is non-empty and induces a connected subgraph of G, and
(ii) for $1 \le i < j < k \le m$, if G has an edge between P_i and P_k and an edge between P_j and P_k, then it also has an edge between P_i and P_j.

A part P_i is a *leaf* of the partition if there is no $j > i$ such that G contains an edge between P_i and P_j. By (i), $G/\mathcal{P}$ is a minor of G. Hence, if G is K_t-minor-free, then $G/\mathcal{P}$ has clique number at most $t-1$. By (ii), the graph $G/\mathcal{P}$ obtained by contracting each of the parts is chordal, and consequently, it has treewidth at most $t-2$. This property makes chordal partitions an important tool in lifting the properties of graphs of bounded treewidth to all proper minor-closed classes.

Of course, every graph has a chordal partition: Just take $\mathcal{P}$ to be the partition of $V(G)$ to vertex sets of the components of G. Clearly, such a trivial partition is useless in the study of the properties of G. Hence, a key point is to be able to choose the partition so that each of the parts has an additional application-specific property π. In general, this is achieved using the following algorithm.

We start with the trivial chordal partition into components, and refine it while preserving the following invariant: Every non-leaf part satisfies the property π. Suppose that $\mathcal{P} = P_1, \ldots, P_m$ is the currently considered partition and there exists $i \in [m]$ such that the part P_i (necessarily a leaf) does not satisfy the property π. Let $i_1 < i_2 < \ldots < i_d < i$ be the indices such that G contains an edge between P_{i_j} and P_i for $j \in [d]$; by (ii), G contains an edge between P_{i_j} and P_{i_k} for all distinct $j, k \in [d]$. For $j \in [d]$, let $A_j \subseteq P_i$ consist of the vertices incident with an edge with the other end in P_{i_j}. We choose a non-empty set $Q \subset P_i$ so that

($\star$) $G[Q]$ is connected, Q intersects $A_1, \ldots, A_d$, and Q satisfies the property π.

Note that the last condition implies that $Q \neq P_i$. We replace P_i in $\mathcal{P}$ by Q followed by the vertex sets of the components of $G[P_i] - Q$ in any order.

© The Author(s), under exclusive license to Springer Nature Switzerland AG 2025
Z. Dvořák, *Graph Minors*, Springer Monographs in Mathematics,
https://doi.org/10.1007/978-3-031-87469-7_13

This gives a chordal partition: Q was chosen to satisfy (i). If K is the vertex set of a component of $G[P_i] - Q$, then the edges of G from K may only end in Q and the sets P_{i_j} for (not necessarily all) $j \in [d]$, and since Q intersects A_j, the graph G also has an edge between Q and P_{i_j}; it follows that (ii) holds. Finally, the components of $G[P_i] - Q$ form leaves in the new version of the partition $\mathcal{P}$, and thus the invariant that π holds for all non-leaf parts is preserved.

Since each iteration of the described procedure increases the number of parts of $\mathcal{P}$, after at most $|G|$ iterations we end up with a chordal partition where each part satisfies the property π. Of course, this assumes that for the property π, the choice of the subset Q satisfying ($\star$) is always possible. Let us remark that π does not need to depend only on Q, but also on the parts $P_{j_1}, \ldots, P_{j_d}$ and the set $\bigcup_{k \geq i} P_k$, since these sets do not change by further iterations of the refinement procedure. Of course, the procedure will not succeed for every property π; we discuss this in more detail at the end of the following section.

Note that the described algorithm can be executed on any graph, not necessarily forbidding any minor. The only place where the applications exploit the fact that G is K_t-minor-free is in concluding that $G/\mathcal{P}$ has treewidth at most $t - 2$. This separation results in rather clear and simple arguments.

Let us now describe two applications of chordal partitions arising from particular choices of π.

13.1 Fractional Chromatic Number

A ***fractional coloring*** of a graph G is a function φ that to each vertex assigns a set of colors of measure one (from some measure space, say $\mathbb{R}$ with Lebesgue measure) such that $\varphi(u) \cap \varphi(v) = \emptyset$ for every edge uv of G. The *span* of φ is the measure of $\bigcup_{v \in V(G)} \varphi(v)$. The ***fractional chromatic number*** $\chi_f(G)$ of G is the infimum of the spans of fractional colorings of G. Clearly, one way to obtain a fractional coloring is to start with an ordinary proper coloring ψ using $k = \chi(G)$ colors, then assign to each vertex v the interval $[\psi(v), \psi(v) + 1)$. The span of the resulting fractional coloring is equal to k. It follows that $\chi_f(G) \leq \chi(G)$.

The fractional chromatic number is one of the best studied variants of the chromatic number; see e.g. [6] for a development of its theory. Reed and Seymour [5] used chordal partitions to show that Hadwiger's conjecture holds for fractional chromatic number up to a factor of two.

Theorem 13.1 *For every positive integer t, every K_{t+1}-minor-free graph has fractional chromatic number at most* $2t$.

In contrast, the current best bound [1] for the ordinary chromatic number is $O(t \log \log t)$, via a significantly more involved argument.

To prove Theorem 13.1, we need a reformulation of the definition of the fractional chromatic number. Let $\mathcal{I}(G)$ be the set of all non-empty independent sets of graph G. Given a fractional coloring φ of a graph G, for each $I \in \mathcal{I}(G)$, define x_I to be the measure of the set

$$\bigcap_{v\in I}\varphi(v)\setminus\bigcup_{v\in V(G)\setminus I}\varphi(v),$$

i.e., of the set of colors that appear in G exactly on I; note that the condition $\varphi(u)\cap\varphi(v)=\emptyset$ for every edge uv implies that each color appears on an independent set. Observe that the values $\{x_I : I\in\mathcal{I}(G)\}$ determine the fractional coloring φ up to a measure-preserving transformation of colors. In particular, a fractional coloring φ' can be recovered from $\{x_I : I\in\mathcal{I}(G)\}$ as follows: To each independent set $I\in\mathcal{I}(G)$, assign an interval S_I of measure x_I so that the assigned intervals are pairwise disjoint. For each $v\in V(G)$, let

$$\varphi'(v)=\bigcup_{I\in\mathcal{I}(G):v\in I} S_I.$$

This correspondence shows that the fractional chromatic number is equal to the solution of the following linear program (P).

$$\begin{aligned} x_I &\geq 0 && \text{for } I\in\mathcal{I}(G)\\ \sum_{I\in\mathcal{I}(G):v\in I} x_I &\geq 1 && \text{for } v\in V(G)\\ \text{minimize} \sum_{I\in\mathcal{I}(G)} x_I&. \end{aligned}$$

Let us remark that this in particular implies that we can replace "infimum" by "minimum" in the definition of the fractional chromatic number, and that $\chi_f(G)$ is a rational number. By linear programming duality, we can equivalently solve the following program (D).

$$\begin{aligned} y_v &\geq 0 && \text{for } v\in V(G)\\ \sum_{v\in I} y_v &\leq 1 && \text{for } I\in\mathcal{I}(G)\\ \text{maximize} \sum_{v\in V(G)} y_v&. \end{aligned}$$

The assignment $\{y_v : v\in V(G)\}$ such that $\sum_{v\in I} y_v\leq 1$ for every independent set I can be viewed as a fractional relaxation of a clique; indeed, if K is a clique in G, then the characteristic function of K satisfies these constraints. Thus, the duality states that the fractional number of G is equal to the size of the largest fractional clique in G.

Another interpretation is in terms of the *weighted independence number*. Given an assignment $w : V(G) \to \mathbb{R}_0^+$ of non-negative weights to vertices of G, let $\alpha_w(G)$ be the maximum of $w(I)$ over all independent sets I in G.

Lemma 13.2 *Let G be a graph and let r be a positive real number. The fractional chromatic number of G is at most r if and only if for every assignment $w : V(G) \to \mathbb{R}_0^+$ of non-negative weights to vertices of G,*

$$\alpha_w(G) \geq \frac{w(G)}{r}.$$

Proof If $\chi_f(G) \leq r$, then there exists a solution $\{x_I : I \in \mathcal{I}(G)\}$ of (P) with $\sum_{I\in\mathcal{I}(G)} x_I \leq r$. Using the constraints of (P), we get

$$\begin{aligned}
\alpha_w(G) &= \max_{I\in\mathcal{I}(G)} w(I) \geq \max_{I\in\mathcal{I}(G)} w(I)\cdot\frac{1}{r}\sum_{I\in\mathcal{I}(G)} x_I\\
&\geq \frac{1}{r}\sum_{I\in\mathcal{I}(G)} x_I w(I) = \frac{1}{r}\sum_{I\in\mathcal{I}(G)}\sum_{v\in I} x_I w(v)\\
&= \frac{1}{r}\sum_{v\in V(G)} w(v) \sum_{I\in\mathcal{I}(G):v\in I} x_I \geq \frac{1}{r}\sum_{v\in V(G)} w(v)\\
&= \frac{w(G)}{r}.
\end{aligned}$$

Conversely, suppose that $\chi_f(G) > r$, and thus there exists a solution $\{y_v : v \in V(G)\}$ of (D) with $\sum_{v\in V(G)} y_v > r$. Let $w(v) = y_v$ for every $v \in V(G)$, so that $w(G) > r$. Using the constraints of (D), we have

$$w(I) = \sum_{v\in I} y_v \leq 1 < \frac{w(G)}{r}$$

for every $I \in \mathcal{I}(G)$, and thus $\alpha_w(G) < \frac{w(G)}{r}$. □

Consequently, Theorem 13.1 is equivalent to the following claim.

Theorem 13.3 *For every positive integer t, every K_{t+1}-minor-free graph G, and every assignment $w : V(G) \to \mathbb{R}_0^+$ of non-negative weights to vertices of G,*

$$\alpha_w(G) \geq \frac{w(G)}{2t}.$$

Given an assignment $w : V(G) \to \mathbb{R}_0^+$ of non-negative weights to vertices of a graph G, a *w-egg* is a non-empty set $S \subseteq V(G)$ containing a subset Y such that Y is an independent set in G and $w(Y) \geq \frac{w(S)}{2}$; the set Y is called a *yolk* of S. We apply the chordal partition construction algorithm from the beginning of the chapter for the following property of a set P_j in the chordal partition $\mathcal{P} = P_1, \ldots, P_m$ of G:

π_{egg}: P_j is a w-egg, and there is no w-egg $S \subseteq \bigcup_{k \geq j} P_k$ such that $G[S]$ is connected and $P_j \subsetneq S$.

Thus, we need to prove the following lemma, asserting that in the step ($\star$) of the algorithm, we can choose a set $Q \subseteq P_i$ satisfying the property π_{egg}. The lemma only asserts that Q can be chosen to be a w-egg; the condition that there is no larger w-egg is ensured by taking a maximal superset of Q that induces a connected w-egg (since the part P_i in the statement of the lemma is a leaf, every subset of $\bigcup_{k \geq i} P_k$ inducing a connected subgraph and containing $Q \subseteq P_i$ is contained in P_i).

Lemma 13.4 *Let w be an assignment of non-negative weights to vertices of a graph G. Let $\mathcal{P} = P_1, \ldots, P_m$ be a chordal partition of G such that the property π_{egg} holds for every $j \in [m]$ such that P_j is not a leaf. Let $i \in [m]$ be an index such that P_i is a leaf. Let $i_1 < i_2 < \ldots < i_d < i$ be the indices such that G contains an edge between P_{i_k} and P_i for $k \in [d]$. For $k \in [d]$, let $A_k \subseteq P_i$ consist of the vertices incident with an edge with the other end in P_{i_k}. There exists a w-egg $Q \subset P_i$ such that $G[Q]$ is connected and Q intersects $A_1, \ldots, A_d$.*

Proof Let Q be a minimal non-empty subset of P_i such that $G[Q]$ is connected and Q intersects $A_1, \ldots, A_d$ (such a set exists, since $G[P_i]$ is connected). We claim that $G[Q]$ is bipartite; then the heavier part Y of the bipartition of $G[Q]$ is an independent set satisfying $w(Y) \geq \frac{w(Q)}{2}$, and thus Q is a w-egg, as desired.

Hence, suppose for a contradiction that $G[Q]$ is not bipartite, and thus it contains an odd cycle K. Consider any vertex $v \in V(K)$, and let Q_v be the vertex set of the component of $G[Q] - v$ containing $K - v$. By the minimality of Q, there exists $k_v \in [d]$ such that A_{k_v} does not intersect Q_v. Let P_v be a shortest path from A_{k_v} to v in $G[Q]$. Note that P_v is an induced path only intersecting K in v, G has no edge between $P_v - v$ and $K - v$, and P_v intersects A_{k_v} exactly in its starting vertex. Let us remark that it is possible that P_v consists of only of the vertex v, in case that $v \in A_{k_v}$.

Since K is an odd cycle, there exists an edge $vv' \in E(K)$ such that the lengths of the paths P_v and $P_{v'}$ have the same parity. Since there is no path in $G[Q] - v$ from A_{k_v} to $v' \in V(K - v)$, we conclude that $k_{v'} \neq k_v$, the paths P_v and $P_{v'}$ are disjoint, and G contains no edge between $P_v - v$ and $P_{v'} - v'$. Therefore, the concatenation P of P_v, the edge vv', and the reversal of $P_{v'}$ is an induced path in $G[Q]$ starting in A_{k_v}, ending in $A_{k_{v'}}$, and otherwise disjoint from $A_{k_v} \cup A_{k_{v'}}$. Moreover, since the lengths of P_v and $P_{v'}$ have the same parity, the path P has odd length.

Let $P = v_1 v_2 \ldots v_{2n}$, where $v_1 \in A_{k_v}$. Let $Y_1 = \{v_1, v_3, \ldots, v_{2n-1}\}$ and $Y_2 = \{v_2, v_4, \ldots, v_{2n}\}$. By symmetry, we can assume that $w(Y_2) \geq w(Y_1)$, and thus $w(Y_2) \geq \frac{w(P)}{2}$. Let Y_v be the yolk of the part P_{k_v}. Since P only intersects A_{k_v} in v_1 and P is an induced path in G, $Y_v \cup Y_2$ is an independent set in G. Moreover, $w(Y_v \cup Y_2) \geq \frac{w(P_{k_v})}{2} + \frac{w(P)}{2} = \frac{w(P_{k_v} \cup V(P))}{2}$. Thus $G[P_{k_v} \cup V(P)]$ is a connected w-egg. This contradicts the maximality in the property π_{egg} for k_v, since $P_{k_v} \cup V(P) \subseteq P_{k_v} \cup P_i \subseteq \bigcup_{k \geq k_v} P_k$. □

Consequently, the algorithm from the beginning of the chapter gives us a chordal partition such that each part is a w-egg.

Corollary 13.5 *Let G be a graph. For every assignment w of non-negative weights to vertices of G, there exists a chordal partition $\mathcal{P} = P_1, \ldots, P_m$ of G such that P_i is a w-egg for every $i \in [m]$.*

The proof of the factor-2 fractional Hadwiger's conjecture now easily follows.

Proof of Theorem 13.3 Let $\mathcal{P} = P_1, \ldots, P_m$ be a chordal partition of G such that P_i is a w-egg for every $i \in [m]$, which exists by Corollary 13.5. The graph $G/\mathcal{P}$ is chordal and has clique number at most t, and thus it is t-colorable; that is, there exists a function $\varphi : [m] \to [t]$ such that $\varphi(i) \neq \varphi(j)$ whenever G contains an edge between P_i and P_j. There exists a color $c \in [t]$ such that

$$w\left(\bigcup_{i \in \varphi^{-1}(c)} P_i\right) \geq \frac{w(G)}{t}.$$

For $i \in [m]$, let Y_i be the yolk of P_i; then

$$w\left(\bigcup_{i \in \varphi^{-1}(c)} Y_i\right) \geq \frac{1}{2} w\left(\bigcup_{i \in \varphi^{-1}(c)} P_i\right) \geq \frac{w(G)}{2t},$$

and $\bigcup_{i \in \varphi^{-1}(c)} Y_i$ is an independent set in G. □

Given the simplicity of the proof, one might hope to get more from the argument. If we actually got a chordal partition where each part induces a bipartite subgraph, this would show that K_{t+1}-minor-free graphs have ordinary chromatic number at most $2t$, a major breakthrough towards Hadwiger's conjecture. This seems promising, as in Lemma 13.4, the set Q indeed turns out to induce a bipartite subgraph. However, this crucially uses the assumption that the adjacent parts of the decomposition are maximal w-eggs, and taking the maximal superegg of Q to preserve this property may result in a set inducing a non-bipartite subgraph. We can restrict ourselves to the type of maximization actually needed in the proof, repeated addition of an odd-length path P with only the first vertex v_1 having neighbors in Q, but this does not fix the issue, since we have no constraints on the neighborhood of v_1 in Q. And indeed, Scott et al. [7] quashed the hopes in this direction rather thoroughly.

Theorem 13.6 (Scott et al. [7]) *There exists a constant $\gamma > 0$ such that for every integer $t \geq 6$, there exists a graph G_t with the following properties: The treewidth of G_t is at most $t-1$ and for any partition $\mathcal{P}$ of $V(G_t)$ into connected non-empty parts such that $G_t/\mathcal{P}$ is perfect, there exists a part $P \in \mathcal{P}$ such that $\omega(G_t[P]) \geq \gamma t^{1/3}$.*

Since chordal graphs are perfect, this excludes the possibility of K_{t+1}-minor-free graphs having a chordal partition with all parts inducing subgraphs of chromatic number $o(t^{1/3})$. Hence, this approach cannot give a bound better than $O(t^{4/3})$ for the chromatic number of K_{t+1}-minor-free graphs. This is rather far from the current best bound of $O(t \log\log t)$.

13.2 Degeneracy and Its Generalizations

A path Q in a graph G is ***geodesic*** if it is a shortest path between its ends. Let $\mathcal{P} = P_1, \ldots, P_m$ be a chordal partition of G. Consider $i \in [m]$, let $J_i \subseteq [i-1]$ consist of the indices $j < i$ such that G contains an edge between P_i and P_j, and for $j \in J_i$, let $A_{i,j} \subseteq P_i$ consist of the vertices incident with edges with the other end in P_j. A ***predecessor transversal*** of P_i is any inclusionwise-minimal set $S \subseteq P_i$ such that S intersects $A_{i,j}$ for every $j \in J_i$. A chordal partition $\mathcal{P} = P_1, \ldots, P_m$ is *geodesic* if the following condition holds for every $i \in [m]$:

π_{geo}: There exist a vertex $v_i \in P_i$, a predecessor transversal S_i of P_i, and a system $\{Q_u : u \in S_i\}$ of paths in $G[P_i]$ such that

$$P_i = \{v_i\} \cup \bigcup_{u \in S_i} V(Q_u)$$

and for every $u \in S_i$, the ends of the path Q_u are v_i and u, and Q_u is geodesic in $G\left[\bigcup_{k \geq i} P_k\right]$.

Recall that if G is K_t-minor-free, then $|J_i| \leq t-2$, and thus P_i is covered by at most $t-2$ geodesic paths.

Using the algorithm from the beginning of the chapter, it is easy to see that geodesic chordal partitions exist: In the step ($\star$), choose $v_i \in P_i$ and a predecessor transversal $S_i \subseteq P_i$ arbitrarily, and let Q be the union of $\{v_i\}$ and the vertex sets of shortest paths from v_i to vertices of S_i in $G[P_i]$.

Geodesic chordal partitions turn out to be especially useful in bounding various degeneracy-type parameters of K_t-minor-free graphs. First, let us note that they imply the following relaxation of Theorem 1.2.

Lemma 13.7 *Every K_t-minor-free graph is $3t^2$-degenerate, and thus has average degree at most $6t^2$.*

Proof It suffices to show that every K_t-minor-free graph G has a vertex of degree at most $3t^2$. Let $\mathcal{P} = P_1, \ldots, P_m$ be a geodesic chordal partition of G. For $i \in [m]$, let v_i, S_i, and $\{Q_u : u \in S_i\}$ be as in π_{geo}, and note that since $G/\mathcal{P}$ has clique number at most $t-1$, we have $|S_i| \leq t-2$ for every $i \in [m]$. We claim that the vertex v_m has degree at most $3t^2$.

Since the paths Q_u for $u \in S_m$ are geodesic in $G[P_m]$ and start in v_m, it follows that v_m has only one neighbor in each of them; i.e., v_m has at most $|S_m|$ neighbors in $G[P_m]$. Let J be the set of at most $t-2$ indices $i \in [m-1]$ such that v_m has a neighbor in P_i. For any $i \in J$ and $u \in S_i$, the path Q_u is geodesic in $G[P_i \cup P_m]$, and thus v_m has at most three neighbors in Q_u, as otherwise we could shortcut Q_u through v_m. Consequently, v_m has at most

$$|S_m| + \sum_{i \in J} 3|S_i| \leq t - 2 + 3(t-2)^2 \leq 3t^2$$

neighbors in total. □

The quadratic bound is not too bad for a rather simple argument. More importantly, the geodesic partition argument gives the best available bounds for many generalizations of degeneracy.

Consider a linear ordering $\prec$ of vertices of a graph G and let r be a positive integer. A vertex u is *r-reachable* from a vertex v if $u \preceq v$ and there exists a path Q in G from v to u of length at most r such that all internal vertices of Q appear after v in the ordering $\prec$; let $R_{\prec,r}(v)$ be the set of vertices that are r-reachable from v in the ordering. The vertex u is *weakly r-reachable* from v if the internal vertices of Q appear after u in the ordering $\prec$; let $W_{\prec,r}(v)$ be the set of vertices that are weakly r-reachable from v. The ***r-coloring number*** $\mathrm{col}_{\prec,r}(G)$ and the ***weak r-coloring number*** $\mathrm{wcol}_{\prec,r}(G)$ of the ordering are $\max_{v \in V(G)} |R_{\prec,r}(v)|$ and $\max_{v \in V(G)} |W_{\prec,r}(v)|$, respectively. The *$r$-coloring number* $\mathrm{col}_r(G)$ (resp. the *weak r-coloring number* $\mathrm{wcol}_r(G)$) of G is the minimum of the r-coloring numbers (resp. the weak r-coloring numbers) over all orderings $\prec$ of $V(G)$.

Note that $\mathrm{col}_1(G) = \mathrm{wcol}_1(G)$ is equal to the degeneracy of G plus one, while $\mathrm{col}_r(G) \leq \mathrm{wcol}_r(G)$ in general. These generalized coloring numbers have various applications in algorithmic design, see e.g. [2]. They are also related to various other graph parameters:

- Coloring vertices of a graph in the optimal order $v_1, \ldots, v_n$ for the 2-coloring number, greedily assigning the smallest color not used on $R_2(v_i) \setminus \{v_i\}$ when v_i is colored, gives an ***acyclic coloring*** of G using at most $\mathrm{col}_2(G)$ colors, i.e., a proper coloring such that the union of any two color classes induces a forest.
- Coloring vertices of a graph in the optimal order $v_1, \ldots, v_n$ for the weak 2-coloring number, greedily assigning the smallest color not used on $W_2(v_i) \setminus \{v_i\}$ when v_i is colored, gives a ***star coloring*** of G using at most $\mathrm{wcol}_2(G)$ colors, i.e., a proper coloring such that the union of any two color classes induces a disjoint union of stars.
- It is easy to see that $\mathrm{col}_n(G) = \mathrm{tw}(G) + 1$ and $\mathrm{wcol}_n(G) = \mathrm{td}(G)$.

Moreover, generalized coloring numbers play an important role in the theory of graph classes with bounded expansion [4].

Van den Heuvel et al. [8] used chordal partition to obtain good bounds on the generalized coloring numbers of K_t-minor-free graphs. First, let us give bounds for chordal graphs shown in [3]. A linear ordering $\prec$ of the vertex set of a graph G is an ***elimination ordering*** if for every $v \in V(G)$, the set $\{u : u \prec v, uv \in E(G)\}$ induces a clique. It is well known that a graph is chordal if and only if it has an elimination ordering. Moreover, elimination orderings have the following key property.

Observation 13.8 *Let G be a chordal graph with elimination ordering $\prec$. For a vertex $v \in V(G)$, let $B_{\prec,v} = \{u : u \prec v, uv \in E(G)\}$ and let $K_{\prec,v}$ be the component of $G - B_{\prec,v}$ containing v. Then $v \preceq x$ for every $x \in V(K_{\prec,v})$*

Proof Suppose for a contradiction that $G - B_{\prec,v}$ contains a path from v to a vertex $x \prec v$, and let Q be a shortest such path. Since $x \prec v$ and $x \notin B_{\prec,v}$, the vertex x is not a neighbor of v, and thus Q has length at least two. Since Q is shortest, every internal vertex of Q is larger than v in $\prec$. Let y be the maximum internal vertex of Q in the ordering $\prec$, and let u_1 and u_2 be the neighbors of y in Q. Then $u_1, u_2 \prec y$, and since $\prec$ is an elimination ordering, $u_1u_2 \in E(G)$. However, then $Q - y + u_1u_2$ is a shorter path from v to x, which is a contradiction. □

We are now ready to give the bounds on the generalized coloring numbers of elimination orderings.

Lemma 13.9 *Let G be a chordal graph with elimination ordering $\prec$. If G has clique number at most $k+1$, then for every positive integer r, $\mathrm{col}_{\prec,r}(G) \leq k+1$ and $\mathrm{wcol}_{\prec,r}(G) \leq \binom{r+k}{k}$.*

Proof Let Q be a path in G from a vertex $v \in V(G)$ to a vertex $u \prec v$ such that $v \prec x$ for every internal vertex x of Q. Then $Q - u$ is disjoint from $B_{\prec,v}$, and thus it is a path in $K_{\prec,v}$. Since $u \prec v$, Observation 13.8 implies that u is not in $K_{\prec,v}$, and since $K_{\prec,v}$ is a component of $G - B_{\prec,v}$, it follows that $u \in B_{\prec,v}$. Therefore, for every positive integer r, we have $R_{\prec,r}(v) \subseteq B_{\prec,v} \cup \{v\}$. Since $\prec$ is an elimination ordering, $B_{\prec,v} \cup \{v\}$ forms a clique in G, and thus $|R_{\prec,r}(v)| \leq k+1$. We conclude that $\mathrm{col}_{\prec,r}(G) \leq k+1$.

Consider now any vertex $u \in W_{\prec,r}(v)$, that is, $u \preceq v$ and there exists a path $Q = v_0v_1 \dots v_p$ in G, where $v_0 = v$, $v_p = u$, $p \leq r$, and $u \prec v_i$ for $i \in [p-1]$. Choose a shortest such path Q; then Q is induced. Moreover, observe that $v_0 \succ v_1 \succ \dots \succ v_p$: Indeed, otherwise we would have $v_{i-1}, v_{i+1} \prec v_i$ for some $i \in [p-1]$ and $v_{i-1}v_{i+1}$ would be an edge, since $\prec$ is an elimination ordering. Let us furthermore choose such a path Q so that the tuple $(v_1, \dots, v_{p-1})$ is lexicographically minimal, where the comparison in each coordinate is according to $\prec$. We claim that the following condition holds:

($\star$) For every $i \in \{0, 1, \dots, p-1\}$ and every vertex $x \in B_{\prec,v_i}$, either

- $v_{i+1} \preceq x$, or
- $x \in B_{\prec,v_j}$ for every $j \in \{i+1, \dots, p\}$.

Indeed, suppose that $x \prec v_{i+1}$, and let $k \in \{i+1, \dots, p\}$ be maximal such that $x \prec v_k$. Since $\prec$ is an elimination ordering and x is adjacent to v_i, we conclude that x is adjacent to v_{i+1}. Repeating this observation, x is also adjacent to $v_{i+2}, \dots, v_k$. Therefore, we have $x \in B_{\prec,v_j}$ for every $j \in \{i+1, \dots, k\}$. If $k < p$, then since Q is an induced path, we have $x \neq v_{k+1}$. Since $\prec$ is an elimination ordering and $x, v_{k+1} \prec v_k$, we have $xv_{k+1} \in E(G)$. But then we can replace Q by the path

$v_0 \dots v_i x v_{k+1} \dots v_p$, which is either shorter than Q, or (when $k = i + 1$) has the same length as Q but is lexicographically smaller. This is a contradiction, and thus $k = p$ and the second outcome of ($\star$) holds.

For a vertex $v \in V(G)$ and integers ℓ and m, let $n_{\ell,m}(v)$ be the number of paths $Q = v_0v_1 \dots v_p$ such that $v_0 = v$, $v_0 \succ v_1 \succ \dots \succ v_p$, $p \leq \ell$, there exist at least m distinct vertices $y \in B_{\prec,v}$ such that $y \prec v_p$, and Q satisfies ($\star$). If $|B_{\prec,v}| \leq m$, this is only possible if $p = 0$, and thus $n_{\ell,m}(v) \leq 1$; in particular, this is always the case if $m = k$. Hence, suppose that $|B_{\prec,v}| > m$. Then the paths $Q = v_0v_1 \dots v_p$ satisfying the conditions described above fall into one of the following cases:

- There are at least $m + 1$ distinct vertices $y \in B_{\prec,v}$ such that $y \prec v_p$.
- The $(m + 1)$-th smallest vertex w in $B_{\prec,v}$ satisfies $v_p \preceq w$. In this case, ($\star$) implies $v_1 \preceq w$. Since all m elements of $B_{\prec,v}$ smaller than w are also smaller than v_p, and thus distinct from v_1, it follows that $w = v_1$. Since $B_{\prec,v}$ is a clique, the m distinct vertices $u \in B_{\prec,v}$ such that $u \prec v_p$ also belong to $B_{\prec,v_1} = B_{\prec,w}$, and the path $Q - v$ starting in w satisfies the conditions described above for $\ell - 1$.

Therefore, we have

$$n_{\ell,m}(v) \leq n_{\ell,m+1}(v) + n_{\ell-1,m}(w).$$

Since $n_{\ell,k}(v) \leq 1$ and $n_{0,m}(v) \leq 1$ for every $v \in V(G)$, ℓ, and m, a straightforward inductive argument shows that

$$n_{\ell,m}(v) \leq \binom{\ell + k - m}{k - m}$$

for every $v \in V(G)$, ℓ and m. Therefore,

$$|W_{\prec,r}(v)| \leq n_{r,0}(v) \leq \binom{r + k}{k}$$

for every $v \in V(G)$, and thus $\mathrm{wcol}_r(G) \leq \binom{r+k}{k}$. □

In particular, this has the following consequence.

Corollary 13.10 *Let G be a graph. For every positive integer r, $\mathrm{col}_r(G) \leq \mathrm{tw}(G) + 1$ and $\mathrm{wcol}_{\prec,r}(G) \leq \binom{r+\mathrm{tw}(G)}{\mathrm{tw}(G)}$.*

Proof Let $k = \mathrm{tw}(G)$ and let (T, β) be a tree decomposition of G of width k. Let G' be the graph obtained from G by adding edges between all vertices in $\beta(x)$ for every $x \in V(T)$. Then (T, β) is a tree decomposition of G' such that every bag induces a clique, and thus G' is chordal. Moreover, G' has treewidth k, and thus $\omega(G') \leq k + 1$. By Lemma 13.9, we have $\mathrm{col}_r(G) \leq \mathrm{col}_r(G') \leq k + 1$ and $\mathrm{wcol}_r(G) \leq \mathrm{wcol}_r(G') \leq \binom{r+k}{k}$. □

The bounds from Lemma 13.9 can be lifted to proper minor-closed classes using the following property of geodesic chordal partitions.

Observation 13.11 *Let $\mathcal{P} = P_1, \ldots, P_m$ be a geodesic chordal partition of a K_t-minor-free graph G. Suppose that $1 \le i \le j \le m$ and v is a vertex of P_j. Then at most $(t-2)(2r+1)$ vertices of P_i are at distance at most r from v in $G\left[\bigcup_{k\ge i} P_k\right]$.*

Proof Since the partition is geodesic and G is K_t-minor-free, P_i is covered by the vertex sets of at most $t-2$ paths which are geodesic in $G\left[\bigcup_{k\ge i} P_k\right]$. The vertex v can be at distance at most r from at most $2r+1$ vertices of each of the paths, as otherwise we could shortcut the path through v. □

We are now ready to give the bound on the generalized coloring numbers.

Lemma 13.12 *For every positive integer r, every K_t-minor-free graph G has r-coloring number at most $t^2(2r+1)$ and weak r-coloring number at most $2(r+t)^t$.*

Proof Let $\mathcal{P} = P_1, \ldots, P_m$ be a geodesic chordal partition of G. For $i \in [m]$, let p_i be the vertex of $G/\mathcal{P}$ corresponding to P_i. Then the ordering $\prec_0$ such that $p_i \prec p_j$ if and only if $i < j$ is an elimination ordering of $G/\mathcal{P}$. Moreover, since G is K_t-minor-free, we have $\omega(G/\mathcal{P}) \le t-1$. For every $i \in [m]$ and every vertex $v \in P_i$, let us define $\pi(v) = p_i$. For a path $Q = v_1 v_2 \ldots v_p$ in G, the *projection* of Q is the walk in $G/\mathcal{P}$ obtained from $\pi(v_1)\pi(v_2)\ldots\pi(v_p)$ by suppressing consecutive appearances of the same vertex.

Let $\prec$ be a linear ordering of $V(G)$ such that for $1 \le i < j \le m$, all vertices $u \in P_i$ and $v \in P_j$ satisfy $u \prec v$. The relative ordering of vertices belonging to the same part of $\mathcal{P}$ is arbitrary. Consider vertices $u, v \in V(G)$ such that $u \in P_i$ and $v \in P_j$ for some $i \le j$. If $u \in R_{\prec,r}(v)$, then there exists a path Q of length at most r in G such that $v \prec x$ for every internal vertex x of Q; observe that the projection of Q shows that $p_i \in R_{\prec_0,r}(p_j)$. Similarly, if $u \in W_{\prec,r}(v)$, then $p_i \in W_{\prec_0,r}(p_j)$.

Lemma 13.9 and Observation 13.11 then imply that

$$\mathrm{col}_{\prec,r}(G) \le (t-2)(2r+1)\,\mathrm{col}_{\prec_0,r}(G/\mathcal{P}) \le (t-2)(2r+1)(t-1) \le t^2(2r+1),$$

and

$$\begin{aligned}\mathrm{wcol}_{\prec,r}(G) &\le (t-2)(2r+1)\,\mathrm{wcol}_{\prec_0,r}(G/\mathcal{P}) \le (t-2)(2r+1)\binom{r+t-2}{t-2}\\ &\le 2(r+t)^t.\end{aligned}$$

□

Let us remark that the bounds on the generalized chromatic numbers can be improved by a better analysis of the same argument; see [8] for details. We are going to see more applications of geodesic chordal partitions in Sects. 14.3 and 16.2.

References

1. M. Delcourt, L. Postle, Reducing linear Hadwiger's conjecture to coloring small graphs (2021). ArXiv, 2108.01633

2. Z. Dvořák, Constant-factor approximation of domination number in sparse graphs. Eur. J. Combin. **34**, 833–840 (2013)
3. M. Grohe, S. Kreutzer, R. Rabinovich, S. Siebertz, K. Stavropoulos, Coloring and covering nowhere dense graphs. SIAM J. Discrete Math. **32**, 2467–2481 (2018)
4. J. Nešetřil, P. Ossona de Mendez, *Sparsity (Graphs, structures, and algorithms).* Algorithms and Combinatorics, vol. 28 (Springer, Berlin, 2012)
5. B. Reed, P.D. Seymour, Fractional colouring and Hadwiger's conjecture. J. Combin. Theory Ser. B **74**, 147–152 (1998)
6. E. Scheinerman, D. Ullman, *Fractional graph theory* (Dover Publications Inc., Mineola, 2011)
7. A. Scott, P.D. Seymour, D.R. Wood, Bad news for chordal partitions. J. Graph Theory **90**, 5–12 (2019)
8. J. van den Heuvel, P. Ossona de Mendez, D. Quiroz, R. Rabinovich, S. Siebertz, On the generalised colouring numbers of graphs that exclude a fixed minor. Eur. J. Combin. **66**, 129–144 (2017)

Chapter 14
Chromatic Number

Indisputably, Hadwiger's conjecture (that every K_{t+1}-minor-free graph is t-colorable) has been one of the most influential problems shaping the graph minors theory. For any $t \geq 4$, it is stronger than the Four Color Theorem: For any planar graph G, observe that K_{t+1} is not a minor of the graph obtained from G by adding $t-4$ universal vertices, and any t-coloring of this graph restricts to a 4-coloring of G. And indeed, Hadwiger's conjecture for $t \in \{4, 5\}$ has been shown to be equivalent to the Four Color Theorem.

The case $t = 4$ follows from Wagner's characterization of K_5-minor-free graphs [19]: Such graphs are obtained by clique-sums from planar graphs and copies of the Wagner graph W_8, the 8-vertex Möbius ladder. Since W_8 is 3-colorable and clique-sums cannot increase the chromatic number, the claim easily follows. For $t = 5$, a much more involved argument by Robertson et al. [14] shows that a minimum counterexample is necessarily obtained from a planar graph by adding a single apex vertex, again reducing the problem to the Four Color Theorem. Hadwiger's conjecture is widely open for all other values of t; indeed, we do not even know whether K_7-minor-free graphs are 7-colorable, much less 6-colorable!

For the reasons we discussed in the introduction to this part of the book, the chromatic number of proper minor-closed classes is an area where the Minor Structure Theorem is largely irrelevant. For a long time, the only general bound on the chromatic number of K_t-minor-free graphs was based on the tight bound $O(t\sqrt{\log t})$ on their density. This bound dates back to the works of Kostochka [8] and Thomason [16], though a much simpler argument was found recently by Alon et al. [1]. Given the form of the bound, it should be unsurprising that the proofs are probabilistic in nature.

In 2019, the density barrier was broken by Norin et al. [13], starting a period of quick progress on this problem, culminating in the current best bound of $O(t \log \log t)$ given by Delcourt and Postle [3]. It does not seem to be possible to improve the bound further without significant new ideas, as we discuss in Sect. 14.2;

© The Author(s), under exclusive license to Springer Nature Switzerland AG 2025

Z. Dvořák, *Graph Minors*, Springer Monographs in Mathematics,
https://doi.org/10.1007/978-3-031-87469-7_14

we would need a much better understanding of the chromatic number of small K_t-minor-free graphs, say with $O(t \operatorname{polylog} t)$ vertices.

14.1 Density

Before that, let us talk about the density argument in more detail. Let $f(k)$ be the minimum integer such that every simple graph of average degree at least $f(k)$ contains K_k as a minor. Mader [11] proved that $f(k) = O(k \log k)$ and that a random-graph based construction shows that $f(k) = \Omega(k\sqrt{\log k})$. To get some intuition for why this is the case, consider the random graph $G = G(n, 1/2)$ with n vertices and with each pair of vertices being joined by an edge independently with probability $1/2$; the average degree of this graph is with high probability close to $n/2$. For any partition of $V(G)$ to parts $X_1, \ldots, X_k$, the expected number of pairs of these parts with no edge between them is

$$\sum_{1\le i<j\le k} 2^{-|X_i||X_j|} \ge \binom{k}{2}\left(\prod_{i<j} 2^{-|X_i||X_j|}\right)^{1/\binom{k}{2}} = \binom{k}{2} 2^{-\frac{1}{\binom{k}{2}}\sum_{i<j}|X_i||X_j|}$$
$$\ge \binom{k}{2} 2^{-(n/k)^2}.$$

Thus, if $n = \varepsilon k\sqrt{\log_2 k}$, then the expected number of non-adjacent pairs is $\Omega(k^{2-\varepsilon^2})$, i.e., rather large. A more complicated concentration argument shows that with high probability, it is not possible to partition $V(G)$ into k parts such that there is an edge between any two of them, implying that K_k is not a minor of G (the random graph G is almost surely connected, and thus a model of K_k in G could be extended to cover the whole graph).

Kostochka [8] and Thomason [16] independently strengthened Mader's upper bound, showing that $f(k) = \Theta(k\sqrt{\log k})$. Finishing this progression of results, Thomason [17] determined the exact value of the multiplicative constant α such that $f(k) = (\alpha + o(1))k\sqrt{\log k}$ (see Theorem 1.2). To showcase some of the ideas of these arguments, let us give a relatively short proof that $f(k) = \Theta(k\sqrt{\log k})$ based on [1]. First, let us note that every graph contains a dense minor with roughly the same average degree.

Lemma 14.1 *Let d be a positive integer. If a simple graph G has average degree at least $2d$, then G has*

- *a minor F_1 of minimum degree at least d with at most $2d$ vertices, and*
- *a minor F_2 of minimum degree at least $\frac{3}{5}d$, connectivity more than $\frac{2}{5}d$, and with at most $2d$ vertices.*

Proof Let H be a smallest non-null minor of G of average degree at least $2d$, i.e., such that $\|H\| \geq d|H|$. Note that this implies that $E(H) \neq \emptyset$, and thus $|H| \geq 2$.

If an edge e of H is contained in t triangles, then $\|H/e\| = \|H\| - t - 1$. By the minimality of H, we have

$$\|H\| - t - 1 = \|H/e\| \leq d|H/e| - 1 = d|H| - d - 1 \leq \|H\| - d - 1.$$

Therefore, $t \geq d$, i.e., every edge of H is in at least d triangles.

By the minimality of H, we cannot delete any edge from H without violating the inequality $\|H\| \geq d|H|$, and thus $\|H\| = d|H|$. Hence, H has average degree exactly $2d$, and thus it contains a vertex v of degree at most $2d$. Let F_1 be the subgraph of H induced by the neighbors of v. Clearly $|F_1| \leq 2d$. Moreover, for every $u \in V(F_1)$, the edge vu is contained in at least d triangles, and thus u has degree at least d in F_1.

If F_1 has connectivity more than $\frac{2}{5}d$, then we can let $F_2 = F_1$. Otherwise, consider a vertex separation (A, B) of F_1 of order at most $\frac{2}{5}d$ such that $A \not\subseteq B$ and $B \not\subseteq A$. By symmetry, we can assume $|B| \geq |A|$. Let $F_2 = F_1 - B$; we have

$$|F_2| = |A \setminus B| \leq \tfrac{1}{2}(|F_1| - |A \cap B|) \leq d - \tfrac{1}{2}|A \cap B|.$$

On the other hand, F_2 has minimum degree

$$\delta(F_2) \geq \delta(F_1) - |A \cap B| \geq d - |A \cap B| \geq \tfrac{3}{5}d.$$

If $S \subseteq V(F_2)$ is a cut in F_2, then a vertex of the smallest component C of $F_2 - S$ has less than $|C| + |S| \leq \frac{1}{2}(|F_2| - |S|) + |S| = \frac{1}{2}(|F_2| + |S|)$ neighbors in F_2, and thus

$$2(d - |A \cap B|) \leq 2\delta(F_2) < |F_2| + |S| \leq d - \tfrac{1}{2}|A \cap B| + |S|,$$

and

$$|S| > d - \tfrac{3}{2}|A \cap B| > d - \tfrac{3}{2} \cdot \tfrac{2}{5}d = \tfrac{2}{5}d.$$

Therefore, F_2 has connectivity more than $\frac{2}{5}d$. □

We are also going to need the following easy observation on diameter of graphs of large minimum degree.

Observation 14.2 *Let c be a positive integer and let G be a connected graph with n vertices and with minimum degree δ. If $c(\delta + 1) > n$, then G has diameter less than* $3(c - 1)$.

Proof Suppose for a contradiction that $P = v_0v_1 \ldots v_{3(c-1)}$ is a shortest path between vertices v_0 and $v_{3(c-1)}$ of G. The closed neighborhoods of v_0, v_3, …,

$v_{3(c-1)}$ are pairwise disjoint and each of them consists of at least $\delta + 1$ vertices. Hence, we have $n \geq (\delta + 1)c$, which is a contradiction. □

Finally, we are going to need the following lemma with an easy probabilistic proof.

Lemma 14.3 *Let $c \geq 4$, $k \geq 5$, and n be integers and let G be a graph with at most n vertices and with minimum degree $\delta \geq n/c$. Let $A_1, \ldots, A_k$ be subsets of $V(G)$ of size at most $ne^{-\sqrt{\log k}}$. Then there exists a non-empty set $B \subseteq V(G)$ of size at most $2c\sqrt{\log k}$ such that $B \not\subseteq A_i$ for $i \in [k]$ and all but at most $ne^{-\sqrt{\log k}}$ vertices of G have a neighbor in B.*

Proof Let $s = \lceil c(\sqrt{\log k} + 1) \rceil$; note that $s \leq c\sqrt{\log k} + c + 1 \leq 2c\sqrt{\log k}$, since $c \geq 4$ and $k \geq 5$. Choose vertices $v_1, \ldots, v_s \in V(G)$ independently uniformly at random with repetitions allowed, and let B be the set of these vertices. For a vertex $v \in V(G)$, the probability that none of the chosen vertices is in the neighborhood of v is at most

$$(1 - \delta/n)^s \leq (1 - 1/c)^{c(\sqrt{\log k}+1)} \leq e^{-\sqrt{\log k}-1} < \tfrac{1}{2}e^{-\sqrt{\log k}}.$$

Let A be the set of vertices with no neighbor in B; the expected size of A is less than $\frac{1}{2}ne^{-\sqrt{\log k}}$, and by Markov's inequality,

$$\Pr[|A| > ne^{-\sqrt{\log k}}] < \tfrac{1}{2}.$$

Now, for $i \in [k]$, the probability that $B \subseteq A_i$ is

$$\left(\frac{|A_i|}{n}\right)^s \leq e^{-s\sqrt{\log k}} \leq e^{-c\log k} \leq \frac{1}{k^c} \leq \frac{1}{5k}.$$

Hence, by the union bound, the probability that B satisfies all the conditions from the statement of the lemma is at least $1 - \frac{1}{2} - k \cdot \frac{1}{5k} > 0$. □

We are now ready to prove the following weaker version of Theorem 1.2.

Theorem 14.4 *Let $k \geq 5$ be an integer. Every graph G of average degree at least $2700\lceil k\sqrt{\log k}\rceil$ contains K_k as a minor.*

Proof Let $m = \lceil k\sqrt{\log k}\rceil$. By Lemma 14.1, G has a minor F of minimum degree at least $810m$, connectivity more than $540m$, and with $n \leq 2700m$ vertices. We are going to choose pairwise disjoint sets $C_1, \ldots, C_k \subseteq V(F)$ forming a model of K_k in F such that the following conditions are satisfied for each $i \in [k]$:

- $F[C_i]$ is connected and there is an edge between C_i and C_j in F for every $j \in [i-1]$,
- $|C_i| \leq 540\sqrt{\log k}$, and
- the set A_i of vertices of $V(F) \setminus \bigcup_{j=1}^{i-1} C_j$ with no neighbor in C_i has size at most $ne^{-\sqrt{\log k}}$.

Suppose that we have already selected such sets $C_1, \ldots, C_{p-1}$ for some $p \in [k]$ and let $A_1, \ldots, A_{p-1}$ be the corresponding sets. Let us finish the argument by showing how to choose C_p.

Let $Z = C_1 \cup \ldots \cup C_{p-1}$ and note that $|Z| \leq 540k\sqrt{\log k} \leq 540m$. Hence, $F - Z$ is a connected graph with at most n vertices and with minimum degree at least $270m \geq n/10$. By Lemma 14.3 with $c = 10$ and the sets $A_1 \setminus Z, \ldots, A_{p-1} \setminus Z$, $\emptyset, \ldots, \emptyset$, there exists a non-empty set $B \subseteq V(F - Z)$ of size at most $20\sqrt{\log k}$ such that

- $B \not\subseteq A_j$ (and thus there exists an edge between C_j and B) for each $j \in [p-1]$, and
- the set A'_p of vertices of $V(F - Z)$ with no neighbor in B has size at most $ne^{-\sqrt{\log k}}$.

By Observation 14.2, $F - Z$ has diameter less than 27. Let v_0 be an arbitrary vertex of B, and let C_p be the union of the vertex sets of shortest paths from v_0 to all vertices of B in $F - Z$. We have

$$|C_p| \leq 27|B| \leq 540\sqrt{\log k}.$$

Adding the shortest paths ensures that $F[C_p]$ is connected. Since $A_p \subseteq A'_p$, it follows that the set C_p satisfies the required invariants. □

Let us remark that the constant 2700 from Theorem 14.4 can be significantly improved by a tighter analysis (see [1] for details) and by proving the bound only for large values of k.

Of course, for small values of k, the general bounds (even the asymptotically tight one from Theorem 1.2) are far from the best possible. One way to construct relatively dense K_k-minor-free graphs is to add $n - k + 2$ pairwise non-adjacent apex vertices to K_{k-2}; this graph has n vertices and

$$(k-2)(n-k+2) + \binom{k-2}{2} = (k-2)n - \binom{k-1}{2} \tag{14.1}$$

edges. Thus, the average degree of these graphs is $2(k-2) - O(1/n)$.

- For $k \leq 4$, the K_k-minor-free graphs are exactly the graphs of treewidth at most $k-2$. This easily implies that the maximum number of edges of such K_k-minor-free n-vertex graphs is given by (14.1). Moreover, every K_k-minor-free graph with $k \leq 4$ has minimum degree is at most $k-2$, roughly half of this bound on their average degree.
- By Wagner's theorem [19], K_5-minor-free graphs are obtained by clique-sums from planar graphs and from the 3-regular Wagner graph. From this, an easy inductive argument shows that K_5-minor-free n-vertex graph has at most $3n - 6$ edges, matching the bound from (14.1). Here, the minimum degree of K_5-minor-free graphs "catches up" to the average degree, since there are planar graphs of minimum degree 5.

A direct way to show this bound is as follows: Suppose for a contradiction that there exists a K_5-minor-free graph with $n \geq 5$ vertices and at least $3n - 5$ edges, and let G be such a graph with $|G| + \|G\|$ minimum. Since $G \neq K_5$ and $3 \cdot 5 - 5 = \binom{5}{2}$, we have $n \geq 6$. Every edge of G is in at least three triangles, as contraction of an edge in $t \leq 2$ triangles would result in a K_5-minor-free graph G' with $\|G'\| = \|G\| - t - 1 \geq 3(n-1) - 5 = 3|G'| - 5$. Consequently, the neighbors of every vertex induce a subgraph of minimum degree at least three, and in particular G has minimum degree at least four. Furthermore, we can assume that G has exactly $3n-5$ edges, as otherwise we can delete an edge from G. Hence, the average (and minimum) degree of G is less than six. If a vertex $v \in V(G)$ has degree four, then its neighborhood is a clique and $K_5 \subseteq G$. If v has degree five, then its neighborhood induces a complement of a matching. Thus, there exist distinct neighbors x_1, x_2, y_1, and y_2 such that all neighbors of v except possibly for x_i and y_i for $i \in [2]$ are pairwise adjacent. Then the contraction of the edge x_1x_2 turns the closed neighborhood of v to the clique K_5. In either case, we obtain a contradiction.

- The lower bound from (14.1) also matches the upper bound for $k \in \{6, 7\}$, as shown by Mader [11] through more complicated versions of the preceding argument for K_5.
- This is no longer the case for $k = 8$, as was again pointed out by Mader: The graph $K_{2,2,2,2,2}$ (the complement of a matching of size five) is K_8-minor-free, since we cannot perform more than two contractions to keep the number of vertices at least 8, but that leaves a pair of non-adjacent vertices. However, it has $40 = 6 \cdot 10 - \binom{7}{2} + 1$ edges. Moreover, any n-vertex graph obtained from the copies of $K_{2,2,2,2,2}$ by gluing them on cliques of size five has $6n - 20 = 6n - \binom{7}{2} + 1$ edges. Jørgensen [7] proved that every K_8-minor-free n-vertex graph is either one of these graphs (called $(K_{2,2,2,2,2}, 5)$-cockades), or has at most $6n - 21$ edges.
- For $k = 9$, Song and Thomas [15] similarly characterized all K_9-minor-free graphs whose density exceeds (14.1). In particular, they show that the maximum number of edges of a K_9-minor-free n-vertex graph is $7n - 27 = 7n - \binom{8}{2} + 1$.

As there are K_k-minor-free graphs of average degree $\Omega(k\sqrt{\log k})$ and we can take their disjoint copies to boost the number of vertices, the analogous bound of form $(k-2)n + o(n)$ is false for sufficiently large k; it is not clear what is the smallest value of k for which this is the case. Seymour and Thomas conjectured that for every k, every sufficiently large $(k-2)$-connected K_k-minor-free graph has at most $(k-2)n - \binom{k-1}{2}$ edges, and consequently, that every $(k-2)$-connected K_k-minor-free graph has at most $(k-2)n + O_k(1)$ edges. By Theorem 10.1, the claim holds for k-connected graphs.

14.2 Beyond the Density

Let us now give a high-level overview of the ideas needed to improve the bound on the chromatic number beyond the one coming from the degeneracy.

Random graphs do not give (or even come close to giving) counterexamples to Hadwiger's conjecture: As we have argued, the random graph $G_{n,1/2}$ has Hadwiger number $\Theta(n/\sqrt{\log n})$, but its chromatic number is known to be $\Theta(n/\log n)$. Still, they highlight one issue in trying to improve the bounds on the chromatic number of K_t-minor-free graphs: It is hard to draw any structural conclusions from the assumption that a graph with relatively few vertices does not have a K_t-minor. The best bounds on the chromatic number of such graphs available to us are based on the bounds on the size of their independent sets. Theorem 13.3 in particular implies that every n-vertex K_{t+1}-minor-free graph has an independent set of size at least $\frac{n}{2t}$ (let us remark that the constant 2 in this claim can be slightly improved, see [2]). This has the following consequence on the chromatic number of small graphs.

Lemma 14.5 *Let t be a positive integer and let G be a graph with γt vertices. If G is K_{t+1}-minor-free, then $\chi(G) \leq t + \lceil 2t \log \gamma \rceil$.*

Proof Let $m = \lceil 2t \log \gamma \rceil$. Let $G_0 = G$. For $i = 1, 2, \ldots, m$, let A_i be a largest independent set in G_{i-1} and let $G_i = G_{i-1} - A_i$. Note that $\chi(G) \leq m + |G_m|$, since we can color each of the sets $A_1, \ldots, A_m$ by a single distinct color, then assign a separate color to each vertex of G_m. Hence, it suffices to bound $|G_m|$. Theorem 13.3 implies that for $i \in [m]$, we have $|A_i| \geq \frac{1}{2t}|G_{i-1}|$, and thus $|G_i| \leq \left(1 - \frac{1}{2t}\right)|G_{i-1}|$. Therefore,

$$|G_m| \leq \left(1 - \frac{1}{2t}\right)^m |G| \leq e^{-\frac{m}{2t}} \gamma t \leq e^{-\log \gamma} \gamma t = t.$$

Therefore, $\chi(G) \leq t + m$, as required. □

Note that the bound from Lemma 14.5 beats the $\chi(G) = O(t\sqrt{\log t})$ bound for graphs with at most $t \exp(\sqrt{\log t})$ vertices, which gives us some room for structural arguments.

Kühn and Osthus [9] proved that a graph of girth at least five and average degree d contains a minor of minimum degree $\Omega(d^{3/2})$. This implies Hadwiger's conjecture for graphs of girth at least five for sufficiently large forbidden minor K_{t+1}: If a graph G is not t-colorable, then it contains a subgraph of minimum degree at least t. Hence, if G has girth at least five, then it has a minor of minimum degree $\Omega(t^{3/2})$, which implies the presence of K_{t+1} (and in fact even of substantially larger clique minors) by Theorem 1.2.

The proof of this "density increment" result is by contracting small subgraphs chosen carefully so that not too many parallel edges arise. It is natural to try a similar argument even without the girth restriction; and it turns out that this only fails when the graph has small dense subgraphs (in which any contraction leads to many edges being lost).

Theorem 14.6 (Delcourt and Postle [3]) *There exists a positive integer c such that the following claim holds. For all integers $k \geq t \geq 1$, if G is a K_t-minor-free graph of average degree at least ck, then G contains a subgraph H of minimum degree at least k with at most $c^2 t \log^3 t$ vertices.*

Note that using Lemma 3.5, we can even assume that H is k-connected, or, using Theorem 3.4, k-linked. Together with Lemma 14.5, we can show that if G has chromatic number $\Omega(\log\log t)$, it actually contains many vertex-disjoint dense subgraphs.

Corollary 14.7 *For any function $g(t) = \text{polylog}(t)$, there exists a constant γ such that the following claim holds for every integer $t \geq 16$. Let G be a K_t-minor-free graph. If $\chi(G) \geq \gamma t \log\log t$, then G contains at least $g(t)$ pairwise vertex-disjoint $\lfloor t \log\log t \rfloor$-linked subgraphs.*

Proof Let m be maximum such that G contains m pairwise vertex-disjoint $\lfloor t \log\log t \rfloor$-linked subgraphs $H_1, \ldots, H_m$, each with $O(t \log^3 t)$ vertices, and suppose for a contradiction that $m < g(t)$. Let $S = V(H_1) \cup \ldots \cup V(H_m)$; we have $|S| \leq t\,\text{polylog}(t)$. By Lemma 14.5, $\chi(G[S]) = O(t \log\log t)$, and if γ is sufficiently large, this implies $\chi(G[S]) \leq \chi(G)/2$. Hence, $\chi(G - S) \geq \chi(G)/2 \geq \lfloor \frac{\gamma}{2} t \log\log t \rfloor$, and thus $G - S$ has a subgraph of minimum degree at least $\lfloor \frac{\gamma}{2} t \log\log t \rfloor$. By Theorem 14.6 with $k = \lfloor t \log\log t \rfloor$ and the following discussion, we conclude that G contains a $\lfloor t \log\log t \rfloor$-linked subgraph with $O(t \log^3 t)$ vertices disjoint from S, which is a contradiction. □

Let us now try to argue that every K_t-minor-free graph G is $tf(t)$-colorable, for some $f(t) = \Omega(\log\log t)$. Suppose for a contradiction this is not the case, and let G be a minimum counterexample. Then G is minor-χ-critical, and by Lemma 10.20 and Theorem 3.4, G is $\Omega(tf(t))$-linked. By Corollary 14.7, G contains pairwise vertex-disjoint $t \log\log t$-linked subgraphs $H_1, \ldots, H_m$, where $m \geq \text{polylog}(t)$. Theorem 1.2 then shows that each of these subgraphs contains as a minor the clique with $\Omega(t \log\log t/\sqrt{\log t})$ vertices. The argument then can be finished by linking these smaller minors to form a minor of K_t in G, obtaining a contradiction; of course, achieving this requires further non-trivial ideas.

In any case, we see that Lemma 14.5 is currently a bottleneck: For a K_t-minor-free graph G with say $t\sqrt{\log t}$ vertices, there is no hope for gaining anything by the density increment argument, since there are K_t-minor-free graphs of average degree $\Omega(t\sqrt{\log t})$; and Lemma 14.5 only shows that G is $O(t \log\log t)$-colorable. Thus, to get closer to Hadwiger's conjecture, we would need to get a better understanding of such small graphs. As shown in [3], this is actually the only obstruction to proving Linear Hadwiger's conjecture.

14.3 Relaxed Colorings

As we are quite far from being able to prove (or disprove) Hadwiger's conjecture, it is natural to consider its variants in terms of less restrictive types of coloring. Two popular options are *defective coloring*, where each color class is required to induce a subgraph of bounded maximum degree, and *clustered coloring*, where each connected monochromatic subgraph is required to have bounded size.

More precisely, a (not necessarily proper) coloring φ of a graph G has *defect at most d* if for every color c, every vertex of color c has at most d neighbors of color c. It has *clustering at most κ* if for every color c, every connected subgraph formed by vertices of color c has at most κ vertices. Note that a coloring is proper if and only if it has defect 0 and clustering 1. Moreover, a coloring with clustering at most κ clearly has defect at most $\kappa - 1$.

We say that G is *k-colorable with defect d* (resp., with *clustering κ*) if G has a coloring with defect at most d (resp. clustering at most κ) using at most k colors. In general, we want to minimize the number of colors, with the defect or clustering playing secondary role. Of course, this does not make sense for single finite graphs, since every graph G is 1-colorable with clustering $|G|$. However, this is not an issue for graph classes.

- The ***defective chromatic number*** $\chi_\Delta(\mathcal{G})$ of a graph class $\mathcal{G}$ is the minimum integer k such that for some d, every graph from $\mathcal{G}$ is k-colorable with defect d.
- The ***clustered chromatic number*** $\chi_\star(\mathcal{G})$ of a graph class $\mathcal{G}$ is the minimum integer k such that for some κ, every graph from $\mathcal{G}$ is k-colorable with clustering κ.

Edwards et al. [6] proved that Hadwiger's conjecture holds in the defective chromatic number setting. A simpler proof of this result (as well as a quite good bound on the clustered chromatic number) based on chordal partitions was found in [18].

Theorem 14.8 *For every integer $t \geq 2$, the class of K_{t+1}-minor-free graphs has defective chromatic number at most t and clustered chromatic number at most $2t$.*

Proof Let G be a K_{t+1}-minor-free graph and let $\mathcal{P} = P_1, \ldots, P_m$ be a geodesic chordal partition of G (see Sect. 13.2). Since G is K_{t+1}-minor-free, the graph $G/\mathcal{P}$ has treewidth at most $t-1$, and thus it is t-colorable. Hence, there exists a coloring $\psi : [m] \to [t]$ such that for any distinct $i, j \in [m]$, if G contains an edge between P_i and P_j, then $\psi(i) \neq \psi(j)$.

Moreover, by π_{geo}, for each $i \in [m]$ there exists a vertex $v_i \in P_i$ such that P_i is covered by the vertex sets of at most $t-1$ geodesic paths in $G[P_i]$ starting in v_i. Since the paths are geodesic, each vertex of P_i has at most three neighbors in each of them and at most two in the one containing it, and thus $G[P_i]$ has maximum degree at most $3t-4$. Hence, the coloring φ_1 that gives for each $i \in [m]$ the color $\psi(i)$ to all vertices of P_i is a t-coloring of G with defect $3t-4$.

Consider now a coloring φ_2 that for each $i \in [m]$ gives the color $2\psi(i)$ to vertices of P_i whose distance from v_i in $G[P_i]$ is even and the color $2\psi(i)-1$ to those whose distance is odd. Note that if two vertices of P_i of the same color are adjacent, then they necessarily have the same distance from v_i. Since P_i is covered by the vertex sets of at most $t-1$ geodesic paths, it follows that any monochromatic connected subgraph has at most $t-1$ vertices. Hence, φ_2 is a $2t$-coloring of G with clustering $t-1$. □

The bound on the defective chromatic number from Theorem 14.8 is tight: For a positive integer d, let $G_{2,d} = K_{1,d}$, and for $t \geq 3$, let $G_{t,d}$ be obtained from d copies of $G_{t-1,d}$ by adding a universal vertex. By induction on t, it is easy to show that $G_{t,d}$ is K_{t+1}-minor-free, since it is obtained from disjoint union of copies of $G_{t-1,d}$ (which are K_t-minor-free by the induction hypothesis) by adding an apex vertex. Moreover, every coloring of $G_{t,d}$ by at most $t-1$ colors has defect at least d: By symmetry, we can assume that the universal vertex u has color $t-1$. If every copy of $G_{t-1,d}$ contains a vertex of color $t-1$, then u has at least d neighbors of color $t-1$. Otherwise, there is a copy of $G_{t-1,d}$ colored by at most $t-2$ colors, and the coloring has defect at least d by the induction hypothesis. Since the class of K_{t+1}-minor-free graphs contains $\{G_{t,d} : d \geq 1\}$, we conclude that it does not have defective chromatic number less than t.

The clustered chromatic number of K_{t+1}-minor-free graphs turns out to be also t, as was shown recently [4] (another proof was announced a few years earlier [5], but without full details). Both proofs are based on the Minor Structure Theorem; the one in [4] uses a simpler product structure version of the theorem which we discuss in Chap. 15. Here, let us show just one of the ideas going to these proofs.

A ***k-island*** in a graph G is a non-empty set S of vertices of G such that each vertex of S has less than k neighbors outside of S. Islands can be used to bound the clustered chromatic number, as shown in following observation.

Observation 14.9 *Let k and κ be positive integers. If every induced subgraph of a graph G contains a k-island of size at most κ, then G is k-colorable with clustering κ.*

Proof We prove the claim by induction on the number of vertices of G. If $|G| \leq \kappa$, then we can color all vertices of G by color 1. Otherwise, let S be a k-island in G of size at most κ. By the induction hypothesis, $G - S$ has a coloring φ by k colors with clustering at most κ. Extend φ to a k-coloring of G by, for each vertex $v \in S$, assigning to v the smallest color that is not used on its neighbors outside of S. Such a color exists, since v has less than k neighbors outside of S. Note that any new monochromatic connected subgraphs are contained in S, and thus they have at most κ vertices. Hence, the resulting coloring has clustering at most κ. □

An easy way to obtain an island is by reducing a set of vertices incident with few edges. For a set $S \subseteq V(G)$, let $e_G(S)$ be the number of edges of G with at least one end in S.

Observation 14.10 *Let k be a positive integer and let S_0 be a non-empty set of vertices of a graph G. If $e_G(S_0) < k|S_0|$, then G contains a k-island $S \subseteq S_0$.*

Proof Let S be a minimal subset of S_0 such that $e_G(S) < k|S|$; clearly, S is non-empty. If S were not a k-island, then there would exist a vertex $v \in S$ with $d \geq k$ neighbors outside of S. But then

$$e_G(S \setminus \{v\}) = e_G(S) - d < k|S| - k = k|S \setminus \{v\}|,$$

contradicting the minimality of S. □

From that, we get the following conclusion for classes with sublinear separators.

Theorem 14.11 *Let k and n_0 be positive integers and let $\delta > 0$ be a real number. Let $\mathcal{G}$ be a hereditary class of graphs with $O(n^{1-\varepsilon})$-separators for some positive $\varepsilon \le 1$. If every graph $G \in \mathcal{G}$ with $n \ge n_0$ vertices has less than $(k-\delta)n$ edges, then $\chi_\star(\mathcal{G}) \le k$.*

Proof Let $t = \lceil k/\delta \rceil$. By Lemma 12.5, there exists $\kappa_0 = O(t^{1/\varepsilon})$ such that the following claim holds: For every graph $G \in \mathcal{G}$ with n vertices, there exists a set $R \subseteq V(G)$ of size at most n/t such that every component of $G - R$ has at most κ_0 vertices. Let $\kappa = \max(\kappa_0, n_0)$.

Suppose that $n \ge n_0$, and let $S_1, \ldots, S_m$ be the vertex sets of the components of $G - R$. Then

$$\sum_{i=1}^{m} e_G(S_i) \le \|G\| < (k-\delta)n \le k(n - n/t) \le k|V(G-R)| = k \cdot \sum_{i=1}^{m} |S_i|.$$

Hence, there exists $i \in [m]$ such that $e_G(S_i) < k|S_i|$, and by Observation 14.10, a subset of S_i forms a k-island. Hence, G contains a k-island of size at most κ_0.

Since $\mathcal{G}$ is hereditary, we conclude that every induced subgraph of G with at least n_0 vertices contains a k-island of size at most κ_0. On the other hand, if G' is an induced subgraph of G with less than n_0 vertices, then $V(G')$ is a k-island in G' of size less than n_0. Therefore, every induced subgraph of G contains a k-island of size at most κ. By Observation 14.9, we conclude that every graph $G \in \mathcal{G}$ is k-colorable with clustering κ. □

Let us mention some interesting graph classes for which this theorem applies:

- Every n-vertex graph drawn on a surface Σ of Euler genus at most g has at most $3n + O(g)$ edges. Hence, we can apply Theorem 14.11 with $k = 4$, $\delta = 1/2$, and $n_0 = \Theta(g)$ and conclude that the class of graphs drawn on Σ has clustered chromatic number at most 4. Thus, in the clustered setting, the Four Color Theorem extends to all surfaces.

  Let us remark this is the best possible even for planar graphs: Let $G'_{2,\kappa}$ be the path with κ vertices, and for $k \ge 3$, let $G'_{k,\kappa}$ be obtained from κ copies of $G'_{k-1,\kappa}$ by adding a universal vertex. It is easy to see that every $(k-1)$-coloring of $G'_{k,\kappa}$ has clustering at least κ. Moreover, the graph $G'_{4,\kappa}$ is planar for every $\kappa \ge 1$. Hence, the class of planar graphs does not have clustered chromatic number less than four.
- Every n-vertex triangle-free graph drawn on a surface Σ of Euler genus at most g has at most $2n + O(g)$ edges. Hence, we can apply Theorem 14.11 with $k = 3$, $\delta = 1/2$, and $n_0 = \Theta(g)$ and conclude that the class of triangle-free graphs drawn on Σ has clustered chromatic number at most 3. Thus, in the clustered setting, the Grötzsch theorem extends to all surfaces. Again, it is not hard to see

that three is the right value of the clustered chromatic number even for planar triangle-free graphs.

- Consider the class of K_{t+1}-minor-free graphs for $t \leq 8$. As we have seen in Sect. 14.1, the maximum number of edges of an n-vertex K_{t+1}-minor-free graph for these small values of t is less than $(t-1)n$. Hence, by Theorem 14.11 with $k = t$ and $\delta = n_0 = 1$, we see that their clustered chromatic number is at most t, giving us these first few cases of the clustered Hadwiger's theorem essentially for free.

Consider a proper minor-closed class $\mathcal{G}$ of graphs. The maximum chromatic number of graphs in $\mathcal{G}$ is certainly at least the size of the largest complete graph in $\mathcal{G}$. This is not the case for defective and clustered chromatic number; e.g., for any positive integer s, the class of graphs with all components of size at most s has clustered chromatic number one, but contains K_s. It is natural to ask whether there exists a characterization of the defective and clustered chromatic number for proper minor-closed classes. I.e., for a fixed positive integer k, can we describe a class $\mathcal{F}_k$ (resp. $\mathcal{F}'_k$) of graphs such that every proper minor-closed class $\mathcal{G}$ satisfies $\chi_\Delta(\mathcal{G}) \geq k$ (resp. $\chi_\star(\mathcal{G}) \geq k$) if and only if $\mathcal{F}_k \subseteq \mathcal{G}$ (resp. $\mathcal{F}'_k \subseteq \mathcal{G}$)?

For the defective chromatic number, this was completely answered in [10]: One can take $\mathcal{F}_k = \{G_{k,d} : d \in \mathbb{N}\}$, for the graphs $G_{k,d}$ discussed after Theorem 14.8. For the clustered chromatic number, the problem seems even more complicated. Such a characterization would of course imply that K_{t+1}-minor-free graphs have clustered chromatic number at most t, which was only proved recently and the proofs are not simple. However, see [12] for a partial result. A more detailed discussion of many other interesting topics relating to the defective and clustered coloring can be found in a survey by Wood [20].

References

1. N. Alon, M. Krivelevich, B. Sudakov, Complete minors and average degree: a short proof. J. Graph Theory **103**, 599–602 (2023)
2. J. Balogh, A. Kostochka, Large minors in graphs with given independence number. Discrete Math. **311**, 2203–2215 (2011)
3. M. Delcourt, L. Postle, Reducing linear Hadwiger's conjecture to coloring small graphs (2021). ArXiv, 2108.01633
4. V. Dujmović, L. Esperet, P. Morin, D.R. Wood, Proof of the clustered Hadwiger conjecture (2023). ArXiv, 2306.06224
5. Z. Dvořák, S. Norin, Islands in minor-closed classes. I. Bounded treewidth and separators. ArXiv, 1710.02727 (2017)
6. K. Edwards, D. Kang, J. Kim, S. Oum, P.D. Seymour, A relative of Hadwiger's conjecture. SIAM J. Discrete Math. **29**, 2385–2388 (2015)
7. L. Jørgensen, Contractions to K_8. J. Graph Theory **18**, 431–448 (1994)
8. A. Kostochka, Lower bound of the hadwiger number of graphs by their average degree. Combinatorica **4**, 307–316 (1984)
9. D. Kühn, D. Osthus, Minors in graphs of large girth. Random Struct. Algorithms **22**, 213–225 (2003)

10. C. Liu, Defective coloring is perfect for minors (2022). ArXiv, 2208.10729
11. W. Mader, Homomorphiesäze für Graphen. Math. Ann. **178**, 154–168 (1968)
12. S. Norin, A. Scott, P.D. Seymour, D.R. Wood, Clustered colouring in minor-closed classes. Combinatorica **39**, 1387–1412 (2019)
13. S. Norin, L. Postle, Z. Song, Breaking the degeneracy barrier for coloring graphs with no kt minor. Adv. Math. **422**, 109020 (2023)
14. N. Robertson, P.D. Seymour, R. Thomas, Hadwiger's conjecture for K_6-free graphs. Combinatorica **13**, 279–361 (1993)
15. Z. Song, R. Thomas, The extremal function for K_9 minors. J. Combin. Theory Ser. B **96**, 240–252 (2006)
16. A. Thomason, An extremal function for contractions of graphs. Math. Proc. Camb. Philos. Soc. **95**, 261–265 (1984)
17. A. Thomason, The extremal function for complete minors. J. Combin. Theory Ser. B **81**, 318–338 (2001)
18. J. van den Heuvel, D.R. Wood, Improper colourings inspired by Hadwiger's conjecture. J. Lond. Math. Soc. **98**, 129–148 (2018)
19. K. Wagner, Über eine Eigenschaft der ebenen Komplexe. Math. Ann. **114**, 570–590 (1937)
20. D.R. Wood, Defective and clustered graph colouring. Electron. J. Combin. **DS23** (2018)

Chapter 15
Product Structure

The Minor Structure Theorem combines several technical ingredients (graphs on surfaces, vortices, tree decompositions, ...), all of which are necessary to obtain an approximate description of graphs from proper minor-closed classes, in the sense that for every graph H, there exists a graph H' such that

(i) every H-minor-free graph has the described structure, and
(ii) every graph with this structure is H'-minor-free.

The resulting structure theorem is somewhat unwieldy and difficult to apply. However, could the result be simplified by sacrificing (ii), obtaining a structure theorem for a wider family of graphs, yet so that many of the properties (similarity to graphs on surfaces, sublinear separators, ...) are preserved? In this chapter, we investigate a recent approach along these lines, via embedding into products of simple basic graphs. Although the proof of the existence of this product description uses the Minor Structure Theorem, the resulting structural description is much simpler. Moreover, the properties exposed by the product structure description turned out to be crucial in resolving several long-standing problems for proper minor-closed classes.

The ***strong product*** $G \boxtimes H$ of graphs G and H is the graph with vertex set $V(G) \times V(H)$, with distinct vertices (u_1, v_1) and (u_2, v_2) adjacent if and only if $u_2 \in N_G[u_1]$ and $v_2 \in N_H[v_1]$, i.e., they are equal or adjacent in both coordinates. Thus, the strong product of two paths is the grid with all diagonals. A fuzzy idea about generic planar graphs might be that they are somewhat similar to such grids; e.g., in Corollary 2.34, we have seen that every planar graph is a minor of a sufficiently large grid. There are a number of issues with this intuition; for one, the strong product of two paths has maximum degree 8, while planar graphs can have unbounded maximum degree. But even if we restrict ourselves to planar graphs of maximum degree at most three, we run into a problem: Let $n_{G,r}[v]$ denote the number of vertices of a graph G at distance at most r from v. In the strong product Q of paths, we have $n_{Q,r}[v] \leq (2r+1)^2$, while the root x of a complete binary

© The Author(s), under exclusive license to Springer Nature Switzerland AG 2025

Z. Dvořák, *Graph Minors*, Springer Monographs in Mathematics,
https://doi.org/10.1007/978-3-031-87469-7_15

tree T of depth d satisfies $n_{T,r}[x] = 2^{r+1} - 1$ for every $r \leq d$. Hence, even very special planar graphs such as trees are not subgraphs of the strong product of two (or any bounded number of) paths. Interestingly, the superpolynomial growth of the neighborhoods is the only issue.

Theorem 15.1 (Krauthgamer and Lee [10]) *For all positive integers c and d, there exists an integer $t = O_c(d \log d)$ such that the following claim holds. If G is a graph such that $n_{G,r}[v] \leq cr^d$ for every vertex $v \in V(G)$ and every positive integer r, then G is a subgraph of the strong product of t paths.*

In this theorem, the paths of course have to be sufficiently long, say of length equal to the diameter of G. Rather surprisingly, Dujmović et al. [3] demonstrated that planar graphs also admit product structure; the issue with exponential growth of neighborhoods is addressed by including a tree-like factor in the product.

Theorem 15.2 (Dujmović et al. [3]) *For every planar graph G, there exists a path P and a graph H of treewidth at most 8 such that $G \subseteq P \boxtimes H$.*

The bound on the treewidth was subsequently improved to six [14]. Moreover, as we are going to discuss in the rest of the chapter, similar results were obtained for other graph classes, generally derived from ones avoiding minors or other obstructions.

15.1 Product Structure in Graphs on Surfaces

It will be more convenient to obtain the graph G as a subgraph of the strong product $P \boxtimes F \boxtimes K_a$ of three graphs, a path P, a graph F of bounded treewidth, and a clique with a vertices. This implies a result of form given in Theorem 15.2, since $\mathrm{tw}(F \boxtimes K_a) = a(\mathrm{tw}(F) + 1) - 1$. What is needed to show the existence of such a product representation? Let $\lambda : V(G) \to \mathbb{N}$ be a layering of vertices of G and let $\mathcal{Q}$ be a partition of vertices of G, where the parts do not necessarily induce connected subgraphs. The *λ-width* of the partition $\mathcal{Q}$ is the maximum of $|\lambda^{-1}(i) \cap Q|$ over $i \in \mathbb{N}$ and $Q \in \mathcal{Q}$.

Observation 15.3 *The following conditions are equivalent for graphs F and G and a positive integer a:*

- *$G \subseteq P \boxtimes F \boxtimes K_a$, where P is a path.*
- *There exists a layering λ and a partition $\mathcal{Q}$ of vertices of G of λ-width at most a such that $G/\mathcal{Q} \subseteq F$.*

Proof Suppose λ is a layering of G and $\mathcal{Q}$ is a partition of vertices of G of λ-width at most a such that $G/\mathcal{Q} \subseteq F$. Let $f : V(G) \to [a]$ be an arbitrary function injective on $\lambda^{-1}(i) \cap Q$ for every $i \in \mathbb{N}$ and $Q \in \mathcal{Q}$. For each $v \in V(G)$, let $q(v)$ be the vertex of $G/\mathcal{Q}$ resulting from the contraction of the part of $\mathcal{Q}$ containing v. Let $P = x_0 x_1 \ldots x_m$ be a path such that $\lambda^{-1}(i) = \emptyset$ for every $i > m$, and let $y_1, \ldots, y_a$

be the vertices of the clique of size a. Then the mapping of every vertex $v \in V(G)$ to $(x_{\lambda(v)}, q(v), y_{f(v)})$ shows that G is a subgraph of $P \boxtimes F \boxtimes K_a$.

Conversely, if $g : V(G) \to V(P \boxtimes F \boxtimes K_a)$ is an embedding of G as a subgraph of $P \boxtimes F \boxtimes K_a$, then the projection of g to the first coordinate gives a layering λ of G, the partition Q of $V(G)$ according to the second coordinate satisfies $G/Q \subseteq F$, and because there are only a possible values in the third coordinate, the λ-width of Q is at most a. □

Thus, we aim to reduce the treewidth by contracting subsets which are "orthogonal" to a layering λ, in the sense that each of them intersects each layer in a small number of points. In the basic case that λ is a BFS layering, this can be ensured by contracting paths in a BFS spanning tree. A useful property of this choice of the contracted paths is that they are geodesic. Indeed, such partitions in planar graphs were first developed by Pilipczuk and Siebertz [12] to achieve this property, and only later shown in [3] to lead to the product structure. Interestingly, the argument uses the following well-known result.

Lemma 15.4 (Sperner's Lemma) *Let G be a plane graph with the outer face bounded by a cycle C and all other faces of length three. Let $\varphi : V(G) \to [3]$ be a (not necessarily proper) coloring of G. If $C[\varphi^{-1}(i)]$ is a non-empty path for every $i \in [3]$, then G contains an internal face whose vertices are colored* 1, 2, *and* 3.

The desired result will be a consequence of the following lemma applied to a BFS spanning tree. A path in a rooted spanning tree T of a graph G is *vertical* if it is a subpath of a path from a leaf of T to the root. A subgraph H of G is *a-vertical* if for some $a' \leq a$, there exist pairwise vertex-disjoint vertical paths $P_1, \ldots, P_{a'} \subseteq H$ such that $V(H) = V(P_1) \cup \ldots \cup V(P_{a'})$. Note that if H is a path, then every subpath of H is also a-vertical. A set of vertices of G is *a-vertical* if it is the vertex set of an a-vertical subgraph.

Lemma 15.5 *Let G be a simple plane triangulation and let T be a rooted spanning tree of G. There exists a partition Q of $V(G)$ into* 3*-vertical parts such that* $\mathrm{tw}(G/Q) \leq 3$.

Proof Without loss of generality, we can assume that T is rooted in a vertex incident with the outer face of G. For a cycle C in G, let G_C denote the subgraph of G drawn in the closed disk bounded by C. We are going to prove the following claim by induction on the number of vertices of G_C:

($\star$) Let C be a cycle of G and let $X \subseteq E(C)$ be a non-empty set of size at most three such that each component of $C - X$ is 2-vertical. Let Q'_C be the partition of $V(C)$ to the vertex sets of components of $C - X$. Then there exists a partition $Q_C \supseteq Q'_C$ of $V(G_C)$ into 3-vertical parts such that $\mathrm{tw}(G_C/Q_C) \leq 3$.

The claim of the lemma then follows by applying ($\star$) to the triangle C bounding the outer face of G, with $X = E(C)$.

The claim is trivial in case that $G_C = C$, since then we can let $\mathcal{Q}_C = \mathcal{Q}'_C$ and let the tree decomposition consist of just a single bag. Hence, suppose that $G_C \neq C$. Moreover, we can assume that $|\mathcal{Q}'_C| = |X| = 3$: Since G is simple, C has length at least three. In case that $|X| \leq 2$, we can add arbitrary edges of C to X to increase its size to three, prove ($\star$), then merge the parts of $\mathcal{Q}'_C$ back. This does not increase the treewidth of $G_C/\mathcal{Q}_C$, since the merged parts form a clique.

Let P_1, P_2, and P_3 be the components of $C - X$. Consider a vertex $v \in V(G_C)$. Since the root r of T is incident with the outer face of G, the path from v to r in T intersects C; let P_v be the vertical path from v to the first intersection with C, and let $\varphi(v) \in [3]$ be the index such that the last vertex of P_v belongs to $P_{\varphi(v)}$. By Lemma 15.4, there exists a triangle $R = v_1v_2v_3$ bounding an internal face of G_C such that $\varphi(v_i) = i$ for $i \in [3]$.

For $i \in [3]$, let C_i be the cycle in $(C \cup P_{v_{i-1}} \cup P_{v_{i+1}} \cup v_{i-1}v_{i+1}) - V(P_{v_i})$ distinct from C, where $v_0 = v_3$ and $v_4 = v_1$; in case that $v_{i-1}v_{i+1}$ is an edge of C, let C_i denote this edge, instead. Let Q_i be the subpath of C_i formed by the edges not in C. Observe that $|X \cap E(C_i)| = 1$. Let X_i consist of the edge of $X \cap E(C_i)$ and of the first and the last edge of Q_i (which can be the same edge). Clearly each component of $C_i - X_i$ is 2-vertical, as it is either a subpath of P_1, P_2, or P_3, or a subpath of $P_{v_{i-1}} \cup P_{v_{i+1}} \cup v_{i-1}v_{i+1}$. Let $\mathcal{Q}'_{C_i}$ be the partition of $V(C_i)$ to the vertex sets of the components of $C_i - X_i$. If C_i is a cycle, then by the induction hypothesis, there exists a partition $\mathcal{Q}_{C_i} \supseteq \mathcal{Q}'_{C_i}$ of $V(G_{C_i})$ into 3-vertical parts such that $\mathrm{tw}(G_{C_i}/\mathcal{Q}_{C_i}) \leq 3$. Note that the vertices corresponding to $\mathcal{Q}'_{C_i}$ form a clique K_i in $G_{C_i}/\mathcal{Q}_{C_i}$. In case that C_i is the edge $v_{i-1}v_{i+1}$, we simply let $\mathcal{Q}_{C_i} = \mathcal{Q}'_{C_i} = \{\{v_{i-1}\}, \{v_{i+1}\}\}$.

Let $Q = (R \cup P_{v_1} \cup P_{v_2} \cup P_{v_3}) - V(C)$; clearly, the subgraph Q is 3-vertical. Let

$$\mathcal{Q}_C = \{V(Q)\} \cup \mathcal{Q}'_C \cup \bigcup_{i=1}^{3} (\mathcal{Q}_{C_i} \setminus \mathcal{Q}'_{C_i}).$$

Thus $\mathcal{Q}_C$ is a partition of $V(G_C)$ to 3-vertical parts. Let K be the clique in $G_C/\mathcal{Q}_C$ of size four formed by the vertices corresponding to $V(Q)$ and the sets in $\mathcal{Q}'_C$. Observe that $G_C/\mathcal{Q}_C$ is obtained from K and the graphs $G_{C_i}/\mathcal{Q}_{C_i}$ for $i \in [3]$ by clique-sums on the cliques K_i. Therefore,

$$\mathrm{tw}(G_C/\mathcal{Q}_C) \leq \max(\mathrm{tw}(K), \mathrm{tw}(G_{C_1}/\mathcal{Q}_{C_1}), \mathrm{tw}(G_{C_2}/\mathcal{Q}_{C_2}), \mathrm{tw}(G_{C_3}/\mathcal{Q}_{C_3})) = 3,$$

as required. □

If T is a BFS spanning tree of a connected graph G and λ is its BFS layering, then for each $i \in \mathbb{N}$, each vertical path in T intersects $\lambda^{-1}(i)$ in at most one vertex, and thus $|Q \cap \lambda^{-1}(i)| \leq a$ for every a-vertical set Q. Hence, Lemma 15.5 together with Observation15.3 give the following consequence.

Corollary 15.6 *Every simple planar graph G is a subgraph of the product $P \boxtimes F \boxtimes K_3$, where P is a path and F is a graph of treewidth at most three, and thus also of the product $P \boxtimes H$, where $H = F \boxtimes K_3$ is a graph of treewidth at most* 11.

Proof Without loss of generality, we can assume that G is a plane triangulation, as we can add edges to its faces to triangulate it otherwise. Let Q be the partition of $V(G)$ obtained by Lemma 15.5 applied with a BFS spanning tree T. Let λ be the layering of G according to the distance from the root of T. Then $F = G/Q$ has treewidth at most three, and since each part of Q is 3-vertical, the partition Q has λ-width at most three. The claim then follows from Observation 15.3. □

The bound of 8 rather than 11 in Theorem 15.2 follows by a slightly more careful analysis of the process to obtain the tree decomposition in Lemma 15.5: Recall that while Q is 3-vertical, each component of $C - X$ is 2-vertical. Hence, if we consider the refinement of Q_C where each part is broken into vertical parts, every bag will only consist of $3 + 2|X| \le 9$ parts.

Moreover, it is simple to generalize the result to graphs on surfaces.

Theorem 15.7 *Every simple graph G of Euler genus g is a subgraph of the product $P \boxtimes F \boxtimes K_a$, where P is a path, F is a graph of treewidth at most four, and $a =$* $\max(3, 2g)$. *Hence, it is also a subgraph of the product $P \boxtimes H$, where $H = F \boxtimes K_a$ is a graph of treewidth less than* $5\max(3, 2g)$.

Proof Again, without loss of generality, we can assume that G is a triangulation of a surface Σ of Euler genus g (we need to be a bit careful, since just adding edges to triangulate the faces could result in parallel edges; however, we can also add vertices to G to deal with this issue). Let T_0 be a BFS spanning tree of G and λ the layering of G according to the distance from the root of T_0.

We use the idea of interdigitating trees that we discussed in Sect. 2.12: Let $G^\star$ be the dual graph to G and let A be the set of the edges of the dual that correspond to the edges of T_0. Then $G^\star - A$ is connected, and thus it has a spanning tree T_1. Let Z be the set of edges $e \in E(G) \setminus E(T_0)$ whose duals do not belong to $E(T_1)$, and recall that $|Z| = g$. Let Z' be the set of vertices incident with the edges of Z.

Let us cut the surface Σ along the edges of the subgraph $R = T_0 + Z$. The resulting surface Δ is obtained from the triangular faces of G by gluing them along the edges whose duals belong to the tree T_1, and thus Δ is a closed disk. The boundary of Δ corresponds to a closed walk in G traversing each edge of R twice. Let us now perform the following reduction: If R has a vertex v of degree one different from the root of T_0 (necessarily $v \notin Z'$), then the two appearances of the incident edge e in the boundary of Δ are necessarily consecutive and traversed in opposite directions. We delete v from R and glue the two segments of the boundary of Δ corresponding to e. Let us repeat this reduction as long as possible.

In the end, the resulting graph R' is the union of Z and of the vertical paths in T_0 from Z' to the root, and cutting Σ along R' still turns the surface into a disk Δ'. The graph $G - V(R')$ is drawn in the disk Δ', and thus it can be drawn in the plane so that all vertices adjacent in G to the vertices of R' are incident with the outer face of $G - V(R')$.

Let T be the tree obtained from T_0 by contracting R' to a single root vertex r. Let G' be a simple plane triangulation obtained from $G - V(R')$ by adding r and the incident edges of T and then triangulating the non-triangular faces; then T is a spanning tree of G'. By Lemma 15.5, there exists a partition Q' of $V(G')$ into 3-vertical (with respect to T) parts such that $\text{tw}(G'/Q') \leq 3$. Let Q be obtained from Q' by removing r from the part that contains it; then Q is a partition of $V(G) \setminus V(R')$ to 3-vertical (with respect to T_0) parts such that $\text{tw}((G - V(R'))/Q) \leq 3$.

Let Q_0 be obtained from Q by adding the part $V(R')$. Note that $T_0[V[R']]$ is the union of $|Z'| \leq 2g$ vertical paths in T_0, and thus it is $2g$-vertical. The graph G/Q_0 is obtained from $(G - V(R'))/Q$ by adding a single vertex, and thus $\text{tw}(G/Q_0) \leq 4$.

Since λ is the BFS-layering of T_0, Q_0 has λ-width at most $\max(3, 2g)$. The result then follows from Observation 15.3. □

15.2 Beyond Embedded Graphs

Going one step further along the lines of the Minor Structure Theorem, it is not hard to see that adding vortices still preserves the existence of the product structure.

Lemma 15.8 *For all non-negative integers g, d, and m, there exists an integer t such that every graph G drawn on a surface of Euler genus g up to m vortices of width at most d is a subgraph of the product $P \boxtimes H$, where P is a path and H is a graph of treewidth at most t.*

Proof Let us give just a quick outline: Let G_0 be the surface part of G and let F_1, …, F_m be the vortices. Without loss of generality, we can assume that ∂F_i is a cycle in G_0 for $i \in [m]$. Let G_1 be obtained from G_0 by adding a vertex r adjacent to the vertices in $\bigcup_{i=1}^{m} \partial F_i$; then G_1 has Euler genus at most $g + 2m$. Let λ be the BFS layering of G_1 according to the distance from r. The argument from the proof of Theorem 15.7 gives us a partition Q_1' of $V(G_1)$ of λ-width $O(g + m)$ such that $\text{tw}(G_1/Q_1') \leq 4$. By the argument from the proof of Observation 15.3, this also implies that there exists a partition Q_1 of $V(G_1)$ of λ-width one such that $\text{tw}(G_1/Q_1) = O(g + m)$.

Let Q_0 be obtained from Q_1 by removing r from the part that contains it. Let Q be obtained from Q_0 by, for each $i \in [m]$, adding the vertices of $V(F_i) \setminus \partial F_i$ as singleton parts. We also extend λ to G by letting $\lambda(v) = 0$ for every $v \in \bigcup_{i=1}^{m} (V(F_i) \setminus \partial F_i)$. Then Q has λ-width $O(g + m)$. Moreover, G/Q is obtained from G_0/Q_0 by adding the vortices. Note that for each $i \in [m]$, we have $\partial F_i \subseteq \lambda^{-1}(1)$, and since Q_0 has λ-width one, the cycle ∂F_i can be viewed as a subgraph of G_0/Q_0. As in the proof of Lemma 7.5, we conclude that $\text{tw}(G/Q) = O(d\,\text{tw}(G_0/Q_0)) = O(d(g + m))$. The claim then follows from Observation 15.3. □

Moreover, it is also possible to handle clique-sums, based on the following lemma.

Lemma 15.9 *Let G and H be graphs such that $G \subseteq P \boxtimes H$ for a path P. Let K be a clique in G. For any function $\mu : V(K) \to \{0, 1\}$ and any partition $\mathcal{Q}_K$ of $V(K)$, there exists a layering λ of G extending μ and a partition $\mathcal{Q} \supseteq \mathcal{Q}_K$ of $V(G)$ of λ-width at most* $\max(4, |K|)$ *such that* $\text{tw}(G/\mathcal{Q}) \leq \text{tw}(H) + |K|$.

Proof Since $G \subseteq P \boxtimes H$, Observation 15.3 implies that there exists a layering λ_0 of G and a partition $\mathcal{Q}_0$ of $V(G)$ of λ_0-width one such that $G/\mathcal{Q}_0 = H$. Since K is a clique, there exists a non-negative integer k such that $V(K) \subseteq \lambda_0^{-1}(\{k, k+1\})$. Let

$$\lambda(v) = \begin{cases} 0 & \text{if } v \in \mu^{-1}(0) \\ 1 & \text{if } v \in \lambda_0^{-1}(\{k-1, k, k+1, k+2\}) \setminus \mu^{-1}(0) \\ i \geq 2 & \text{if } v \in \lambda_0^{-1}(\{k-i, k+1+i\}). \end{cases}$$

Note that λ is a layering of G extending μ, and since each layer of λ is contained in at most four layers of λ_0, the partition $\mathcal{Q}_0$ has λ-width at most four.

Let $\mathcal{Q}$ be obtained from $\mathcal{Q}_0$ by removing the vertices of K from the parts that contain them and adding $\mathcal{Q}_K$. The λ-width of $\mathcal{Q}$ clearly is at most $\max(4, |K|)$. Moreover, $G/\mathcal{Q}$ is obtained from a subgraph of $G/\mathcal{Q}_0 = H$ by adding $|\mathcal{Q}_K| \leq |K|$ vertices, and thus $\text{tw}(G/\mathcal{Q}) \leq \text{tw}(H) + |K|$. □

We say that a graph G has a *clique-friendly* (path $\boxtimes$ tw$_k$ $\boxtimes K_a$)*-structure* if for every clique K in G, every function $\mu : V(K) \to \{0, 1\}$ and every partition $\mathcal{Q}_K$ of $V(K)$, there exists a layering λ of G extending μ and a partition $\mathcal{Q} \supseteq \mathcal{Q}_K$ of $V(G)$ of λ-width at most a such that $\text{tw}(G/\mathcal{Q}) \leq k$. Lemma 15.9 has the following consequence.

Corollary 15.10 *Let G be a graph and t a positive integer such that $G \subseteq P \boxtimes H$ for a path P and a graph H of treewidth at most t. Then G has clique-friendly (path $\boxtimes$ tw$_{3t+2}$ $\boxtimes K_{2t+2}$)-structure.*

Proof For any clique K in G, there exists a vertex $v \in V(P)$ such that $v \boxtimes H$ contains at least half of the clique K. The subgraph $v \boxtimes H$ is isomorphic to H, and thus $\frac{1}{2}|K| \leq \text{tw}(H) + 1 \leq t + 1$. The claim then follows from Lemma 15.9. □

As usual, the point of introducing such a technical notion is that it is exactly preserved by clique-sums.

Lemma 15.11 *Suppose that G_1 and G_2 are graphs with a clique-friendly (path $\boxtimes$ tw$_k$ $\boxtimes K_a$)-structure. If G is a clique-sum of G_1 and G_2, then G also has a clique-friendly (path $\boxtimes$ tw$_k$ $\boxtimes K_a$)-structure.*

Proof Let K be any clique in G, and consider any function $\mu : V(K) \to \{0, 1\}$ and any partition $\mathcal{Q}_K$ of $V(K)$. By symmetry, we can assume that K is contained in G_1. Hence, there exists a layering λ_1 of G_1 extending μ and a partition $\mathcal{Q}_1 \supseteq \mathcal{Q}_K$ of $V(G_1)$ of λ_1-width at most a such that $\text{tw}(G_1/\mathcal{Q}_1) \leq k$.

Let K' be the common clique of G_1 and G_2 on which we perform the clique-sum to obtain G. Let b be the minimum of λ_1 on K', and let $\mu'(v) = \lambda_1(v) - b$ for every $v \in V(K')$. Note that $\mu'(v) \in \{0, 1\}$ for every $v \in V(K')$, since the clique K' is necessarily contained in two consecutive layers of the layering λ_1. Let $Q_{K'}$ be the partition of $V(K')$ according to Q_1. There exists a layering λ_2 of G_2 extending μ' and a partition $Q_2 \supseteq Q_{K'}$ of $V(G_2)$ of λ_2-width at most a such that $\operatorname{tw}(G_2/Q_2) \le k$. Then $Q = Q_1 \cup (Q_2 \setminus Q_{K'}) \supseteq Q_K$ is a partition of $V(G)$. Moreover, G/Q is a clique-sum of G_1/Q_1 and G_2/Q_2 on the clique $K'/Q_{K'}$, and thus $\operatorname{tw}(G/Q) \le k$.

Let $\lambda(v) = \lambda_1(v)$ for $v \in V(G_1)$ and $\lambda(v) = \lambda_2(v) + b$ for $v \in V(G_2)$; note that both definitions coincide for $v \in V(K')$. Then λ is a layering of G extending μ, and clearly Q has λ-width at most a. □

Putting the results together, the existence of the product structure is preserved by clique-sums.

Corollary 15.12 *Let t be a positive integer. Let G be a clique-sum of graphs G_1, $\dots$, G_m. Suppose that for $i \in [m]$, the graph G_i is a subgraph of the product $P_i \boxtimes H_i$, where P_i is a path and H_i is a graph of treewidth at most t. Then $G \subseteq P \boxtimes H$ for a path P and a graph H of treewidth at most* $6(t+1)^2$.

Proof By Corollary 15.10, G_i has a clique-friendly (path $\boxtimes$ $\operatorname{tw}_{3t+2} \boxtimes K_{2t+2}$)-structure for each $i \in [m]$. By Lemma 15.11, so does G. By Observation 15.3, this implies that $G \subseteq P \boxtimes F \boxtimes K_{2t+2}$ for a path P and a graph F of treewidth at most $3t + 2$. The claim follows by letting $H = F \boxtimes K_{2t+2}$. □

Theorem 10.5 together with Lemma 15.8 and Corollary 15.12 then implies that apex-minor-free graphs have product structure.

Corollary 15.13 *For every apex graph F, there exists an integer t such that every F-minor-free graph is a subgraph of the product $P \boxtimes H$ for a path P and a graph H of treewidth at most t.*

Unfortunately, this claim does not extend to F-minor-free graphs for non-apex graphs F. Note that if $G \subseteq \text{path} \boxtimes H$ and G has a universal vertex, then $G \subseteq P_3 \boxtimes H$, and thus $\operatorname{tw}(G) \le 3\operatorname{tw}(H) + 2$. Thus, a sufficiently large "grid plus one vertex" graph G is not a subgraph of path $\boxtimes$ H for any graph H of bounded treewidth, and such graphs G are F-minor-free for every graph F with $a(F) \ge 2$. Still, we can at least formulate the relaxation of the Minor Structure Theorem with simplified basic pieces. We say that a graph G has ((path $\boxtimes$ tw_t) + K_a)*-structure* if there exists a set $A \subseteq V(G)$ of size at most a such that $G - A \subseteq P \boxtimes H$ for a path P and a graph H of treewidth at most t.

Theorem 15.14 *For every graph H, there exist integers a and t such that every H-minor-free graph can be expressed as a clique-sum of graphs of* ((*path* $\boxtimes$ tw_t) + K_a)*-structure.*

Let us remark that while the product structure does not deal well with apex vertices, it is quite robust otherwise. E.g., graphs that are drawn on a fixed surface still can be expressed as a product of a path and bounded treewidth graph if we allow a bounded number of crossings on each edge [5]. This and many other results follow from the fact that the product structure is preserved by taking shallow minors.

Theorem 15.15 (Hickingbotham and Wood [9]) *Suppose G is a subgraph of the product $P \boxtimes H \boxtimes K_a$ for a path P and a graph H. Let G' be obtained from G by contracting pairwise vertex-disjoint subgraphs, each of radius at most r. Then G' is a subgraph of the product $P' \boxtimes H' \boxtimes K_{a(2r+1)^2}$, where P' is a path and H' is a graph of treewidth less than*

$$\binom{\mathrm{tw}(H) + 2r + 1}{2r + 1}.$$

As an example, consider a graph F drawn in the plane with at most one crossing on each edge. Let F_0 be the planar graph obtained from F by replacing the crossings by vertices of degree four, and let G be obtained from F_0 by adding twins of these vertices of degree four. Then a supergraph of F can be obtained from G by contracting stars centered on the vertices of F. Since F_0 is planar, it is a subgraph of $P \boxtimes H \boxtimes K_3$ for a path P and a graph H of treewidth at most 11 by Corollary 15.6. Thus, $G \subseteq F_0 \boxtimes K_2 \subseteq P \boxtimes H \boxtimes K_6$, and by Theorem 15.15 applied with $r = 1$, we conclude that F is a subgraph of the strong product of a path, a graph of treewidth less than $\binom{14}{3}$, and K_{54}.

Moreover, many geometric graph classes (nearest neighbor graphs, bounded degree string graphs, ...) admit product structure [5]. Given this, one might wonder how far can we push this notion, especially in combination with apex vertices and clique-sums. As we are going to see in the following section, product structure implies sublinear separators. Could all classes with (sufficiently) sublinear separators, possibly subject to other natural constraints, admit the kind of decomposition described in Theorem 15.14? One way to see that the answer is negative and to reveal further limitations is through the *comparable box dimension*.

For a graph G, a *box representation in* $\mathbb{R}^d$ is a function ρ mapping vertices of G to axis-aligned boxes in $\mathbb{R}^d$ with pairwise disjoint interiors, such that $uv \in E(G)$ if and only if the boxes $\rho(u)$ and $\rho(v)$ intersect (in their boundaries, i.e., the boxes just touch). Let us remark that quite complicated graphs, even ones without sublinear separators, have a box representation in bounded dimension. For example, if G is obtained from a bipartite graph by subdividing every edge once, then G has a box representation in $\mathbb{R}^3$: The vertices of one part A of the bipartition are represented by narrow boxes that are long in the x-axis direction, the vertices of the other part B by narrow boxes that are long in the y-axis direction and almost touch all the boxes for A, and the vertices created by subdividing the edges are represented by small cubes touching the boxes of the adjacent vertices of A and B.

Thus, let us further restrict the representation: We say that two axis-aligned boxes B_1 and B_2 are *comparable* if a translation of one is a subset of the other one, i.e., if one of the boxes is at most as wide as the other one in all axis directions. A box representation ρ of a graph G is *comparable* if the boxes $\rho(u)$ and $\rho(v)$ are comparable for every $u, v \in V(G)$. The ***comparable box dimension*** of a graph G is the minimum integer d such that G has a comparable box representation in $\mathbb{R}^d$. As shown in [7], graphs of comparable box dimension at most d have $O(n^{1-1/(2d+4)})$-separators. Moreover, every graph that can be expressed as a clique-sum of graphs of ((path $\boxtimes$ tw_t) + K_a)-structure for fixed t and a has bounded comparable box dimension [6]. Finally, Dvořák et al. [7] gave for every function $f(n) = o(n)$ an example of a hereditary class of graphs with $f(n)$-separators but unbounded comparable box dimension, and thus not admitting product structure, even in the sense of Theorem 15.14.

15.3 Applications

We obviously lose a lot of information by replacing the condition of being nearly drawn on a surface by the condition of being a subgraph of the product of a path and a bounded treewidth graph, as in Theorem 15.14. However, for many applications, this is actually an advantage, since it makes it possible to focus on the important properties rather than getting lost in the details.

As a simple example, let us note that product structure easily implies the result of Chap. 7: Every proper minor-closed class is treewidth-fragile. Indeed, we have the following simple observation.

Observation 15.16 *Let G be a subgraph of the product $P \boxtimes H$, where P is a path and the graph H has treewidth at most t. Then for every positive integer k, the graph G has a $((t+1)(k-1)+1)$-low-treewidth coloring by k colors.*

Proof Clearly, it suffices to show that this is the case when $G = P \boxtimes H$. Let $P = v_1 v_2 \ldots v_m$, and let $\gamma(v_i, u) = (i \bmod k) + 1$ for every vertex $(v_i, u) \in V(P \boxtimes H)$. Then for each $i \in [k]$, each component of $G - \gamma^{-1}(i)$ is isomorphic to the product $P' \boxtimes H$ for a path P' with at most $k-1$ vertices, and thus $\mathrm{tw}(G - \gamma^{-1}(i)) \leq (t+1)(k-1)+1$. It follows that γ is a $((t+1)(k-1)+1)$-low-treewidth coloring of G. □

As shown in Sect. 7, treewidth-fragility is preserved by adding a bounded number of apices and taking clique-sums. Note that a graph with ((path $\boxtimes$ tw_t) + K_a)-structure has clique number at most $2(t+1)+a$. Using Observations 7.3 and 7.9 and Lemma 7.11, we get the following conclusion.

Corollary 15.17 *The class of graphs that can be expressed as a clique-sum of graphs of ((path $\boxtimes$ tw_t) + K_a)-structure is treewidth-fragile, with bounding function $f(k) = (t+1)(k+1) + 2a + 1$.*

By Theorem 15.14, this is a strengthening of Theorem 7.12. Furthermore, as we have seen in Sect. 7.2, Corollary 15.17 implies that clique-sums of graphs of ((path $\boxtimes$ tw_t) + K_a)-structure have $O_{a,t}(\sqrt{n})$-separators.

More importantly, the product structure was used to prove results for planar graphs and other minor-closed classes that were not known before. In their paper introducing this tool, Dujmović et al. [3] proved that they have bounded queue number, solving a long-standing open question of Heath et al. [8]. They also used them to bound track numbers and volume of 3-dimensional drawings of such graphs. The product structure tools were later used to obtain space-efficient representations [4] and to bound nonrepetitive chromatic number [2].

To showcase an application, let us consider the last example: A coloring φ of a graph G is *nonrepetitive* if there is no path $v_1 v_2 \dots v_{2m}$ in G such that $\varphi(v_i) = \varphi(v_{m+i})$ for $i \in [m]$. In particular, by considering the case $m = 1$, this coloring must be proper. The ***nonrepetitive chromatic number*** of G is the minimum number of colors needed in a nonrepetitive coloring of G. Thue [13] proved a statement on words that can be rephrased as saying that paths have nonrepetitive chromatic number at most three. More generally, Alon et al. [1] proved that a graph of maximum degree Δ has nonrepetitive chromatic number $O(\Delta^2)$. They also asked whether the nonrepetitive chromatic number of planar graphs is bounded, a natural question that was resolved almost 20 years later using the product structure.

As we are going to see, it is not too hard to prove that graphs of bounded treewidth have bounded nonrepetitive chromatic number. An issue that we need to overcome is that nonrepetitive chromatic number is not preserved by products: Stars have nonrepetitive chromatic number two, but $K_{1,n} \boxtimes K_{1,n}$ contains as a subgraph $K_{n,n}$, a graph of nonrepetitive chromatic number $n + 1$. As usual, we deal with this issue by considering more restrictive notions. A walk $v_1 \dots v_{2m}$ in a graph G is a *stroll* if $v_i \neq v_{m+i}$ for every $i \in [m]$, and it is *relevant* if there exists $i \in [m]$ such that $v_i \neq v_{m+i}$. Thus, all strolls are relevant. Consider a coloring φ of G; a walk is *repetitively colored* if $\varphi(v_i) = \varphi(v_{m+i})$ for every $i \in [m]$. Clearly, every irrelevant walk is repetitively colored. A coloring of G is ***stroll-nonrepetitive*** if no stroll is repetitively colored. A coloring is ***walk-nonrepetitive*** if no relevant walk is repetitively colored. The *stroll-nonrepetitive (resp. walk-nonrepetitive) chromatic number* of a graph is the minimum number of colors in a stroll-nonrepetitive (resp. walk-nonrepetitive) coloring. Clearly, stroll-nonrepetitive coloring is also nonrepetitive. Moreover, walk-nonrepetitivity is a substantially stronger notion than stroll-nonrepetitivity; e.g., $K_{1,n}$ has stroll-nonrepetitive chromatic number two and walk-nonrepetitive chromatic number $n + 1$. Indeed, the two notions relate as follows.

Observation 15.18 *A stroll-nonrepetitive coloring φ of a graph G is walk-nonrepetitive if and only if $\varphi(u) \neq \varphi(v)$ for all distinct vertices $u, v \in V(G)$ with a common neighbor.*

Proof Suppose that φ is walk-nonrepetitive. If u and v are distinct vertices with a common neighbor x, then $uxvx$ is a relevant walk, and thus $\varphi(u) \neq \varphi(v)$.

Conversely, suppose that φ is stroll-nonrepetitive and $\varphi(u) \neq \varphi(v)$ for all distinct vertices $u, v \in V(G)$ with a common neighbor. Consider any walk $W = v_1 \dots v_{2m}$ colored repetitively by φ. Since φ is stroll-nonrepetitive, W is not a stroll, and thus there exists $i \in [m]$ such that $v_i = v_{m+i}$. We claim that W is irrelevant: Otherwise, there exists $j \in [m]$ such that $v_j \neq v_{m+j}$. By choosing the indices i and j with these properties so that $|i - j|$ is minimum, we can assume that $|i - j| = 1$. However, then v_j and v_{m+j} are distinct vertices with a common neighbor $v_i = v_{m+i}$ and $\varphi(v_j) = \varphi(v_{m+j})$, which is a contradiction. Hence, φ is repetitive only on irrelevant walks. □

In particular, walk-nonrepetitive chromatic number can be bounded only for graphs of bounded maximum degree. The definitions of stroll- and walk-nonrepetitive colorings are motivated by the fact that they behave well with respect to strong products. To show this, a further technical observation will be useful. A *lazy walk* in a graph G is a sequence $v_1 \ldots v_t$ of its vertices such that $v_{i+1} \in N_G[v_i]$ for $i \in [t-1]$, i.e., v_{i+1} is either equal or adjacent to v_i. The repetitivity of a coloring and the relevance is defined for lazy walks in the natural way, and a *lazy stroll* is a lazy walk $v_1 \ldots v_{2m}$ such that $v_i \neq v_{m+i}$ for every $i \in [m]$.

Observation 15.19 *Let φ be a coloring of a graph G. If the coloring φ is stroll-nonrepetitive (resp. walk-nonrepetitive), then it is also nonrepetitive on all lazy strolls (resp. relevant lazy walks).*

Proof Assume for now only that φ is a proper coloring of G, and let $W = v_1 \ldots v_{2m}$ be any lazy walk in G on which φ is repetitive. Suppose that there exists $i \in [2m-1]$ such that $v_i = v_{i+1}$.

- Let us first consider the case that $i \neq m$, and by symmetry assume that $i < m$. Since φ is repetitive on W, we have $\varphi(v_{i+m}) = \varphi(v_i) = \varphi(v_{i+1}) = \varphi(v_{i+m+1})$. Since φ is a proper coloring, we also have $v_{i+m} = v_{i+m+1}$. Let W' be the lazy walk obtained from W by removing v_i and v_{m+i} (but keeping v_{i+1} and v_{i+m+1}). Clearly φ is repetitive on W', and W' is relevant or a lazy stroll if and only if W is.
- In case that $i = m$, i.e., $v_m = v_{m+1}$, let W_1 be the lazy walk obtained from W by removing v_m and v_{2m} and let W_2 be the one obtained by removing v_1 and v_{m+1}. Note that φ is repetitive on W_1 and W_2, and if W is a lazy stroll, then so are W_1 and W_2. Moreover, if W is relevant, then W_1 or W_2 is relevant.

By repeating these reductions, we conclude that if φ is repetitive on a lazy stroll or lazy relevant walk, then it is also repetitive on a non-lazy one. □

We can now show the result on products.

Lemma 15.20 *Let φ_1 be a walk-nonrepetitive coloring of a graph P and φ_2 a stroll-nonrepetitive coloring of a graph H. For $(u, v) \in V(P) \times V(H)$, let $\varphi(u, v) = (\varphi_1(u), \varphi_2(v))$. Then φ is a stroll-nonrepetitive coloring of $P \boxtimes H$.*

Proof Consider any walk $W = (u_1, v_1) \ldots (u_{2m}, v_{2m})$ in $P \boxtimes H$ on which φ is repetitive. Then φ_1 is repetitive on the lazy walk $u_1 \ldots u_{2m}$, and since φ_1 is walk-nonrepetitive, this lazy walk is necessarily irrelevant, i.e., $u_i = u_{i+m}$ for every $i \in [m]$. Moreover, φ_2 is repetitive on the lazy walk $v_1 \ldots v_{2m}$, and since φ_2 is stroll-nonrepetitive, this lazy walk is not a lazy stroll. Hence, there exists $i \in [m]$ such that $v_i = v_{i+m}$, and consequently $(u_i, v_i) = (u_{i+m}, v_{i+m})$. Therefore, W is not a stroll. We conclude that φ is stroll-nonrepetitive. □

Let us note that although the notion of stroll-nonrepetitivity is somewhat convoluted, it is necessary to make this argument work.

Thus, to obtain results for graphs with product structure, we need to bound the walk-nonrepetitive chromatic number of paths and the stroll-nonrepetitive chromatic number of graphs of bounded treewidth. For the former, Kündgen and Pelsmajer [11] proved that every path has a walk-nonrepetitive coloring using four colors, and at least four colors are needed for paths with at least six vertices. For the latter, let us apply another useful tool, *shadow-complete layerings*.

Consider a layering λ of a graph G, and let Q be a component of $G[\lambda^{-1}(\{j \in \mathbb{N} : j \geq i\})]$ for a positive integer i. The *shadow* of Q is the subgraph of G induced by the vertices in $\lambda^{-1}(i-1)$ with a neighbor in Q. We say that λ is ***shadow-complete*** if all shadows are cliques. An important example of shadow-complete layerings arises in the context of chordal graphs.

Lemma 15.21 *Let G be a connected chordal graph and let λ be a BFS layering of G. Then λ is shadow-complete. Moreover, $\omega(G[\lambda^{-1}(i)]) < \omega(G)$ for every positive integer i.*

Proof Let v_0 be the vertex of G such that $\lambda(v)$ is the distance between v_0 and v in G for every $v \in V(G)$.

Let i be a non-negative integer and let $K \subseteq G[\lambda^{-1}(i)]$ be a shadow. Therefore, for all distinct $u, v \in V(K)$, there exists a path from u to v in $G[\lambda^{-1}(\{j \in \mathbb{N} : j \geq i\})]$ intersecting $\lambda^{-1}(i)$ exactly in its ends. Let P_1 be a shortest such path. Since λ is the BFS layering according to the distance from v_0, the subgraph $G[\lambda^{-1}(\{0, \ldots, i-1\})$ is connected and u and v both have neighbors in this subgraph; let P_2 be a shortest path from u to v in $G[\lambda^{-1}(\{0, \ldots, i\})]$ intersecting $\lambda^{-1}(i)$ exactly in its ends. Then P_1 and P_2 are induced paths. If u and v were not adjacent, then $P_1 \cup P_2$ would be an induced cycle in G of length at least four, contradicting the assumption that G is chordal. Therefore, K is a clique. It follows that λ is shadow-complete.

Suppose now for a contradiction that $\omega(G[\lambda^{-1}(i)]) = \omega(G)$ for a positive integer i, i.e., the subgraph $G[\lambda^{-1}(i)]$ contains a clique K of size $\omega(G)$. Let $y \in \lambda^{-1}(i-1)$ be a vertex with the largest number of neighbors in K. Since $|K| = \omega(G)$, there exists a vertex $v \in V(K)$ non-adjacent to y. Let z be a neighbor of v in $\lambda^{-1}(i-1)$. Since z has at most as many neighbors in K as y, there exists a neighbor $u \in V(K)$ of y non-adjacent to z. Let P be a shortest path between y and z in $G[\lambda^{-1}(\{0, \ldots, i-1\})]$ intersecting $\lambda^{-1}(i-1)$ only in its ends. Then $P + yuvz$ is an induced cycle in G of length at least four, which is a contradiction. □

Let us now show how shadow-complete layerings can be used to obtain stroll-nonrepetitive colorings.

Lemma 15.22 *Let λ be a shadow-complete layering of a graph G such that for every $i \in \mathbb{N}$, $G[\lambda^{-1}(i)]$ has a stroll-nonrepetitive coloring φ_i using c colors. Then the stroll-nonrepetitive chromatic number of G is at most* $4c$.

Proof Let t be the maximum index such that $\lambda^{-1}(t)$ is non-empty. As we have discussed above, there exists a walk-nonrepetitive coloring ψ of the path $P = 012\ldots t$

using four colors. For each $v \in V(G)$, let us define $\varphi(v) = (\psi(\lambda(v)), \varphi_{\lambda(v)}(v))$, so that φ uses at most $4c$ colors.

Consider a walk $W = v_1 \ldots v_{2m}$ in G on which φ is repetitive. Then $\psi \circ \lambda$ is repetitive on W, and since ψ is a walk-nonrepetitive coloring, the walk $\lambda(v_1) \ldots \lambda(v_{2m})$ must be irrelevant. That is, $\lambda(v_i) = \lambda(v_{i+m})$ for $i \in [m]$. Let s be the minimum of $\lambda(v_i)$ for $i \in [2m]$, and let $i_1 < \ldots < i_{2k}$ be the indices such that $\lambda(v_{i_j}) = s$ for $j \in [2k]$; note that $i_{j+k} = i_j + m$ for every $j \in [k]$. For $j \in [2k-1]$, v_{i_j} and $v_{i_{j+1}}$ are connected by a walk in G with internal vertices contained in $\lambda^{-1}(\{s+1, \ldots, t\})$, and thus v_{i_j} and $v_{i_{j+1}}$ are contained in the same shadow. Since λ is shadow-complete, $W' = v_{i_1} \ldots v_{i_{2k}}$ is a lazy walk in $G[\lambda^{-1}(s)]$. Since $i_{j+k} = i_j + m$ for every $j \in [k]$ and φ is repetitive on W, it is also repetitive on W'. Hence, φ_s is repetitive on W'. Since φ_s is stroll-nonrepetitive, W' is not a stroll, and thus there exists $j \in [k]$ such that $v_{i_j} = v_{i_{j+k}} = v_{i_j+m}$. Therefore, W also is not a stroll, and we conclude that φ is stroll-nonrepetitive. □

Combining Lemmas 15.21 and 15.22, a straightforward inductive argument gives the following conclusion.

Corollary 15.23 *Every chordal graph G has stroll-nonrepetitive chromatic number at most* $4^{\omega(G)}$.

Since every graph of treewidth t is a subgraph of a chordal graph of clique number at most $t+1$, this immediately gives the desired claim about graphs of bounded treewidth.

Corollary 15.24 *Every graph G has stroll-nonrepetitive chromatic number at most* $4^{\mathrm{tw}(G)+1}$.

In combination with Lemma 15.20, this gives the basic fact about graphs with product structure.

Corollary 15.25 *If G is a subgraph of the product* $P \boxtimes H$ *for a path P and a graph H, then G has stroll-nonrepetitive chromatic number at most* $4^{\mathrm{tw}(H)+2}$.

By Corollary 15.6, Theorem 15.7, and Corollary 15.13, this immediately gives upper bounds on the nonrepetitive chromatic number of planar graphs, graphs on surfaces, and apex-minor-free graphs. Moreover, adding apex vertices colored by unique colors preserves stroll-nonrepetitiveness.

Corollary 15.26 *Every graph of* $((\text{path} \boxtimes \mathrm{tw}_t) + K_a)$*-structure has stroll-nonrepetitive chromatic number at most* $4^{t+2} + a$.

To deal with clique-sums, we again leverage shadow-complete layerings, through the following useful lemma. Let (T, β) be a tree decomposition of a graph G; a tree decomposition (T', β') of an induced subgraph G' of G is *dominated* by (T, β) if there exists an injective function $\iota : V(T') \to V(T)$ such that $\beta'(x') = \beta(\iota(x')) \cap V(G')$ for every $x' \in V(T')$. Note that this implies that the torsos of (T', β') are subgraphs of the torsos of (T, β).

Lemma 15.27 *Let G be a connected graph and let (T, β) be a tree decomposition of G of adhesion $b \geq 1$. There exists a layering λ of G such that*

- *for every shadow S, there exist distinct nodes $x, y \in V(T)$ such that $V(S) \subseteq \beta(x) \cap \beta(y)$, and*
- *for every $i \in \mathbb{N}$, $G[\lambda^{-1}(i)]$ has a tree decomposition of adhesion less than b dominated by (T, β).*

Proof Without loss of generality, we can assume that $\beta(x) \not\subseteq \beta(y)$ and $\beta(y) \not\subseteq \beta(x)$ for every distinct $x, y \in V(T)$: Otherwise, there exists an edge $xy \in E(T)$ such that $\beta(x) \subseteq \beta(y)$, and we can instead consider the smaller tree decomposition of adhesion at most b dominated by (T, β) which is obtained by contracting the edge xy, keeping only the bag of y.

Let us choose an arbitrary vertex r of T and let us define $\lambda(v) = 0$ for all vertices $v \in \beta(r)$. For $i = 1, 2, \ldots$, let $\lambda(v) = i$ for every vertex $v \in V(G)$ for which we have not yet selected the layer and for which there exists a node $x \in V(T)$ such that $v \in \beta(x)$ and $\beta(x) \cap \lambda^{-1}(i-1) \neq \emptyset$. Since G is connected, this procedure assigns a layer to every vertex of G. Note that each bag of (T, β) is contained in at most two consecutive layers of λ, and since every edge of G has both ends in one of the bags, it follows that λ is indeed a layering of G.

For a non-negative integer i, let T_i be the subgraph of T induced by the nodes $x \in V(T)$ such that $\max(\lambda(\beta(x))) \leq i$. Observe that T_i is a subtree of T containing r, and that $\lambda^{-1}(\{0, 1, \ldots, i\}) = \bigcup_{x \in V(T_i)} \beta(x)$. Consider any component Q of $G - \lambda^{-1}(\{0, 1, \ldots, i\})$. Since Q is connected and disjoint from $\bigcup_{x \in V(T_i)} \beta(x)$, there exists a component T_Q of $T - V(T_i)$ such that $V(Q) \subseteq \bigcup_{x \in V(T_Q)} \beta(x)$. Let $y \in V(T_Q)$ be the unique node of T_Q with a neighbor $x \in V(T_i)$. Then $\beta(x) \cap \beta(y)$ separates Q from the rest of $\lambda^{-1}(\{0, 1, \ldots, i\})$, and thus the shadow of Q is contained in $G[\beta(x) \cap \beta(y)]$.

Let us also note the following property:

($\star$) Let x be a node of T different from r and let x' be the neighbor of x on the path to r. For every vertex $v \in \beta(x) \setminus \beta(x')$, there exists a vertex $v' \in \beta(x) \cap \beta(x')$ such that $\lambda(v') < \lambda(v)$.

Indeed, let T_0 be the component of $T - xx'$ containing x (and not containing r) and let v_0 be the vertex in $\bigcup_{y \in V(T_0)} \beta(y) \setminus \beta(x')$ with $\lambda(v_0)$ minimum. In particular, $\lambda(v_0) \leq \lambda(v)$. Since $v_0 \notin \beta(x')$, all nodes whose bag contains v_0 belong to T_0. It follows that $\lambda(v_0) > 0$, as otherwise we would have $v_0 \in \beta(r)$. Thus, the definition of λ implies that there exists a node $x_0 \in V(T_0)$ with $v_0 \in \beta(x_0)$ and a vertex $v' \in \beta(x_0)$ such that $\lambda(v') = \lambda(v_0) - 1$. By the minimality of $\lambda(v_0)$, we conclude that $v' \in \beta(x')$. Since the node x is on the path between x' and x_0, $v' \in \beta(x) \cap \beta(x')$ is a vertex with $\lambda(v') < \lambda(v_0) \leq \lambda(v)$, as desired.

Let us now consider the tree decomposition (T, β_i) of $G[\lambda^{-1}(i)]$ defined by letting $\beta_i(x) = \beta(x) \cap \lambda^{-1}(i)$ for each $x \in V(T)$. Clearly, (T, β_i) is dominated by (T, β). This tree decomposition does not necessarily have adhesion less than b; however, as we are going to show next, if $|\beta_i(x) \cap \beta_i(y)| = b$ for distinct $x, y \in V(T)$, then there exists a node $z \in V(T)$ such that $\beta_i(x) \cup \beta_i(y) \subseteq \beta_i(z)$. Hence, after contracting the edges of the tree decomposition (T, β_i) with comparable bags, no nodes x and y with this property remain, and thus the resulting tree decomposition has adhesion less than b.

It remains to prove this property. If $i = 0$, this is clearly the case, since all vertices of $\lambda^{-1}(0)$ are contained in $\beta(r)$. Hence, suppose that $i > 0$. Let z be the node of T nearest to r such that $\beta(x) \cap \beta(y) \subseteq \beta(z)$. Such a node z exists, since all nodes on the path between x and y in T have this property. Note that z separates the nodes x and y from r. Suppose for a contradiction that say $\beta_i(x) \not\subseteq \beta_i(z)$, and thus there exists a vertex $v \in \beta_i(x) \setminus \beta_i(z)$. In particular, $z \neq x$. Let x' be the neighbor of x on the path to r (or equivalently, to z) in T. Note that $\beta(x) \cap \beta(y) \subseteq \beta(x')$. Since $|\beta(x) \cap \beta(x')| \leq b$ and $|\beta_i(x) \cap \beta_i(y)| = b$, we have $\beta(x) \cap \beta(x') = \beta_i(x) \cap \beta_i(y)$, and thus $\lambda(v') = i$ for every $v' \in \beta(x) \cap \beta(x')$. This contradicts $(\star)$. □

A tree decomposition (T, β) of a graph G is *adhesion-complete* if $G[\beta(x) \cap \beta(y)]$ is a clique for all distinct $x, y \in V(T)$. For such a decomposition, the layering from Lemma 15.27 is shadow-complete. By Lemma 15.22, we obtain the following consequence.

Corollary 15.28 *Let (T, β) be a tree decomposition of a graph G of adhesion $b \geq 0$. If every torso of this tree decomposition has stroll-nonrepetitive chromatic number at most c, then G has stroll-nonrepetitive chromatic number at most $4^b c$.*

Proof We prove the claim by induction on b. If $b = 0$, then G is the disjoint union of graphs $G[\beta(x)]$ for $x \in V(T)$, and each of the graphs has stroll-nonrepetitive chromatic number at most c by the assumptions. Hence, suppose that $b > 0$.

We can add to G a clique on $\beta(x) \cap \beta(y)$ for all distinct $x, y \in V(T)$, since that does not change the torsos of the decomposition. Hence, we can assume that (T, β) is adhesion-complete. Moreover, we can assume that G is connected, as otherwise we can consider each component of G separately. By Lemma 15.22, there exists a shadow-complete layering λ of G such that for each $i \in \mathbb{N}$, the subgraph $G[\lambda^{-1}(i)]$ has a tree decomposition of adhesion at most $b - 1$ whose torsos are subgraphs of the torsos of (T, β). By the induction hypothesis, $G[\lambda^{-1}(i)]$ has stroll-nonrepetitive chromatic number at most $4^{b-1}c$. The claim then follows by Lemma 15.22. □

Note that the assumptions of Corollary 15.28 are satisfied whenever G is a clique-sum of graphs of stroll-nonrepetitive chromatic number at most c and clique number at most b. Hence, Corollary 15.26 has the following consequence.

Corollary 15.29 *Every graph obtained from graphs with ((path $\boxtimes$ tw_t) + K_a)-structure by clique-sums has stroll-nonrepetitive chromatic number at most*

$$4^{2t+2+a}(4^{t+2} + a).$$

Theorem 15.14 now gives the desired conclusion for proper minor-closed classes.

Corollary 15.30 *For every graph H, there exist a constant c_H such that every H-minor-free graph has (stroll-)nonrepetitive chromatic number at most c_H.*

References

1. N. Alon, J. Grytczuk, M. Hałuszczak, O. Riordan, Nonrepetitive colorings of graphs. Random Struct. Algorithms **21**, 336–346 (2002)
2. V. Dujmović, L. Esperet, G. Joret, B. Walczak, D.R. Wood, Planar graphs have bounded nonrepetitive chromatic number. Adv. Combin. **5** (2020)
3. V. Dujmović, G. Joret, P. Micek, P. Morin, T. Ueckerdt, D.R. Wood, Planar graphs have bounded queue-number. J. ACM **67**, 1–38 (2020)
4. V. Dujmović, L. Esperet, C. Gavoille, G. Joret, P. Micek, P. Morin, Adjacency labelling for planar graphs (and beyond). J. ACM **68**, 1–33 (2021)
5. V. Dujmović, P. Morin, Wood, D.R. (2023). Graph product structure for non-minor-closed classes. J. Combin. Theory Ser. B **162**, 34–67 (2023)
6. Z. Dvořák, D. Gonçalves, A. Lahiri, J. Tan, T. Ueckerdt, On comparable box dimension, in *38th international symposium on computational geometry (SoCG 2022)*. Leibniz International Proceedings in Informatics (LIPIcs), vol. 224 (Schloss Dagstuhl – Leibniz-Zentrum für Informatik, Wadern, 2022), pp. 38:1–38:14
7. Z. Dvořák, R. McCarty, S. Norin, Sublinear separators in intersection graphs of convex shapes. SIAM J. Discrete Math. **35**, 1149–1164 (2021)
8. L. Heath, F. Leighton, A. Rosenberg, Comparing queues and stacks as machines for laying out graphs. SIAM J. Discrete Math. **5**, 398–412 (1992)
9. R. Hickingbotham, D.R. Wood, Shallow minors, graph products and beyond planar graphs (2021). ArXiv, 2111.12412
10. R. Krauthgamer, J. Lee, The intrinsic dimensionality of graphs. Combinatorica **27**, 551–585 (2007)
11. A. Kündgen, M. Pelsmajer, Nonrepetitive colorings of graphs of bounded tree-width. Discrete Math. **308**, 4473–4478 (2008)
12. M. Pilipczuk, S. Siebertz, Polynomial bounds for centered colorings on proper minor-closed graph classes. J. Combin. Theory Ser. B **151**, 111–147 (2021)
13. A. Thue, Über unendliche Zeichenreihen. Norske Vid. Selsk. Skr. I. Mat. Nat. Kl. Christiania **7**, 1–22 (1906)
14. T. Ueckerdt, D.R. Wood, W. Yi, An improved planar graph product structure theorem. Electron. J. Combin. **29**, P2.51 (2022)

Chapter 16
Iterated Layerings

We have already seen a number of times that considering a bounded number of layers in a suitably chosen layering (often simply a BFS layering) of a graph reveals important properties useful both in theoretical arguments and in design of algorithms. Of course, a major limitation is that every layering of a graph with a universal vertex (or more generally, bounded radius) has a small number of layers, and thus the amount of information that we can gain in this way is limited. In a sense, as we are going to see in Sect. 16.1, if we iterate BFS layering in a graph avoiding a fixed minor, we always eventually run into this limitation. However, this fact by itself turns out to have applications in areas such as graph flows and metric embeddings.

A way around this limitation is to intersperse the layering actions with removal of "central" vertices. It is not hard to see that for every graph H and every positive integer r, any H-minor-free graph G can be reduced to nothing by a bounded number of vertex removals and restrictions to r consecutive layers of a layering. An issue reducing the applicability of this result is that the number of iterations turns out to be substantially larger than r, and thus actually a majority of the graph may turn out to be thrown away in the layering steps of the process. Somewhat surprisingly, this can be circumvented by increasing r in each iteration. The resulting decomposition enables us to generalize Baker's powerful method for design of approximation algorithms to all proper minor-closed classes.

16.1 Iterated BFS Layerings

Let G be a connected graph, and for positive integers k and r, let $G = G_1, \ldots, G_{k+1}$ be a sequence of induced subgraphs of G, such that for $i \in [k]$, G_{i+1} is a connected component of a subgraph induced by the union of r consecutive layers in a BFS layering λ_i of G_i. In this case, we say that G_{k+1} is obtained from G by *k-iterated*

© The Author(s), under exclusive license to Springer Nature Switzerland AG 2025
Z. Dvořák, *Graph Minors*, Springer Monographs in Mathematics,
https://doi.org/10.1007/978-3-031-87469-7_16

r-*layering*. A key observation of Klein et al. [6] is that in a $K_{k,k}$-minor-free graph, the diameter of $V(G_{k+1})$ in G is bounded. Let us emphasize that the distance is measured in the whole graph G, rather than in the subgraph G_{k+1}.

Theorem 16.1 *If G_{k+1} is obtained from a $K_{k,k}$-minor-free connected graph G by k-iterated r-layering, then the distance between any vertices of G_{k+1} in G is at most $6k^2r$.*

Proof Suppose for a contradiction that G_{k+1} contains vertices u and v at distance more than $6k^2r$ in G. Let P be a path between u and v in G_{k+1}, and for $j \in [k]$, let v_j be a vertex of P at distance $6(j-1)kr$ from u in G. Observe that such vertices exist, since the distances of consecutive vertices of P differ by at most one. By the triangle inequality, we have $d_G(v_{j_1}, v_{j_2}) \geq 6kr$ for all distinct $j_1, j_2 \in [k]$.

For $i \in [k]$, let G_i and λ_i be as in the definition of k-iterated r-layering. Let T_i be a BFS spanning tree of G_i corresponding to λ_i. Let d_i be the index such that G_{i+1} is a subgraph of $G_i[\lambda_i^{-1}(\{d_i, \ldots, d_i + r - 1\})]$. For $i = k, k-1, \ldots, 0$, we are going to construct a model μ_i of $K_{k,k-i}$ in G such that the following conditions hold, denoting by $X = \{x_1, \ldots, x_k\}$ and Y_i the two parts of $K_{k,k-i}$:

(a) For each vertex $y \in Y_i$, $\mu_i(y) \subseteq G_{i+1}$.
(b) If $i > 0$, then for each $j \in [k]$, $\mu_i(x_j)$ is a subgraph of G_i consisting of vertices at distance (in G_i) less than $3r(k+1-i)$ from v_j. Moreover, $\mu_i(x_j)$ consists of a subgraph of G_{i+1} and a vertical path $P_{i,j}$ in T_i ending in a vertex $b_{i,j}$ with $\lambda_i(b_{i,j}) = d_i - 2r$.

The construction is started for $i = k$ by letting $\mu_k(x_j)$ for each $j \in [k]$ consist of the vertical path $P_{k,j}$ in T_k starting in v_j and ending in the vertex $b_{k,j}$ with $\lambda_k(b_{k,j}) = d_k - 2r$. The condition (b) is satisfied, since these paths have length less than $3r$.

Consider now i such that $0 \leq i \leq k-1$. The model μ_i is obtained as follows, see Fig. 16.1 for an illustration. Let Y_i be obtained from Y_{i+1} by adding a new vertex y_i, and let $\mu_i(y) = \mu_{i+1}(y)$ for every $y \in Y_{i+1}$ and let $\mu_i(y_i) = T_{i+1}[\lambda_{i+1}^{-1}(\{0, \ldots, d_{i+1} - 2r - 1\})]$. By (a) and (b), the vertices of the subgraphs forming μ_{i+1} are contained in $V(G_{i+2}) \subseteq \lambda_{i+1}^{-1}(\{d_{i+1}, \ldots, d_{i+1} + r - 1\})$ or in $\lambda_{i+1}^{-1}(\{d_{i+1} - 2r, d_{i+1} - 2r + 1, \ldots\})$, and thus they are vertex-disjoint from $\mu_i(y_i)$. Moreover, for each $j \in [k]$, the vertex $b_{i+1,j} \in V(\mu_{i+1}(x_j))$ has a neighbor in $\mu_i(y_i)$.

If $i = 0$, the construction is finished by letting $\mu_0(x_j) = \mu_1(x_j)$ for each $j \in [k]$. If $i > 0$, for each $j \in [k]$, we obtain $\mu_i(x_j)$ as follows. Let $m_{i,j}$ be the vertex of $P_{i+1,j}$ with $\lambda_{i+1}(m_{i,j}) = d_{i+1} - r$. Let $\mu_i(x_j)$ consist of $\mu_{i+1}(x_j)$ and the vertical path $P_{i,j}$ in T_i starting in $m_{i,j}$ and ending in the vertex $b_{i,j}$ with $\lambda_i(b_{i,j}) = d_i - 2r$. Note that this path has length less than $3r$, and since $\mu_{i+1}(x_j)$ consist of vertices at distance less than $3r(k-i)$ from v_i, it follows that $\mu_i(x_j)$ consists of vertices at distance less than $3r(k+1-i)$ from v_j. Since $d_G(v_{j_1}, v_{j_2}) \geq 6kr$ for distinct $j_1, j_2 \in [k]$, this ensures that $\mu_i(x_{j_1})$ and $\mu_i(x_{j_2})$ are vertex-disjoint. Moreover, note that for each $j \in [k]$, the path $P_{i,j}$ intersects $\lambda_i^{-1}(\{d_i, \ldots, d_i + r - 1\}) \supseteq V(G_{i+1})$

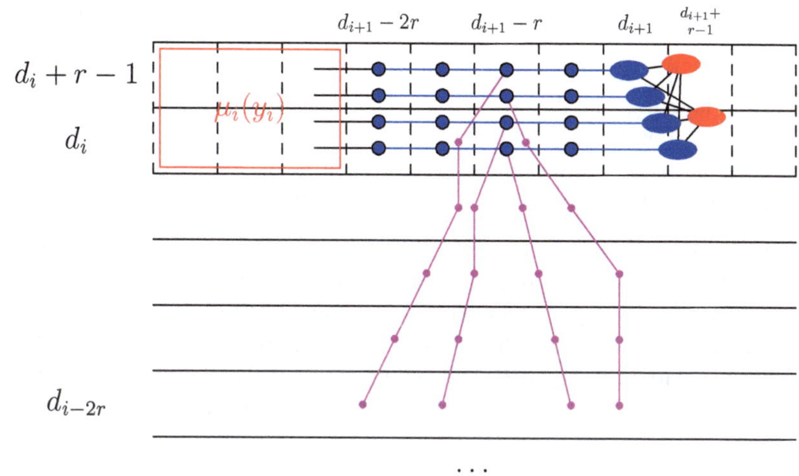

Fig. 16.1 The construction from the proof of Theorem 16.1, with $r = 2$ and $k = 4$. The subgraphs $\mu_{i+1}(Y_{i+1})$ are shown in red, the subgraphs $\mu_{i+1}(X)$ in blue. The paths added in $\mu_i(X)$ are drawn in purple

at most in the first r vertices, and thus this intersection is contained in layers of λ_{i+1} at distance at most $r-1$ from $m_{i,j}$, i.e., in $\lambda_{i+1}^{-1}(\{d_{i+1}-2r+1,\ldots,d_{i+1}-1\})$. Hence, $\mu_i(x_j)$ is also disjoint from $\mu_i(y)$ for all $y \in Y_i$.

For $i = 0$, we conclude that $K_{k,k}$ is a minor of G, which is a contradiction. □

The fact that the diameter in Theorem 16.1 is linear in r gives us an interesting result even in the case of graphs with non-uniform edge lengths. Given a graph G and an assignment ℓ of non-negative lengths to its edges, let $d_{G,\ell}(u, v)$ denote the ℓ-distance between the vertices u and v of G, i.e., the minimum of $\sum_{e\in E(P)} \ell(e)$ over all paths P in G between u and v.

Corollary 16.2 *Let $k \geq 3$ be an integer, let G be a $K_{k,k}$-minor-free graph and let $\ell : E(G) \to \mathbb{N}$ be an assignment of lengths to edges of G. Let $L = \sum_{e\in E(G)} \ell(e)$. For every $r \geq 1$, there exists a set X of at most kL/r edges of G such that for every component K of $G - X$, all vertices $u, v \in V(K)$ satisfy $d_{G,\ell}(u, v) \leq 6k^2r$.*

Proof Let G' be the graph obtained from G by contracting all edges of length 0 and for every other edge e, subdividing the edge $(\ell(e) - 1)$ times. Note that $\|G'\| = L$ and that G' is also $K_{k,k}$-minor-free. Clearly, it suffices to find a set X' at most $k\|G'\|/r$ edges of G' such that for every component K' of $G' - X'$, all vertices $u, v \in V(K')$ satisfy $d_{G'}(u, v) \leq 6k^2r$; then, we can let X consist of the edges of G such that the corresponding paths in G' intersect X'.

Let $G_0 = G'$. For $i = 1,\ldots,k$, proceed as follows. Choose a function $\lambda_i : V(G_{i-1}) \to \mathbb{N}$ so that the restriction of λ_i to each component of G_{i-1} is a BFS layering. Choose an integer $b \in [r]$ uniformly at random and let X_i be the set of edges $uv \in E(G_{i-1})$ such that $\lambda(u) \equiv b \pmod r$ and $\lambda(v) \equiv b+1 \pmod r$. Let $G_i = G_{i-1} - X_i$.

Let $X = \bigcup_{i=1}^{k} X_i$. Then each component of $G' - X$ is obtained from G' by k-iterated r-layering, and thus by Theorem 16.1, we have $d_{G'}(u, v) \leq 6k^2 r$ for any vertices u and v in the same component. Moreover, for each edge $e \in E(G')$, the probability that $e \in X_i$ is at most $1/r$ for each i, and thus $\mathrm{E}[|X|] \leq k\|G'\|/r$. Consequently, we have $|X| \leq k\|G'\|/r$ with non-zero probability. □

A basic example of how this result can be applied comes in the setting of multicommodity flows. For simplicity of presentation, let us consider the uniform-demand uniform-capacity case. Let G be a graph and let $\mathcal{P}(G)$ be the set of all paths in G with at least one edge. The paths $P \in \mathcal{P}(G)$ are directed, i.e., they have a distinguished starting vertex $s(P)$ and ending vertex $t(P)$. A ***multicommodity flow*** in G is a function $f : \mathcal{P}(G) \to \mathbb{R}_0^+$ such that for every edge e,

$$\sum_{P \in \mathcal{P}(G), e \in E(P)} f(P) \leq 1.$$

The interpretation is that over each path P starting in a vertex u and ending in a vertex v, the amount $f(P)$ of a commodity is transported from u to v, where the commodities for distinct pairs (u, v) are different, and the amount transported over each edge cannot exceed its capacity 1. For distinct $u, v \in V(G)$, let

$$f(u, v) = \sum_{P \in \mathcal{P}(G), s(P)=u, t(P)=v} f(P)$$

be the total amount of the commodity transferred from u to v. We aim to maximize the *throughput* of f, defined as

$$\min_{u, v \in V(G), u \neq v} f(u, v).$$

I.e., we want to maximize the amount a that can be transferred between every pair of vertices simultaneously. Thus, the problem corresponds to the following linear program:

$$\begin{aligned}
x_P &\geq 0 && \text{for } P \in \mathcal{P}(G) \\
\sum_{P \in \mathcal{P}(G), e \in E(P)} x_P &\leq 1 && \text{for } e \in E(G) \\
\sum_{P \in \mathcal{P}(G), s(P)=u, t(P)=v} x_P &\geq a && \text{for distinct } u, v \in V(G) \\
&\text{maximize } a
\end{aligned}$$

Note that the program has superpolynomial number of variables, and thus we cannot solve it straightforwardly. Let us instead consider the dual program:

$$\ell_e \geq 0 \qquad \text{for } e \in E(G)$$

$$d_{u,v} \geq 0 \qquad \text{for distinct } u, v \in V(G)$$

$$\sum_{e \in E(P)} \ell_e \geq d_{s(P),t(P)} \qquad \text{for } P \in \mathcal{P}$$

$$\sum_{\text{distinct } u,v \in V(G)} d_{u,v} = 1$$

$$\text{minimize } \sum_{e \in E(G)} \ell_e$$

The last two conditions can be replaced by minimizing

$$\frac{\sum_{e \in E(G)} \ell_e}{\sum_{\text{distinct } u,v \in V(G)} d_{u,v}}.$$

That is, over all length assignments $\ell : E(G) \to \mathbb{R}_0^+$, letting $L = \sum_{e \in E(G)} \ell(e)$, we want to maximize

$$T = \frac{1}{L} \sum_{u,v \in V(G)} d_{G,\ell}(u, v),$$

and the maximum throughput is then equal to $1/T$. Let us remark that in this dual form, the linear program can be solved in polynomial time using the methods that only require a separation oracle, rather than an explicit representation of the program.

We are interested in obstructions to a large-throughput multicommodity flow. A natural obstruction comes from cuts: For a non-empty set $U \subsetneq V(G)$, let $e_G(U, \overline{U})$ denote the number of edges of G with exactly one end in U, and let the *ratio* of U be defined as

$$t(U) = \frac{e_G(U, \overline{U})}{|U| \cdot |(|G| - |U|)}.$$

Observe that any multicommodity flow in G has throughput at most $\frac{1}{2}t(U)$. A *minimum-ratio* cut in G is one with $t(U)$ minimum. In the single-commodity case, the size of a maximum flow and the capacity of a minimum cut coincide. Analogously, it is natural to ask how far apart the maximum throughput and the minimum cut ratio can be in the multicommodity case. Leighton and Rao [7] proved that in general n-vertex graphs, they can differ by a factor of at most $O(\log n)$. This bound cannot be improved: If G is a 3-regular graph, then each vertex is at distance $\Omega(\log n)$ from $\Omega(n)$ other vertices and yet G has $O(n)$ edges, implying that the maximum throughput is $O\left(\frac{1}{n\log n}\right)$. Moreover, if G is an expander, then $t(U) = \Theta(1/n)$ for every non-empty $U \subsetneq V(G)$.

However, Corollary 16.2 enables us to reduce this factor to $O(k^3)$ in $K_{k,k}$-minor-free graphs.

Theorem 16.3 *Let $k \geq 3$ be an integer, let G be a $K_{k,k}$-minor free graph with at least two vertices, and let a be the maximum throughput of a multicommodity flow in G. Then there exists a non-empty set $U \subsetneq V(G)$ such that $t(U) \leq 384k^3a$.*

Proof Let $n = |G|$. Suppose for a contradiction that $t(U) > 384k^3a$ for every non-empty $U \subsetneq V(G)$. In particular, if $|U| \leq n/2$, then

$$e_G(U, \overline{U}) = t(U)|U|(n - |U|) \geq t(U)|U|n/2 > 192k^3an|U|. \tag{16.1}$$

Let v be a vertex of G of minimum degree; for $U = \{v\}$, we have $e_G(U, \overline{U}) = \deg v \leq \frac{2\|G\|}{n}$, and together with (16.1), this gives

$$a < \frac{\|G\|}{96k^3n^2}. \tag{16.2}$$

Let $\ell : E(G) \to \mathbb{R}_0^+$ be a length assignment such that, letting $L = \sum_{e\in E(G)} \ell(e)$, we have $\frac{1}{L}\sum_{u,v\in V(G)} d_{G,\ell}(u, v) = \frac{1}{a}$. Without loss of generality, we can assume that $L = \|G\|$, since we can scale all lengths by the same factor. Let $\ell'(e) = \lceil \ell(e)\rceil$ for every edge e; then $L' = \sum_{e\in E(G)} \ell'(e) < L + \|G\| = 2\|G\| = 2L$, and $L' \geq L$. Consequently,

$$\frac{1}{2a} = \frac{1}{2L}\sum_{u,v\in V(G)} d_{G,\ell}(u, v) \leq \frac{1}{L'}\sum_{u,v\in V(G)} d_{G,\ell'}(u, v).$$

Let $r' = \frac{L'}{48k^2an^2}$; by (16.2), $r' > 2k > 1$. Let us apply Corollary 16.2 for ℓ' and $r = \lfloor r'\rfloor$. Thus, there exists a set $X \subseteq E(G)$ of size at most kL'/r such that each component of $G - X$ has diameter at most $6k^2r \leq 6k^2r'$ in the $d_{G,\ell'}$ metric. Let $\mathcal{U}$ be the system of the vertex sets of the components of $G - X$.

Suppose first that $|U| \leq n/2$ for every $U \in \mathcal{U}$. Using (16.1), we obtain

$$|X| = \frac{1}{2}\sum_{U\in\mathcal{U}} e_G(U, \overline{U}) > 96k^3an\sum_{U\in\mathcal{U}} |U| = 96k^3an^2.$$

However, by the choice of r', we have

$$|X| \leq \frac{kL'}{r} \leq \frac{2kL'}{r'} = 96k^3an^2.$$

This is a contradiction. Hence, there exists a unique set $U_0 \in \mathcal{U}$ of size more than $n/2$. Let a vertex $v_0 \in U_0$ be chosen arbitrarily. Note that

$$\begin{aligned}\frac{L'}{2a} &\le \sum_{u,v\in V(G)} d_{G,\ell'}(u,v) \le \sum_{u,v\in V(G)} (d_{G,\ell'}(v_0,u) + d_{G,\ell'}(v_0,v)) \\ &= 2n \sum_{v\in V(G)} d_{G,\ell'}(v_0,v),\end{aligned}$$

and thus

$$\sum_{v\in V(G)} d_{G,\ell'}(v_0,v) \ge \frac{L'}{4an}.$$

Since U_0 has diameter at most $6k^2r'$ in the $d_{G,\ell'}$ metric, we have

$$\begin{aligned}\frac{L'}{4an} &\le \sum_{v\in V(G)} d_{G,\ell'}(v_0,v) \le 6k^2r'n + \sum_{v\in V(G)} d_{G,\ell'}(U_0,v) \\ &= \frac{L'}{8an} + \sum_{v\in V(G)} d_{G,\ell'}(U_0,v),\end{aligned}$$

and thus

$$\sum_{v\in V(G)} d_{G,\ell'}(U_0,v) \ge \frac{L'}{8an}.$$

For $i \ge 1$, let U_i be the set of vertices of G at $d_{G,\ell'}$-distance at least i from U_0; we have $|U_i| \le n - |U_0| < n/2$. Using (16.1), we conclude that

$$\frac{L'}{8an} \le \sum_{v\in V(G)} d_{G,\ell'}(U_0,v) = \sum_{i\ge 1} |U_i| < \frac{1}{192k^3an} \sum_{i\ge 1} e_G(U_i,\overline{U_i}).$$

However, note that an edge $e \in E(G)$ has exactly one end in U_i for at most $\ell'(e)$ values of i, and thus

$$\sum_{i\ge 1} e_G(U_i,\overline{U_i}) \le \sum_{e\in E(G)} \ell'(e) = L'.$$

This is a contradiction. □

Observe that the proof of Theorem 16.3 can be turned into an algorithm that in polynomial time returns an $O(k^3)$-approximation of the minimum-ratio cut. An interesting consequence is that we can approximate the minimum size of a balanced edge cut in a $K_{k,k}$-minor-free graph, subject to an additional relaxation in the balance of the cut. Let β be a real number such that $\frac{1}{2} \le \beta \le 1$ and let U be a set of vertices of an n-vertex graph G. If $(1-\beta)n \le |U| \le \beta n$, then the set of edges of G with exactly one end in U is a *β-balanced edge-cut* in G. Let $\mathrm{bec}_\beta(G)$ denote the minimum size of a β-balanced edge-cut in G.

Corollary 16.4 *There exists a polynomial-time algorithm that, given a positive integer k and a $K_{k,k}$-minor-free graph G, returns a $\frac{3}{4}$-balanced edge-cut in G of size at most $O(k^3) \cdot \mathrm{bec}_{2/3}(G)$.*

Proof Let $n = |G|$, $s = \mathrm{bec}_{2/3}(G)$, and let U be a set of vertices of G such that $\frac{n}{3} \le |U| \le \frac{2n}{3}$ and $e(U, \overline{U}) = s$. Since we can replace U by $V(G) \setminus U$, we can assume that $|U| \le \frac{n}{2}$. Consider any induced subgraph G' of G with $n' \ge \frac{3}{4}n$ vertices, and let $U' = V(G') \cap U$. Note that $|U'| \ge |U| - \frac{n}{4} \ge \frac{n}{12}$ and $n' - |U'| \ge \frac{3n}{4} - \frac{n}{2} = \frac{n}{4}$. Thus,

$$t_{G'}(U') = \frac{e_{G'}(U', \overline{U'})}{|U'|(n' - |U'|)} \le \frac{e_G(U, \overline{U})}{n^2/48} = \frac{48s}{n^2}.$$

Hence, the maximum throughput of a multicommodity flow in G' is at most $\frac{1}{2}t_{G'}(U') \le \frac{24s}{n^2}$. Consequently, the algorithm based on Theorem 16.3 returns for any such induced subgraph G' an edge-cut of ratio at most ck^3s/n^2, where $c = 9216$.

Let us proceed as follows. Let $G_1 = G$. For $i = 1, 2, \ldots$, if $|G_i| < \frac{3n}{4}$, then stop and let $m = i$. Otherwise, let $U_i \subsetneq V(G_i)$ be a set with $t_{G_i}(U_i) \le ck^3s/n^2$, which we can find in polynomial time as described above. Since we could exchange U_i for its complement, we can assume that $|U_i| \le |G_i|/2$. Let $G_{i+1} = G_i - U_i$.

At the end of the process, let $R = V(G) \setminus V(G_m) = \bigcup_{i=1}^{m-1} U_i$. Since $|G_m| < \frac{3n}{4}$, we have $|R| \ge \frac{n}{4}$. On the other end, $|G_{m-1}| \ge \frac{3n}{4}$ and

$$|R| = |V(G) \setminus V(G_{m-1})| + |U_{m-1}| \le n - \tfrac{1}{2}|G_{m-1}| \le \tfrac{5n}{8}.$$

Hence, the edges with exactly one end in R form a $\frac{3}{4}$-balanced edge-cut in G. Moreover,

$$\begin{aligned} e_G(R, \overline{R}) &\le \sum_{i=1}^{m-1} e_{G_i}(U_i, \overline{U_i}) \le \sum_{i=1}^{m-1} t_{G_i}(U_i)|U_i|n \\ &\le \sum_{i=1}^{m-1} \frac{ck^3s}{n}|U_i| = \frac{ck^3s|R|}{n} \le ck^3s, \end{aligned}$$

as required. □

Let us remark that using similar (though somewhat more involved) ideas, one can also approximate the minimum size of a balanced vertex cut [4, 5].

16.2 Baker-amenable Classes

Layerings where the unions of small numbers of consecutive layers have restricted structure (e.g., bounded treewidth) are useful in design of approximation algorithms using Baker's technique [1]. Let us illustrate this technique on the domination number. Actually, both to showcase the power of this technique and for a later convenience, let us consider a technical generalization: For an integer s, an instance of the *s-request dominating set problem* is a triple $(G, M, \mathcal{R})$, where G is a graph, $M \subseteq V(G)$, and $\mathcal{R}$ consists of at most s subsets of $V(G)$. The goal is to find the minimum size $\gamma(G, M, \mathcal{R})$ of a set $S \subseteq V(G)$ such that

- each vertex of $M \setminus S$ has a neighbor in S, and
- for each $R \in \mathcal{R}$, $R \cap S \neq \emptyset$.

Thus, the domination number $\gamma(G)$ of a graph G is equal to $\gamma(G, V(G), \emptyset)$.

Lemma 16.5 *Let s be an integer and $\varepsilon \geq 0$ a real number, and let $\mathcal{G}_0$ be a class of graphs such that there exists a polynomial-time $(1+\varepsilon)$-approximation algorithm for s-request dominating set problem for graphs from $\mathcal{G}_0$. Let $r \geq 3$ be an integer and let λ be a layering of a graph G such that the subgraph of G induced by any r consecutive layers belongs to $\mathcal{G}_0$. Then we can in polynomial time for any s-request dominating set problem instance $(G, M, \mathcal{R})$ return a solution of size at most $(1 + \frac{2}{r-2})(1+\varepsilon)\gamma(G, M, \mathcal{R})$.*

Proof Let $n = |G|$, and let m be the largest integer such that $\lambda^{-1}(m) \neq \emptyset$; clearly, we can assume that $m \leq rn$, since any gap of more than $r-1$ empty layers can be shrunk without affecting the subgraphs induced by r consecutive layers.

For each $i \in \mathbb{N}$, let $V_i = \lambda^{-1}(\{i-r+1, \ldots, i\})$ and $M_i = M \cap \lambda^{-1}(\{i-r+2, \ldots, i-1\})$, and for each subset $\mathcal{R}' \subseteq \mathcal{R}$, let $\mathcal{R}'_i = \{R \cap V_i : R \in \mathcal{R}'\}$. First, using the algorithm from the assumptions, we find for each $i \in \mathbb{N}$ such that $V_i \neq \emptyset$ and for each subset $\mathcal{R}' \subseteq \mathcal{R}$ a $(1+\varepsilon)$-approximate solution $S_{i,\mathcal{R}'}$ to the instance $(G[V_i], M_i, \mathcal{R}'_i)$. In case that the instance does not have solution, i.e., $\mathcal{R}'_i$ contains an empty set, we let $S_{i,\mathcal{R}'} = \bot$. Note that the total number of vertices of these instances is at most $2^s rn$, and thus this can be performed in polynomial time.

Now, we stitch these partial solutions together using dynamic programming. For $i = 1, 2, \ldots, m+r-2$, let $V_{\leq i} = \lambda^{-1}(\{0, 1, \ldots, i\})$ and $M_{\leq i} = M \cap \lambda^{-1}(\{0, \ldots, i-1\})$, and for each $\mathcal{R}' \subseteq \mathcal{R}$, let $\mathcal{R}'_{\leq i} = \{R \cap V_{\leq i} : R \in \mathcal{R}'\}$. If $i \leq r-2$, then let $S_{\leq i,\mathcal{R}'} = S_{i,\mathcal{R}'}$. If $i \geq r-1$, then let $S_{\leq i,\mathcal{R}'}$ be the smallest of the sets $S_{\leq i-r+2,\mathcal{R}'_1} \cup S_{i,\mathcal{R}'_2}$ for $\mathcal{R}'_1, \mathcal{R}'_2 \subseteq \mathcal{R}'$ such that $\mathcal{R}'_1 \cup \mathcal{R}'_2 = \mathcal{R}'$, $S_{\leq i-r+2,\mathcal{R}'_1} \neq \bot$, and $S_{i,\mathcal{R}'_2} \neq \bot$; or $\bot$ if no such choice of $\mathcal{R}'_1$ and $\mathcal{R}'_2$ exists. Note that if $S_{\leq i,\mathcal{R}'} \neq \bot$, then it is a valid solution to the instance $(G[V_{\leq i}], M_{\leq i}, \mathcal{R}'_{\leq i})$; in particular, any vertex in $M_{\leq i}$ belongs to $M_{\leq i-r+2}$ or to M_i, and thus it is guaranteed to be dominated. Moreover, all the sets can be found in time $O((m+r)8^s n) = O(8^s(n+r)^2)$.

The algorithm outputs the smallest of the sets $S_{\leq m+1,\mathcal{R}}, \ldots, S_{\leq m+r-2,\mathcal{R}}$, all of which are valid solutions to the instance $(G, M, \mathcal{R})$. Let us now argue about the approximation ratio of this algorithm. Let O be an optimal solution and for $i \in \mathbb{N}$, let $O_i = O \cap V_i$ and let $\mathcal{R}^{(i)} = \{R \in \mathcal{R} : R \cap O_i \neq \emptyset\}$; thus, O_i is a valid solution to the instance $(G[V_i], M_i, \mathcal{R}_i^{(i)})$, and consequently

$$|S_{i,RR^{(i)}}| \leq (1+\varepsilon)|O_i|.$$

Clearly, there exists $t \in [r-2]$ such that

$$|O \cap \lambda^{-1}(\{i-1, i : i \equiv m+t \pmod{r-2}\})| \leq \frac{2}{r-2}|O|.$$

Consequently,

$$\sum_{i \equiv m+t (\text{mod } r-2)} |O_i| \leq \left(1 + \frac{2}{r-2}|O|\right).$$

Therefore,

$$\begin{aligned} |S_{\leq m+t,\mathcal{R}}| &\leq \sum_{i \equiv m+t (\text{mod } r-2)} |S_{i,\mathcal{R}^{(i)}}| \leq (1+\varepsilon) \sum_{i \equiv m+t (\text{mod } r-2)} |O_i| \\ &\leq \left(1 + \frac{2}{r-2}\right)(1+\varepsilon)|O|, \end{aligned}$$

showing that the returned set has size at most $\left(1 + \frac{2}{r-2}\right)(1+\varepsilon)\gamma(G, M, \mathcal{R})$. □

As a basic application, we have seen in Corollary 2.74 that if G is a graph drawn on a fixed surface, then for any fixed r, the graphs in $\mathcal{G}_0$ have bounded treewidth, and thus the s-request dominating set problem for these graphs can be solved exactly (with $\varepsilon = 0$). Hence, since we can choose r to be arbitrarily large, this gives a polynomial-time approximation scheme for the domination number for graphs of bounded Euler genus.

As we have already seen, the grid plus universal vertex graphs show that K_6-minor-free graphs G do not necessarily have a layering such that the union of any three layers has bounded treewidth. However, the result from the previous section suggests that having bounded radius, i.e., all vertices of G being close to a single vertex v_0, might be the only issue that needs to be overcome. What if we allowed deletion of a bounded number of vertices? As illustrated in the following lemma, this operation can also be handled in the design of approximation algorithms.

Lemma 16.6 *Let s be an integer and $\varepsilon \geq 0$ a real number, and let $\mathcal{G}_0$ be a class of graphs such that there exists a polynomial-time $(1+\varepsilon)$-approximation algorithm for $(s+1)$-request dominating set problem for graphs from $\mathcal{G}_0$. Let v_0 be a vertex of a graph G such that $G - v_0 \in \mathcal{G}_0$. Then we can in polynomial time for any s-request dominating set problem instance $(G, M, \mathcal{R})$ return a solution of size at most $(1+\varepsilon)\gamma(G, M, \mathcal{R})$.*

Proof We simply try two possibilities depending on whether v_0 is in the solution or not. Let $M_1 = M \setminus N[v_0]$ and let $\mathcal{R}_1 = \{R \in \mathcal{R} : v_0 \notin R\}$. Let $M_2 = M \setminus \{v_0\}$, and let $\mathcal{R}_2 = \{R \setminus \{v_0\} : R \in \mathcal{R}\}$ if $v_0 \notin M$, and $\mathcal{R}_2 = \{R \setminus \{v_0\} : R \in \mathcal{R}\} \cup \{N(v_0)\}$ if $v_0 \in M$. For $i \in [2]$, let S_i be the $(1+\varepsilon)$-approximate solution to the instance $(G - v_0, M_i, \mathcal{R}_i)$ returned by the algorithm from the assumptions. Observe that both $S_1 \cup \{v_0\}$ and S_2 (if not equal to $\perp$) are solutions to the instance $(G, M, \mathcal{R})$: If v_0 is in the dominating set, then we only need to further dominate M_1 and intersect the sets of $\mathcal{R}$ that do not contain v_0. Otherwise, we need to dominate M_2 and intersect $R \setminus \{v_0\}$ for each $R \in \mathcal{R}$; and in case that $v_0 \in M$, we also need to dominate v_0, forcing us to additionally intersect $N(v_0)$ with the solution. We return the smaller of the sets $S_1 \cup \{v_0\}$ and S_2.

Let O be an optimal solution. If $v_0 \in O$, then $O \setminus \{v_0\}$ is a solution to $(G - v_0, M_1, \mathcal{R}_1)$, and thus $|S_1| \le (1+\varepsilon)(|O|-1)$, and $|S_1 \cup \{v_0\}| \le (1+\varepsilon)(|O|-1)+1 \le (1+\varepsilon)|O|$. Otherwise, O is a solution to $(G - v_0, M_2, \mathcal{R}_2)$, and $|S_1| \le (1+\varepsilon)|O|$. In either case, the returned set has size at most $(1+\varepsilon)\gamma(G, M, \mathcal{R})$. □

This motivates the following definition. For a (finite or infinite) sequence $\mathbf{r} = r_1, r_2, \ldots$, let $\text{head}(\mathbf{r}) = r_1$ and $\text{tail}(\mathbf{r}) = r_2, \ldots$ For a finite sequence $\mathbf{r}$ of positive integers, we say that a graph G is $\mathbf{r}$-*Baker-amenable* if

[empty] $V(G) = \emptyset$, or

[delete] $\mathbf{r}$ is non-empty and there exists a vertex $v_0 \in V(G)$ such that $G - v_0$ is $\text{tail}(\mathbf{r})$-Baker-amenable, or

[layer] $\mathbf{r}$ is non-empty and there exists a layering λ of G such that every subgraph of G induced by the union of $\text{head}(\mathbf{r})$ consecutive layers is $\text{tail}(\mathbf{r})$-Baker-amenable.

Implicitly, we are going to assume that we can also find the vertex v_0 or the layering λ in polynomial time; this will be the case for graphs from proper minor-closed classes. Combining Lemmas 16.5 and 16.6, we easily obtain the following conclusion by induction on m.

Corollary 16.7 *For any finite sequence $\mathbf{r} = r_1, \ldots, r_m$ of integers such that $r_i \ge 3$ for $i \in [m]$, there exists a polynomial-time $\prod_{i=1}^{m}\left(1 + \frac{2}{r_i - 2}\right)$-approximation algorithm for domination number in $\mathbf{r}$-Baker-amenable graphs.*

We say that a graph class $\mathcal{G}$ is ***Baker-amenable*** if for every infinite sequence $\mathbf{r}$ of positive integers, there exists a finite prefix $\mathbf{r}_0$ of $\mathbf{r}$ such that every graph from $\mathcal{G}$ is $\mathbf{r}_0$-Baker-amenable.

Corollary 16.8 *Let $\mathcal{G}$ be a Baker-amenable class of graphs. For every $\varepsilon > 0$, there exists a polynomial-time $(1+\varepsilon)$-approximation for domination number in graphs from $\mathcal{G}$.*

Proof Note that

$$\prod_{i \ge 1}\left(1 + \frac{1}{ci^2}\right) \le \exp\left(\frac{1}{c}\sum_{i \ge 1}\frac{1}{i^2}\right) = \exp\left(\frac{\pi^2}{6c}\right),$$

and thus we can select $c \geq 2$ as a sufficiently large integer so that this product is at most $1+\varepsilon$. Let $\mathbf{r} = r_1, r_2, \ldots$, where $r_i = 2ci^2 + 2$; then

$$\prod_{i \geq 1} \left(1 + \frac{2}{r_i - 2}\right) \leq 1 + \varepsilon.$$

Let $\mathbf{r}_0$ be a finite prefix of $\mathbf{r}$ such that every graph from $\mathcal{G}$ is $\mathbf{r}_0$-Baker-amenable. The claim follows by applying Corollary 16.7 for $\mathbf{r}_0$. □

Of course, domination number is not the only graph parameter that can be efficiently approximated in Baker-amenable graph classes; indeed, every minimization or maximization problem expressible in the first-order logic has a polynomial-time approximation scheme in this setting [2].

Why do we need the somewhat technical definition of $\mathbf{r}$-Baker-amenability with the sequence $\mathbf{r}$, and cannot just work with a fixed integer r giving the number of layers? The problem we run into here is that the number m of iterations for graphs from proper minor-closed classes turns out to depend on r, and in fact to be much larger than r. Thus, e.g., in Corollary 16.7, we would only get a $\left(1 + \frac{2}{r-2}\right)^m$-approximation algorithm, with $\left(1 + \frac{2}{r-2}\right)^m$ constant but much larger than 1.

One might wonder whether this could not be fixed by accounting separately for the [layer] and [delete] steps, but that is not the case, as even the number of [layer] steps can be too large. Moreover, there are arguments where [delete] steps in a strategy used to show Baker-amenability of a graph H are turned into a [layer] step followed by several [delete] steps in a strategy for a related graph G, and such a separate accounting would make this significantly more complicated.

Another aspect the reader might wonder about is that the base case for Baker-amenability are graphs with no vertices, rather than those of bounded treewidth. As we are going to show, bounded treewidth graphs are Baker-amenable, and thus stopping at bounded treewidth would not make the notion more general. Moreover, stopping at the null graph potentially makes it possible to use the notion for algorithmic problems that are not a priori known to be approximable on graphs of bounded treewidth.

The definition of Baker-amenability of a graph class allows the sequence $\mathbf{r}$ to grow arbitrarily fast, and thus it may seem implausible at first that any nontrivial graph class is Baker-amenable. To allay this concern at least a bit, let us prove Baker-amenability of forests.

Observation 16.9 *Let $\mathbf{r} = r_1, \ldots, r_m$ be a sequence of positive integers. If $m > 2r_1$, then every forest is $\mathbf{r}$-Baker-amenable. Hence, the class of forests is Baker-amenable.*

Proof Let F be a forest, and choose a root in each component of F arbitrarily. Let λ be the BFS layering according to the distance from the roots, and consider any subgraph F_1 of F induced by r_1 consecutive layers of λ. For $i = 1, \ldots, r_1$, we proceed as follows:

- Let λ_i be a layering of F_i that puts the vertices of each component to a single layer, with the layers containing different components at least r_{2i} layers apart. Consider any subgraph F_i' induced by r_{2i} consecutive layers, i.e., a component of F_i.
- Let v_i be the unique vertex of F_i' with $\lambda(v_i)$ minimum, and let $F_{i+1} = F_i' - v_i$.

Observe that for each $i \in [r_1]$, the forest F_i has depth at most $r_1 - i + 1$, and F_{r_1+1} has no vertices. Thus, by reverse induction, F_i' is $\text{tail}^{2i}(\mathbf{r})$-Baker-amenable and F_i is $\text{tail}^{2i-1}(\mathbf{r})$-Baker-amenable for $i \in [r_1]$, and F is $\mathbf{r}$-Baker-amenable. □

As can be seen from the proof of this observation, Baker-amenability can be seen as a kind of a game: In each step, we can either delete a vertex, or choose a layering. In the latter case, an "enemy" chooses the given number of consecutive layers, determining the subgraph considered in the next turn. We win if we destroy the graph in a bounded number of steps, or more precisely, without exhausting the sequence $\mathbf{r}$.

Moreover, observe that in the proof we deleted only quite specific vertices, those currently minimum in the ordering according to the distance from the root. It will be quite important for the proof of Baker-amenability of proper minor-closed classes to enforce this property. An ***ordered graph*** is a pair $(G, \prec)$, where $\prec$ is a linear ordering of vertices of the graph G. For a finite sequence $\mathbf{r}$ of positive integers, we say that the ordered graph $(G, \prec)$ is $\mathbf{r}$-*Baker-amenable* if

[empty] $V(G) = \emptyset$, or

[delete] $\mathbf{r}$ is non-empty and $(G - v_0, \prec)$ is $\text{tail}(\mathbf{r})$-Baker-amenable for the minimum vertex v_0 in the ordering $\prec$, or

[layer] $\mathbf{r}$ is non-empty and there exists a layering λ such that for each $i \in \mathbb{N}$, the ordered graph $(G[\lambda^{-1}(\{i, \ldots, i + \text{head}(\mathbf{r}) - 1\})], \prec)$ is $\text{tail}(\mathbf{r})$-Baker-amenable.

A class $\mathcal{G}$ of ordered graphs is *Baker-amenable* if for every infinite sequence $\mathbf{r}$ of positive integers, there exists a finite prefix $\mathbf{r}_0$ of $\mathbf{r}$ such that every ordered graph $(G, \prec) \in \mathcal{G}$ is $\mathbf{r}_0$-Baker-amenable. As another simple ingredient, let us note the following observation.

Observation 16.10 *Let $\mathbf{r}$ be a finite sequence of positive integers, let $(G, \prec)$ be an ordered graph, and let $(G', \prec)$ be a subgraph of $(G, \prec)$. If $(G, \prec)$ is $\mathbf{r}$-Baker-amenable, then so is $(G', \prec)$.*

Proof We mimic the strategy showing the $\mathbf{r}$-Baker-amenability of $(G, \prec)$ on $(G', \prec)$. This is possible, since any layering of a graph is also a layering of its subgraphs. Moreover, in a [delete] step where we remove the minimum vertex v_0 from G, if v_0 is not contained in the subgraph, then we "skip" the step by using a [layer] step with all vertices contained in a single layer, instead. □

In the following arguments, we are going to focus on destroying a part Q of an (ordered) graph G by restricting to consecutive layers of carefully chosen layerings (of the whole graph), and it will be convenient to ignore the deletions from the part of the graph outside of Q caused by this process. That is, if the [layer] operation results in the graph $G - L$, we instead prefer to work with the graph $G - (L \cap V(Q))$. Observation 16.10 shows that we can do this, as it is sufficient to show Baker-amenability of the supergraph. Thus, we allow ourselves to use the following variant of the [layer] operation:

[layer'] **r** is non-empty and there exists a layering λ of G and a set $Q \subseteq V(G)$ such that for each $i \in \mathbb{N}$, the ordered graph $(G - \{v \in Q : \lambda(v) < i \text{ or } \lambda(v) \geq i + \text{head}(\mathbf{r})\}, \prec)$ is $\text{tail}(\mathbf{r})$-Baker-amenable.

We can now show a key lemma for inductive arguments. A linear ordering $\prec$ of vertices of a graph G *extends* a layering λ of G if for every $u, v \in V(G)$, if $u \prec v$, then $\lambda(u) \leq \lambda(v)$.

Lemma 16.11 *Let $\mathcal{G}$ and $\mathcal{G}_0$ be classes of ordered graphs, where $\mathcal{G}_0$ is Baker-amenable. Suppose that for every ordered graph $(G, \prec) \in \mathcal{G}$, there exists a shadow-complete layering λ_0 such that $\prec$ extends λ_0 and $(G[\lambda_0^{-1}(i)], \prec) \in \mathcal{G}_0$ for each $i \in \mathbb{N}$. Then $\mathcal{G}$ is Baker-amenable.*

Proof Let us first prove this for a layering λ_0 with only two layers, i.e., with $\text{img}(\lambda_0) = \{0, 1\}$. Consider any infinite sequence **r** of positive integers, and let $\mathbf{r}_1$ be a finite prefix of **r** such that every ordered graph in $\mathcal{G}_0$ is $\mathbf{r}_1$-Baker-amenable. Let $\mathbf{r}'$ be the suffix of **r** after $\mathbf{r}_1$, and let $\mathbf{r}_2$ be a finite prefix of $\mathbf{r}'$ such that every ordered graph in $\mathcal{G}_0$ is $\mathbf{r}_2$-Baker-amenable.

We claim that every ordered graph $(G, \prec) \in \mathcal{G}$ is $\mathbf{r}_1\mathbf{r}_2$-Baker-amenable. Indeed, we first mimic the strategy for the $\mathbf{r}_1$-Baker-amenability of the ordered subgraph $(G[\lambda_0^{-1}(0)], \prec) \in \mathcal{G}_0$. Suppose we have already reduced $(G, \prec)$ to an induced subgraph $(G', \prec)$, such that $G'[\lambda_0^{-1}(1)] = G[\lambda_0^{-1}(1)]$:

- If the [delete] action is used on $(G'[\lambda_0^{-1}(0)], \prec)$, then we also use the [delete] action on $(G', \prec)$; this removes the same vertex, since $\prec$ extends λ_0.
- Suppose now that a [layer] action is used on $(G'[\lambda_0^{-1}(0)], \prec)$ with a layering λ. Since λ_0 is shadow-complete, each component K of $G'[\lambda_0^{-1}(1)]$ has neighbors only in a clique C_K in $G'[\lambda_0^{-1}(0)]$. Since λ is a layering of $G'[\lambda_0^{-1}(0)]$, there exists $i_K \in \mathbb{N}$ such that $V(C_K) \subseteq \lambda^{-1}(\{i_K, i_K + 1\})$. We extend λ to a layering of G' by letting $\lambda(v) = i_K$ for each vertex v belonging to a component K of $G'[\lambda_0^{-1}(1)]$. We then use the [layer'] action on $(G', \prec)$ with this extended version of λ and with $Q = \lambda_0^{-1}(0) \cap V(G')$.

Once $G[\lambda_0^{-1}(0)]$ is reduced to an empty graph, if needed we exhaust the rest of $\mathbf{r}_1$ using [layer'] actions with any layering and $Q = \emptyset$. Then, we simply follow the strategy for the $\mathbf{r}_2$-Baker-amenability of the ordered subgraph $(G[\lambda_0^{-1}(1)], \prec)$, which belongs to $\mathcal{G}_0$.

For a positive integer ℓ, let $\mathcal{G}_\ell$ consist of $\mathcal{G}_0$ and of the ordered graphs $(G, \prec) \in \mathcal{G}$ which have a shadow-complete layering λ_0 with the properties described in the statement of the lemma and with $\max(\text{img}(\lambda_0)) - \min(\text{img}(\lambda_0)) < \ell$. That is, $\mathcal{G}_1 = \mathcal{G}_0$, and we have just shown that $\mathcal{G}_2$ is Baker-amenable. We can now show by induction on ℓ that all classes $\mathcal{G}_\ell$ are Baker-amenable: Suppose that $(G, \prec) \in \mathcal{G}_\ell$ for some $\ell \geq 3$. Without loss of generality, we can assume that $\mathrm{img}(\lambda_0) = \{0, 1, \ldots, \ell-1\}$. Let λ_1 be defined by setting $\lambda_1(v) = 0$ if $\lambda_0(v) = 0$ and $\lambda_1(v) = 1$ otherwise; then λ_1 is a shadow-complete layering of G, the ordering $\prec$ extends λ_1, and $(G[\lambda_1^{-1}(i)], \prec) \in \mathcal{G}_{\ell-1}$ for $i \in \{0, 1\}$. Since $\mathcal{G}_{\ell-1}$ is Baker-amenable by the induction hypothesis, the claim follows by the argument for the 2-layer case, with $\mathcal{G}_{\ell-1}$ playing the role of $\mathcal{G}_0$.

Finally, let us consider the general case that the number of layers of λ_0 is not bounded. Consider any infinite sequence $\mathbf{r}$ of positive integers, let $\ell = \mathrm{head}(\mathbf{r})$, and let $\mathbf{r}_1$ be a finite prefix of $\mathrm{tail}(\mathbf{r})$ such that every ordered graph from $\mathcal{G}_\ell$ is $\mathbf{r}_1$-Baker-amenable. Then clearly every ordered graph $(G, \prec) \in \mathcal{G}$ is $\ell\mathbf{r}_1$-Baker-amenable: We first take the [layer] action with λ_0; this reduces $(G, \prec)$ to its subgraph induced by ℓ consecutive layers of λ_0, and thus belonging to G_ℓ. We then just follow the strategy showing $\mathbf{r}_1$-Baker-amenability of this subgraph. Therefore, $\mathcal{G}$ is indeed Baker-amenable. □

We say that a layering λ of a graph G is ***umbral*** if it is shadow-complete, $\omega(G[\lambda^{-1}(i)]) < \omega(G)$ for every positive integer i, and $\lambda^{-1}(0)$ is an independent set. A linear ordering $\prec$ of $V(G)$ is *umbral* if either $E(G) = \emptyset$, or $\prec$ extends an umbral layering λ of G and the ordering $\prec$ restricted to $G[\lambda^{-1}(i)]$ is umbral for every $i \in \mathbb{N}$.

Lemma 15.21 applied separately to each component shows that every chordal graph has an umbral layering. It is easy to see that every chordal graph has an umbral ordering. Indeed, the following stronger claim holds. Given a (not necessarily umbral) ordering $<$ of $V(G)$ and a vertex $v \in V(G)$, let $G_{\geq v}$ denote the connected component of $G[\{u \in V(G) : u \geq v\}]$ containing v.

Lemma 16.12 *Let G be a chordal graph and let $<$ be an elimination ordering of G. Then there exists an umbral ordering $\prec$ of G such that $G_{\geq v} = G_{\succeq v}$ for every $v \in V(G)$.*

Proof We prove the claim by induction on $\omega(G)$; the basic case $\omega(G) = 1$ is trivial, as we can let $\prec$ be $<$. Hence, suppose that $\omega(G) \geq 2$. Let R be the set consisting of the minimum vertices of the components of G in the ordering $<$, and let λ be the BFS layering of G according to the distance from R. By Lemma 15.21, λ is an umbral layering of G. For $i \in \mathbb{N}$, let $G^i = G[\lambda^{-1}(i)]$ and let $<^i$ be the restriction of $<$ to $\lambda^{-1}(i)$; by the induction hypothesis, there exists an umbral ordering $\prec^i$ of G^i such that $G^i_{\geq^i v} = G^i_{\succeq^i v}$ for each $v \in \lambda^{-1}(i)$.

Let us now define the ordering $\prec$: For distinct $u, v \in V(G)$, we have $u \prec v$ if either $\lambda(u) < \lambda(v)$, or $\lambda(u) = \lambda(v) = i$ and $u \prec^i v$. Then $\prec$ extends the umbral layering λ and the restriction $\prec^i$ of $\prec$ to $G[\lambda^{-1}(i)]$ is umbral for every $i \in \mathbb{N}$, and thus $\prec$ is an umbral ordering.

We claim that $\prec$ preserves the ordering of adjacent vertices, i.e., if $uv \in E(G)$ and $u < v$, then $u \prec v$. We prove this by induction on the clique number. Let Q be the component of G containing v and let $r \in R$ be the minimum vertex of Q in the ordering $<$. Let K be the set of neighbors of v preceding it in the ordering

$<$. Since $<$ is an elimination ordering, the set K separates v from r, and thus $\lambda(v) = 1 + \min \lambda(K)$. Moreover, K is a clique, and since $u \in K$, we have $\lambda(u) \leq 1 + \min \lambda(K) \leq \lambda(v)$. If $\lambda(u) < \lambda(v)$, then $u \prec v$. If $\lambda(u) = \lambda(v) = i$, then $u \prec v$ by the induction hypothesis for G^i and $\prec^i$.

It follows that for each $v \in V(G)$, the sets of neighbors of v preceding it in $<$ and in $\prec$ coincide, and in particular, $\prec$ is also an elimination ordering of G. Hence, the fact that $G_{\geq v} = G_{\succeq v}$ for every $v \in V(G)$ is a consequence of the following claim: If $\sqsubset$ is an elimination ordering of G, then $x \in V(G_{\sqsupseteq v})$ if and only if there exists a path $x = x_0 x_1 \ldots x_m = v$ in G such that $x_i \sqsubset x_{i-1}$ for $i \in [m]$. The "if" part is trivial. For the "only if" part, we claim that any induced path from x to v in $G_{\sqsupseteq v}$ has this property. Indeed, since v is the minimum vertex of this component in the ordering $\sqsubset$, if the path did not have this property, then there would exist $i \in [m-1]$ such that $x_{i-1}, x_{i+1} \sqsubset x_i$; but then $x_{i-1}x_{i+1} \in E(G)$, since $\sqsubset$ is an elimination ordering. □

The following claim easily follows using Lemma 16.11 by induction on ω. In the basic case $\omega = 1$, we simply use one [layer] action to restrict ourselves to a component of the graph, which is an isolated vertex, then a [delete] action to remove it.

Corollary 16.13 *For every positive integer ω, the class of umbrally-ordered chordal graphs of clique number at most ω is Baker-amenable.*

Since every graph G is a subgraph of a chordal graph of clique number $\text{tw}(G) + 1$, together with Lemma 16.12 and Observation 16.10 this implies that graphs of bounded treewidth are Baker-amenable. Using geodesic chordal partitions, this is easy to generalize to all proper minor-closed classes.

Theorem 16.14 *For every positive integer k, the class of K_k-minor-free graphs is Baker-amenable.*

Proof Let $\mathcal{G}_k$ be the class of umbrally-ordered chordal graphs of clique number less than k, which is Baker-amenable by Corollary 16.13. Consider any infinite sequence $\mathbf{r}$ of positive integers. Let $\mathbf{r}_0 = \mathbf{r}$, and for $i = 1, 2, \ldots$, let $a_i = \text{head}(\mathbf{r}_{i-1})$ and $\mathbf{r}_i = \text{tail}^{1+ka_i}(\mathbf{r}_{i-1})$. Let $\mathbf{a} = a_1, a_2, \ldots$, and let $\mathbf{a}'$ be a finite prefix of $\mathbf{a}$ such that every ordered graph from $\mathcal{G}_k$ is $\mathbf{a}'$-Baker-amenable. Let m be the length of $\mathbf{a}'$, and let $\mathbf{r}'$ be the prefix of $\mathbf{r}$ up to (and including) $\mathbf{r}_m$.

We claim that every K_k-minor-free graph G is $\mathbf{r}'$-Baker-amenable. In Sect. 13.2, we have seen that G has a geodesic chordal partition $\mathcal{P} = P_1, \ldots, P_t$. Since $G' = G/\mathcal{P}$ is a minor of G, it has clique number less than k. Moreover, letting $\prec$ be the ordering of $V(G')$ such that $P_i \prec P_j$ if and only if $i < j$,

- $\prec$ is an elimination ordering of the chordal graph G', and
- for every $i \in [t]$, there exist a vertex $v_i \in P_i$, a predecessor transversal S_i of P_i, and a system $Q_i = \{Q_u : u \in S_i\}$ such that

$$P_i = \{v_i\} \cup \bigcup_{u \in S_i} V(Q_u)$$

and for every $u \in S_i$, Q_u is a path in $G[P_i]$ from v_i to u geodesic in $G[\bigcup V(G'_{\succeq P_i})]$.

By Lemma 16.12, we can without loss of generality assume that $\prec$ is an umbral ordering. To show $\mathbf{r}'$-Baker-amenability of G, we follow the strategy for $\mathbf{a}'$-Baker-amenability of G', preserving the following invariant: For $i \in [m]$, when we are emulating the i-th step of the strategy for G', we have already consumed the prefix of $\mathbf{r}'$ till a_i (exclusive) in the strategy for G, and we have erased from G exactly the union of the sets of $\mathcal{P}$ corresponding to the erased vertices of G'. The strategy is emulated as follows:

- If the i-th step of the strategy for G' is a [layer] step with a layering λ', we perform a [layer] step on G with the layering λ defined by letting $\lambda(v) = \lambda'(P)$ for every vertex v belonging to a part P of $\mathcal{P}$. We then perform ka_i [layer'] steps with an arbitrary layering and $Q = \emptyset$ to skip till a_{i+1} in $\mathbf{r}'$.
- If the i-th step is a [delete] step, then let P_j be the erased part, the smallest one in the ordering $\prec$. Let λ be the BFS layering of the current subgraph starting from v_j (vertices in the components not containing v_j are assigned layer 0). since the paths in Q_j are geodesic and cover P_j, the restriction of λ to $G[P_j]$ is also the BFS layering from v_j in this subgraph. Hence, after performing the [layer'] step with λ and $Q = P_j$, the set P_j is reduced to at most a_i vertices from each of the paths of Q_j. Since $|Q_j| \leq k$, at most $a_i k$ vertices of P_j remain. We erase them by performing at most $a_i k$ [delete] actions, possibly followed by [layer'] steps with an arbitrary layering and $Q = \emptyset$ to skip till a_{i+1} in $\mathbf{r}'$.

□

Thus, approximation algorithms using Baker's technique can be naturally applied to all graphs from proper minor-closed classes. A downside is that the length of the prefix of $\mathbf{r}$ taken in the proof of Theorem 16.14, or even in Corollary 16.13, is impractically large. This may not be immediately obvious; if $\mathbf{r}$ is a constant sequence with all elements equal to a, then the length of the taken segment is actually polynomial in a. However, as we have seen, this is not very useful for the design of approximation algorithms: We need to consider sequences $\mathbf{r} = r_1, r_2, \ldots$ such that $\prod_{i \geq 1}(1+1/r_i)$ is finite (and close to 1), and in particular, we need $r_i \gg i$. For such sequences, it turns out that the length of the prefix taken by the strategy underlying the proof of Corollary 16.13 for chordal graphs of clique number ω is larger than $2 \uparrow^{\omega-2} r_1$, where $a \uparrow^0 b = ab$ and for $i \geq 1$,

$$a \uparrow^i b = \underbrace{a \uparrow^{i-1} a \uparrow^{i-1} \ldots \uparrow^{i-1} a}_{b \text{ times}},$$

associated to the right. It is not clear whether a more reasonable number of steps suffices.

Let us remark that as can be seen in the following lemma, Baker-amenability implies fractional treewidth-fragility, but much better bounds for fractional treewidth-fragility can be obtained by a direct analysis of the argument underlying Corollary 16.13; see [3] for details.

Lemma 16.15 *For every Baker-amenable class $\mathcal{G}$, there exists a function f such that the following claim holds. For every graph $G \in \mathcal{G}$, every assignment w of non-negative weights to vertices of G, and every positive integer k, there exists a set $S \subseteq V(G)$ such that $w(S) \le \frac{1}{k}w(G)$ and $\mathrm{td}(G - S) \le f(k)$.*

Proof For a positive integer k, let c be the minimum positive integer such that

$$\sum_{i \ge 1} \frac{1}{ci^2} \le 1/k.$$

Let $\mathbf{r} = r_1, r_2, \ldots$, where $r_i = ci^2$, and let $\mathbf{r}'$ be a finite prefix of $\mathbf{r}$ such that every graph $G \in \mathcal{G}$ is $\mathbf{r}'$-Baker-amenable. We define $f(k)$ to be the length of $\mathbf{r}$.

To prove the lemma, it suffices to show the following: If a graph G is $\mathbf{a}$-Baker-amenable for a finite sequence $\mathbf{a} = a_1, \ldots, a_m$ of positive integers, then for every assignment w of non-negative weights to vertices of G, there exists a set $S \subseteq V(G)$ such that $w(S) \le \left(\sum_{i=1}^{m} \frac{1}{a_i}\right) w(G)$ and $\mathrm{td}(G - S) \le m$. We prove this claim by induction on m. If $m = 0$, then $V(G) = \emptyset$ and the claim holds with $S = \emptyset$. Consider now the case $m \ge 1$. Since G is $\mathbf{a}$-Baker-amenable, one of the following possibilities holds.

- There exists $v_0 \in V(G)$ such that $G - v_0$ is tail($\mathbf{a}$)-Baker-amenable. Hence, by the induction hypothesis, there exists $S \subseteq V(G - v_0)$ such that

$$w(S) \le \left(\sum_{i=2}^{m} \frac{1}{a_i}\right) w(G - v_0) \le \left(\sum_{i=1}^{m} \frac{1}{a_i}\right) w(G)$$

  and $\mathrm{td}(G - (S \cup \{v_0\})) \le m - 1$. Then $\mathrm{td}(G - S) \le m$.
- There exists a layering λ of G such that the subgraph of G induced by any a_1 consecutive layers is tail($\mathbf{a}$)-Baker-amenable. For $b \in [a_1]$, let $S_b = \{v \in V(G) : \lambda(v) \equiv b \pmod{a_1}\}$. There exist $b \in [a_1]$ such that $w(S_b) \le \frac{1}{a_1} w(G)$; fix such a value b. Let $\gamma = \sum_{i=2}^{m} \frac{1}{a_i}$. For any $j \in \mathbb{Z}$ such that $j \equiv b + 1 \pmod{a_1}$, let $G_j = G[\lambda^{-1}(\{j, j+1, \ldots, j+a_1-2\})]$; then G_j is tail($\mathbf{a}$)-Baker-amenable, and by the induction hypothesis, there exists a set $S_j \subseteq V(G_j)$ such that $w(S_j) \le \gamma w(G_j)$ and $\mathrm{td}(G_j - S_j) \le m - 1$. Let S' be the union of these sets for all values of j; then $G - (S_b \cup S')$ is the disjoint union of the graphs $G_j - S_j$ for all values of j, and thus it has treedepth at most $m - 1$. Moreover, $w(S_b \cup S') \le \frac{1}{a_1} w(G) + \gamma w(G - S_b) \le \left(\frac{1}{a_1} + \gamma\right) w(G)$, as required.

□

References

1. B.S. Baker, Approximation algorithms for NP-complete problems on planar graphs. J. ACM **41**, 153–180 (1994)
2. Z. Dvořák, Baker game and polynomial-time approximation schemes, in *Proceedings of the thirty-first annual ACM-SIAM symposium on discrete algorithms, SODA '20* (Society for Industrial and Applied Mathematics, Philadelphia, 2020), pp. 2227–2240
3. Z. Dvořák, J. Sereni, On fractional fragility rates of graph classes. Electron. J. Combin. **27**, P4.9 (2020)
4. U. Feige, M. Mahdian, Finding small balanced separators, in *Proceedings of the thirty-eighth annual ACM symposium on Theory of computing* (ACM, 2006), pp. 375–384
5. U. Feige, M. Hajiaghayi, J. Lee, Improved approximation algorithms for minimum weight vertex separators. SIAM J. Comput. **38**, 629–657 (2008)
6. P. Klein, S. Plotkin, S. Rao, Excluded minors, network decomposition, and multicommodity flow, in *Proceedings of the twenty-fifth annual ACM symposium on Theory of computing* (1993), pp. 682–690
7. F. Leighton, S. Rao, An approximate max-flow min-cut theorem for uniform multicommodity flow problems with applications to approximation algorithms, in *FOCS*, vol. 88 (1988), pp. 422–431

Chapter 17
Isomorphism Testing

The complexity of testing whether two graphs are isomorphic is one of the most important open problems in algorithmic graph theory, as well as complexity theory in general. There are indications that it is not NP-hard; otherwise, the polynomial hierarchy collapses to its second level [13], which is widely believed not to be the case. On the other hand, despite of decades of research, a polynomial-time algorithm for the problem is not known. In a recent breakthrough, Babai [1] improved the bound on the complexity of isomorphism testing for general graphs to quasi-polynomial; more precisely, he gave an algorithm that decides whether two n-vertex graphs are isomorphic in time $\exp(O(\log^3 n))$. Note that this makes it even more unlikely that graph isomorphism is NP-hard, as otherwise all problems in NP would be solvable in quasi-polynomial time.

On the other hand, graph isomorphism is known to be decidable in polynomial time for graphs restricted to a number of special graph classes, e.g., interval graphs [2], graphs of bounded maximum degree [9], and most relevantly for us, graphs avoiding a forbidden minor [12] or a topological minor [5].

It is possible to base the graph isomorphism test for proper minor-closed classes on the structure theorem, see e.g. [4] for a systematic (though not the simplest possible) development. Indeed, as a consequence of the structure theorem, Grohe [4] proved the following interesting result:

Theorem 17.1 *For every graph H, there exists a positive integer k such that the k-dimensional Weisfeiler-Lehman algorithm decides isomorphism of H-minor-free graphs.*

The *k-dimensional Weisfeiler-Lehman algorithm* is a standard labeling refinement algorithm used as a heuristic for graph isomorphism or a subroutine in more sophisticated algorithms, and can be implemented in time $O(n^{k+1}\log n)$, see [8]. Thus, testing the isomorphism of H-minor-free graphs can be done by a rather simple algorithm (agnostic of the Minor Structure Theorem), though as the constant k derives from the structure theorem, this is not relevant in practice.

© The Author(s), under exclusive license to Springer Nature Switzerland AG 2025
Z. Dvořák, *Graph Minors*, Springer Monographs in Mathematics,
https://doi.org/10.1007/978-3-031-87469-7_17

Even before the Minor Structure Theorem was developed, a polynomial-time algorithm for isomorphism of graphs in proper minor-closed classes was given by Ponomarenko [12], combining group-theoretic techniques with relatively simple observations on the structure of graphs avoiding a fixed clique minor. The time complexity of his algorithm for n-vertex K_k-minor-free graphs is $O(n^{\text{poly}(k)})$. Using the ideas of Babai's quasipolynomial algorithm [1], Grohe et al. [7] recently developed another algorithm avoiding the structure theorem with time complexity $O(n^{\text{polylog}(k)})$. In the rest of this chapter, we outline some of the ideas, giving a simpler algorithm with time complexity $O(n^{\text{poly}(k)})$. We are actually going to show that this algorithm works for graphs avoiding K_k as a topological minor, as this comes out for free from the argument.

17.1 Weisfeiler-Lehman Algorithm

Weisfeiler-Lehman algorithm is a simple preprocessing routine used in many graph isomorphism algorithms. It generalizes simple degree-based observations on graph isomorphism, such as the following: Suppose that G_1 and G_2 are isomorphic. Then for every $d \in \mathbb{N}$, the graphs G_1 and G_2 have the same number of vertices of degree d. Moreover, for every $d \in \mathbb{N}$ and $m_1, m_2, \ldots \in \mathbb{Z}_0^+$, the graphs G_1 and G_2 have the same number of vertices of degree d which have exactly m_1 neighbors of degree one, m_2 neighbors of degree two, etc. And so on, taking larger and larger neighborhoods into account.

Weisfeiler and Lehman observed that such arguments can be generalized in the following coloring refinement procedure. Let $\varphi_0 : V(G) \to C_0$ be an initial coloring of vertices of a graph G; often, the trivial coloring assigning to all vertices the same color is used. We proceed in rounds, refining this coloring. Consider $r \geq 1$ and let $\varphi_{r-1} : V(G) \to C_{r-1}$ be the coloring from round $r-1$, where the colors $c_1, c_2, \ldots, c_t$ of C_{r-1} have a fixed ordering. For each vertex $v \in V(G)$ and each color c, let $n(v, c)$ be the number of neighbors of v of color c. As the color $\varphi_r(v)$, let us use the tuple $(\varphi(v), n(v, c_1), \ldots, n(v, c_t))$. We order the colors (tuples) used by φ_r lexicographically.

Since the colorings are defined in terms of the properties of the graph (independently of labels of vertices, and not making any arbitrary choices), we obtain the following easy observation. We say that an isomorphism $f : V(G_1) \to V(G_2)$ *preserves* colorings φ_1 and φ_2 of G_1 and G_2 if $\varphi_1(v) = \varphi_2(f(v))$ for every $v \in V(G_1)$

Observation 17.2 *Let $\varphi_{1,0} : V(G_1) \to C_0$ and $\varphi_{2,0} : V(G_2) \to C_0$ be colorings of graphs G_1 and G_2 and let $\varphi_{1,r}$ and $\varphi_{2,r}$ be their refinements after r rounds of the Weisfeiler-Lehman algorithm. Then every isomorphism of G_1 and G_2 that preserves $\varphi_{1,0}$ and $\varphi_{2,0}$ also preserves $\varphi_{1,r}$ and $\varphi_{2,r}$.*

The color refinement is typically repeated until the coloring stabilizes, i.e., until φ_{r-1} and φ_r partition the vertex set to the same color classes; we say that the resulting coloring is *stable*. Of course, in a practical implementation, we would not represent the colors by long tuples; rather, each new color is assigned an integer to represent it. Moreover, it turns out to be easier not to perform the whole round at once: We just need to locate a pair of colors c_1 and c_2 such that not all vertices of color c_1 have the same number of neighbors of color c_2, then refine the color class c_1 based on the numbers of such neighbors. With additional optimizations and careful handling of implementation details, the Weisfeiler-Lehman algorithm can be implemented in time $O((n+m)\log n)$ for graphs with n vertices and m edges [3].

In a basic application for the isomorphism testing, we start with coloring of all vertices of G_1 and G_2 by the same color. We then run the Weisfeiler-Lehman algorithm on them, obtaining their colorings $\varphi_{1,r}$ and $\varphi_{2,r}$. There are two good outcomes:

- If there exists a color $c \in C_r$ such that $|\varphi_{1,r}^{-1}(c)| \neq |\varphi_{2,r}^{-1}(c)|$, then no isomorphism preserves $\varphi_{1,r}$ and $\varphi_{2,r}$. Since every isomorphism preserves the initial coloring, Observation 17.2 implies that G_1 and G_2 are not isomorphic.
- If every color class of the final colorings $\varphi_{1,r}$ and $\varphi_{2,r}$ is trivial (i.e., has size exactly one), then the graphs are isomorphic, via the isomorphism mapping each vertex of G_1 to the vertex of G_2 of the same color.

If neither of these extremes happens, the outcome may at least restrict the set of bijections that we need to consider for isomorphism. Of course, there are graphs for which this approach does not help much or even at all. For example, if G is regular, then the initial monochromatic coloring is already stable.

A usual way around this issue is through vertex individualization: Suppose that we have reached stable colorings φ_1 and φ_2 of G_1 and G_2 and there still are non-trivial color classes. Deterministically choose a color c such that $|\varphi_1^{-1}(c)| = |\varphi_2^{-1}(c)| > 1$, say one with $|\varphi_1^{-1}(c)|$ minimum, and subject to that with c smallest. Let $v_1 \in \varphi_1^{-1}(c)$ be chosen arbitrarily and let $\varphi_2^{-1}(c) = \{v_{2,1}, \ldots, v_{2,t}\}$. Let φ_1' be obtained from φ_1 by giving v_1 a new color c_0, and for $i \in [t]$, let $\varphi_{2,i}'$ be obtained from φ_2 by giving $v_{2,i}$ the color c_0. Clearly, an isomorphism $f : V(G_1) \to V(G_2)$ preserving φ_1 and φ_2 maps v_1 to a vertex of $\varphi_2^{-1}(c)$, and thus there exists a unique $i \in [t]$ such that f also preserves φ_1' and $\varphi_{2,i}'$. Thus, to decide whether such an isomorphism exists, it suffices to check for each $i \in [t]$ whether there exists an isomorphism of G_1 and G_2 preserving φ_1' and $\varphi_{2,i}'$.

Hence, this reduces the isomorphism test for (G_1, φ_1) and (G_2, φ_2) to t isomorphism tests of (G_1, φ_1') and $(G_2, \varphi_{2,1}'), \ldots, (G_2, \varphi_{2,t}')$, which can be handled recursively. Importantly, coloring a vertex by the unique color c_0 opens up new possibilities for the color refinement. Although the resulting algorithm does not run in polynomial time in general, it works rather well as a practical heuristic.

For special graph classes, we may be able to argue that after each specialization and Weisfeiler-Lehman refinement, the smallest non-trivial color class (if any) of the current coloring φ has bounded size. For example, this is the case for connected graphs of maximum degree at most d: Because we performed the specialization, there is a color c_1 whose color class has size one. Suppose that there is a color c_2

whose color class has size more than one. Since the graph is connected, we can choose the colors c_1 and c_2 so that the unique vertex v_1 of color c_1 has a neighbor v_2 of color c_2. Because the coloring is stable and v_2 has a neighbor of color c_1, all vertices of $\varphi^{-1}(c_2)$ must have a vertex of color c_1, i.e., they all must be adjacent to v_1. Therefore, $|\varphi^{-1}(c_2)| \leq \deg v_1 \leq d$. In this case, the time complexity of the described algorithm is at most $d^n \operatorname{poly}(n)$, since the specialization can of course be performed at most n times. This already is a bit of an improvement from the naive $n! \operatorname{poly}(n)$ algorithm testing all permutations of vertices.

17.2 Small Cuts from Stable Colorings

Ideally, we would like to obtain a similar bound on the minimum size of a non-trivial color class in graphs avoiding a fixed (topological) minor. We are not going to be able to achieve this, but as we are about to show, the only reason why this can fail is the existence of small vertex cuts.

Theorem 17.3 *Let k be a positive integer and let G be a graph which does not contain K_k as a topological minor. Let $\varphi : V(G) \to C$ be a stable coloring of G and let*

$$X = \bigcup_{c \in C : |\varphi^{-1}(c)|=1} \varphi^{-1}(c)$$

be the union of the trivial color classes. Then

- *there exists a color $c \in C$ such that $2 \leq |\varphi^{-1}(c)| \leq k^3$, or*
- *$|N(V(K)) \cap X| \leq k - 1$ for every component K of $G - X$.*

To prove this claim, suppose for a contradiction that $\varphi : V(G) \to C$ is a stable coloring such that all color classes are either trivial or have size at least k^3, and that there exists a component K of $G-X$ such that $|N(V(K)) \cap X| \geq k$. In this situation, we need to find K_k as a topological minor in G.

Let G/φ be the graph with vertex set C and with two colors c_1 and c_2 adjacent if and only if G contains an edge $v_1 v_2$ such that $\varphi(v_1) = c_1$ and $\varphi(v_2) = c_2$. Let C_K be the set of colors that appear on K and let C_X be the set of colors appearing on $N(V(K)) \cap X$. Note that $C_K \cap C_X = \emptyset$, since the X is the union of the trivial color classes. Since K is a connected component of $G - X$, the graphs $(G/\varphi)[C_K]$ and $(G/\varphi)[C_K \cup C_X]$ are connected, and thus there exists a tree $T_0 \subseteq (G/\varphi)[C_K \cup C_X]$ whose leaves are formed by the vertices of C_X.

A tree $T' \subseteq G$ is a *lift* of T_0 if $\varphi \restriction V(T')$ is an isomorphism from T' to T_0, i.e., the vertex set of T' contains a single vertex $v_c \in \varphi^{-1}(c)$ for each color $c \in C_K \cup C_X$ and no other vertices, and $v_{c_1} v_{c_2} \in E(T')$ if and only if $c_1 c_2 \in E(T_0)$. We are going to show that G contains $m = \binom{k}{2}$ lifts $T_1, \ldots, T_m$ of T_0 intersecting exactly in their leaves in $N(V(K)) \cap X$. We can then use a path in each of the lifts to represent one of the edges of the topological minor of K_k in G.

The key property that we are going to use to obtain these lifts is that the subgraph of G between any two color classes is biregular, which follows from the assumption that the coloring φ is stable (a bipartite graph F with specified bipartition (A_1, A_2) is ***biregular*** if there exist integers a_1 and a_2 such that $\deg v = a_i$ for every $i \in [2]$ and $v \in A_i$). We are first going to deal with the important special case of finding pairwise vertex-disjoint paths in chains of biregular graphs, in a somewhat more general setting describing how a set of already existing paths can be enlarged. For a graph H and disjoint sets $A, B \subset V(H)$, let $H(A, B)$ be the bipartite subgraph of H with bipartition (A, B) containing exactly the edges of H between A and B. Let $\mathcal{A} = A_1, \ldots, A_n$ be a sequence of pairwise disjoint sets of vertices of H. We say that this sequence is a *biregular chain* if for $i \in [n-1]$, the graph $H(A_i, A_{i+1})$ is biregular and has non-empty edge set; the *width* of the chain is $\min_{i \in [n]} |A_i|$. An *$\mathcal{A}$-monotone path* in H is a path consisting of a vertex of A_1, a vertex of A_2, ..., and a vertex of A_n in order. We need the following observation on the proportion of vertices of A_n that can be cut off by a small vertex cut.

Lemma 17.4 *Let $\mathcal{A} = A_1, \ldots, A_n$ be a biregular chain of width d in a graph H. For any sets $Y \subseteq A_1$ and $B \subseteq \bigcup \mathcal{A}$, the number of vertices of A_n reachable from Y by an $\mathcal{A}$-monotone path disjoint from B is at least $|A_n| \cdot \left(\frac{|Y|}{|A_1|} - \frac{|B|}{d}\right)$.*

Proof Let us prove this claim by induction on n. For $n = 1$, the number of reachable vertices is at least $|Y| - |B| = |A_n| \cdot \left(\frac{|Y|}{|A_1|} - \frac{|B|}{d}\right)$, since $|A_1| = |A_n| = d$. Hence, suppose that $n \geq 2$. Let $\mathcal{A}' = A_1, \ldots, A_{n-1}$, let $P \subseteq A_{n-1}$ be the set of vertices of A_{n-1} reachable from Y by an $\mathcal{A}'$-monotone path disjoint from $B' = B \setminus A_n$, and let d' be the width of $\mathcal{A}'$. By the induction hypothesis,

$$|P| \geq |A_{n-1}| \cdot \left(\frac{|Y|}{|A_1|} - \frac{|B'|}{d'}\right) \geq |A_{n-1}| \cdot \left(\frac{|Y|}{|A_1|} - \frac{|B'|}{d}\right).$$

Let $a_1, a_2 \geq 1$ be the integers such that for $j \in [2]$, each vertex of A_{n-2+j} has a_j neighbors in A_{n+1-j}. Note that the number of edges between A_{n-1} and A_n is $|A_{n-1}|a_1 = |A_n|a_2$, and thus $|A_n| = \frac{a_1}{a_2}|A_{n-1}|$. Similarly, by counting the edges between P and the neighbors of vertices of P in A_n, we have $|N(P) \cap A_n| \geq \frac{a_1}{a_2}|P|$. Therefore, the number of vertices of A_n reachable from Y by an $\mathcal{A}$-monotone path disjoint from B is at least

$$\begin{aligned}
|N(P) \cap A_n| - |B \cap A_n| &\geq \frac{a_1}{a_2}|P| - (|B| - |B'|) \\
&\geq \frac{a_1}{a_2}|A_{n-1}| \cdot \left(\frac{|Y|}{|A_1|} - \frac{|B'|}{d}\right) - (|B| - |B'|) \\
&= |A_n| \cdot \left(\frac{|Y|}{|A_1|} - \frac{|B'|}{d}\right) - (|B| - |B'|)
\end{aligned}$$

$$\geq |A_n| \cdot \left(\frac{|Y|}{|A_1|} - \frac{|B'|}{d} \right) - \frac{|A_n|}{d}(|B| - |B'|)$$
$$= |A_n| \cdot \left(\frac{|Y|}{|A_1|} - \frac{|B|}{d} \right).$$

□

Using Lemma 17.4, we can show that an existing system of $\mathcal{A}$-monotone paths can be enlarged in many different ways. For sets $Y \subseteq A_1$ and $Z_0 \subseteq A_n$, let $\tau_{\mathcal{A}}(Y, Z_0)$ be the set of vertices $z \in A_n \setminus Z_0$ such that H contains $|Z_0| + 1$ pairwise vertex-disjoint $\mathcal{A}$-monotone paths from Y to $Z_0 \cup \{z\}$.

Lemma 17.5 *Let $(A_1, \ldots, A_n)$ be a biregular chain of width d in a graph H. Let Y be a subset of A_1 and Z_0 subset of A_n such that H contains $|Z_0|$ pairwise vertex-disjoint $\mathcal{A}$-monotone paths from Y to Z_0. Then*

$$\frac{|\tau_{\mathcal{A}}(Y, Z_0)|}{|A_n|} \geq \frac{|Y|}{|A_1|} - \frac{|Z_0|}{d}.$$

Proof Let $\mathbf{F}$ be the directed graph with vertex set $\bigcup \mathcal{A}$ and with edge set consisting exactly of the of edges (u, v) such that $uv \in E(H)$ and $u \in A_i$ and $v \in A_{i+1}$ for some $i \in [n-1]$. Note that $\mathcal{A}$-monotone paths in H are exactly the directed paths in $\mathbf{F}$ from A_1 to A_n. In particular, $\mathbf{F}$ contains a strict (Y, Z_0)-linkage $\mathcal{P}_0$ of size $|Z_0|$ by the assumptions of the lemma.

Let $M = A_n \setminus \tau_{\mathcal{A}}(Y, Z_0)$; note that $Z_0 \subseteq M$. The graph $\mathbf{F}$ does not contain a (Y, M)-linkage of size $|Z_0|+1$, as otherwise by Lemma 5.2, it would contain a strict $(Y, Z_0 \cup \{z\})$-linkage of size $|Z_0| + 1$ for some $z \in M$, showing that $z \in \tau_{\mathcal{A}}(Y, Z_0)$. By Menger's theorem, there exists a set B of at most $|Z_0|$ vertices such that every path in $\mathbf{F}$ (and thus also every $\mathcal{A}$-monotone path in H) from Y to M intersects B. That is, every vertex of A_n reachable from Y by an $\mathcal{A}$-monotone path disjoint from B belongs to $A_n \setminus M = \tau_{\mathcal{A}}(Y, Z_0)$, and by Lemma 17.4, there are at least

$$|A_n| \cdot \left(\frac{|Y|}{|A_1|} - \frac{|B|}{d} \right) \geq |A_n| \cdot \left(\frac{|Y|}{|A_1|} - \frac{|Z_0|}{d} \right)$$

such vertices. □

Let us now use this observation to extend a system of lifts of trees. For a rooted tree T, a vertex v of T is a *terminal* if it is the root, a leaf of T, or a vertex of T of degree at least three. Let $b(T)$ be the number of non-root terminals of T.

Lemma 17.6 *Let d be a positive integer, let $\varphi : V(G) \to C$ be a stable coloring of a graph G and let $T \subseteq G/\varphi$ be a rooted tree such that $|\varphi^{-1}(c)| \geq d$ for every $v \in V(T)$. Let $T_1, \ldots, T_m$ be pairwise vertex-disjoint lifts of T in G. Let c be a terminal of T, let T^c be the subtree of T rooted in c, and let $Z_c = \varphi^{-1}(c) \cap \bigcup_{i=1}^m V(T_i)$. Let τ_c consists of Z_c and the vertices $z \in \varphi^{-1}(c) \setminus Z_c$ such that G contains pairwise vertex-disjoint lifts $T'_1, \ldots, T'_{m+1}$ of T^c intersecting $\varphi^{-1}(c)$ exactly in $Z_c \cup \{z\}$. Then*

$$\frac{|\tau_c|}{|\varphi^{-1}(c)|} \geq 1 - \frac{mb(T^c)}{d}.$$

Proof We prove the claim by induction on the depth of c in T, starting from the leaves. If c is a leaf, then $\tau_c = \varphi^{-1}(c)$ and $b(T^c) = 0$, and thus

$$\frac{|\tau_c|}{|\varphi^{-1}(c)|} = 1 = 1 - \frac{mb(T^c)}{d}.$$

Hence, suppose that c is not a leaf. Let $c_1, \ldots, c_p$ be the terminals of T^c that are reachable from c by paths $P_1, \ldots, P_p$ in T^c whose internal vertices are not terminals. Consider any $i \in [p]$, and let $\mathcal{A}_i$ be the biregular chain in G formed by the color classes of the colors forming the path P_i. Consider any vertex

$$z \in \bigcap_{i=1}^{p} \tau_{\mathcal{A}_i}(\tau_{c_i}, Z_c).$$

We claim that $z \in \tau_c$: Note that for $i \in [p]$, the lifts $T_1, \ldots, T_m$ contain m pairwise vertex-disjoint $\mathcal{A}_i$-monotone paths from Z_{c_i} to Z_c. By the definition of $\tau_{\mathcal{A}_i}$, there exists a system $\mathcal{P}_i$ of $m+1$ pairwise vertex-disjoint $\mathcal{A}_i$-monotone paths from τ_{c_i} to $Z_c \cup \{z\}$. By Lemma 5.2, we can assume that the paths of $\mathcal{P}_i$ start in $Z_{c_i} \cup \{z_i\}$ for a vertex $z_i \in \tau_{c_i} \setminus Z_{c_i}$. By the definition of τ_{c_i}, there exist pairwise vertex-disjoint lifts $T_1^{c_i}, \ldots, T_{m+1}^{c_i}$ of T^{c_i} in G intersecting $\varphi^{-1}(c_i)$ in $Z_{c_i} \cup \{z_i\}$. The union of the lifts $T_j^{c_i}$ for $i \in [p]$ and $j \in [m+1]$ and of the paths of $\mathcal{P}_i$ for $i \in [p]$ gives us $m+1$ pairwise vertex-disjoint lifts of T^c in G intersecting $\varphi^{-1}(c)$ exactly in $Z_c \cup \{z\}$, as desired.

Hence, we only need to bound the size of the set $\bigcap_{i=1}^{p} \tau_{\mathcal{A}_i}(\tau_{c_i}, Z_c)$. By the induction hypothesis, we have $\frac{|\tau_{c_i}|}{|\varphi^{-1}(c_i)|} \geq 1 - \frac{mb(T^{c_i})}{d}$. Hence, by Lemma 17.5,

$$\frac{|\tau_{\mathcal{A}_i}(\tau_{c_i}, Z_c)|}{|\varphi^{-1}(c)|} \geq \frac{|\tau_{c_i}|}{|\varphi^{-1}(c_i)|} - \frac{m}{d} \geq 1 - \frac{m(b(T^{c_i})+1)}{d}.$$

Note that $b(T^c) = \sum_{i=1}^{p}(b(T^{c_i})+1)$. Consequently,

$$\frac{|\tau_c|}{|\varphi^{-1}(c)|} \geq \frac{\bigcap_{i=1}^{p} |\tau_{\mathcal{A}_i}(\tau_{c_i}, Z_c)|}{|\varphi^{-1}(c)|} \geq 1 - \sum_{i=1}^{p} \frac{m(b(T^{c_i})+1)}{d} = 1 - \frac{mb(T^c)}{d}.$$

□

From this, it is easy to conclude the existence of many disjoint lifts.

Corollary 17.7 *Let m_0 be a positive integer, let $\varphi : V(G) \to C$ be a stable coloring of a graph G, let $T \subseteq G/\varphi$ be a tree with at most k leaves, and let $d = (2k+1)m_0$.*

If $|\varphi^{-1}(c)| \geq d$ *for every* $v \in V(T)$*, then* G *contains* m_0 *pairwise vertex-disjoint lifts of* T.

Proof Root the tree T in an arbitrary vertex c; then $b(T) < 2k$. Let $m \leq m_0$ be maximum integer such that G contains m pairwise vertex-disjoint lifts $T_1, \ldots, T_m$ of T, and suppose for a contradiction that $m < m_0$. Let Z_c be the set of vertices in which these lifts intersect $\varphi^{-1}(c)$. With the notation as in Lemma 17.6, the maximality of m implies that $\tau_c = Z_c$. However, then Lemma 17.6 implies that

$$\frac{m}{d} \geq \frac{|Z_c|}{|\varphi^{-1}(c)|} = \frac{|\tau_c|}{|\varphi^{-1}(c)|} \geq 1 - \frac{mb(T^c)}{d} > 1 - \frac{2mk}{d},$$

which is a contradiction, since $d = (2k+1)m_0 > (2k+1)m$. □

We are now ready to prove the bound on the size of cuts arising from stable colorings with no small non-trivial color classes.

Proof of Theorem 17.3 Suppose for a contradiction that $\varphi : V(G) \to C$ is a stable coloring such that all classes are either trivial or have size at least k^3, and that there exists a component K of $G - X$ and distinct vertices $v_1, \ldots, v_k \in X$ with neighbors in K. Let C_K be the set of colors that appear on K; since K is connected, $(G/\varphi)[C_K]$ is connected. For $i \in [k]$, let u_i be a neighbor of v_i in K and let $c_i = \varphi(u_i)$ (the colors $c_1, \ldots, c_k$ are not necessarily distinct). Since $v_i \in X$, the color class of v_i has size one, and since the subgraph of G between v_i and $\varphi^{-1}(c_i)$ is biregular, note that v_i is adjacent to all vertices of $\varphi^{-1}(c_i)$.

Let T be a minimal subtree of $(G/\varphi)[C_K]$ containing $c_1, \ldots, c_k$; clearly, T has at most k leaves. Let $m_0 = \binom{k}{2}$. Since $(2k+1)m_0 < k^3$, Corollary 17.7 implies that G contains pairwise vertex-disjoint lifts $T_1, \ldots, T_{m_0}$ of T. For every $1 \leq i < j \leq k$, choose a distinct lift T_a and let $P_{i,j}$ be a path in G consisting of an edge from v_i to a vertex v_i' of T_a of color c_i, an edge from v_j to a vertex v_j' of T_a of color c_j, and a path in T_a between v_i' and v_j'. Then $\bigcup_{1 \leq i < j \leq k} P_{i,j}$ is a subdivision of K_k, contradicting the assumption that G does not contain K_k as a topological minor. □

17.3 Handling the Small Cuts

Because of the result from the previous section, we need a way to deal with the bounded-size cuts. We are going to do so through dynamic programming, precomputing the information about isomorphism between the pieces cut of by bounded-size cuts starting from the smallest pieces.

More precisely, a ***linearly bordered graph*** is a bordered graph G with its boundary ∂G having a fixed linear ordering. For two linearly bordered graphs G_1 and G_2, a function $f : V(G_1) \to V(G_2)$ is their isomorphism if f is the isomorphism of the underlying graphs and f maps vertices of ∂G_1 to vertices of

∂G_2 in order (and in particular, $|\partial G_1| = |\partial G_2|$). A ***fragment*** of a graph G is a linearly bordered graph G' such that

- the underlying graph of G' is an induced subgraph of G,
- $G' - \partial G'$ is connected and every vertex of $\partial G'$ has a neighbor in $G' - \partial G'$, and
- vertices of $G' - \partial G'$ do not have neighbors in $V(G) \setminus V(G')$ in G.

That is, $\partial G'$ is a minimal cut separating $G' - \partial G'$ from the rest of G. A *t-fragment* is a fragment G' with $|\partial G'| \le t$. Note that if G is an n-vertex graph, then G has less than $(n+1)^{t+1}$ t-fragments, since each sequence of at most t vertices of G gives rise to at most n fragments. For a set $X \subseteq V(G)$, an *X-fragment* of G is a fragment G' such that $G' - \partial G'$ is a component of $G - X$ and $\partial G' \subseteq X$.

Instead of just determining whether two graphs G_1 and G_2 are isomorphic, we are going to determine for every two $(k-1)$-fragments G'_1 of G_1 and G'_2 of G_2 whether they are isomorphic (when G_1 and G_2 are connected, this process also computes the answer to the original problem of whether G_1 and G_2 are isomorphic, since G_1 and G_2 with empty boundaries are $(k-1)$-fragments). Clearly, the number of such fragment isomorphism tests is at most $(n+1)^{2k}$. Of course, we only need to test fragments of the same size, i.e., with $|G'_1| = |G'_2|$, and we are going to compute this information in order of increasing size of the fragments.

In this way, we divide the fragments into isomorphism classes. In practice, a more efficient method to achieve this is to produce a ***canonical encoding*** of the fragments, i.e., a way of encoding which is guaranteed to assign the same code to fragments if and only if they are isomorphic. The codes then directly identify the isomorphism classes. Hence, we only need to call the canonical encoding procedure for at most $(n+1)^k$ fragments of G_1 and G_2, rather than test the isomorphism for all pairs of the fragments. As this adds another layer of complexity, we are not going to go into the details of how such canonical encodings can be obtained.

We now have all ingredients to obtain an isomorphism test for connected n-vertex K_k-topological-minor-free graphs with time complexity $k^{3n}\,\mathrm{poly}(k)$. As outlined above, we need to describe how to test the isomorphism of $(k-1)$-fragments G'_1 and G'_2 with the same number $n_1 \le n$ of vertices, assuming that we have already precomputed the isomorphism information for the smaller $(k-1)$-fragments. Let $t = |\partial G'_1| = |\partial G'_2|$. We can simply use brute force if $n_1 \le k-1$. Otherwise, we are going to go over all (at most $n_1^{2(k-t)}$) choices of ways how to extend the boundaries $\partial G'_1$ and $\partial G'_2$ to length k and for each of them check whether the resulting linearly bordered graphs G''_1 and G''_2 are isomorphic.

Let $\partial G''_1 = u_1, \ldots, u_k$ and $\partial G''_2 = v_1, \ldots, v_k$. We start with the colorings $\varphi_{1,0}$ and $\varphi_{2,0}$ of G''_1 and G''_2 that for $i \in [k]$ assign to the vertices u_i and v_i the color i and assign the color 0 to all other vertices. We then run the Weisfeiler-Lehman algorithm. If the resulting stable colorings contain a non-trivial class of size at most k^3, we apply individualization to vertices of this class, branching into at most k^3 new instances as described at the end of Sect. 17.1.

Thus, the problem eventually reduces to at most $k^{3n_1}\,\mathrm{poly}(n_1)$ instances of verifying whether there exists an isomorphism of G''_1 and G''_2 preserving their stable colorings φ_1 and φ_2 with no non-trivial classes of size at most k^3. Let X_1 and X_2 be the unions of trivial color classes of φ_1 and φ_2 and let f be the function mapping the vertices of X_1 to vertices of X_2 of the same color. By Theorem 17.3, for $i \in [2]$, every X_i-fragment F of G_i'' is a $(k-1)$-fragment. Since $\partial G_i' \subseteq X_i$, F is also a fragment of G_i. Moreover, since $|X_i| \geq |\partial G_i''| = k$, we have $|F| < |G_i''| = |G_i'|$. Therefore, we have already precomputed the isomorphism relation between such fragments of G_1 and G_2. To verify whether f extends to an isomorphism of G_1'' and G_2'', it suffices to check whether the following condition holds for each $(\leq k-1)$-tuple B of distinct vertices of X_1: The number of X_1-fragments F_1 of G_1'' with $\partial F_1 = B$ and the number of X_2-fragments F_2 of G_2'' with $\partial F_2 = f(B)$ of each isomorphism class is the same.

17.4 Group Theoretic Ingredients

To improve the time complexity to polynomial, we need tools from the algorithmic group theory. Specifically, we are going to consider the ***set-of-strings isomorphism problem***. Given a finite index set I and an *alphabet* (another set) C, an *I-string* is a function $s : I \to C$. For a permutation γ of I, s^γ is the I-string such that $s^\gamma(i) = s(\gamma^{-1}(i))$ for each $i \in I$; i.e., the character from each position i is moved to the position $\gamma(i)$. A permutation γ of I is an isomorphism of two multisets S_1 and S_2 of I-strings if $S_2 = \{s^\gamma : s \in S_1\}$. The instance of the set-of-strings isomorphism problem consists of a finite index set I, a group $\mathfrak{G}$ of permutations of I, and two finite multisets S_1 and S_2 of I-strings; the output should be the set M of all isomorphisms of S_1 and S_2 belonging to $\mathfrak{G}$. Note that if M is non-empty, then it can be expressed as $\gamma_0\mathfrak{H}$, where $\gamma_0 \in \mathfrak{G}$ is an isomorphism of S_1 and S_2 and the subgroup $\mathfrak{H} \leq \mathfrak{G}$ consists of all automorphisms of S_1.

It might not be immediately clear to the reader how to efficiently represent the input group $\mathfrak{G}$ and the output group $\mathfrak{H}$, whose sizes may be up to $n!$, where $n = |I|$. This is done by specifying a set of ***generators*** of the group: For a set Q of permutations of I, let $\langle Q \rangle$ be the set of all permutations that can be obtained from them by repeated compositions. Since I is finite, this automatically includes the inverses of all the elements: Composing any permutation with itself repeatedly eventually results in the identity permutation. Then $\langle Q \rangle$ is the group *generated* by Q, and Q is a set of generators of this group. Any group $\mathfrak{G}$ has a set of generators, since we can take $Q = \mathfrak{G}$, but minimal sets of generators are of course more interesting. Recall that the *index* $|\mathfrak{G}|/|\mathfrak{H}|$ of the subgroup $\mathfrak{H}$ of a group $\mathfrak{G}$ is an integer by Lagrange's theorem. If $Q = \{q_1, \ldots, q_m\}$ is a minimal set of generators, then $\{\mathrm{id}\} < \langle\{q_1\}\rangle < \langle\{q_1, q_2\}\rangle < \ldots < \langle Q \rangle = \mathfrak{G}$, and thus $|\mathfrak{G}| \geq 2^m$. Consequently, every group $\mathfrak{G}$ of permutations of I can be represented by a system of at most $\log_2 |\mathfrak{G}| \leq \log_2 n! = O(n \log n)$ generators. See [14] for algorithmic aspects of performing operations on groups in this representation.

It is not known whether the set-of-strings isomorphism problem can be solved in polynomial-time for an arbitrary group $\mathfrak{G}$, or in the most natural case of $\mathfrak{G} = \mathrm{Sym}(I)$ being the group of all permutations of I. However, this is known to be the case for a useful family of groups, originally introduced by Luks [9] in order to solve the isomorphism problem for graphs of bounded maximum degree: For a positive integer d, let Sym_d be the group of all permutations of a d-element set. For a normal subgroup $\mathfrak{H}$ of a group $\mathfrak{G}$, let $\mathfrak{G}/\mathfrak{H}$ be the corresponding quotient group. Let $\hat{\Gamma}_d$ denote the class of all groups $\mathfrak{G}$ with the following property: Every non-trivial subgroup $\mathfrak{G}'$ of $\mathfrak{G}$ has a proper normal subgroup $\mathfrak{H}$ such that $\mathfrak{G}'/\mathfrak{H}$ is isomorphic to a subgroup of Sym_d. Note that every subgroup of a group in $\hat{\Gamma}_d$ clearly also belongs to $\hat{\Gamma}_d$.

Theorem 17.8 (Miller [10], Neuen [11]) *For every positive integer d, there exists a polynomial-time algorithm to solve the set-of-strings isomorphism problem when the input group is in $\hat{\Gamma}_d$.*

Let us remark that the size of the input instance is measured in the size n of the index set I and the number m of strings; the size of the alphabet can clearly be assumed to be bounded by nm and the group $\mathfrak{G}$ can be given by $O(n \log n)$ generators. Miller [10] only considers the problem of hypergraph isomorphism, but there is a simple transformation between the problems (see e.g. [11, Theorem 4.1]). The time complexity of Miller's algorithm is $O((n+m)^{O(d)})$, while Neuen [11] improves the complexity to $O((n+m)^{\mathrm{polylog}(d)})$. A nice exposition of the main ideas going into these algorithms can be found in [6].

Let us now describe the consequences of Theorem 17.8 that we need. The first one shows that we win as soon as we can show that there is an isomorphism belonging to a group in $\hat{\Gamma}_d$.

Lemma 17.9 *For every positive integer d, there exists a polynomial-time algorithm that given graphs G_1 and G_2 on the same set of vertices V and a group $\mathfrak{G} \in \hat{\Gamma}_d$ of permutations of V, either decides that no isomorphism of G_1 and G_2 belongs to $\mathfrak{G}$, or returns a permutation $\gamma_0 \in \mathfrak{G}$ and the subgroup $\mathfrak{H} \leq \mathfrak{G}$ such that $\gamma_0\mathfrak{H}$ is the set of all isomorphisms of G_1 and G_2 belonging to $\mathfrak{G}$.*

Proof For $i \in \{1, 2\}$, let $s_i : \binom{V}{2} \to \{0, 1\}$ be the string such that for every $\{u, v\} \in \binom{V}{2}$, $s_i(\{u, v\}) = 1$ if and only if $uv \in E(G_i)$. For each permutation $\gamma \in \mathfrak{G}$, let γ' be the permutation of $\binom{V}{2}$ such that $\gamma'(\{u, v\}) = \{\gamma(u), \gamma(v)\}$ for every $\{u, v\} \in \binom{V}{2}$. Let $\mathfrak{G}' = \{\gamma' : \gamma \in \mathfrak{G}\}$. Note that $\mathfrak{G}'$ is a group isomorphic to $\mathfrak{G}$ via the isomorphism $\gamma \mapsto \gamma'$, and thus $\mathfrak{G}' \in \hat{\Gamma}_d$. Moreover, γ is an isomorphism of G_1 and G_2 if and only if $s_1^{\gamma'} = s_2$. Therefore, the answer is obtained by running the algorithm from Theorem 17.8 for $I = \binom{V}{2}$, $\mathfrak{G}'$, $\{s_1\}$, and $\{s_2\}$, then translating the returned permutation γ_0' and subgroup $\mathfrak{H}' \leq \mathfrak{G}'$ to $\mathfrak{G}$ via the isomorphism of $\mathfrak{G}$ and $\mathfrak{G}'$. □

The next result will be used at the end of the argument, when we need to match up the pieces separated by small cuts.

Lemma 17.10 *For all positive integers k and d, there exists a polynomial-time algorithm as follows. The input consists of*

- *graphs G_1 and G_2,*
- *sets $X_1 \subseteq V(G_1)$ and $X_2 \subseteq V(G_2)$ such that for $i \in [2]$, all X_i-fragments of G_i are k-fragments,*
- *the partition π of X_1-fragments of G_1 and X_2-fragments of G_2 to isomorphism classes,*
- *a group $\mathfrak{G} \in \hat{\Gamma}_d$ of permutations of X_1, and*
- *a bijection $\kappa : X_1 \to X_2$.*

The output consists of the set M of bijections from $\kappa\mathfrak{G}$ which extend to an isomorphism of G_1 and G_2; if M is non-empty, it is represented as $\gamma_0\mathfrak{H}$, where $\gamma_0 \in \kappa\mathfrak{G}$ and $\mathfrak{H} \leq \mathfrak{G}$.

Proof The proof is analogous to the proof of Lemma 17.9, except that we need to match up the $(\leq k)$-tuples instead of just pairs. To simplify the notation, let us rename for each $x \in X_1$ the vertex $\kappa(x)$ of X_2 to x, so that $X_1 = X_2 = X$ and $\kappa = \text{id}$. Using Lemma 17.9, we can additionally restrict the group $\mathfrak{G}$ so that it only contains isomorpisms of $G[X_1]$ and $G[X_2]$.

Let Z be the set of all $(\leq k)$-tuples of distinct vertices of X. For $B \in Z$ and $i \in [2]$, let the i-color of B consist of the tuple $(n_{i,Q} : Q \in \pi)$, where for each isomorphism class Q of π, $n_{i,Q}$ is the number of X-fragments G' of G_i such that $G' \in Q$ and $\partial G' = B$. For $i \in [2]$, let s_i be the Z-string assigning to each tuple $B \in Z$ the i-color of B. For $\gamma \in \mathfrak{G}$, let γ' be the permutation of Z mapping each tuple $(x_1, \ldots, x_a)$ to the tuple $(\gamma(x_1), \ldots, \gamma(x_a))$, and observe that γ extends to an isomorphism of G_1 and G_2 if and only if $s_1^{\gamma'} = s_2$. Let $\mathfrak{G}' = \{\gamma' : \gamma \in \mathfrak{G}\}$ and note that $\gamma \mapsto \gamma'$ is an isomorphism of $\mathfrak{G}$ and $\mathfrak{G}'$. Hence, the result is obtained by running the algorithm from Theorem 17.8 for $I = Z$, $\mathfrak{G}'$, $\{s_1\}$, and $\{s_2\}$ and translating it to $\mathfrak{G}$ via the isomorphism of the groups. □

Finally, we are going to study the effect of the Weisfeiler-Lehman refinement applied to colorings that do not necessarily arise directly from the Weisfeiler-Lehman algorithm, but just respect the isomorphisms in a weaker sense. A coloring $\varphi : V(G) \to C$ of a linearly bordered graph is *boundary-respecting* if for each $v \in \partial G$, the color $\varphi(v)$ is only used on v. Given such a coloring, let G/φ be the linearly bordered graph obtained by contracting the color classes and with the boundary $\partial(G/\varphi) = \varphi(\partial G)$. A bijection $f : V(G_1) \to V(G_2)$ is *compatible* with boundary-respecting colorings $\varphi_1 : V(G_1) \to C_1$ and $\varphi_2 : V(G_2) \to C_2$ of linearly bordered graphs G_1 and G_2 if for every $u, v \in V(G_1)$, $\varphi_2(f(u)) = \varphi_2(f(v))$ if and only if $\varphi_1(u) = \varphi_1(v)$. In that case, we define $f_{\varphi_1,\varphi_2} : C_1 \to C_2$ to be the function that maps every color $c \in C_1$ to the common color of the vertices in $f(\varphi_1^{-1}(c))$. We say that φ_1 and φ_2 *respect the isomorphisms* of G_1 and G_2 if every isomorphism of G_1 and G_2 is compatible with φ_1 and φ_2.

Observation 17.11 *Let $\varphi_1 : V(G_1) \to C_1$ and $\varphi_2 : V(G_2) \to C_2$ be boundary-respecting colorings of linearly bordered graphs G_1 and G_2 respecting their isomorphisms. Let ψ_1 and ψ_2 be obtained from φ_1 and φ_2 by running one round* *of the Weisfeiler-Lehman algorithm. Then ψ_1 and ψ_2 also respect the isomorphisms of G_1 and G_2.*

Proof Consider any isomorphism $f : V(G_1) \to V(G_2)$ and vertices $u, v \in V(G_1)$ such that $\psi_1(u) \neq \psi_1(v)$. If $\varphi_1(u) \neq \varphi_1(v)$, then $\varphi_2(f(u)) \neq \varphi_2(f(v))$, and thus $\psi_2(f(u)) \neq \psi_2(f(v))$.

Suppose now that $\varphi_1(u) = \varphi_1(v)$. Since $\psi_1(u) \neq \psi_1(v)$, by the definition of the Weisfeiler-Lehman algorithm round, there exists a color $c \in C_1$ such that $|N(u) \cap \varphi_1^{-1}(c)| \neq |N(v) \cap \varphi_1^{-1}(c)|$. Since f is compatible with φ_1 and φ_2, there exists a color $c' \in C_2$ such that $f(\varphi_1^{-1}(c)) = \varphi_2^{-1}(c')$. Since f is an isomorphism of G_1 and G_2, we have

$$\begin{aligned}|N(f(u)) \cap \varphi_2^{-1}(c')| &= |f(N(u)) \cap f(\varphi_1^{-1}(c))| = |N(u) \cap \varphi_1^{-1}(c)| \\ &\neq |N(v) \cap \varphi_1^{-1}(c)| = |N(f(v)) \cap \varphi_2^{-1}(c')|,\end{aligned}$$

and thus $\psi_2(f(u)) \neq \psi_2(f(v))$ by the definition of the Weisfeiler-Lehman algorithm round.

The same argument for f^{-1} shows that if $\psi_2(f(u)) \neq \psi_2(f(v))$, then $\psi_1(u) \neq \psi_1(v)$. □

So far, we have only needed the special case of Theorem 17.8 for a single string. The final application uses the full power of the theorem. Let $\varphi_1 : V(G_1) \to C_1$ and $\varphi_2 : V(G_2) \to C_2$ be boundary-respecting colorings of linearly bordered graphs G_1 and G_2 respecting their isomorphisms. For a positive integer d, an *isomorphism d-enclosure* of $(G_1, \varphi_1, G_2, \varphi_2)$ is a pair $(\kappa, \mathfrak{G})$, where $\kappa : C_1 \to C_2$ is a bijection and $\mathfrak{G} \in \hat{\Gamma}_d$ is a group of permutations of C_1 such that $f_{\varphi_1,\varphi_2} \in \kappa\mathfrak{G}$ for every isomorphism $f : V(G_1) \to V(G_2)$.

Lemma 17.12 *For very positive integer d, there exists a polynomial-time algorithm as follows. The input consists of*

- *linearly bordered graphs G_1 and G_2,*
- *boundary-respecting colorings $\varphi_1 : V(G_1) \to C_1$ and $\varphi_2 : V(G_2) \to C_2$ respecting the isomorphisms of G_1 and G_2, and*
- *an isomorphism d-enclosure $(\kappa_0, \mathfrak{G}_0)$ of $(G_1, \varphi_1, G_2, \varphi_2)$.*

For $i \in [2]$, let $\psi_i : V(G_i) \to C_i'$ be obtained from φ_i by running one round of the Weisfeiler-Lehman algorithm. The algorithm either decides that G_1 and G_2 are not isomorphic, or returns an isomorphism d-enclosure $(\kappa, \mathfrak{G})$ of $(G_1, \psi_1, G_2, \psi_2)$.

Proof (⇸) By replacing φ_2 by $\kappa_0^{-1} \circ \varphi_2$, we can assume that $C_1 = C_2 = C$ and that κ_0 is the identity function.

For $i \in [2]$ and colors $a \in C_i'$ and $c \in C$, let $n_{i,a,c}$ be the number of neighbors of each vertex of color a in $\varphi_i^{-1}(c)$, which is the same for all such vertices by the definition of the Weisfeiler-Lehman algorithm round. Let $\delta_{i,a,c} = 1$ if $\psi_i^{-1}(a) \subseteq \varphi_i^{-1}(c)$ and $\delta_{i,a,c} = 0$ otherwise. For every $i \in [2]$ and every color $a \in C_i'$, let $s_{i,a}$ be the C-string such that $s_{i,a}(c) = (\delta_{i,a,c}, n_{i,a,c})$ for each color $c \in C$. Let $S_i = \{s_{i,a} : a \in C_i'\}$.

Suppose that $f : V(G_1) \to V(G_2)$ is an isomorphism, and let $\gamma = f_{\varphi_1,\varphi_2}$. Note that $\gamma \in \mathfrak{G}_0$, since $(\mathrm{id}, \mathfrak{G}_0)$ is an isomorphism d-enclosure. Consider any color $a \in C'_1$ and the color $a' = f_{\psi_1,\psi_2}(a) \in C'_2$. Let $c_0 \in C$ be the color such that $\psi_1^{-1}(a) \subseteq \varphi_1^{-1}(c_0)$ and let $c'_0 = \gamma(c_0)$. For any vertex $v \in \psi_2^{-1}(a')$, we have $\psi_1(f^{-1}(v)) = a$, and thus $\varphi_1(f^{-1}(v)) = c_0$ and $\varphi_2(v) = c'_0$; hence, $\psi_2^{-1}(a') \subseteq \varphi_2^{-1}(c'_0)$. Thus, for every color $c \in C$, we have

$$\delta_{1,a,c} = 1 \Leftrightarrow c = c_0 \Leftrightarrow \gamma(c) = c'_0 \Leftrightarrow \delta_{2,a',\gamma(c)} = 1.$$

Similarly, since f is an isomorphism compatible with φ_1 and φ_2, we conclude that $n_{1,a,c} = n_{2,a',\gamma(c)}$. Therefore, $s_{1,a}^{\gamma} = s_{2,a'}$. Since $f_{\psi_1,\psi_2} : C'_1 \to C'_2$ is a bijection, we conclude that γ is an isomorphism of S_1 and S_2.

Let us apply the algorithm from Theorem 17.8 with $I = C$, $\mathfrak{G}_0$, S_1, and S_2. If the output is empty, then by the previous paragraph, we conclude that G_1 and G_2 are not isomorphic. Otherwise, let $\gamma_0\mathfrak{H}$ with $\gamma_0 \in \mathfrak{G}_0$ and $\mathfrak{H} \le \mathfrak{G}_0$ be the result of the algorithm. Note that the definition of the Weisfeiler-Lehman algorithm round implies that for $i \in [2]$, the strings in S_i are pairwise different. Hence, for every $\theta \in \mathfrak{H}$, there exists a unique bijection $g_\theta : C'_1 \to C'_2$ such that $s_{1,a}^{\gamma_0\theta} = s_{2,g_\theta(a)}$ for every $a \in C'_1$. Let us define $\overline{\theta} : C'_1 \to C'_1$ as $\overline{\theta} = g_{\mathrm{id}}^{-1} g_\theta$. Note that for $\theta_1, \theta_2 \in \mathfrak{H}$ and $a \in C'_1$,

$$\begin{aligned} s_{1,a}^{\gamma_0\theta_1\theta_2} &= s_{1,a}^{\gamma_0\theta_1\gamma_0^{-1}\gamma_0\theta_2} = \left(s_{1,a}^{\gamma_0\theta_2}\right)^{\gamma_0\theta_1\gamma_0^{-1}} = \left(s_{2,g_{\theta_2}(a)}\right)^{\gamma_0\theta_1\gamma_0^{-1}} \\ &= \left(s_{2,g_{\mathrm{id}}(\overline{\theta_2}(a))}\right)^{\gamma_0\theta_1\gamma_0^{-1}} = \left(s_{1,\overline{\theta_2}(a)}^{\gamma_0}\right)^{\gamma_0\theta_1\gamma_0^{-1}} \\ &= s_{1,\overline{\theta_2}(a)}^{\gamma_0\theta_1} = s_{2,g_{\theta_1}(\overline{\theta_2}(a))}.v \end{aligned}$$

Therefore, $g_{\theta_1\theta_2} = g_{\theta_1}\overline{\theta_2}$, and consequently $\overline{\theta_1\theta_2} = \overline{\theta_1}\,\overline{\theta_2}$. It follows that $\mathfrak{G} = \{\overline{\theta} : \theta \in \mathfrak{H}\}$ is a group isomorphic to $\mathfrak{H}$, and since $\mathfrak{H} \le \mathfrak{G}_0 \in \hat{\Gamma}_d$, we also have $\mathfrak{G} \in \hat{\Gamma}_d$.

We claim that $(g_{\mathrm{id}}, \mathfrak{G})$ is an isomorphism d-enclosure of $(G_1, \psi_1, G_2, \psi_2)$, and thus we can let $\kappa = g_{\mathrm{id}}$. Indeed, consider any isomorphism $f : V(G_1) \to V(G_2)$. As we have argued above, f_{φ_1,φ_2} is an isomorphism of S_1 and S_2, and thus there exists $\theta \in \mathfrak{H}$ such that $f_{\varphi_1,\varphi_2} = \gamma_0\theta$. For each color $a \in C'_1$, we have

$$s_{1,a}^{f_{\varphi_1,\varphi_2}} = s_{1,a}^{\gamma_0\theta} = s_{2,g_\theta(a)} = s_{2,\kappa(\overline{\theta}(a))}.$$

On the other hand, since f is an isomorphism, the definition of the Weisfeiler-Lehman algorithm round implies that

$$s_{1,a}^{f_{\varphi_1,\varphi_2}} = s_{2,f_{\psi_1,\psi_2}(a)}$$

for every color $a \in C'_1$. It follows that $f_{\psi_1,\psi_2} = \kappa\overline{\theta}$, and thus $f_{\psi_1,\psi_2} \in \kappa\mathfrak{G}$ as required. □

17.5 Isomorphism Testing

Let k be a fixed positive integer. We are now ready to describe a polynomial-time algorithm to test the isomorphism of two graphs that do not contain K_k as a topological minor. As outlined in Sect. 17.3, we are actually going to compute the isomorphism relation among all $(k-1)$-fragments of the input graphs, in the increasing order of the size of the fragments, thus reducing the problem to the following one: We are given

- linearly bordered graphs G_1 and G_2 with $|\partial G_1| = |\partial G_2| = k$, not containing K_k as a topological minor, and
- letting

$$\mathcal{H} = \{H : i \in [2], H \text{ is a } (k-1)\text{-fragment of } G_i, \partial G_i \cap V(H) \subseteq \partial H\},$$

the partition π of $\mathcal{H}$ to isomorphism classes.

We need to determine whether G_1 and G_2 are isomorphic. Let $d = k^3$. We are going to maintain

- boundary-respecting colorings $\varphi_1 : V(G_1) \to C_1$ and $\varphi_2 : V(G_2) \to C_2$ respecting the isomorphisms of G_1 and G_2, and
- an isomorphism d-enclosure $(\kappa, \mathfrak{G})$ of $(G_1, \varphi_1, G_2, \varphi_2)$.

Initially, for $i \in [2]$, let φ_i be the coloring that for each $j \in [k]$ assigns the color j to the j-th vertex of ∂G_i, and the color 0 to all vertices of $V(G_i) \setminus \partial G_i$. Note that these colorings respect the isomorphisms of G_1 and G_2, and that f_{φ_1,φ_2} is the identity function for every isomorphism f of G_1 and G_2. Hence, we can let $\kappa = \text{id}$ and $\mathfrak{G} = \{\text{id}\}$. We now repeatedly refine the colorings and update the isomorphism d-enclosure, as follows:

- By replacing φ_2 by $\kappa^{-1}\varphi_2$ if necessary, we can assume that $C_1 = C_2 = C$ and $\kappa = \text{id}$.
- We can also assume that for every $c \in C$ and $\gamma \in \mathfrak{G}$, we have $|\varphi_1^{-1}(c)| = |\varphi_2^{-1}(\gamma(c))|$. Indeed, f_{φ_1,φ_2} has this property for any isomorphism f of G_1 and G_2. Moreover, we can restrict $\mathfrak{G}$ to the permutations with this property by using the algorithm from Theorem 17.8 with $I = C$, $\mathfrak{G}$, and $S_1 = \{s_1\}$ and $S_2 = \{s_2\}$ for the strings s_1 and s_2 such that $s_i(c) = |\varphi_i^{-1}(c)|$ for $i \in \{1, 2\}$ and $c \in C$.
- If the colorings φ_1 and φ_2 are not stable, then let ψ_1 and ψ_2 be their refinements obtained by one round of the Weisfeiler-Lehman algorithm. Using Lemma 17.12, we either decide that G_1 and G_2 are not isomorphic, or obtain an isomorphism d-enclosure $(\kappa', \mathfrak{G}')$ of $(G_1, \psi_1, G_2, \psi_2)$. We replace φ_1 by ψ_1, φ_2 by ψ_2, and $(\kappa, \mathfrak{G})$ by $(\kappa', \mathfrak{G}')$ and repeat.
- Hence, assume that φ_1 and φ_2 are stable. Suppose that there exists an integer t such that $2 \le t \le d$ and at least one color class of φ_1 has size t. Let $T \subset C$ consist of the colors $c \in C$ such that $|\varphi_1^{-1}(c)| = t$. Note that since $\kappa = \text{id}$ and

the permutations in $\mathfrak{G}$ preserve the color class sizes, we also have $T = \{c \in C : |\varphi_2^{-1}(c)| = t\}$, and all permutations from $\mathfrak{G}$ map T to itself.

For each color $c \in T$, let $N_c = [t]$, and for each color $c \in C \setminus T$, let $N_c = \{1\}$. Let $C' = \bigcup_{c \in C}(\{c\} \times N_c)$. For $i \in [2]$, let $\psi_i : V(G_i) \to C'$ be obtained from φ_i by, for each $c \in T$, coloring each vertex of $\varphi_i^{-1}(c)$ by a distinct color from $\{c\} \times N_c$, and for each $c \in C \setminus T$ coloring all vertices of $\varphi_i^{-1}(c)$ by the color $(c, 1)$. Clearly, the colorings ψ_1 and ψ_2 respect the isomorphisms of G_1 and G_2.

For $\gamma \in \mathfrak{G}$ and a system $\Gamma = \{\Gamma_c \in \mathrm{Sym}(N_c) : c \in C\}$ of permutations, let $\omega_{\gamma,\Gamma}$ be the permutation of C' which for each $c \in C$ and $c' \in N_c$ maps (c, c') to $(\gamma(c), \Gamma_c(c'))$. Observe that if f is an isomorphism of G_1 and G_2, then there exists such a system Γ for which $f_{\psi_1,\psi_2} = \omega_{f_{\varphi_1,\varphi_2},\Gamma}$. Moreover, for $\gamma_1, \gamma_2 \in \mathfrak{G}$ and systems Γ^1 and Γ^2, we have

$$\omega_{\gamma_1,\Gamma^1}\omega_{\gamma_2,\Gamma^2} = \omega_{\gamma,\Gamma},$$

where $\gamma = \gamma_1\gamma_2$ and

$$\Gamma = \{\Gamma^1_{\gamma_2(c)}\Gamma^2_c : c \in C\}.$$

Therefore, the set

$$\mathfrak{G}' = \left\{\omega_{\gamma,\Gamma} : \gamma \in \mathfrak{G}, \Gamma \in \prod_{c \in C} \mathrm{Sym}(N_c)\right\}$$

of all permutations of this form is a group (a subgroup of the wreath product of $\mathfrak{G}$ and Sym_t). Luks [9] proved that since $\mathfrak{G} \in \hat{\Gamma}_d$ and $t \le d$, this group $\mathfrak{G}'$ also belongs to $\hat{\Gamma}_d$. Hence, we can replace φ_1 by ψ_1, φ_2 by ψ_2, and $(\kappa, \mathfrak{G})$ by $(\mathrm{id}, \mathfrak{G}')$ and repeat.

Note that in this step, the group theory tools in essence allow us to keep track of all possible individualizations of vertices whose color is in T without having to branch.

Finally, suppose that none of the refinement steps applies, and thus φ_1 and φ_2 are stable and every color class of either of them has size 1 or more than d. Let $J \subseteq C$ consist of the colors whose color class (in either of the colorings) has size 1. For $i \in [2]$, let X_i be the union of the trivial color classes of φ_i and let $\iota_i : J \to X_i$ be the function mapping each color $c \in J$ to the unique vertex of G_i of color c. By Theorem 17.3, the X_i-fragments of G_i are $(k-1)$-fragments. Let $\mathfrak{G}_0'$ be the group of the restrictions of the permutations in $\mathfrak{G}$ to J, and let $\mathfrak{G}_0 = \{\iota_1\gamma\iota_1^{-1} : \gamma \in \mathfrak{G}_0'\}$ be the corresponding group of permutations of X_1. Let $\kappa_0 = \iota_2\iota_1^{-1}$ be the bijection mapping vertices of X_1 to the vertices of X_2 of the same color. Since $(\mathrm{id}, \mathfrak{G})$ is an isomorphism d-enclosure, observe that for any isomorphism f of G_1 and G_2, the restriction of f to X_1 belongs to $\kappa_0\mathfrak{G}_0$. Hence, to determine whether there exists such an isomorphism, it suffices to check whether the output of the algorithm from Lemma 17.10 for G_1, G_2, X_1, X_2, π, $\mathfrak{G}_0$, and κ_0 is non-empty.

As usual, we are skimming over some of the implementation details; most importantly, one needs to work out how to implement the necessary permutation group operations (wreath product, restriction to an orbit, ...) in polynomial time.

References

1. L. Babai, Graph isomorphism in quasipolynomial time (2016). *ArXiv*, 1512.03547
2. K. Booth, G. Lueker, A linear time algorithm for deciding interval graph isomorphism. J. ACM **26**, 183–195 (1979)
3. A. Cardon, M. Crochemore, Partitioning a graph in $O(|A| \log_2 |V|)$. Theor. Comput. Sci. **19**, 85–98 (1982)
4. M. Grohe, *Descriptive complexity, canonisation, and definable graph structure theory*. Lecture Notes in Logic (Cambridge University Press, Cambridge, 2017)
5. M. Grohe, D. Marx, Structure theorem and isomorphism test for graphs with excluded topological subgraphs. SIAM J. Comput. **44**, 114–159 (2015)
6. M. Grohe, D. Neuen, Recent advances on the graph isomorphism problem (2020). ArXiv, 2011.01366
7. M. Grohe, D. Neuen, D. Wiebking, Isomorphism testing for graphs excluding small minors. SIAM J. Comput. **52**, 238–272 (2023)
8. N. Immerman, E. Lander, *Describing graphs: A first-order approach to graph canonization* (Springer, Berlin, 1990)
9. E. Luks, Isomorphism of graphs of bounded valence can be tested in polynomial time. J. Comput. Syst. Sci. **25**, 42–65 (1982)
10. G.L. Miller, Isomorphism of graphs which are pairwise k-separable. Inf. Control. **56**, 21–33 (1983)
11. D. Neuen, Hypergraph isomorphism for groups with restricted composition factors. ACM Trans. Algorithms (TALG) **18**, 1–50 (2022)
12. I.N. Ponomarenko, The isomorphism problem for classes of graphs that are invariant with respect to contraction. Zap. Nauchn. Sem. Leningrad. Otdel. Mat. Inst. Steklov. (LOMI) **174**, 147–177 (1988)
13. U. Schöning, Graph isomorphism is in the low hierarchy. J. Comput. Syst. Sci. **37**, 312–323 (1988)
14. Á. Seress, *Permutation group algorithms*, vol. 152 (Cambridge University Press, Cambridge, 2003)

Chapter 18
Sources

The proof of the existence of sublinear separators in graphs with a forbidden minor from Chap. 12 is almost verbatim taken from [1]. The applications and arguments from Sect. 12.2 date back to the seminal papers of Lipton and Tarjan [11, 12].

The notion of chordal partitions presented it in Chap. 13 comes from the fractional chromatic number argument of Reed and Seymour [13], though the fact that it can be applied more generally was highlighted in [15].

Chapter 14 draws on multiple sources. The density bound is a slightly simplified version of the argument of [2]. The ideas of the presented in Sect. 14.2 are drawn from [3]. Section 14.3 is based on [14] and [6], as well as Wood's survey of the defective and clustered chromatic number [17].

The chapter on product structure is largely based on [4] and other related papers; see also [7] for a (nowadays slightly outdated) survey. The application for the non-repetitive chromatic number is based on Wood's survey [16].

The presentation for the iterated BFS layering and its application to multicommodity flows is inspired by [9] and [10]. Results on Baker-amenability are taken from [5].

The isomorphism algorithm given in Chap. 17 is a simplified version (with substantially worse time complexity) of the algorithm of Grohe et al. [8].

References

1. N. Alon, P.D. Seymour, R. Thomas, A separator theorem for graphs with an excluded minor and its applications, in *Proceedings of the twenty-second annual ACM symposium on Theory of computing* (ACM, 1990), pp. 293–299
2. N. Alon, M. Krivelevich, B. Sudakov, Complete minors and average degree: A short proof. J. Graph Theory **103**, 599–602 (2023)
3. M. Delcourt, L. Postle, Reducing linear Hadwiger's conjecture to coloring small graphs (2021). ArXiv, 2108.01633

© The Author(s), under exclusive license to Springer Nature Switzerland AG 2025

Z. Dvořák, *Graph Minors*, Springer Monographs in Mathematics,
https://doi.org/10.1007/978-3-031-87469-7_18

4. V. Dujmović, G. Joret, P. Micek, P. Morin, T. Ueckerdt, D.R. Wood, Planar graphs have bounded queue-number. J. ACM **67**, 1–38 (2020)
5. Z. Dvořák, Baker game and polynomial-time approximation schemes, in *Proceedings of the thirty-first annual ACM-SIAM symposium on discrete algorithms*, SODA '20 (Society for Industrial and Applied Mathematics, Philadelphia, 2020), pp. 2227–2240
6. Z. Dvořák, S. Norin, Islands in minor-closed classes. I. Bounded treewidth and separators (2017). ArXiv, 1710.02727
7. Z. Dvořák, T. Huynh, G. Joret, C. Liu, D.R. Wood, Notes on graph product structure theory. *2019–20 MATRIX Annals* (2021), pp. 513–533
8. M. Grohe, D. Neuen, D. Wiebking, Isomorphism testing for graphs excluding small minors. SIAM J. Comput. **52**, 238–272 (2023)
9. P. Klein, S. Plotkin, S. Rao, Excluded minors, network decomposition, and multicommodity flow, in *Proceedings of the twenty-fifth annual ACM symposium on Theory of computing* (1993), pp. 682–690
10. F. Leighton, S. Rao, An approximate max-flow min-cut theorem for uniform multicommodity flow problems with applications to approximation algorithms, in *FOCS*, vol. 88 (1988), pp. 422–431
11. R. Lipton, R. Tarjan, A separator theorem for planar graphs. SIAM J. Appl. Math. **36**, 177–189 (1979)
12. R. Lipton, R. Tarjan, Applications of a planar separator theorem. SIAM J. Comput. **9**, 615–627 (1980)
13. B. Reed, P.D. Seymour, Fractional colouring and Hadwiger's conjecture. J. Combin. Theory, Ser. B **74**, 147–152 (1998)
14. J. van den Heuvel, D.R. Wood, Improper colourings inspired by Hadwiger's conjecture. J. Lond. Math. Soc. **98**, 129–148 (2018)
15. J. van den Heuvel, P. Ossona de Mendez, D. Quiroz, R. Rabinovich, S. Siebertz, On the generalised colouring numbers of graphs that exclude a fixed minor. Eur. J. Combin. **66**, 129–144 (2017)
16. D.R. Wood, Nonrepetitive graph colouring. Electron. J. Combin. **DS24** (2012)
17. D.R. Wood, Defective and clustered graph colouring. Electron. J. Combin. **DS23** (2018)

Name Index

A
Alon, N., 287, 307, 331

B
Babai, L., 359, 360
Baker, B.S., 59, 231, 339, 347, 355

C
Chekuri, C., 28
Chuzhoy, J., 28

D
Delcourt, M., 5, 307, 313
Dujmović, V., 204, 322, 331
Dvořák, Z., 22, 259, 278, 330

E
Edwards, K., 315
Erdős, P., 46, 47, 51, 149, 182

G
Geelen, J.F., 74
Grohe, M., 233, 234, 359, 360, 377
Gurski, F., 74

H
Hadwiger, H., v, 5, 14, 212, 256, 276, 278, 285, 296, 300, 307, 313–315, 318
Hickingbotham, R., 329

J
Jørgensen, L.K., 255, 312

K
Kawarabayashi, K., vi, 91, 278, 282–285, 287
Klein, P., 340
Kostochka, A.V., 307, 308
Krauthgamer, R., 322
Kühn, D., 313
Kündgen, A., 333
Kuratowski, K., v

L
Lee, J.R., 322
Leighton, T., 343
Lipton, R.J., 57, 59, 377
Luks, E.M., 369, 374

M
Mader, W., 88, 139, 167, 203, 205, 308, 312
Marx, D., 233, 234
Mayhew, D., 91
Miller, G.L., 369

N
Neuen, D., 369
Norin, S., 22, 97, 255, 283, 307

© The Editor(s) (if applicable) and The Author(s), under exclusive license to Springer Nature Switzerland AG 2025
Z. Dvořák, *Graph Minors*, Springer Monographs in Mathematics,
https://doi.org/10.1007/978-3-031-87469-7

O
Osthus, D., 313

P
Pelsmajer, M.J., 333
Pilipczuk, M., 323
Plotkin, S.A., 340
Ponomarenko, I., 360
Pósa, L., 46, 47, 51
Postle, L., 5, 307, 313

R
Rao, S., 340, 343
Reed, B.A., 278, 282, 283, 287, 296, 377
Robertson, N., v, vii, 7, 9, 17, 24, 25, 28, 33, 51, 71, 86, 91, 105, 112, 166, 184, 186–188, 204, 307

S
Schrijver, A., 105
Scott, A., 300
Seymour, P.D., v, vii, 7, 9, 17, 24–28, 33, 51, 65, 71, 86, 91, 105, 112, 166, 184, 186–188, 204, 296, 312, 377
Siebertz, S., 323
Song, Z.-X., 312
Sperner, E., 323
Szekeres, G., 149, 182

T
Tan, Z., 28
Tarjan, R.E., 57, 59, 377
Thomas, R., vi, 26–28, 88, 94, 97, 255, 259, 278, 283, 285, 312
Thomason, A., 5, 307, 308
Thue, A., 331

V
van den Heuvel, J., 302

W
Wagner, K., v, vi, 7, 255, 307, 311
Wanke, E., 74
Whitney, H., 188
Wollan, P., vi, 88, 91, 285
Wood, D.R., 318, 377

Index

Symbol
G-normal, 4
H-avoiding, 17
λ-rearrangement, 193
$\mathcal{T}$-central, 182
d-degenerate, 18
k-unbreakable, 20
m-linked, 83
s-separators, 290
w-balanced cut, 19

A
Acyclic coloring, 211, 302
Adhesion, 12
Anchor, 256
Animal, 196
Apex graph, 228
Apex number, 141, 164
Apex vertices, 17
Arrangement, 160
Atom, 107

B
Baker-amenable, 349
Basic connectivity condition, 102
Battlefield, 122
Bfs layering, 209
Biregular graph, 363
Bisegregation, 179
Bordered graph, 16
Boundary linkage, 94
Bounded expansion, 218
Bramble, 26
Branch decomposition, 70
Branch vertices, 6
Branchwidth, 70
Bridge, 150

C
Cactus, 64
Canonical encoding, 367
Carvingwidth, 72
Cell, 158
Cellular, 4
Centered coloring, 78
Chordal partition, 295
Clearing a zone, 115
Clique-sum, 13
Cliquewidth, 74
Clustered chromatic number, 315
Cographs, 74
Coloring number, 302
Comparable box dimension, 329
Confined subgraph, 63
Conformal tangle, 34
Connectivity function, 70
Consistent arrangement, 180
Contraction, 228
Controlling a model, 134
Crooked transaction, 179
Cross, 153
Crosscap transaction, 181
Cycle decompositions, 12

© The Editor(s) (if applicable) and The Author(s), under exclusive license to Springer Nature Switzerland AG 2025
Z. Dvořák, *Graph Minors*, Springer Monographs in Mathematics,
https://doi.org/10.1007/978-3-031-87469-7

D
Defective chromatic number, 315
Domination number, 167
Double-cross transaction, 181
Drawing, 3
Drawn up to outgrowths, 17
Drawn up to vortices, 17

E
Edgewidth, 4
Elimination forest, 77
Elimination ordering, 302
Erdős–Pósa property, 46
Essentially 4-connected, 14
Euler genus, 3
Extremal constraint, 109
Eye, 137

F
Faces, 4
Feedback vertex set, 47, 79
Flat region, 164
Folio, 171
Forbidden minor, 7
Fractional chromatic number, 296
Fractional coloring, 296
Fractional treewidth-fragility, 215
Fragment, 367

G
General decomposition, 12
Generalized Euler's formula, 4
Geodesic, 301
Grid, 25
Group generators, 368

H
Hadwiger number, 5
h-index, 233
Horn, 140

I
I-arc, 102
Interdigitating tree, 57
Internally linked, 269
Irrelevant vertex, 86
Island, 316

J
Jump, 137

L
Large-degree-outerplanar, 234
Layering, 209
Lean tree decomposition, 94
Leap transaction, 181
Linear Hadwiger's conjecture, 276
Linearly bordered graph, 366
Linkage, 41
Linkage critical, 90
Linked decomposition, 94
Locally bounded treewidth, 167

M
Medial graph, 72
Minor, 2
Minor-closed, 7
Minor-χ-critical, 276
Model, 5
Multicommodity flow, 342

N
Natural model, 187
Nearly drawn, 17
Nearly-k-apex, 278
Neighborhood complexity, 73
Nonrepetitive chromatic number, 331

O
Ordered graph, 351
Orientation of separations, 22

P
Partially triangulated grid, 228
Path decompositions, 12
Pathwidth, 12, 75
Predecessor transversal, 301
Pretangle, 23
Projection of a segregation, 158

R
Radial graph, 53, 61
Rankwidth, 72
Representativity, 4

Respectful tangle, 61
Rigid, 188
Rooted minor, 6
Rooted tree decomposition, 186
Root partition, 101
Root separator, 186
Rural, 154

S
Segregation, 158
Semi-solid, 158
Semi-strong vortical decomp., 261
Separation, 2
Set-of-strings isomorphism, 368
Shadow-complete layering, 333
Slope, 63
Society, 16
Solid, 160
Span, 190
Spider, 137
Spread arrangement, 187
Star coloring, 302
Star decompositions, 12
Strict linkage, 122
Strictly outerplanar, 63
Stroll-nonrepetitive, 331
Strong product, 321
Strong vortical decomposition, 260
Surface, 3
Surface part, 17
Surface tangle, 187

T
Tangle, 23
Tangle number, 71
Tie, 108
Topological feasibility, 101
Topological minor, 6
Topological model, 6
Topological realization, 101
Torso, 13
Transaction, 177
Tree decompositions, 12
Treedepth, 78
Treewidth, 12
Treewidth-fragility, 210
Twinwidth, 74

U
Umbral layering, 353

V
Vertex cover size, 79
Vertex separation, 2
Vortex face, 187
Vortex-like society, 177
Vortex-local, 256
Vortical decomposition, 16

W
Walk-nonrepetitive, 331
Wall, 31
Warp, 264
Weak coloring number, 302

Z
Zone, 113

MIX
Papier aus verantwortungsvollen Quellen
Paper from responsible sources
FSC® C105338

If you have any concerns about our products,
you can contact us on
ProductSafety@springernature.com

In case Publisher is established outside the EU,
the EU authorized representative is:
Springer Nature Customer Service Center GmbH
Europaplatz 3, 69115 Heidelberg, Germany

Printed by Libri Plureos GmbH
in Hamburg, Germany